Biomass and Bioenergy Solutions for Climate Change Mitigation and Sustainability

Ashok Kumar Rathoure
M/s Akone Services, India

Shankar Mukundrao Khade
Ajeenkya D.Y. Patil University, India

IGI Global
PUBLISHER of TIMELY KNOWLEDGE

A volume in the Advances in Environmental Engineering and Green Technologies (AEEGT) Book Series

Published in the United States of America by
IGI Global
Engineering Science Reference (an imprint of IGI Global)
701 E. Chocolate Avenue
Hershey PA, USA 17033
Tel: 717-533-8845
Fax: 717-533-8661
E-mail: cust@igi-global.com
Web site: http://www.igi-global.com

Library of Congress Cataloging-in-Publication Data

Names: Rathoure, Ashok K., 1983- editor. | Khade, Shankar, 1988- editor.
Title: Biomass and bioenergy solutions for climate change mitigation and sustainability / Ashok Rathoure and Shankar Khade, editors.
Description: Hershey, PA : Engineering Science Reference, an imprint of IGI Global, [2022] | Includes bibliographical references and index. | Summary: "'Biomass and bioenergy solutions for climate change mitigation and sustainability' highlights the challenges of energy conservation and the current scenario of existing fossil fuel and uses. It discusses how pollution potential of burning fossil fuel and the depletion of fossil fuel is a major issue for energy generation, thus propelling humanity toward bioenergy solutions"-- Provided by publisher.
Identifiers: LCCN 2022007805 (print) | LCCN 2022007806 (ebook) | ISBN 9781668452691 (h/c) | ISBN 9781668452707 (s/c) | ISBN 9781668452714 (ebook)
Subjects: LCSH: Biomass energy.
Classification: LCC TP339 .B5469 2022 (print) | LCC TP339 (ebook) | DDC 662/.88--dc23/eng/20220520
LC record available at https://lccn.loc.gov/2022007805
LC ebook record available at https://lccn.loc.gov/2022007806

This book is published in the IGI Global book series Advances in Environmental Engineering and Green Technologies (AEEGT) (ISSN: 2326-9162; eISSN: 2326-9170)

British Cataloguing in Publication Data
A Cataloguing in Publication record for this book is available from the British Library.

For electronic access to this publication, please contact: eresources@igi-global.com.

Advances in Environmental Engineering and Green Technologies (AEEGT) Book Series

Sang-Bing Tsai
Zhongshan Institute, University of Electronic Science and Technology of China, China & Wuyi University, China
Ming-Lang Tseng
Lunghwa University of Science and Technology, Taiwan
Yuchi Wang
University of Electronic Science and Technology of China Zhongshan Institute, China

ISSN:2326-9162
EISSN:2326-9170

Mission

Growing awareness and an increased focus on environmental issues such as climate change, energy use, and loss of non-renewable resources have brought about a greater need for research that provides potential solutions to these problems. Research in environmental science and engineering continues to play a vital role in uncovering new opportunities for a "green" future.

The **Advances in Environmental Engineering and Green Technologies (AEEGT)** book series is a mouthpiece for research in all aspects of environmental science, earth science, and green initiatives. This series supports the ongoing research in this field through publishing books that discuss topics within environmental engineering or that deal with the interdisciplinary field of green technologies.

Coverage

- Industrial Waste Management and Minimization
- Radioactive Waste Treatment
- Contaminated Site Remediation
- Biofilters and Biofiltration
- Air Quality
- Electric Vehicles
- Cleantech
- Pollution Management
- Sustainable Communities
- Green Technology

IGI Global is currently accepting manuscripts for publication within this series. To submit a proposal for a volume in this series, please contact our Acquisition Editors at Acquisitions@igi-global.com or visit: http://www.igi-global.com/publish/.

The Advances in Environmental Engineering and Green Technologies (AEEGT) Book Series (ISSN 2326-9162) is published by IGI Global, 701 E. Chocolate Avenue, Hershey, PA 17033-1240, USA, www.igi-global.com. This series is composed of titles available for purchase individually; each title is edited to be contextually exclusive from any other title within the series. For pricing and ordering information please visit http://www.igi-global.com/book-series/advances-environmental-engineering-green-technologies/73679. Postmaster: Send all address changes to above address.

Titles in this Series

For a list of additional titles in this series, please visit: http://www.igi-global.com/book-series/advances-environmental-engineering-green-technologies/73679

Transcending Humanitarian Engineering Strategies for Sustainable Futures
Yiannis Koumpouros (University of West Attica, Greece) Angelos Georgoulas (University of West Attica, Greece) and Georgia Kremmyda (University of Warwick, UK)
Engineering Science Reference • © 2023 • 315pp • H/C (ISBN: 9781668456194) • US $270.00

Handbook of Research on Building Greener Economics and Adopting Digital Tools in the Era of Climate Change
Patricia Ordóñez de Pablos (The University of Oviedo, Spain)
Engineering Science Reference • © 2022 • 391pp • H/C (ISBN: 9781668446102) • US $350.00

Improving Integrated Pest Management for Crop Protection and Food Security
Muhammad Haseeb (Florida A&M University, USA) Lambert H.B. Kanga (Florida A&M University, USA) and Jawwad A. Qureshi (University of Florida, USA)
Engineering Science Reference • © 2022 • 315pp • H/C (ISBN: 9781799841111) • US $215.00

Geoscientific Investigations From the Indian Antarctic Program
Neloy Khare (Ministry of Earth Sciences, India)
Engineering Science Reference • © 2022 • 305pp • H/C (ISBN: 9781668440780) • US $250.00

Handbook of Research on Green Technologies for Sustainable Management of Agricultural Resources
R.S. Sengar (Sardar Vallabhbhai Patel University of Agriculture and Technology, India) Reshu Chaudhary (Sardar Vallabhbhai Patel University of Agriculture and Technology, India) and H.S. Bhadauriya (Jiwaji University, Gwalior, India)
Engineering Science Reference • © 2022 • 637pp • H/C (ISBN: 9781799884347) • US $325.00

Urban Sustainability and Energy Management of Cities for Improved Health and Well-Being
Roberto Alonso González-Lezcano (Universidad CEU San Pablo, Spain)
Engineering Science Reference • © 2022 • 281pp • H/C (ISBN: 9781668440308) • US $250.00

Handbook of Research on Water Sciences and Society
Ashok Vaseashta (International Clean Water Institute, USA & Transylvania University of Brasov, Romania & Academy of Sciences of Moldova, Moldova) Gheorghe Duca (Institute of Chemistry, Moldova State University, Moldova & Academy of Science of Moldova and Romanian Academy, Romania.) and Sergey Travin (Federal Research Center for Chemical Physics and N. N. Semenov of the Russian Academy of Sciences, Russia)
Engineering Science Reference • © 2022 • 728pp • H/C (ISBN: 9781799873563) • US $515.00

IGI Global
PUBLISHER of TIMELY KNOWLEDGE

701 East Chocolate Avenue, Hershey, PA 17033, USA
Tel: 717-533-8845 x100 • Fax: 717-533-8661
E-Mail: cust@igi-global.com • www.igi-global.com

Editorial Advisory Board

Table of Contents

Preface xvi

Acknowledgment xxiii

Chapter 1
Bioenergy: Social, Economic, and Environmental Impacts 1
Shweta Arun Avhad, Savitribai Phule Pune University, India

Chapter 2
Life Cycle Assessment of Biofuels: Challenges and Opportunities 22
Majorie Moraa Nyasani, Machakos University, Kenya
Victor Odhiambo Shikuku, Kaimosi Friends University, Kenya

Chapter 3
Entrepreneurial Opportunities In Bioenergy 32
Prashant Kumar, Department of Bioinformatics, Kalinga University, Raipur, India
Sunil Kumar Verma, B.N. College of Engineering and Technology, Lucknow, India

Chapter 4
Biofuel Policies in India: An Assessment of Policy Barriers 44
Sunil Kumar Verma, B.N. College of Engineering and Technology, Lucknow, India
Prashant Kumar, Department of Bioinformatics, Kalinga University, Raipur, India

Chapter 5
Green House Gases: Challenges, Effect, and Climate Change 65
Kannadhasan S., Study World College of Engineering, India
Nagarajan R., Gnanamani College of Technology, India

Chapter 6
Biofuels From Bio-Waste and Biomass 75
Kondapalli Vamsi Krishna, Christ University (Deemed), India
Sompalli Bhavana, Christ University (Deemed), India
Koushik Koujalagi, Christ University (Deemed), India
Alok Malaviya, Christ University (Deemed), India

Chapter 7
Biogas: Renewable Natural Gas 119
Bela Khiratkar, Ajeenkya D.Y. Patil University, India
Shankar Mukundrao Khade, Ajeenkya D.Y. Patil University, India
Abhishek Dutt Tripathi, Banaras Hindu University, India

Chapter 8
Algae as Superfood 129
Shital Uddhav Giri, Ajeenkya D.Y. Patil University, India
Namdev Gopal Krishna Hadapad, Ajeenkya D.Y. Patil University, India
Aditya Akhade, Ajeenkya D.Y. Patil University, India
Parth Bhilare, Ajeenkya D.Y. Patil University, India

Chapter 9
Biofuels From Macroalgae: A Sustainable Alternative to Conventional Energy Resources 148
Debraj Biswal, Government General Degree College at Mangalkote, India
Dipanwita Sarkar (Paria), Chandernagore College, India

Chapter 10
A Source of Future Energy: Gas Hydrates 170
Anupama Kumari, Indian Institute of Technology, Roorkee, India
Mukund Madhaw, Independent Researcher, India
C. B. Majumder, Indian Institute of Technology, Roorkee, India

Chapter 11
Recent Advancements in Microalgae-Biofuel Generation Employing Nano-Additives 188
Mahesh Pattabhiramaiah, Bangalore University, India
Bhargavi Rajarathinam, Centre for Applied Genetics, Department of Zoology, Bangalore University, Bangalore, India
Shanthala Mallikarjunaiah, Bangalore University, India

Chapter 12
Macroalgae-Based Bioethanol 206
Sakshi Kolpe, Ajeenkya D.Y. Patil University, India
Shankar Mukundrao Khade, Ajeenkya D.Y. Patil University, India
Shivraj Nile, Zhejiang Chinese Medical University, Hangzhou, China

Chapter 13
Microalgae Biofuels: Challenges and Potential 217
Naresh Tanwer, Maharshi Dayanand University, India
Vaishali Arora, Maharshi Dayanad University, India
Priyanka Bumbra, Maharshi Dayanand University, India
Kiran Grewal, Maharshi Dayanand University, India
Jitender Singh Laura, Maharshi Dayanand University, India
Babita Khosla, Maharshi Dayanand University, India

Chapter 14
Microbial Cellulase in the Production of Second Generation Biofuels: State-of-the-Art and Beyond ... 233
Jovana Trbojević-Ivić, Innovative Centre, Faculty of Chemistry, Belgrade, Ltd., Serbia

Chapter 15
Comparative Studies on Neem and Jatropha Oil-Derived Biodiesels ... 258
Sunil Kulkarni, Gharda Institute of Technology, Lavel, India
Ajaygiri Goswami, University Institute of Chemical Technology, Jalgaon, India
Ghayas Usmani, University Institute of Chemical Technology, Jalgaon, India

Chapter 16
Characterization and Comparison of Various Blends of Honge Oil Methyl Ester (Biodiesel) With Diesel Fuel ... 274
Sunil Kulkarni, Gharda Institute of Technology, Lavel, India
Ajaygiri K. Goswami, University Institute of Chemical Technology, Jalgaon, India
Ghayas A. Usmani, University Institute of Chemical Technology, North Maharashtra University, Jalgaon, India

Chapter 17
Explicit Conceptual Design Approach to Adapt a Biomass-Fed Anaerobic Digester and Status Indicators in Semi-Arid Areas ... 291
Yusto Mugisha Yustas, Sokoine University of Agriculture, Tanzania

Chapter 18
Pretreatment of Lignocellulosic Biomass and 2G Ethanol ... 322
Hitesha J. Panchal, Parul University, Vadodara, India
Krishan Kumar, Parul University, Vadodara, India

Compilation of References ... 340

About the Contributors ... 403

Index ... 410

Detailed Table of Contents

Preface xvi

Acknowledgment xxiii

Chapter 1
Bioenergy: Social, Economic, and Environmental Impacts 1
Shweta Arun Avhad, Savitribai Phule Pune University, India

As we know that escalating population demand and technological advancement lead us towards a new world dominion power, but with that, there is huge pressure on resources due to an increase in consumption at all levels globally. This leads to revamping environmental motifs all over the world and non-renewable sources are estimated to be not rehabilitating soon. In this case, bio-energy is energy presented as a major opportunity to address the issues of climate change by reducing carbon emissions. This chapter explore bioenergy.

Chapter 2
Life Cycle Assessment of Biofuels: Challenges and Opportunities 22
Majorie Moraa Nyasani, Machakos University, Kenya
Victor Odhiambo Shikuku, Kaimosi Friends University, Kenya

To address energy security and climate-change concerns, substitutes are needed for petroleum-based transportation fuels. In addition to electricity and natural gas, biofuels are emerging as an important class of substitutes. The promotion of biofuels as energy is mainly driven by the perspective of oil depletion, concerns about energy security, and global warming. However, results published in the past few decades present conflicting pictures regarding the energy efficiency and sustainability of biofuels. To adequately address these real or apparent contradictions, a life cycle assessment (LCA) of biofuels is indispensable. "Life cycle" refers to all stages of a process: from the cradle, that is, raw material extraction through manufacturing, distribution, and use to ultimate disposal. This chapter discusses the benefits and limitations of biofuels from an LCA standpoint. The discrepancies and strengths in LCA of biofuels are critically analyzed and opportunities are highlighted. LCA of biofuels is shown to be critical to informing policy and for practical application.

Chapter 3
Entrepreneurial Opportunities In Bioenergy 32
Prashant Kumar, Department of Bioinformatics, Kalinga University, Raipur, India
Sunil Kumar Verma, B.N. College of Engineering and Technology, Lucknow, India

Entrepreneurism is vital as entrepreneurs push technological change and innovation, consequently generating economic growth. Bio-energy entrepreneurism is the focal point of this chapter, along with entrepreneurial motivations, opportunities, and market drivers. Market drivers include government policies, trends in energy use, and progression of the energy sector into more renewables. Several factors influence entrepreneurs, which involves several factors ranging from the environment to economics, pioneering spirit, and social factors. With the help of entrepreneurial motivations and market drivers, individuals can easily visualize and plan all the bio-energy sector opportunities. Bio-energy entrepreneurs moderate the market of transformation of biomass to energy. The bio-energy field is filled with stories of failures, successes, and everything in between. There is a marginal difference between a failure story and a success story in the modern bio-energy sector, which narrow down to timing.

Chapter 4
Biofuel Policies in India: An Assessment of Policy Barriers 44
Sunil Kumar Verma, B.N. College of Engineering and Technology, Lucknow, India
Prashant Kumar, Department of Bioinformatics, Kalinga University, Raipur, India

Energy is one of the most precious and demanded commodities among various industries and consumers to sustain the current lifestyle. Energy is a crucial element, which unswervingly influences the country's economic development. Numerous methods are adopted to reduce global warming, embracing clean energy from wind, solar, and biomass sources. This chapter speaks about the current situation of energy demand, the innovations in biofuel sources, and the obstacles regarding the commercialization and production of microalgal biofuel.

Chapter 5
Green House Gases: Challenges, Effect, and Climate Change 65
Kannadhasan S., Study World College of Engineering, India
Nagarajan R., Gnanamani College of Technology, India

Changes in weather patterns over decades or longer periods of time are referred to as climate change. Natural and human forces both contribute to climate change. Since the Industrial Revolution, humans have contributed to climate change through emitting greenhouse gases and aerosols, as well as changing land use, leading global temperatures to rise. More storms, floods, droughts, and rising sea levels may result from increasing global temperatures and the melting of ice sheets, sea ice, and glaciers. The sun's rays supply energy to the whole world. GHGs are required to trap heat and maintain a temperature that permits life to flourish on the planet. The greenhouse effect is a natural and necessary occurrence for life on Earth to survive. If the greenhouse effect did not exist, the world would be around 33°C cooler right now. Humans have contributed to the increase in atmospheric GHGs via rising fossil fuel usage and deforestation in recent decades. Increased GHG emissions have been the primary cause of global warming during the last century.

Chapter 6
Biofuels From Bio-Waste and Biomass 75
Kondapalli Vamsi Krishna, Christ University (Deemed), India
Sompalli Bhavana, Christ University (Deemed), India
Koushik Koujalagi, Christ University (Deemed), India
Alok Malaviya, Christ University (Deemed), India

The planet's limited natural fossil fuel reserves are anticipated to be very soon owing to massive usage. Biofuels would be a critical alternative source that may reduce global warming and CO2 emissions. The food-versus-fuel dilemma is, however, one of the key drawbacks of first-generation biofuels like corn ethanol, sugarcane ethanol, etc. Cellulose and hemicellulose, the primary constituents of lignocellulosic feedstocks, could be reduced to sugars by either thermochemical/biological processes before being fermented to generate biofuels. However, owing to structural heterogeneity, more complicated operational techniques are required before the production technology can be commercialized, and several challenges must be addressed. This chapter provided an assessment of various feedstocks, availability, various processing techniques, obstacles, and current technical developments in the generation of biofuels from biomass.

Chapter 7
Biogas: Renewable Natural Gas 119
Bela Khiratkar, Ajeenkya D.Y. Patil University, India
Shankar Mukundrao Khade, Ajeenkya D.Y. Patil University, India
Abhishek Dutt Tripathi, Banaras Hindu University, India

Biogas is a renewable natural gas used in production of energy which is generated by breakdown of organic matter by anaerobic digestion. It is a composition of methane, carbon dioxide, and some other gases in small quantities. The most common natural resources for production of biogas are aquatic sediments, animal waste, crop residues, wastewater sludge, and many others. Some other sources are landfills, water lagoons, etc., which are a result of human activity. There are several technologies used for biogas production, mainly biodigesters, wastewater treatment plants, and landfill gas recovery systems. Production of biogas is highly dependent on the feedstock availability and the policy support by the government.

Chapter 8
Algae as Superfood 129
Shital Uddhav Giri, Ajeenkya D.Y. Patil University, India
Namdev Gopal Krishna Hadapad, Ajeenkya D.Y. Patil University, India
Aditya Akhade, Ajeenkya D.Y. Patil University, India
Parth Bhilare, Ajeenkya D.Y. Patil University, India

With the increasing population, there are problems like the shrinking of arable land, pollution, and feeding the entire growing population high-quality food and reaching consumer demand, and the list goes on. Novel superfoods are rich in nutrition and produced with the most sustainable methods, including algae, specifically microalgae. Microalgae have been studied for decades. Microalgae are one of the largest and most poorly understood organisms on planet earth, but in recent years, the research has been increased for the search of renewable and sustainable energy sources. Microalgae don't require much land for cultivation and thus do not compete with conventional agricultural land. Microalgae can double their size in 24 hours. CO2, solar rays, inorganic nutrients, and water are the basic requirements for them to grow. There are microalgae and macroalgae (e.g., seaweed) that grow in saline and hypersaline water. This chapter presents the promising potential of algae (microalgae and macroalgae) as a superfood.

Chapter 9
Biofuels From Macroalgae: A Sustainable Alternative to Conventional Energy Resources............. 148
Debraj Biswal, Government General Degree College at Mangalkote, India
Dipanwita Sarkar (Paria), Chandernagore College, India

The steep rise in human population together with the simultaneous growth of urbanization and industrialization has triggered overuse of fossil fuels in the recent years. However, they are non-renewable and release carbon dioxide causing global warming. In this context, biofuels come as excellent substitutes because they are renewable and eco-friendly. They are obtained from biological products and include biodiesel, biomethane, bioethanol, and biohydrogen. They are classified into first, second, and third generation fuels depending on the biological feedstocks being used. Of these, the third-generation fuels, obtained from marine resources like algae, have attracted special research interests because they do not rival the food crop production or depend on land and freshwater for their cultivation. However, the potentialities of macroalgae as a biofuel resource have not been investigated thoroughly. The chapter tries to outline all the aspects of biofuel production from macroalgae pointing out the scopes for future research.

Chapter 10
A Source of Future Energy: Gas Hydrates.. 170
Anupama Kumari, Indian Institute of Technology, Roorkee, India
Mukund Madhaw, Independent Researcher, India
C. B. Majumder, Indian Institute of Technology, Roorkee, India

Natural gas hydrate (NGH) has emerged as a future source of energy for the world that can fulfill the future energy demand. The natural gas hydrates hold twice the amount of energy than all available fossil fuels. At present, the energy demand is growing, but these are not available commercially. Various field applications on the dissociation of the natural gas hydrates have been performed till now, and the output of these field applications confirms the possibility of dissociation of the natural gas hydrates. At distinct locations in the world, the huge deposits of hydrates have been identified, but the safe and economic dissociation of gas hydrates is not known yet. Hence, the evolution of new and inexpensive, suitable methods and technologies are required for the dissociation of gas hydrates. In this chapter, the authors reviewed several articles to compare the different dissociation techniques of gas hydrates and their field applications. The advantages and disadvantages of these production methods have also been discussed.

Chapter 11
Recent Advancements in Microalgae-Biofuel Generation Employing Nano-Additives.................... 188
Mahesh Pattabhiramaiah, Bangalore University, India
Bhargavi Rajarathinam, Centre for Applied Genetics, Department of Zoology, Bangalore University, Bangalore, India
Shanthala Mallikarjunaiah, Bangalore University, India

Biofuels have gained significant interest as an alternative fuel in recent years owing to their environmental sustainability, cost-effectiveness, and the ability to blend with traditional fuels like gasoline without requiring engine modifications. The use of microalgae for biofuel production is universally preferred due to its energy efficiency and environmental sustainability; current research is geared towards enhancing the production of microalgae-biofuel from the initial stages to the final product as it is a cost-effective fuel option. The use of different types of nano-additives at different stages of microalgae cultivation and incorporation into biofuel produced noteworthy improvements in the final product. This chapter focuses on the prospective uses of nano-additives in microalgae cultivation, microalgal biomass conversion to biofuels, and biofuel combustion improvement for revolutionary advancements in biofuel technology.

Chapter 12
Macroalgae-Based Bioethanol 206
Sakshi Kolpe, Ajeenkya D.Y. Patil University, India
Shankar Mukundrao Khade, Ajeenkya D.Y. Patil University, India
Shivraj Nile, Zhejiang Chinese Medical University, Hangzhou, China

The chapter discusses the methods effective for the extraction of useful energy from the macroalgae biomass including liquefaction, anaerobic digestion, fermentation to biobutanol, trans-esterification to biodiesel, pyrolysis, direct combustion, fermentation to bioethanol, and gasification. However, if the algae are suited for the production of biodiesel, they can be studied from the content of their triacylglycerols (TAGs). Due to having high fatty acid content, they have a high conversion rate to biodiesel, and the lack of sulphur, phosphorus, and nitrogen also aids in the conversion. This chapter highlights the limitations and suitability of macroalgae for the conversion process in reference to chemical composition, process optimization, and cost effectiveness. It is concluded that bio-oils and bioethanol produced from wet macroalgae are considered over biodiesel production because of high lipid content of microalgae biomass. Moreover, the chapter considers electricity production from the dry mass as it would turn profitable, and this can be achieved from fast-growing macroalgae like "Ulva."

Chapter 13
Microalgae Biofuels: Challenges and Potential 217
Naresh Tanwer, Maharshi Dayanand University, India
Vaishali Arora, Maharshi Dayanad University, India
Priyanka Bumbra, Maharshi Dayanand University, India
Kiran Grewal, Maharshi Dayanand University, India
Jitender Singh Laura, Maharshi Dayanand University, India
Babita Khosla, Maharshi Dayanand University, India

Development of sustainable energy resources is a major concern in front of the scientific community and stakeholders in this field. Invariably, fossil fuels are utilized to meet the major part of energy demand all over the world, but their stocks are limited. Burning of fossil fuels also causes significant environmental pollution that results in drastic climate change. There is an emergent need of an alternative to fulfill the current energy demand in a sustainable way along with sorting out the pollution issues. Microbial biofuels could be the most appropriate choice in this regard because it bears enormous potential that needs to be exploited. But it has some challenging issues to employ it to meet out the current energy demands. This chapter illustrates the microbial biofuels, harvesting techniques, different process, and potential to utilize them as sustainable energy sources along with considerable challenges that need to be addressed to make it an eco-friendly and economic fuel choice to pave the path of sustainable development.

Chapter 14
Microbial Cellulase in the Production of Second Generation Biofuels: State-of-the-Art and Beyond 233
Jovana Trbojević-Ivić, Innovative Centre, Faculty of Chemistry, Belgrade, Ltd., Serbia

Bioethanol from inedible cellulose-rich biomass is the most promising candidate to replace fast depleting, environmentally polluting fossil fuels. Hydrolysis of cellulose to glucose is the crucial step in its biotransformation to bioethanol. Enzymatic hydrolysis is favored over acid hydrolysis, as enzymes are eco-friendly biocatalysts with high substrate specificity and superior catalytic efficiency in mild reaction conditions. Complete hydrolysis of cellulose is achieved by cellulase. Higher cellulase production yield, stability, and catalytic efficiency are the main attentive points for the successful implementation in industrial production of bioethanol. This chapter will highlight general characteristics of microbial cellulases and their role in the bioconversion of cellulose to biofuels, economic sustainability of cellulose-based biofuels, and the latest innovations in cellulase immobilization as the most comprehensive strategy for improvement of enzyme stability, activity, and reusability for cost-effective large-scale application.

Chapter 15
Comparative Studies on Neem and Jatropha Oil-Derived Biodiesels........ 258
Sunil Kulkarni, Gharda Institute of Technology, Lavel, India
Ajaygiri Goswami, University Institute of Chemical Technology, Jalgaon, India
Ghayas Usmani, University Institute of Chemical Technology, Jalgaon, India

Sustainable development is a key aspect of modern-day development. Reducing fuel cost and emissions is considered as the most important research area in sustainable development. Biodiesel has the potential to reduce the requirement of petroleum diesel if blended in suitable proportion. Various feedstocks are being explored for biodiesel production. Oils derived from neem and jatropha can be utilized for biodiesel production. In the current investigation, optimum parameters for the production, namely catalyst proportion, temperature, and oil-to-alcohol ratio were optimized. Experiments were carried out at different blend proportions to study fuel properties, namely kinematic viscosity, aniline point, diesel index, flash and fire points, specific and API gravity, cetane number, and ASTM distillation characteristics.

Chapter 16
Characterization and Comparison of Various Blends of Honge Oil Methyl Ester (Biodiesel) With Diesel Fuel 274
Sunil Kulkarni, Gharda Institute of Technology, Lavel, India
Ajaygiri K. Goswami, University Institute of Chemical Technology, Jalgaon, India
Ghayas A. Usmani, University Institute of Chemical Technology, North Maharashtra University, Jalgaon, India

The Honge oil, known as pongamia oil, is one of the most widely used lamp oils around the world. It is also used in leather tanning, in soap making, and as a lubricant. The Honge oil methyl ester can be used as an alternative fuel for diesel. The transesterification of methanol and Honge oil in the presence of base catalyst results in Honge oil methyl ester. In the current work, the Honge oil methyl-ester known as biodiesel was produced. It was mixed with petroleum diesel fuel in different proportions. Various physical and petroleum properties were investigated for Honge oil biofuel and also for various diesel and ester mixtures. The optimum parameters were temperature was 65oC, catalyst quantity, 0.4 grams and oil: alcohol ratio, 1:6. The blends of 10%, 20%, and 30% compositions were most suitable as potential alternative fuel. Different proportions of biodiesel ester in diesel fuel have been also compared with petroleum diesel fuel to identify a suitable blend having good engine fuel potential.

Chapter 17
Explicit Conceptual Design Approach to Adapt a Biomass-Fed Anaerobic Digester and Status Indicators in Semi-Arid Areas 291
Yusto Mugisha Yustas, Sokoine University of Agriculture, Tanzania

The mitigation of adverse climate change requires holistic strategies. The strategies include the proper handling of biomass wastes such as cow dung. The handling of the wastes has to be efficient using appropriately designed anaerobic digesters. In addition, the operating status of these digesters has to be monitored for detection of any fault of the digesters. Unfortunately, existing approaches for designing and monitoring the working condition of digesters have limitations. This chapter presents the innovative and explicit conceptual design approach of adapting the biomass-fed anaerobic digester and the digester operating-status-indicators in semi-arid areas. The approach is recommended to the renewable energy system designers. Finally, the appropriate indicators are recommended to be applied by the biogas plant operators or users.

Chapter 18
Pretreatment of Lignocellulosic Biomass and 2G Ethanol 322
Hitesha J. Panchal, Parul University, Vadodara, India
Krishan Kumar, Parul University, Vadodara, India

Rapid depletion of fossil fuel-based energy sources increased the demand for alternate energy sources. Lignocellulosics-based 2G ethanol can be used as an alternative sustainable source that presents in ample amount. Sources of lignocellulose biomass are wood, food-agriculture wastes, and forest residues. Cellulose, hemicellulose, and lignin are the core components of lignocellulosic biomass. Cellulosic and hemicellulosic biomass are enzymatically hydrolyzed to produce the monomer sugar (such as glucose or xylose) which is further converted into ethanol using fermentation process. The presence of lignin provides physical barrier that limit the access of enzymes required for saccharification. Pretreatment helps in removing the lignin from biomass and reducing recalcitrance. Pretreatment can be done by conventional methods, which are chemical, physical, and biological. This study covers the different methods of pretreatment including their disadvantages and benefits along with saccharification and fermentation processes.

Compilation of References 340

About the Contributors 403

Index 410

Preface

This is the right time to tackle the changing climate by making concrete plans and taking necessary actions to regulate the existing responsible factors for changing the climate. This can be only possible with the renewable source of energies. In addition to increase the level of pollutant gases upon burning in the environment, the near future depletion of existing fuels made the researchers to think on the alternate sources to mitigate the problems associated.

This issue is mainly concerned with this phenomenon and must be regarded as the commencement of a more extensive efforts for understanding public policy-making and public administration in the bioenergy. This era sees the growth of bioenergy, introducing new products, inclinations, procedures, tactics, and theories daily, thereby changing the environmental issues.

Biofuels produced either from living cell biomass or once live biomass can have efficient energy to acts as an alternative to the existing non-renewable fossil fuels. However, in the production of biofuels efforts are being made worldwide to make the fuels cost-effective. In the first generation, biofuels are derived by fermentation from the edible sources such as edible oils, corn, sugar etc. which limit the use of these sources. In the second generation, the biofuels are obtained from the non-edible sources like wheat straw, rice straw, jatropha seeds, sugarcane bagasse etc. The most critical step followed in this generation is to depolymerize and removal of lignin present in the lignocellulosic biomasses, thereby makes the process costly. Most of the challenges are overcome in the third-generation biofuels, where biofuels are derived from the living cellular biomass i. e. algal cells. The ability of certain algal species to accumulate the lipid inside the cells which can be transesterified into biodiesel. The algae can be cultivated in a small area using relatively cheaper nutrients and limited resources which can be used to produce biofuels.

With considering the global climate change, the several countries have initiated the efforts in mitigating the hazardous effects of pollutants by increasing the dependency on bioenergy.

INSIDE THIS BOOK

In this regard, the first chapter is devoted bioenergy-social, economic, and environmental impacts, which can comprehensively constitute the importance of bioenergy to protect the environment. As we know that escalating population demand and technological advancement leads us towards a new world dominion power, but there is huge pressure on resources due to an increasing consumption at all levels globally. This leads to revamping environmental motifs all over the world and non-renewable sources estimated to be not rehabilitating soon. In this case, bioenergy is energy presented as a major opportunity to address the issues of climate change by reducing carbon emissions. It affects socially by increasing living of standard to creating societal conflict, poverty and transportation cost and job opportunity economically. Environmentally reducing pollution to forest land unavailability concerns. As its major threats of bioenergy have not instigated at greater risk for the world, yet hence it seems to be a good solution for now. The rate of expansion of bioenergy products depends upon a great deal of government policies, straightforward strategy, and foreign investments in developing countries. With taking care of these bioenergy negative aspects also hindered.

The second chapter is based on the challenges and opportunities of biofuels. The results of this study reveal the hurdles that comes in producing the biofuels and replacing the existing fossil fuels. To address energy security and climate-change concerns, substitutes are needed for petroleum-based transportation fuels. In addition to electricity and natural gas, biofuels are emerging as an important class of substitutes. The promotion of biofuels as energy is mainly driven by the perspective of oil depletion, concerns about energy security, and global warming. However, results published in the past few decades present conflicting pictures regarding the energy efficiency and sustainability of biofuels. To adequately address these real or apparent contradictions, a Life cycle assessment (LCA) of biofuels is indispensable. "Life cycle" refers to all stages of a process: from the cradle, that is, raw material extraction through manufacturing, distribution, and use to ultimate disposal. This chapter discusses the benefits and limitations of biofuels from an LCA standpoint. The discrepancies and strengths in LCA of biofuels are critically analyzed and opportunities are highlighted. LCA of biofuels is shown to be critical to informing policy and for practical application.

The third chapter, "Entrepreneurial Opportunities in Bioenergy," claims that the necessity to broaden the scope of bioenergy across the globe. This can be achieved by initiating the new firms to protect the environment from the fossil fuels. One can start the firm in the area of bioenergy for which Government also gives support. Entrepreneurism is vital as entrepreneur's push technological change and innovation, consequently generating economic growth. Bio-energy entrepreneurism is the focal point of this chapter, along with entrepreneurial motivations, opportunities and market drivers. Market drivers include government policies, trends in energy use and progression of the energy sector into more renewables. Several factors influence entrepreneurs, which involves several factors ranging from the environment to economics, pioneering spirit to social factors. With the help of entrepreneurial motivations and market drivers, individuals can easily visualize and plan all the bio-energy sector opportunities. Bio-energy entrepreneurs moderate the market of transformation of biomass to energy. The bioenergy field is filled with stories of failures, successes, and everything in between. There is a marginal difference between a failure story and a success story in the modern bio-energy sector, which narrow down to timing.

The fourth chapter emphasizes the biofuel policies in India and the associated barrier of policy helps to understand the opportunities to initiate the entrepreneurial journey. Energy is one of the most precious and demanded commodities among various industries and consumers to sustain the current lifestyle. Energy is a crucial element, which unswervingly influences the country's economic development. Numerous methods are adopted to reduce global warming, embracing clean energy from wind, solar, and biomass sources. To bring these technologies into reality, it needs to undergo several phases, considering the technology development, environmental sustainability, economic viabilities, and conclusively with the government support in terms of public awareness and effective policies appropriate Research and Development efforts could overawe every variable, but it can be achieved only through an effective policy. Therefore, to begin with, this chapter speaks about the current situation of energy demand, the innovations in biofuel sources, and the obstacles regarding the commercialization and production of microalgal biofuel.

The fifth chapter refers to greenhouse gases, challenges, effect and climate change, puts emphasis on the release of greenhouse gases and their impact on the environment by means of increasing the global warming. Weather patterns alter over decades or longer periods of time, which is referred to as climate change. Natural and human forces both contribute to climate change. Since the Industrial Revolution, humans have contributed to climate change through emitting greenhouse gases and aerosols, as well as changing land use, resulting in a rise in global temperatures. More storms, floods, droughts, and rising sea levels may result from increasing global temperatures, as well as the melting of ice sheets, sea ice, and glaciers. The planet receives energy from the sun's radiation. GHGs are necessary for trapping heat and maintaining a temperature that permits life to flourish on the planet. The greenhouse effect is a natural and necessary phenomenon for life on Earth to survive. If the greenhouse effect didn't exist, the world would be around 33°C cooler right now. Humans have contributed to a rise in atmospheric GHGs as a result of increased fossil fuel use and deforestation in recent decades. The rise in GHGs has been the primary cause of global warming during the last century.

The sixth chapter signifies the production of biofuels from the biomass and biowaste. The cost of production of biofuels can be reduced by utilizing the biowaste and biomass for the production purposes. The planet's limited natural fossil fuel reserves are anticipated to be very soon owing to massive usage. Biofuels would be a critical alternative source that may reduce global warming and CO2 emissions. The food-versus-fuel dilemma is, however, one of the key drawbacks of first-generation biofuels like Corn ethanol, sugarcane ethanol, etc. Cellulose and hemicellulose, the primary constituents of lignocellulosic feedstocks, could be reduced to sugars by thermochemical / biological processes before being fermented to generate biofuels. However, owing to structural heterogeneity, more complicated operational techniques are required before the production technology can be commercialized and several challenges must be addressed. This book chapter provided an assessment of various feedstocks, availability, various processing techniques, obstacles, and current technical developments in the generation of biofuels from biomass.

The seventh chapter emphasizes the biogas generation from the renewable sources under a specified condition. The importance of biogas is in reducing the dependency on the non-renewable gases which can be produced using household wastes too. Biogas is a renewable natural gas used in production of energy which is generated by breakdown of organic matter by anaerobic digestion. It is a composition of methane, carbon dioxide and some other gases in small quantities. The most common natural resources for production of biogas are aquatic sediments, animal waste, crop residues, wastewater sludge and many others. Some other sources are landfills, water lagoons etc. which are a result of human activity. There are several technologies used for biogas production, mainly- biodigesters, wastewater treatment plants and landfill gas recovery systems. Production of biogas is highly dependent on the feedstock availability and the policy support by the government.

The eighth chapter signifies the importance of algal biomass on reducing the malnutrition cases across the globe. The protein rich algal food can be used to enrich the health upon consumption. With the increasing population, there are problems like the shrinking of arable land, pollution and feeding the entire growing population high-quality food and reaching consumer demand and the list goes on. Novel superfoods are rich in nutrition and produced with the most sustainable methods, including Algae, specifically microalgae. Microalgae have been studied for decades. Microalgae are one of the largest and most poorly understood organisms on planet earth but, in recent years the research has been increased for the search of renewable and sustainable energy sources. Microalgae don't require much land for cultivation and thus do not compete with conventional agricultural land. Microalgae can grow in double their size in 24 hours.CO2, solar rays, inorganic nutrients, and water are the basic requirements for them to grow. There are microalgae and macroalgae (eg: seaweed) that grow in saline and hypersaline water. This Chapter presents the promising potentials of algae (microalgae and macroalgae) as a superfood.

The ninth chapter focuses on the production of bioethanol with using macroalgae biomass. Macroalgae are known to accumulate the starch which can be used to accomplish fermentation to produce ethanol as an important biofuel. Steep rise in human population together with the simultaneous growth of urbanization and industrialization has triggered overuse of fossil fuels in the recent years. However, they are non-renewable and release carbon dioxide causing global warming. In this context, biofuels come as excellent substitutes because they are renewable and eco-friendly. They are obtained from biological products and include biodiesel, biomethane, bioethanol and biohydrogen. They are classified into first, second and third generation fuels depending on the biological feedstocks being used. Of these, the third-generation fuels, obtained from marine resources like algae have attracted special research interests because they do not rival the food crop production or depend on land and freshwater for their cultivation. However, the potentialities of macroalgae as a biofuel resource have not been investigated thoroughly. The current chapter tries to outline all the aspects of biofuel production from macroalgae pointing out the scopes for future research.

The tenth chapter signifies the methodology to produce the biogas in the form of gas hydrates. The gas hydrates in future can be efficiently used as an alternative source to the non-renewable gases. This can be produced from the waste biomass to make cost effective process. Natural Gas Hydrate (NGH) is emerged as a future source of energy for the world which can fulfill the future energy demand. The natural gas hydrates hold the twice amount of energy more than all available fossil fuels. At present, the energy demand is growing, but these are not available commercially. Various field applications on the dissociation of the natural gas hydrates have been performed till now and the output of these field applications confirms the possibility of dissociation of the natural gas hydrates. At distinct locations in the world, the huge deposits of hydrates have been identified but the safe and economical dissociation of gas hydrates is not known yet. Hence, the evolution of new and inexpensive, suitable methods and technologies are required for the dissociation of gas hydrates. In this chapter, we reviewed several articles to compare the different dissociation techniques of gas hydrates and their field applications. The advantages and disadvantages of these production method has also been discussed.

The eleventh chapter describes the use of microalgae to accumulate the lipid into them which can be transesterified to fatty acid alkyl esters. The process uses the cost-effective approach to culture the lipid accumulating microalgae on the wastewater and thereby harvesting, extraction, and transesterification of alcohol. Biofuels have gained significant interest as an alternative fuel in recent years owing to their environmental sustainability, cost-effectiveness, and the ability to blend with traditional fuels like gasoline without requiring engine modifications. The use of microalgae for biofuel production is universally preferred due to its energy efficiency and environmental sustainability; Current research is geared towards enhancing the production of microalgae-biofuel from the initial stages to the final product as it is a cost-effective fuel option. The use of different types of nano-additives at different stages of microalgae cultivation and incorporation into biofuel produced noteworthy improvements in the final product. This chapter focuses on the prospective uses of nano-additives in microalgae cultivation, microalgal biomass conversion to biofuels, and biofuel combustion improvement for revolutionary advancements in biofuel technology.

The twelfth chapter emphasizes the macroalgae based ethanol production in which the carbohydrate rich algal biomass is utilized to ferment into ethanol under anaerobic conditions. The chapter discusses the methods effective for the extraction of useful energy from the macroalgae biomass including liquefaction, anaerobic digestion, fermentation to biobutanol, trans-esterification to biodiesel, pyrolysis, direct combustion, fermentation to bioethanol and gasification. However, if the algae are suited for the production of biodiesel can be studied from the content of their triacylglycerols (TAGs). Due to having high fatty acid content they have a high conversion rate to biodiesel, and the lack of sulphur, phosphorus and nitrogen also aids in the conversion. This chapter highlights the limitations and suitability of macroalgae for the conversion process in reference to chemical composition, process optimization and cost effectiveness. It is concluded bio-oils and bioethanol produced from wet macroalgae are considered over biodiesel production because of high lipid content of microalgae biomass. Moreover, the chapter considers electricity production from the dry mass as it would turn profitable and this can be achieved from fast-growing macroalgae like "Ulva".

The thirteenth chapter clarifies the challenges and potential associated with the lipid accumulating microalgal biofuels. Development of sustainable energy resources is a major concern in front of scientific community and stakeholder of this field. Invariably, fossil fuels are utilizing to meet the major part of energy demand all over the world but its stocks are limited. Burning of fossil fuels also causes significant environmental pollution that results in drastic climate changes. There is emergent need of an alternative to fulfill the current energy demand in a sustainable way along with sorting out the pollution issues. Microbial Biofuels could be the most appropriate choice in this regard because it bears enormous potential that needs to be exploited. But it has some challenging issues to employ it to meet out the current energy demands. This chapter illustrates the microbial biofuels, harvesting techniques, different process and potential to utilize it as sustainable energy sources along with considerable challenges that need to be addressed to make it as an eco-friendly and economical fuel choice to pave the path of sustainable development.

The fourteenth chapter signifies the importance of microbial cellulase enzyme in the bioenergy. The cellulase enzyme is used to hydrolyze the lignocellulosic biomass to their monomeric subunits, i.e., glucose which can be used to do fermentation to produce bioethanol. Bioethanol from inedible cellulose - rich biomass is the most promising candidate to replace fast depleting, environmentally polluting fossil fuels. Hydrolysis of cellulose to glucose is the crucial step in its biotransformation to bioethanol. Enzymatic hydrolysis is favored over acid hydrolysis, as enzymes are eco-friendly biocatalysts with high substrate specificity and superior catalytic efficiency in mild reaction conditions. Complete hydrolysis of cellulose is achieved by cellulase. Higher cellulase production yield, stability and catalytic efficiency are the main attentive points for the successful implementation in industrial production of bioethanol. This chapter will highlight general characteristics of microbial cellulases and their role in bioconversion of cellulose to biofuels, economic sustainability of cellulose-based biofuels and the latest innovations in cellulase immobilization as the most comprehensive strategy for improvement of enzyme stability, activity, and reusability for cost-effective large-scale application.

The fifteenth chapter describes the comparative studies on the biodiesel productions using neem and jatropha oils. Sustainable development is key aspect of modern-day development. Reducing fuel cost and emissions is considered as the most important research area in sustainable development. Biodiesel has the potential to reduce the requirement of petroleum diesel if blended in suitable proportion. Various feedstocks are being explored for biodiesel production. Oils derived from neem and jatropha can be utilized for biodiesel production. In the current investigations, Optimum parameters for the production, namely catalyst proportion, temperature, oil to alcohol ratio were optimized. Experiments were carried out at different blend proportions to study fuel properties namely kinematic viscosity, aniline point, diesel index, flash and fire points, specific and API gravity, cetane number and ASTM distillation characteristics.

The sixteenth chapter denotes the characterization and comparison of various blends of Honge oil methyl esters with the diesel fuel in order to reduce the air pollution and fossil fuel dependency. The Honge oil, known as pongamia oil is one of the most widely used oil around the world as lamp oil, in leather tanning, in soap making, and as a lubricant. The Honge oil methyl ester can be used as an alternative fuel for diesel. The transesterification of methanol and Honge oil in presence of base catalyst results in Honge oil methyl ester. In the current work, the Honge oil methyl-ester known as biodiesel was produced. It was mixed with petroleum diesel fuel in different proportions. Various physical and petroleum properties were investigated for Honge oil biofuel and also for various diesel and ester mixtures. The optimum parameters were, temperature was 65 0C, catalyst quantity, 0.4 grams and oil: alcohol ratio, 1:6. The blends of 10%, 20% and 30% compositions were most suitable as potential alternative fuel. Different proportions of biodiesel ester in diesel fuel have been also compared with petroleum diesel fuel to identify suitable blend having good engine fuel potential.

The seventeenth chapter signifies the biomass-fed anaerobic digesters in semi-arid areas. The mitigation of adverse climate change requires holistic strategies. The strategies include the proper handling of biomass wastes such as cow dung. The handling of the wastes has to be efficient using appropriately designed anaerobic digesters. In addition, the operating-status of these digesters has to be monitored for detection of any fault of the digesters. Unfortunately, existing approaches for designing and monitoring the working condition of digesters have limitations. This chapter presents the innovative and explicit conceptual design approach of adapting the biomass-fed anaerobic digester and the digester operating-status-indicators in semi-arid areas. The approach is recommended to the renewable energy system designers. Finally, the appropriate indicators are recommended to be applied by the biogas plant operators or users.

The eighteenth chapter emphasizes the different ways to pretreat the lignocellulosic biomass to remove the lignin so that the access to cellulose is improved efficiently. Rapidly depletion of fossil fuel-based energy sources increased the demand for alternate energy sources. Lignocellulosic based 2G ethanol can be used as an alternative sustainable source that present in ample amount. Sources of lignocellulose biomass are wood, food-agriculture wastes and forest residues. Cellulose, hemicellulose, and lignin are the core components of lignocellulosic biomass. Cellulosic and hemi cellulosic biomass are enzymatically hydrolyzed to produce the monomer sugar (such as glucose or xylose) which is further converted into ethanol using fermentation process. The presence of lignin provides physical barrier that limit the access of enzymes required for saccharification. Pretreatment helps in removing the lignin from biomass and reduce recalcitrance. Pretreatment can be done by conventional methods which are chemical, physical and biological method. This study covers the different methods of pretreatment including their disadvantages and benefits along with saccharification and fermentation processes.

CONCLUSION

The Paris Agreement on climate change and the Sustainable Development Goals (SDGs) create momentum for international joint forces to combat climate change. Bioenergy, which represents a significant type of renewable energy is key to supporting the agreed SDGs, particularly in the context of energy security, sustainability, and economic development. Bioenergy is considered the most problematic area among renewable energy sources – with multi feedstock, technology and diverse energy carriers. Presently, several novel physical, chemical, and biological routes are used to produce renewable bioenergy. In India, crop biomass availability is estimated at around 500 million metric tons per year and can produce energy equivalent to 18,000 megawatts. It shows a bright future in energy production in India if adequately used.

Ashok Kumar Rathoure
M/s Akone Services, India

Shankar Mukundrao Khade
Ajeenkya D. Y. Patil University, India

Acknowledgment

This book is the result of the dedicated effort of numerous individuals. We would like to acknowledge the help of all the people involved in this project and, more specifically, to the authors and reviewers that took part in the review process. Without their support, this book would not have become a reality.

First, we would like to thank each one of the authors for their contributions. Our sincere gratitude goes to the chapter's authors who contributed their time and expertise to this book.

Second, we wish to acknowledge the valuable contributions of the reviewers regarding the improvement of quality, coherence and content presentation of chapters. Most of the authors also served as referees; we highly appreciate their double task.

Additionally, we would like to thank our Editorial Advisory Board.

Our Production Coordinator skillfully managed the production of the textbook. We wish to thank the publisher for the sustained interest shown by him during the entire work. A sincere expression of the thanks goes to the entire staff of IGI Global.

Generous participation of faculties and students to eliminate errors in the text and to eliminate errors in the text and to refine the presentation will be greatly acknowledged.

Last and not least, we beg forgiveness of all those who have been with us over the course of the years and whose names we have failed to mention.

Acknowledgment

Chapter 1
Bioenergy:
Social, Economic, and Environmental Impacts

Shweta Arun Avhad
Savitribai Phule Pune University, India

ABSTRACT

As we know that escalating population demand and technological advancement lead us towards a new world dominion power, but with that, there is huge pressure on resources due to an increase in consumption at all levels globally. This leads to revamping environmental motifs all over the world and non-renewable sources are estimated to be not rehabilitating soon. In this case, bio-energy is energy presented as a major opportunity to address the issues of climate change by reducing carbon emissions. This chapter explore bioenergy.

INTRODUCTION

As we know that escalating population demand and technological advancement lead us towards a new world dominion power, but with that, there is huge pressure on resources due to an increase in consumption at all levels globally. This leads to revamping environmental motifs all over the world and non-renewable sources are estimated to be not rehabilitating soon. In this case, bio-energy is energy presented as a major opportunity to address the issues of climate change by reducing carbon emissions.

BACKGROUND

Bio-energy can be defined as a flexible renewable energy source that can be obtained from organic matter, which includes husbandry, and food waste, forestry sector includes timber, also sewage, and so on. In this, all energy is obtained from a biological source, which produces heat energy (thermal energy), electricity or power, transport biofuel, cooling purposes, etc. Bio-energy makes up nearly 15% of global primary energy consumption substantially from fuel wood and 50% of global renewable energy sources.

DOI: 10.4018/978-1-6684-5269-1.ch001

(Gheewala Shabbir, et al., 2011). Unlike other renewable energy sources, bio-energy spare on a raw material feedstock that can use within energy processes.

BIOMASS

Bioenergy is energy deduced from biomass. It has appertained as a material which is produced by photosynthesis and which is residue from husbandry, forestry, and related diligence, a beast and vegetal composites, algae, fungi, and biological wastes which all are of biological origin and are biodegradable. They are obtained as feedstock in the manufacture of bio-oils, biofuels like petrochemical substances, and other products. The key challenge for bioenergy is arising due to changing environmental substitute patterns that adversely affect the supply chain of biomass, as it is sensitive to environmental impacts. This change in biomass use for energy has an insinuation associated with changes in land use patterns, which affects the subjection of biodiversity to feedstock production.

In this chapter, we are going to discuss the impacts of bioenergy socially, economically, and environmentally.

As its major threats of bioenergy have not been instigated at greater risk for the world, yet hence it seems to be a good solution for now. In short, bioenergy can act as a key part of a future demand solution. According to the IEA Bioenergy report, bioenergy includes efficient and increasing the sustainable use of biomass for biomaterial production. Also, power and thermal energy for transportation purpose as the major opportunity to reduce carbon dioxide emission including clean efficient renewable energy solution this efficient source act as an alternative source of energy for non-renewable source for sustainable bioenergy use (IEA Bioenergy report, 2020). This also helps to step forward to achieve sustainable development goals.

According to IPCC (intergovernmental panel on climate change) 6th assessment report, global temperature speed up in the last 100 years and reach its peak and in the next 20 years, global warming will breach the threshold of 1.5 degree Celsius. If GHG (greenhouse gases) continue to emit global warming, will be above 2 degrees Celsius by the mid-2100s. The rise in sea level started affecting sea-lined countries and Small Island are at greater risk .as per researchers, this bioenergy can act as a stabilizer to control GHG emission and help to reduce extreme weather events. Nevertheless, demand for bioenergy feedstock increases due to bioenergy demand escalating, this feedstock is primarily sowing the seeds of agricultural and forest material. The IPCC determined that the average amount of BECCS(bioenergy with carbon capture storage) in the climate framework requires 25 to 45% of fertile and permanent cropland that is nearly 6,500,000 km square of area up to the next 80 years (Vaughan, A., 2021). Hence, to keep our energy supply pristine, bioenergy must emanate from sustainable resources. This biomass production to derive bioenergy should not affect food material production. Impacts on the environment (including animal welfare, plants, air, water, and ecosystem) must be superintended aptly. LUC i.e. land use consumption has to be sustainable.

IMPACTS OF BIOENERGY

Bioenergy is act as an alternative to fossil fuels, and energy suppliers and increases the energy security of rural communities. Further, it has accommodated sustainably and carefully towards affecting socially, also feedstock affects economically and a limited amount of land, water, the environmental condition affects environmentally. These effects may be positive or maybe negative hence; we can say its impact is having both aspects.

Figure 1.

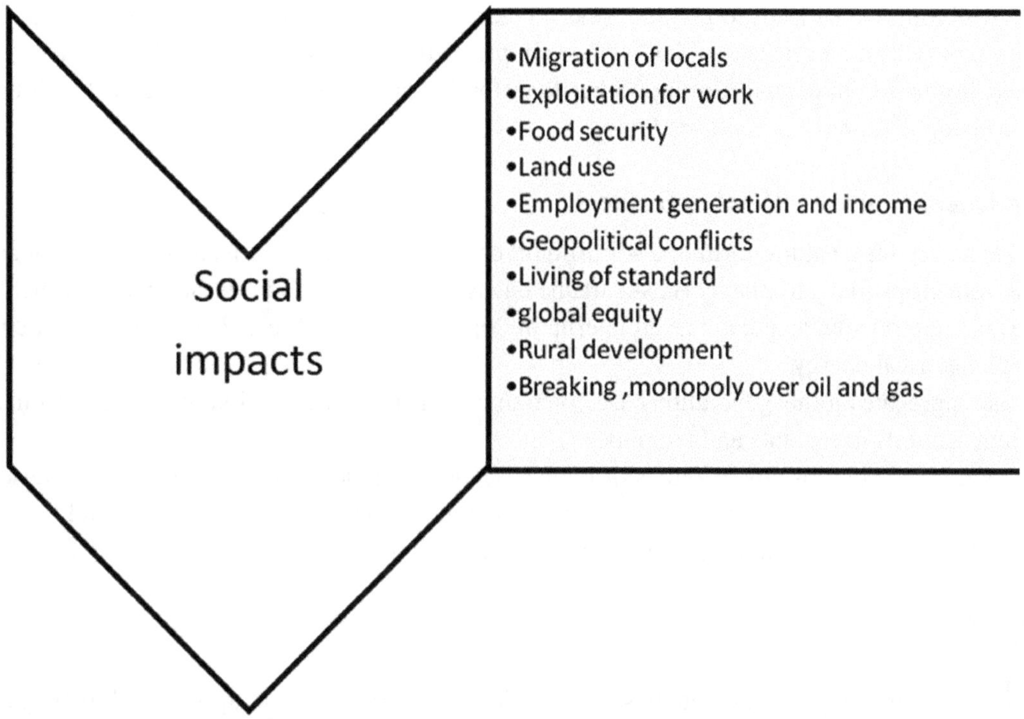

Social Impacts

Bioenergy, which is executed in different forms, has an impact on society. They vary from the merits of new job opportunities, public support, awareness among the people, food security, and water quality to its demerits of societal conflicts arising due to geopolitical regions inflation due to reducing food agriculture land due to changes in land use pattern. There are many factors, that sculpt the rate of change of bioenergy and its consumption demand depends upon local and regional factor influencers. We can see the impacts of bioenergy in a stepwise manner as follows:

1. Poverty Reduction

Uses of bioenergy require biomass, which can obtain from land use. In developing, nation's agriculture is still the dominant sector and has increased population pressure in many such countries. In addition, some agricultural pastureland produces marginal food. However, the introduction of biomass production from following crops like corn, sunflower palm, etc. leads to the production of biofuel such as corn ethanol that is used in transport goods. Even in many countries, petroleum products blended with ethanol are used in the markets.

E.g. For India's ethanol blending program in accordance with national policy on biofuel, the government has set a target of 10% ethanol blending by 2022 to 20% in 2030. This has environmental effects too which we can see in the next part on environmental impacts. As a part of a social impact, this leads to an increase in demand for biofuel, which is a sub-component of bioenergy and causes an increased demand for biomass that leads to giving farmers a side income profit that helps them to reduce poverty and become wealthy. Nevertheless, that increase in prices limits the supply of food products as farmers may shift towards commercial farming rather than food supply. This may also affect the global level prices, causing deficiency in food products.

Then a Question Arises: What Is Biofuel?

Biofuel is a term for a natural form of a solid-liquid or gaseous fuel, which is, arises from biomass processing from crops like jatropha, corn, sugarcane bagasse, algae, etc. and releases energy which acts as a renewable neutral source and act as an alternating option for fossil fuels. They can convert chemical energy to electrical energy.

We can segregate biofuel generation based on four criteria i.e. 1st generation of biofuel is from Starch, sugar, fats, and oil from plants and animals.

2nd, lignin, and cellulose by enzymes or by gasification include municipal solid waste, agricultural residue, forest residue, 3rd from algae, 4th from vegetable oil, and biodiesel. Here Liquid biofuels are biodiesel, bioethanol, bioethanol, bio ethyl tertiary butyl ether, etc.

2. Job Creation

At a societal level, the global climate bash often orientates with inherent support for bioenergy needs. This leads to creating job opportunities for many unemployed generations both from rural to urban levels. At the primary level of agriculture, harvesting bioenergy products from Jatropha curcas is labor intensive and can lead to generating job opportunities and income for rural employers, which are laborers, farmers, suppliers, transporters, etc.

In addition, these are used in Transport biofuels, which act as a fast-growing new energy sector due to their demand. Hence, it will emerge as an important part of energy in the next few years whose hunger even started from now. This increase leads to the development of markets for agricultural components. (BBSRC Bioenergy Dialogue - Report on a pilot public dialogue, Dec. 2013) e.g., INDC (Intended Nationally Determined Contribution) under UNFCCC –in India PANCHAMITRA mission has a target of - 40% energy from renewable sources by 2030.

3. Standard of Living Increases

Bioenergy affects people not only environmentally, but socially well-being too, which helps to reduce casteism between communities within the intercountry or intra countries level. As this energy has, the potential to produce employment that directly affects the income of people mostly to rural middle-income class people to live a free life from a debt gravl income generation. In addition, this energy in the form of renewable energy reduces the high cost of electricity bills from coal and gives eco-friendly energy from available sources without any load shedding effect and unlimited access to energy that indirectly improves their living standard.

This leads to awareness among the people about the need for bioenergy and its social impact on their livelihood. This often indirectly promotes the children and youngsters towards school and college for education. According to the report of UNICEF (United Nations of children's fund), many children mainly from the African continent are still facing problems in education. Due to poverty, hunger, non-availability of standard living due to unemployment, and social conflicts in their barren lands like the Sahel area (including countries of northern Africa like Niger, Mali, Burkina Faso, Chad, and Sudan where the great green wall project established to restore greenery. Which are unfolding their climate change crisis desertification, rainfall variation, land degradation, and erosion) these situations discourage children from education. Hence, in such areas bioenergy production and its distribution can act as a catalyst to trigger the standard of living. This is also applicable to many under-developed and developing countries. Also in developed countries like the U.S.A. bioenergy now seems to be a new solution to deal with the oil crisis and stabilize their currency and power, which in short maintains their living standard.

4. Land Use

Geopolitical Conflicts

Social conflict arises due to the need for land use for primary sources and the need for bioenergy production via agricultural biomass. Conflicts between people asserting clientele rights over lands, this cause inconvenience and problems to any scheduled program.

When it comes to the production of energy from biofuel and needs employers to fulfill company needs may arrange facilities for an employer near the biofuel. Projects cause centralization of the working population with their families in distributed areas or cotters but this increasing population may result in a metathesis reaction of farmers from their traditional land site. This may result in land ownership problems called proprietor conflicts.

Case Study

Forest living tribes, in eastern Asia Penan nomadic tribal community living in Malaysian island Brunei fighting a long conflict with palm oil companies because their territorial boundaries were declining and their natural surviving sources used without permission become illegal, which leads to face environmental impacts over their territorial lands due to energy consumption.

In addition, the greatest impact on the urban socially and economically backward class arises because they do not have enough land for taking advantage of bioenergy projects.

What Is a Metathesis Reaction? How Is It Connected?

Generally, this word metathesis is used in chemical reactions, which is also called a double displacement. When two bonds exchange between each other or to a corresponding substrate it formed as a new product known as a metathesis reaction.

It is represented as - AB + CD ® AC + BD. In this equation, B and C replace their positions.

If we interlink this phenomenon to the social impact on land use consumption due to bioenergy projects or any other such projects we can find out that like the new product form in the above phenomenon by interchanging constituents the farmland can be interchange or conversion take place due to such product implementation.

5. Societal Gap

Such type of projects for bioenergy implementation is strongly favored by the renewable concerning party (here it means environmentalist or public in support of bioenergy) while there is another group present that opposes it and disapproves of the project held at the local level. They are having such reasons to oppose like:

1. Bioenergy projects hold people's traditional land that they (landholders) do not want to give for any purposes.
2. If it has future impacts, it may hazardous for maintaining sustainable balance.
3. It can generate waste.
4. If the implementation is improper may be due to corruption, held under the name of a bioenergy project.

In addition, there is a lack of updated data on bioenergy from the local level to the national level to a global level. Because of idiomatic biomass and laborious challenging technology, it is difficult to gather information at the local level. This internal challenge needs to be in a focus on bioenergy project successions.

6. Political Interference

Increasing conflict or societal gap that emerges as an opportunity for some political leaders to make it a big issue rather than solving or awakening people, they use this topic to collect votes in their favor, which is a harsh reality in society.

Global Equity

A broad range of international organizations ensures sustainable and affordable bioenergy and supports the formation of international collaborations.

In 2017, fossil fuel consumption in Asia is nearly 50% of the global supply. While EUROPE and America account for 47%, hence to tackle future energy crisis as many continents and countries in dependent on primary fossils primary energy supply fossil fuels there is a wide scope of bioenergy, and its sharing among friendly nation increase global equity (Global bioenergy statistics, 2019).

E.g.- ISA international solar alliance including 100+countries is a major bioenergy project from solar technique to generate energy that is renewable, once implemented they have a longer shelf life and no electricity bill this helps to encourage the world for shifting toward eco-friendly technologies and increase its global partnership.

Waste to Energy

Municipal waste includes domestic waste, which is biodegradable and utilized for energy predominantly in urban areas. In this situation combustion of waste product, take place. After processing various times, dry waste collected by municipal corporations is converted into bioenergy. Such energy is used in the generation of electricity, fuel for vehicles, and thermal energy for domestic use. This technology helps to reduce waste and recycle it for social and environmental well-being. However, with that, there is a problem of segregation of this domestic waste. Hence, we can be defined it as ''It is the process of generating energy in the form of electricity from the primary treatment of a waste converting it into a source of fuel. 'It is one type of energy recovery from biowaste.

Figure 2.

7. Breaking Monopoly Over Oil and Gas

Petroleum resources are mainly found geographically in the regions of west Asia like UAE, Saudi Arabia, Oman, and Yemen and they extracted it more than any other country and rise from poor to developed economy. It does not mean that other parts do not have reserves of fossils but they have not been explored yet. Some parts of the American continent hold much amount of natural reserves of fossils but have not been extracted yet in that mount and location finding is still in progress. Although, most countries depend on imports of petroleum products from these countries. In addition, faces crises called oil shocks during regional conflicts, world wars, etc. hence bioenergy is a solution over them to break the monopoly of these countries over oil products.

Economical Impacts

Bioenergy can be a cost-effective solution over costly non-renewable fossil fuels. This can lead to a major economic impact due to its availability biomass production cost, processing cost, then supply chain cost, budget, accelerating transition and creating an alliance, open access consumer, captive power plants, income effect, open access to the consumer. According to a recent study, researchers reveal that increasing investment in the bioenergy sector and its overall processing from production to distribution may step towards a wealthier economy if implemented profoundly. The increasing importance of bioenergy constrains organizations like FAO (Food and Agriculture Organization), and OECD (Organization for Economic Co-operation and Development) to make a framework for sustainable use of bioenergy.

Economical impacts are as follows:

1. Capital Investment

Non-renewable energy resources such as fossil fuels have some lacuna. Like, a slow rate of production leads to generate pressure on a course due to its high demand leads to price inflation and mini oil shocks situations whereas Bioenergy products are less expensive and diverse than fossil fuels like petroleum, diesel, and kerosene, which significantly affects the economy. It has great future capabilities ahead hence; there is a huge chance of increasing investment in bioenergy projects. In addition, this bioenergy feedstock material is harvested locally and they are abundantly available making it a sustainable and self-sufficient source of energy. Hence, it is the center of attraction for investors and helps in economic gain.

E.g. - Bioenergy projects such as the biodiesel industry generated $21.6 billion which greatly and positively affects countries economy of the United States. It approximately created thousands of job opportunities.

2. Industry Engagement

As capital investment starts rising, many different companies meet to discuss, projects and their analysis, which procure exquisite understanding among the companies or industries, forming a cooperation alliance among them.

Figure 3.

3. Price Stability

According to FAO, a larger country produces the most reliable estimates than poorer countries. It is common that larger and wealthier countries to have modern techniques and analysis of estimated bit in case of poorer countries have to depend on outdated surveys and they have to face technical deficiencies. Sustainable bioenergy is beneficial for the price stability of both developed larger countries and poorer countries. As we previously discussed that it generates income source and employment, and help deprived people to overcome the debt trap. Schemes like feed-in tariff (FIT) are helpful for bioenergy elaboration that include domestic to municipal and small businesses to be benefitted from it and also from its social acceptance and awareness about it.

Case Study

Hawaii is a small island in the Pacific Ocean 2000 miles away from the U.S. mainland and is mostly dependent on foreign nations to attain fuels like petroleum products, natural gas, and coal as it lacks when it comes to its own fossil reserves. Which greatly increased their expenditure cost on electricity but they found a resilience strategy. They started to use renewable sources like bioenergy from biomass, this helps to save the economy of the country and make it sustainable energy supply. Investment factor also helps them in such projects.

Figure 4.

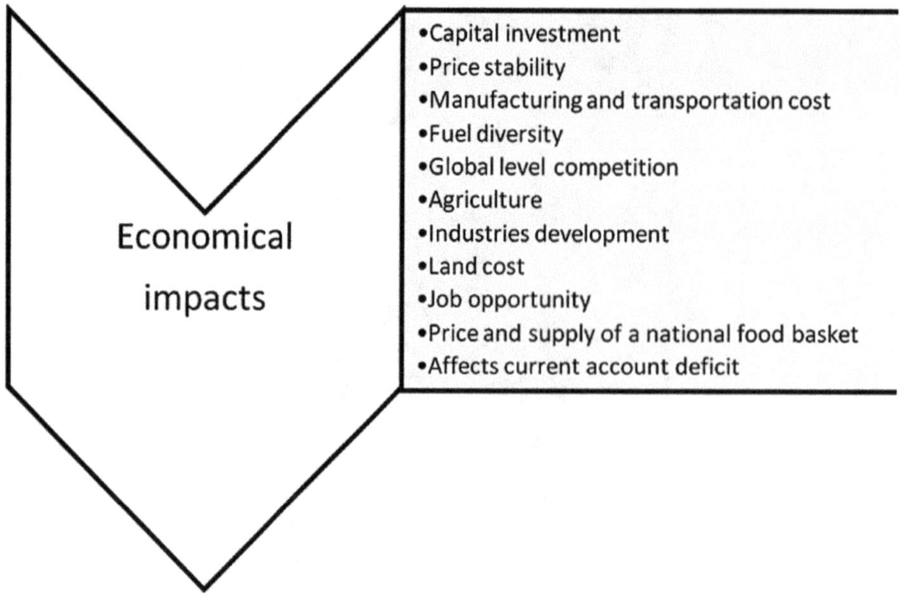

4. Logistic Support and Supply Chain

For the development of bioenergy, it is impossible without including logistic support, which acts as a vertebra for bioenergy gaining. Because energy, produced has to transport to deprived sectors or interchange fossil fuel consumption there is a need for logistic support. From the start of the project, its basic needs, important components, facilities gained from these resources, handling, processing until packing and distribution of materials. Hence, we can say it is a catalyst of energy. This socially empowers people by giving them employment.

Supply chain management, from the implementation of the project, and its location to emission and energy cost need to account for under logistics. Nevertheless, with that, it has negative aspects too. Transportation act as a costly component of bioenergy. It is prominently seen because a measure of bioenergy has lower energy and less volumetric solidity than petroleum products obtained from underground. In some countries, the cost for supply management use is very high than the actual budget, leading to a developing financial crisis. This is because of, a long distance from the location to the distributor, roadways condition, traffic, pipeline management, and waterways having an issue of time consumption due to port management was inappropriate leads to delay loading and unloading in some countries but countries like Philippines exception to that. Bioenergy is the second largest employer globally with approx. 3.2 million people working in the bioenergy supply chain (United Nations Development Programme, 2020).

Now the Question Arises: What Is the Concept of Catalyst?

A substance, which does not take part in a chemical reaction but by its mere presence, can speed up or reduce the speed of any chemical reaction known as a catalyst, and the phenomenon process is called catalysis.

5. Manufacturing Cost

Manufacturers are mostly concerned about cost factors and revenues in any project, bioenergy projects are going from this phase of the cycle. We can say it is a bioenergy production cost. From a bioenergy production point of view, the manufacturing cost is considerably less than that of fossil fuel. However, along with that new implementations have to be made like reserve capacity building has to be increased this is the part of impact assessment. Another impact is different countries have different currencies from the American dollar, and European euro in the west to the yen in Japan, the Cambodian rial, Kuwaiti dinar, and Indian rupees. In these currencies, there is a huge gap in the international market. Which can create an imbalance (there are international organizations to make a balance IMF international monetary fund, which makes the balance of payment BOP) hence, when it comes to importing or exporting feedstock material or basic manufacturing products it may be costly or cheaper. And manufacturing companies find their profits by importing at a cheaper cost but the exporting countries also do not have to be in loss because the purpose of introducing bioenergy is to make sustainable development but along with that it also helps to secure or develop living of standard by giving many employment opportunities and income economically.

Figure 5.

6. Price and Supply of a National Food Basket

According to research, bioenergy affects domestic food supply and affects food baskets due to reversal stipulation in a food market, climate change pattern affecting agriculture hence agronomics cost also fluctuate un accordingly this may cause inflation. Causing inflation greatly influence both international and local level. Many countries have already undergone food crises due to many reasons like financial crises, conflicts among nations or internally, environmental violating influences, etc. Along with that agricultural land also decreasing in such cases globally contribution to bioenergy projects is wracked or seriously affected which may trap the development of bioenergy projects because in such countries they are still facing primary concerns of basic requirements like food shelter and clothes. According to the Global hunger index, which is jointly published by Concern Worldwide and Welthungerhilf. Many African countries like the Central African Republic, Chad, Sierra Leone, Madagascar, Zambia, Yemen, and Sudan, are facing such a crisis. Price volatility also greatly influences their condition. However, bioenergy can be a solution for their problems not completely but leg up and lift them from a crisis. As per researchers, total biofuel consumption rises mainly in countries like China, India, Brazil, the USA, and European countries because these countries were largely dependent on fossil fuels for energy supply or fuel consumption whose prices are relatively high hence their shifting to agricultural feedstock, cellulosic biofuel, etc as another option but it leads to rise food prices too. In such circumstances, markets have to access their physical and fiscal assets. They also have to be transparent functions and efficient to contract and cope with the demand rights of people to oppose or support taken into consideration.

7. Transportation Cost

Despite the increase in the advantage of bioenergy and its growth steadily escalating, transportation is scarcely increasing in terms of bioenergy. This market gives unique opportunities to new startups related to transport facilities that boost the economy of rural and poorer sections as these small businesses have structured their business to give effective transport facilities for biofuels. Transportation of biofuels with greater attention is currently vigorously growing bioenergy sector. However, represent only 2% of total bioenergy. If improvements take place in the transportation sector could lead to quality, the wise better quality of biofuels. it seems to be one of the most feasible ways for the transportation sector and according to sources it reaches 25% in the next 30 years from the current 3%. Increasing global warming and its disastrous nature due to human over technological advancement without an impact assessment and increasing fossil cost leads to an alternate source of bioenergy in the transportation sector (UNDESA, 2006). As bioenergy is mostly produced from sustainable sources like lignocellulose compounds have had a greater impact overall.

Figure 6.

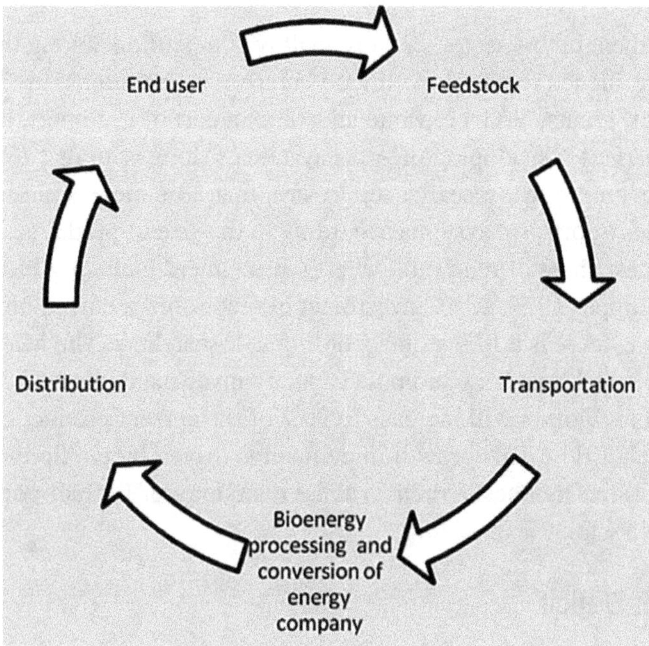

8. Energy Cost

Bioenergy greatly decreases energy costs as we previously see an example of Hawaii. How they stable their energy cost by using bioenergy. The Electricity generated from coal (carbon fossil fuel) along with pollution is price consuming. In addition, petrochemicals are in great demand but many countries do not have their natural reserve hence their importation leads to levy taxation within countries by their governments, which affect society and their living. Reducing this extra energy demand of fossil fuels by altering them with biofuels is a need of the hour. We can see the concept of how this energy is consumed by TFSC-Total final energy consumption here counting an energy content by its total amount of energy transformation and distribution and its consumption in residential, commercial, and agriculture. Gross final energy consumption is a constituent amount of renewable form of energy consumed while energy transportation, its transmission, and distribution. We can also say that Gross final energy consumption=TFEC total final energy consumption + thermal and electrical heat consumption - are used for the non-energy sector (IEA, May 2021). In countries like India, the REC mechanism renewable energy certificate to boost green economy is implemented.

It is a tradable and legal renewable energy credit mechanism where one REC is equivalent to one MWh. This mechanism is associated with the production of renewable energy like bioenergy and encourages the generation of electricity from this source. To deal with challenges associated with it i.e., excess supply over demand REC prices remains close to ground level prices. ESCerts is also a tradable certificate under the PAT (perform achieve trade)-it aims to stimulate energy efficiency. It is a part of the bureau of energy efficiency. Certification schemes to abolish trade barriers, now and in the future

9. Global Level Competition

Due to increased investment in this sector, creates healthy competition among the different nations as it increases income and resilience for rural livelihood enterprises. Economically it has financial and fiscal incentives. Subsidies, grants, VAT i.e. value added tax a part of it, import, and export duties tariffs all act as a financial barrier to developing off-grid systems. Along with that if we look the percentage shows that In Africa according to a research report says that 20% increase in import tariffs (currently at 0%) could reduce sales of off-grid systems according to the size of products. In countries like China, it accounts for the greatest share of renewable energy investment including bioenergy followed by the United States 20% in Europe (19%). Also, investment in renewable seems to be declining in China and India but excluding these, there is a 17% gain according to researchers. The Middle East and Africa accounted for 5%, India 3% and Brazil 2% biomass capacity investment climbed 26% to USD 2.6 billion. In 2030 global demand for biofuel will increase to 90% of the current production of palm oil and other products (Gao Yan, et al, 2011). This growth in demand is driven by public incentives. Like India, its bioethanol blending mission Indonesia already make a strategy of biofuel consumption their biofuel blending increase from 5% to 30% in 2020.

10. Current Account Deficit

The high transaction cost of bioenergy for developing and underdeveloped countries may lead to an increased current account deficit (CAD-simply we can say that current account deficit means if the expenditure is more than its income from various sources and its difference called CAD

E.g., suppose bioenergy production's overall value is 2 million dollars but its consumption by distributing only collects 1 million dollars their difference is one million dollars called a current account deficit. According to economics, less amount of CAD is an indicator of increasing the economy but a large amount decreases it. Bioenergy also influences the GVA of countries which is the gross value added per unit of bioenergy production. The bioenergy sector also leads to the social stock exchange and corporate social responsibility and the digital divide. There are such impacts economically that can happen which indicates both positive and negative aspects.

11. Agricultre

Bioenergy production from 90% solid material which grown in agriculture as biomass, feedstock, and its production started to increase which is a cost-benefit to low land farmers. But to reach its peak near future and to encourage farmers to grow such biomass it's training along with subsidies and insurance in the case of natural calamities or accidental burns on the farm hence to safeguard them such formulas have to be applied. Contesting for agricultural feedstock and its land may increase food prices and farmers and from APMC markets dealers' income it may increase but net buyers will suffer from inflation.

Figure 7.

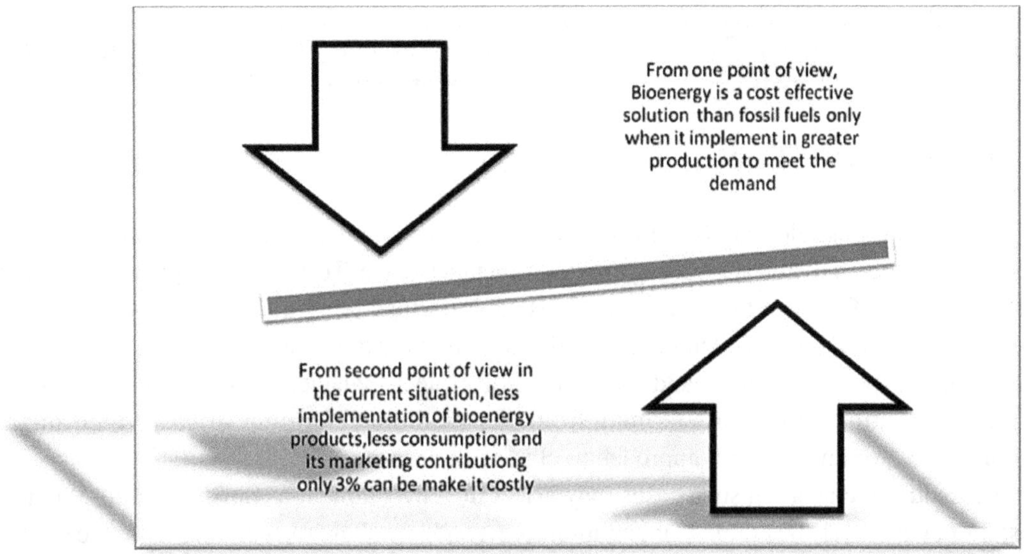

If we talk about its positive aspects, it is economically beneficial to agriculture primary sector and agricultural laborers, but such products also not have negatively affected fertile productive land. Until now, not much such big research has taken for the evaluation of the cost-benefit ratio of the land. Bioenergy seems like grace in many sectors along with social, economical, and mainly environmental concerns but with that, we have research on its long-term impacts by EIA environmental impact assessment so, with taken safety measures we have to ensure that this grace does not convert into a curse for our future generations.

Environmental Impacts

After we have seen the social and economic impacts of bioenergy, let us move on to the environmental impacts of it. From previously seen case studies of the Hawaii area and Sarawak community, we can say that energy security along with global warming and other climatic issues needs new solutions to reduce its hazard. Bioenergy feedstock than carbon fossil fuels may be cost-effective and may contribute to slightly addressing the problem of climate change for sustainable development. Although it seems like a solution it has to highlight that, bioenergy is not a full solution for environmental issues but it can drive as a GHG emission reducer. Contribute to waste treatment in an effective way.

From local condition to various global factors depending upon a type of environmental condition, land, and soil used, its intensity, and more or less productivity such factors affects bioenergy production. An increase in demand for environment-friendly energy like bioenergy also influences feedstock, which indirectly contributes to land use competition among primary agriculture producers that are farmers. They also tend to convert food cropland to energy commercial due to price factors.

Actually, Bioenergy is not a carbon neutral source but can influence such conditions to ensure sustainable development. With the Merits of reduction of GHG (Green House Gas) and the fossil fuels, storage within minimal energy loss, wildlife health and human health, giving agriculturally marginal product income, carbon sequestration, and sustainable source. Bioenergy is also affecting the cons of soil erosion, biodiversity loss, water contamination due to material used in bioenergy agricultural fields, deforestation for lands to grow biomass, etc. There are challenges in the new era of bioenergy implementation. Let us see the impact of bioenergy from an environmental point of view in detail.

1. Land Acquisition or Use

In many countries, degraded lands are used for bioenergy plantation making them reproductive land. Planting trees reduces erosion and ecology restoration takes place. To expand the production and cultivation of biofuels there is a need for a clear well research strategy to regulate land conditions.

The environmental impact of land use includes that in many developing countries small landholding farmers cannot afford its products and materials. For project managers also it is not easy to convince most farmers to introduce biomass farming from their traditional farming. Also, to their migration in case of land need for an industrial approach to change their land.

Indirect land use change in such cases may affect the environment as new agricultural places mean deforestation, which may be arable land, or aural forest and grassland may contaminate the land by using fertilizers and pesticides in such areas. (Bertzky M., et al., 2011).

Figure 8.

2. Soil Quality

Soil is an essential element in the productivity of the land. Due to climatic adverse conditions, poor agricultural practice lowers the fertility of the soil. Major effects of biomass production have not been seen in the products like Jatropha, Palm, Bagasse, algae, etc. Nevertheless, repetition of crops May leads to decrease productivity as nutrients in the soil may be deficient amount. Hence, fodder of macro and micronutrients are given to such farmland to increase yield but most farmers not having actual knowledge of the quantity-wise ratio of such fodder material, which causes to decline in soil fertility due to excessive nutrients.

Agricultural practices may use fertilizers, and artificial insecticides giving rise to soil erosion due to an increase in acidic content and reduction in its water holding capacity and reducing the fertility of soil the effluents which get diffuse in water when in com in contact due to rain, river runoff near agricultural land and underground water leading to water pollution. Hence, we can see the next impact on water.

3. Impact on Water

As we know, the amount of drinkable water on land is 3% from which a large amount is in the form of ice in Polar Regions and only less than 1% of water is available. According to a report by the International Water Management Institute, the rush for liquid biofuels could worsen the water crisis in some countries. In addition, many countries around the world facing water scarcity, and global warming emphasizes it. In such a crisis, bioenergy production could add pressure on existing sources. Pollutants from the soil by agricultural fertilizers and all loaded into water bodies cause pollution. For this, some resource management has to work done efficiently to solve the issue .but on the second side encouraging bioenergy production can reduce the cost of hydraulic renewable from freshwater means energy or electricity generated from freshwater sources to ensure sustainable balance.

4. Forest

Forests hold a large amount of carbon and remove large amounts of carbon from the surrounding atmosphere providing clean air, water, and food, shelter for many wild species hence they play a vital role in sustainable development. In many countries, forestland is replaced by croplands or by large buildings, parks, malls, etc. From a bioenergy point of view, this forestland is replacing due to an increase in demand for biofuels and basic material feedstock which ultimately leads to increased pressure on forestlands. Lungs of the earth ''Amazon forest'' is also not an exception in such case. (whereas still wood is a greater amount of energy sources across the world that act as a biofuel but its cutting and burning leads to pollution and greenhouse gas emission). Shrinking of forestland and savanna-type grassland leads to reduced feed available for animals and local people, which affects their social, environmental, and economic conditions. (Fritsche R. Uwe, et al., 2010)

Loss of forestland or its area leads to the release of more carbon, which influences global warming and biodiversity losses. According to sources, deforestation and poor land cultivation alone released 25% of carbon. (Vaughan, A., 2021). Here implementing the Carbon sequestration method helps to capture such carbon emissions from the environment and helps for better development of a sustainable environment.

Figure 9.

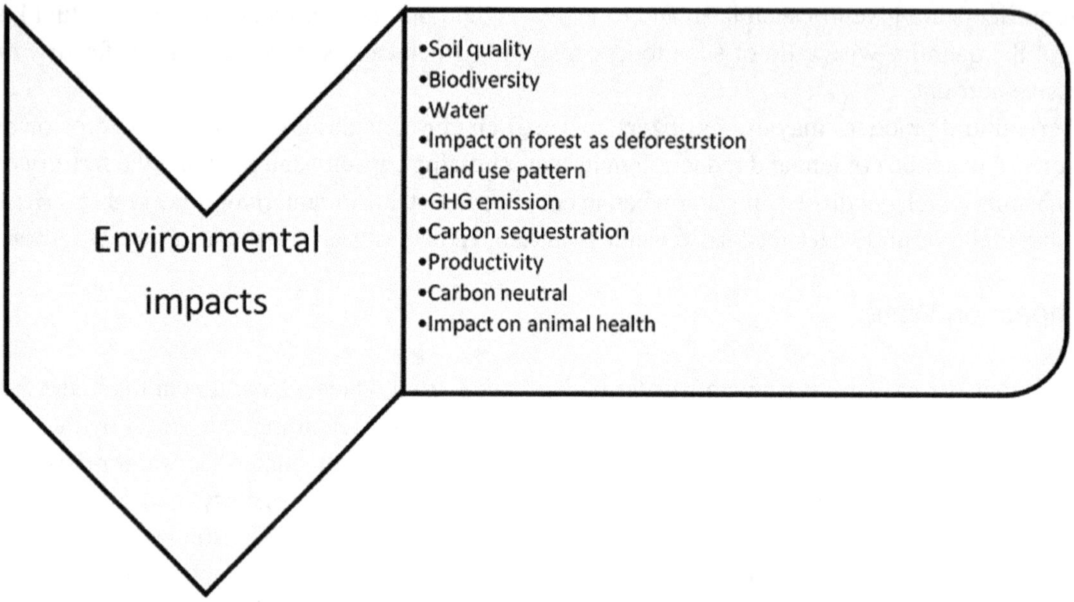

According to the Greenpeace report, biofuel leads to increased GHG emissions due to deforestation of forests and leads to conflicts among communities, between animals and humans, which we called as man-animal conflicts. On the other hand, some scientists argue that bioenergy reduces fossil fuel consumption and emission from it providing wealthy income and jobs to rural peoples.

Deforestation also environmentally affects the carbon cycle and nitrogen fixation.

5. Biodiversity Impact

It has been argued that an increase in feedstock production influences surrounding ecosystems. For example - ICAR NEERI in India implements a Bamboo plantation program to reduce fly ash, which is under processing. In barren and dump land introducing such species increases the greenery of that site like a small forest in less time measurably 3 to 4 years which also helps to increase biodiversity as different small insects, snakes, and bird species are now native to this area.

But on the other hand, deforestation and agricultural practices for bioenergy production reduce the space of animals for their shelter and food, emissions from biorefineries and indirectly creating noise may create disturbance and habitat loss of such species in that area. Here, natural ponds and aquatic life are affected by it.

In addition, biodiversity is under threat due to large-scale monoculture for which a significant amount of bioenergy production takes place.

Figure 10.

6. Emission of Carbon

Carbon is constantly absorbed into the trees and it leaves in the atmosphere when such plants are cut down or die. Bioenergy can be used to reduce carbon emissions from fossil fuels like petroleum, and diesel that emits carbon from vehicles, combustion of it, emission from industries, etc. affecting net carbon emission. BECCS (Bioenergy with carbon capture and storage) could remove large amounts of carbon dioxide from the atmosphere and works as carbon sequestration. However, according to some reports of researchers BECCS could not capture total carbon dioxide from the atmosphere. (Kemper J., 2015). The net energy balance concept arises when the ratio of the primary amount of energy consumed is divided by the total energy produced and to assess the net impact such impact assessment has to be done prior. (Berndes Goran, 2010).

7. Impact On Animals and Human Health

Conversion of forest land to cropland for many projects where bioenergy is also one of it, whose environmental impact also affects many species in the form of habitat loss, their changing behavioral patterns, spawning and mating seasons of animals. Due to noise pollution and excessive involvement and movement associated with creating disturbances among the animals. Animal habitat is changing its pattern. E.g. effluent from bioenergy refineries directly or indirectly dissolves in nearby river ecosystems, pond ecosystem leads to the forceful migration of aquatic animals. (Rupp, S. P., et al., 2012) In addition, increasing toxic content and extra nutrients from croplands in static water develop a HAB that is harmful algal bloom, which is dangerous for an aquatic environment. As they reduce oxygen from reaching the surface and aquatic plants and animals create a condition called a biological oxygen demand also-spawning season and place of fishes fixed from their ancestral part it is an interesting fact about aquatics but effluents and water pollution divert or become a barrier for aquatics. The same condition happens in the case of terrestrial animals.

Bioenergy can have a health impact associated with it. Biomass used for cooking purposes and the unavailability of proper ventilation systems causes indoor pollution which has a long-term effect when exposed for a long time. Potential impacts include respiratory diseases, headaches, etc. however, such impact is not seen in a large amount but it is a point of concern in a rapidly changing world (UNDP and WHO, 2009).

CONCLUSION

In the above whole chapter, we discuss the impacts of bioenergy but along with that, it needs a proper solution to balance carbon levels and to deal with problems associated with it for sustainable development. As we move towards a solution it broader the bioenergy term. There can be a Suggestion as follows:

1. Environmental degradation due to deforestation, land use consumption, soil and water health, etc, monitoring from a status of environmental benefits. Such studies previously gave an idea of future risks and according to that presolution for calamities has to study.
2. Biodiesel made from used cooking oil benefits the health of the environment and the economy linked with it. Such type of new ideas has to be encouraged, also financial support to researchers and projects helps to build green energy corridors in the world.
3. IEA organizations like international energy agencies dealing with proper enactment of energy sources, their applications and impacts ensure reliable, affordable, and clean energy and its worldwide distribution.
4. Green hydrogen usage as a transportation fuel can save us from impacts. Hence green hydrogen mobility project acts on it. Hydrogen produced from renewable sources is called green hydrogen which has no carbon footprint.
5. Global energy review-assesses the direction of energy demand and carbon dioxide emission
6. Climate smart public-private partnership-Climate Resilient Public Private Partnerships: it is a toolkit used to solve the issue related to the following issue. Such type of bioenergy-related bioenergy public-private partnership can influence and spread awareness among the people too.
7. Nature-based solutions UNDP in the 2020 Human development report also called for investment in Nature-based solutions and the assessment of the environment Assessment of the vulnerability of habitats and species is also taken into consideration before implementing bioenergy projects.
8. Implementation of such impact assessment tools effectively mitigate risk linked with bioenergy production from its land use to sustainable production and consumption

The rate of expansion of bioenergy products depends upon a great deal of government policies, straightforward strategy, and foreign investments in developing countries. Taking care of these bioenergy negative aspects is also hindered.

REFERENCES

Bertzky, M., Kapos, V., & Scharlemann, J.P.W. (2011). *Indirect land use change from biofuels production: Implications for biodiversity*. JNCC Report, No 456.

FAO. (2008). *State of Food and Agriculture. Biofuels: prospects, risks, and opportunities. Food and Agricultural Organisation*. FAO.

Goran, B. (2010). Bioenergy land use change and climate change mitigation. IEA Technology Network.

IEA Bioenergy Annual Report. (2020). Under the IEA Framework for International Energy Technology Cooperation, the Executive Committee of each Technology Collaboration Programme (TCP) must produce an Annual Report for IEA Headquarters.

Kemper, J. (2015). Biomass and carbon dioxide capture and storage: A review. *International Journal of Greenhouse Gas Control*, (40), 401–430.

Rupp, S. P., Bies, L., Glaser, A., Kowaleski, C., McCoy, T., Rentz, T., Riffell, S., Sibbing, J., Verschuyl, J., & Wigley, T. (2012). *Effects of bioenergy production on wildlife and wildlife habitat*. Wildlife Society Technical.

UNDESA. (2006). *Indicators of Sustainable Development* (3rd ed.). Guidelines, and Methodologies.

UNDP & WHO. (2009). *The energy access situation in developing countries*. Author.

United Nations Development Programme. (2020). Human development report. United Nations Development Programme.

Uwe, Klaus, Wiegeman, & Herrera. (2010). *FAO Environmental and Natural Resources Service Series, No. 46*. FAO.

Vaughan, A. (2021). Will the IPCC report matter? *New Scientist*, *251*(3348), 8–9. doi:10.1016/S0262-4079(21)01442-1 PMID:34456414

Yan, G., & Skutsch, M. (2011). *A global analysis of deforestation due to biofuel development*. Working Paper 68. CIFOR.

Chapter 2
Life Cycle Assessment of Biofuels:
Challenges and Opportunities

Majorie Moraa Nyasani
Machakos University, Kenya

Victor Odhiambo Shikuku
https://orcid.org/0000-0002-2295-293X
Kaimosi Friends University, Kenya

ABSTRACT

To address energy security and climate-change concerns, substitutes are needed for petroleum-based transportation fuels. In addition to electricity and natural gas, biofuels are emerging as an important class of substitutes. The promotion of biofuels as energy is mainly driven by the perspective of oil depletion, concerns about energy security, and global warming. However, results published in the past few decades present conflicting pictures regarding the energy efficiency and sustainability of biofuels. To adequately address these real or apparent contradictions, a life cycle assessment (LCA) of biofuels is indispensable. "Life cycle" refers to all stages of a process: from the cradle, that is, raw material extraction through manufacturing, distribution, and use to ultimate disposal. This chapter discusses the benefits and limitations of biofuels from an LCA standpoint. The discrepancies and strengths in LCA of biofuels are critically analyzed and opportunities are highlighted. LCA of biofuels is shown to be critical to informing policy and for practical application.

INTRODUCTION

The rising demand for clean, affordable and ecofriendly energy, climate change and global warming concerns, political agitation among nations, and the depletion of fossil-based fuels has necessitated the search and development of alternative sustainable technologies focused on renewable raw materials. Biofuels are considered to curb most, if not all, the aforementioned issues because they serve as an

DOI: 10.4018/978-1-6684-5269-1.ch002

alternative clean fuel with supposed reduced greenhouse gas emissions, the drivers of global warming and climate change (Somerville, 2007). The link between biofuels and climate change is attributed to the fact that biofuels are supposed to be carbon-neutral alternative sources of energy. During photosynthesis, plants get sunlight energy and CO_2 from the atmosphere to form complex chemicals that constitute the plant body. When these plants are burnt, the resulting energy from oxidation is released as heat, CO_2 is given out as well and is recycled again (Somerville, 2007). Therefore, incident solar insolation can be harvested as biomass.

Bioenergy has the potential to offer a significant contribution to clean, low carbon, and secure energy. It is also referred to as a carbon-neutral process. Biofuels are also encouraged to mitigate anticipated shortages of fossil fuels resulting from anticipated depletion of natural resources or geopolitical instability. Despite the merits of biofuels, policy frameworks developed and incentives initiated, issues have been raised concerning other environmental impacts associated with biofuels and their energy efficiency sometimes with conflicting results. As such, sustainability assessment, before adoption, from an environmental, economic and social standpoint is very crucial and this can be achieved through a life cycle assessment (LCA). Life cycle assessment is a methodological tool used to evaluate the requirements and technological impacts, processes, and products to determine their propensity on resource consumption and pollution generation. It refers to all stages of a process that includes: from cradle, that is, raw material extraction, manufacturing, and distribution in the supply chain to grave, i.e. ultimate disposal (Mckone et al., 2011). Life Cycle Assessment is used within the framework of internationally accepted methods as explained in the ISO 14040 and ISO 14044. Figure 1 is a graphical sketch of the different steps in the life cycle of biofuel from the extraction of the raw materials to the end-user.

Figure 1. Graphical Representation of LCA of biofuels
(Hauschild et al., 2017)

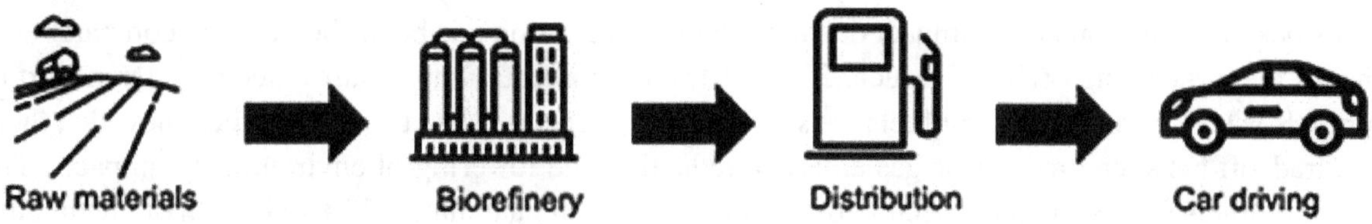

LCA has been used in different fields of study such as wastewater treatment (Gallego-Schmid et al., 2019), solid waste management (Sebastian et al., 2022), the development of construction materials (Liu et al., 2022), food production systems (Lamnatou et al., 2022), and the environmental impacts of processes such as cement production (Ige et al., 2021) among other sectors. These studies underscore the usefulness of LCA in decision-making process and the need for comprehensive LCA of biofuels before adoption.

There are four types of activities that constitute the Life cycle assessment. The initial step entails defining the goal and scope of the assessment. The second activity is to collect life cycle inventory data on materials and energy flows emissions as well as wastes. The third activity entails conducting a life cycle impact assessment (LCIA), which serves to characterize the constituent process impacts. Lastly is the analysis and interpretation of the major findings alongside sensitivity and measurement of uncertainty to influence the decision-making processes (Mckone et al., 2011). Figure 2 is a schematic diagram of the four activities that define a life cycle assessment process.

Figure 2. The general steps of life cycle assessment (LCA)

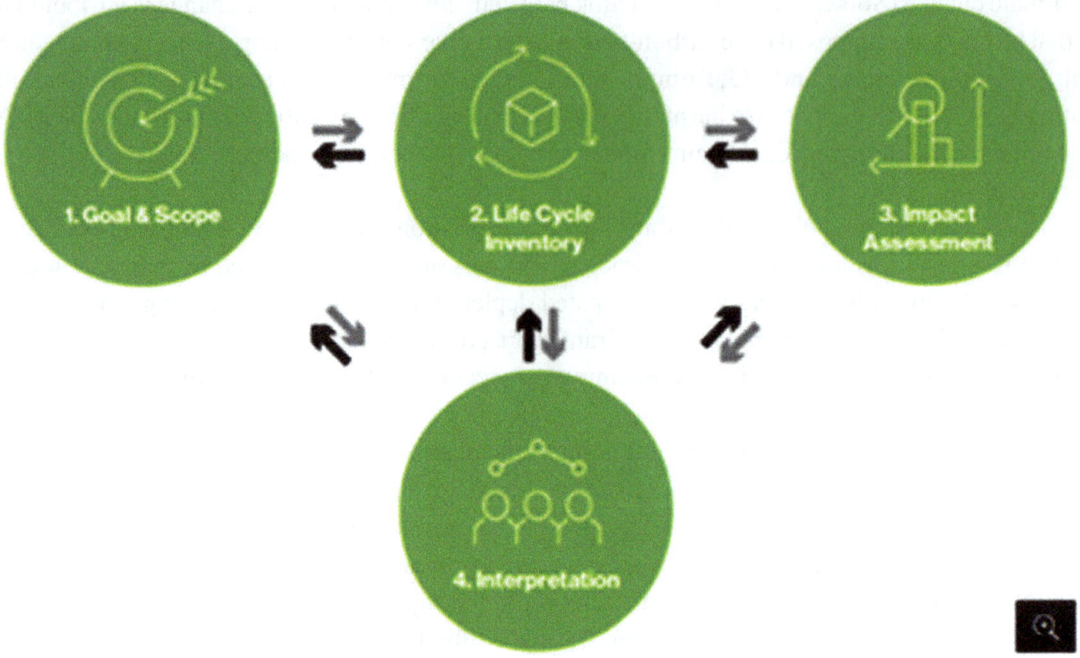

Regardless of the source of the biofuel used, biofuel production can be done in an environmentally friendly way depending on the raw materials used and the production technology utilized. Some important criteria for assessing fuels is their environmental impact and energy efficiency. A greater fraction of the environmental impacts of biofuel are due to the agricultural process of cultivation of the respective feedstock. The environmental impact due agricultural cultivation has been shown to be comparable to that of fuel processing (Zah and Gauch, 2007). The environmental impact attributed to the transport of biofuels is less even when the biofuels are sourced from overseas. Most research studies show that there is a tradeoff between greenhouse gas emission reduction and lowering of environmental impacts. The use of biofuels from sustainable technologies and sources can guarantee a 30% reduction of greenhouse gas emissions which is crucial for climate change mitigation (Zah and Gauch, 2007). A 30% reduction in carbon footprint is appreciable in the search for climate change resilient technologies. However, other studies have suggested that the benefits of biofuels are limited or even absent altogether (Reijnders et al., 2008). The sources of these variances are a subject of debate among researchers (Larson, 2006).

In terms of the environmental impacts of biofuels, tropical agriculture has been identified as a source of air pollution. Combustion of biomass for energy releases large amounts of CO_2, which has been linked with measurable adverse effects on biodiversity. For regions in moderate latitudes, lower yields of crops, intensive use of fertilizer and mechanical tilling of the land cause significant environmental impacts (Zah and Gauch, 2007). Using the LCA tool for environmental impact assessment of biofuels provides fresh evidence that not all biofuels have reduced environmental impacts when compared to the impacts caused by fossil fuels. From research studies, for biofuel production paths, utilization of biogenic wastes and use of grass to wood results in reduced environmental impacts when compared with petrol.

In Europe, use of biofuels has led to reduced greenhouse gas emissions and therefore improving environmental performance. Comparative studies and LCA studies on biofuels and fossil fuels have been done and indicated reduced CO_2 emissions for biofuels. Countries like the United States, Brazil and China are the largest global producing billions of gallons of ethanol as an alternative source of renewable energy. In Europe, production of biodiesel is higher compared to production of ethanol where rapeseed is utilized as the major biofuel feedstock (Nanaki and Koroneos, 2011). European union produced about 5.7 million metric tons of biodiesel with Germany leading with production of 2890 thousand metric tons of biodiesel followed by France. Production of biodiesel is mainly done from oil crops like rapeseed and sunflower oil. Waste cooking oil and animal fats can also be used. The extracted oil is converted to biodiesel through transesterification with an alcohol. The resulting biodiesel can be blended with fossil fuel with about 5-10%. Figure 3 presents a comparison of LCA of biofuel and fossil fuel while Figure 4 shows comparative values of emissions between biofuels and fossil fuels.

Figure 3. Life Cycle of biofuels and Fossil Fuels
(Nanaki & Koroneos, 2011)

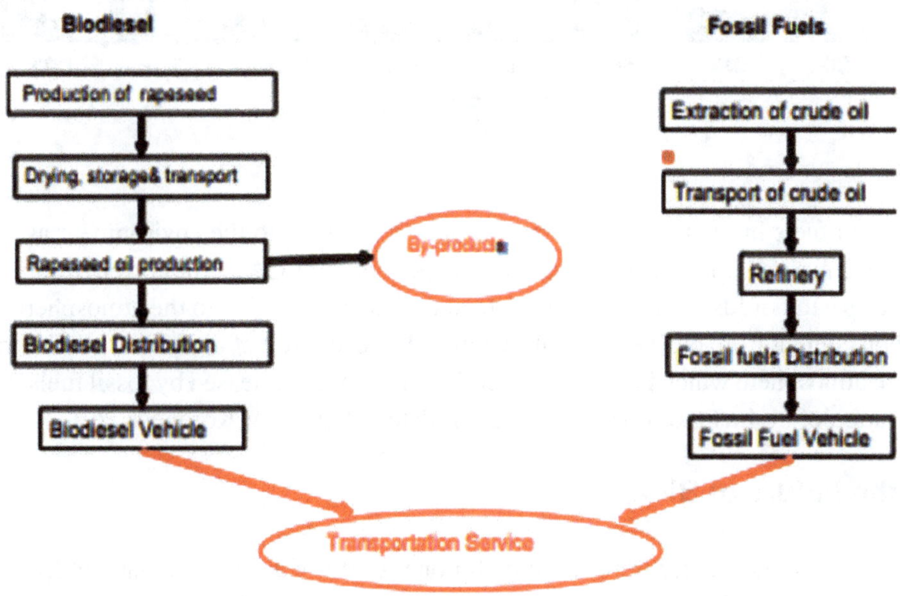

Figure 4. A comparison of emission between biodiesel and diesel
(Nanaki & Koroneos, 2011)

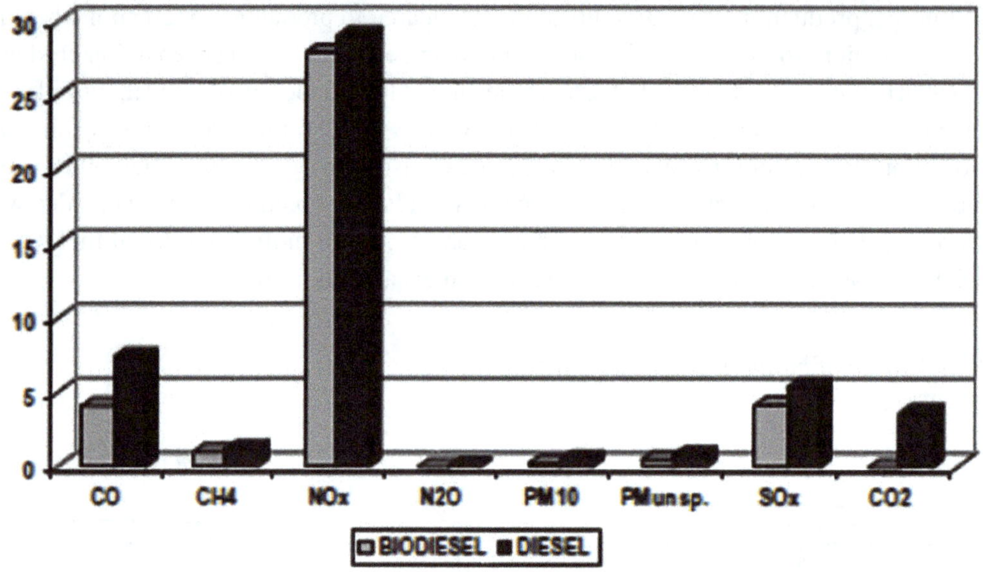

The use of biofuels instead of fossil energy is quite beneficial to the environment as it reduces the greenhouse gas emissions and is also a relatively cheaper option. LCA studies on environmental impacts indicate that every ton of fossil fuel contributes to about 2.8 tons of CO_2 to the atmosphere whereas, the specific carbon content from biofuels is slightly lower. For every ton of biofuel, about 2.4 tons of CO_2 is released to the atmosphere which is slightly lower than the amount released by fossil fuels. Furthermore, during cultivation of biofuel feedstock, CO_2 is assimilated (Nanaki & Koroneos, 2011).

LCA and the Future of Biofuels

Annual crops used for biofuel feedstock account for one-fourth of agricultural land utilized and indirect land use. In Sweden, only 5% of the arable land is being used for biofuel feedstock production (Zah & Gauch, 2007). To increase the production yield of biofuels, the percentage of direct land use for grassland must be increased. However, such a move will increase the biogenic greenhouse gas emissions. Even so, such an increase in biogenic emissions of greenhouse gas can be countered by ensuring efficiency through the entire biofuel production chain. For instance, in the supply chain, during the manufacture of mineral nitrogen fertilizers, oxides of nitrogen can be reduced by fitting in nitrous oxide cleaning equipment.

According to Zah & Gauch (2007), ethanol from wheat biofuel feedstock accounts for 71% climate benefit as compared to fossil fuel utilization if system expansion is applied and this excludes straw. Currently, the byproduct distillers waste is used as biofuel feedstock and it accounts for about 1-2 TWh of bioethanol. This corresponds to about 42 TWh of petrol consumption and a total of 84TWh when diesel is included (Zah & Gauch, 2007). Additionally, the existence of the export market, like in the EU, the total market for distillers' waste that is used as feedstock for biofuels corresponds to ethanol production of 5% of lEurope's petrol consumption. Using the energy allocation method, wheat-based ethanol accounts for 63% of climate benefit excluding straw.

When distillers waste is used as protein feed in biogas production, the climate benefits from the production of biogas account for about 67% when system expansion expansions are utilized. That means digestion residues are assumed to substitute mineral fertilizers. Sugar beet-based ethanol has a climate benefit of 80% and if the energy allocation method is applied, then the climate benefit stands at 74%. In Sweden, sugarcane-based ethanol imported from Brazil accounts for 79% climate benefit and about 77% if the energy allocation method is used. To consider climate benefits and apportioning future contributions, continued expansion, the type of land used for feedstock production will be considered (Zah & Gauch, 2007).

Climate change mitigation due to the use of RME stands at 68% as compared to fossil fuels. If soy meat is substituted by rapeseed, which is a by-product, then climate benefits increase from 64% to 72%. Using the energy allocation method, excluding straw, the climate benefits stand at 53%.biomass from ley crops, sugar beets and maize promotes climate benefit of 86% for ley crops, 85% from sugar beet, and 75% climate benefit from maize biofuel feedstock.it is also worth noting that if methane losses are considered and biogas upgrades applied, then there will be a reduction of climate benefits by 5% points (Zah & Gauch, 2007).

Using manure residues results in climate benefits of 148% using waste from food industries as biofuel feedstock accounts for 119% whereas biofuel feedstock from household waste to generate biogas accounts for a climate benefit of 103% when compared to fossil fuels. The climate benefits in this case exceed 100% due to the indirect benefits gained due to recycling of the nutrients and the reduced utilization of fertilizers and recycling organic matter to the soil.

On the contrary, Hill et al. (2016) explains that the use of biofuels to reduce greenhouse gas emissions tends to encourage high consumption of fuel. Hill et al. (2016) argues that increased biofuel production will lower the fuel prices and therefore the fuel demand will rocket up encouraging additional consumption. This leads to a fuel market rebound effect which end up undermining the strategies of climate change mitigation. High biofuel consumption will lead to increased greenhouse gas emissions. Further, increased biofuel feedstock cultivation will result to indirect land use change which also contributes to increased greenhouse gas emissions.

((Hill et al., 2016)) advises that with right policies, biofuel feedstock production may help mitigate climate change issues like greenhouse gas emissions. Hill et al. (2016) further notes that the use of biofuels as an alternative low carbon source can lower greenhouse gas emissions with right policies in place, such as imposing punitive measures like increased tax on pollution activities rather than lowering fuel prices. High tax on pollution-causing activities will curb the market rebound effect. It will direct the market towards the right direction.

Benefits of LCA for Biofuels

LCA for biofuels is done to give a comprehensive comparison of the biofuels and that of fossil fuels to understand the difference in their respective environmental performance. Regulators, policymakers, and government agencies utilize the LCA tool. For instance, the Greenhouse gas calculator is developed for policymaking using a single impact category, which is global warming. LCA also gives an insight into the significant environmental impacts related to biofuels without having to compare them with fossil fuel impacts. This applies when getting data about the various feedstock's used, information gathered from LCA is also used by the biofuel companies to foster improvement and sustainability and also in research and development(Van Der Voet et al., 2010).

LCA helps in evaluating the environmental impacts of a given product or system. This means that it not only considers the product use of the biofuels when converting chemical energy to kinetic energy in a vehicle's engine, but also considers the industrial and agricultural processes before biofuel is delivered to the fuel tank of the vehicle (Hauschild et al., 2017). LCA helps illustrate that no biofuel is climate neutral because at the industrial process, fossil fuels are used as inputs before the biofuel usage stage. With LCA, one is able to determine that an increase in the demand of biofuels could lead to increased demand for conversion of natural land to meet the biofuel feedstock need. The conversion leads to release of carbon bound in the biomass as well as in the soil in form of Carbon dioxide. A country increasing its production of biofuels contributes indirectly to natural land loss and deforestation.

LCA utilizes a comparative assessment that uses best estimate principle. This is quite beneficial because it allows researchers to have accurate comparisons that are free of bias. This means that equal level of precaution is observed throughout the entire environmental impact assessment. Furthermore, LCA considers a wide range of environmental issues including land use, climate change, fresh water usage, toxicological impacts on human health and aquatic eutrophication. The importance of focusing on the wide range of environmental issues it to eliminate burden shifting to one type of environmental issue which has a potential of increasing other types of environmental impacts (Hauschild et al., 2017).

Moreover, due to the comprehensive LCA, one is able to understand significant effects of biofuels in socio economic status. For instance, cultivation of biofuels releases introduces fertilizer chemicals and pesticides to the soil which contribute to eutrophication as well as toxicity to the aquatic ecosystem. The biofuel feedstock cultivation will also require massive land and substantial amounts of water for irrigation, there is is loss of biodiversity and water scarcity. Additionally, increased cultivation of biofuels contributes to increased food prices (Hauschild et al., 2017).

LCA studies are used to develop policies as well as identifying the crucial areas of contribution an improvement. Recommendations from LCA studies are used in research and development areas which contribute to biofuels sustainability. Data obtained is used for contribution analysis and to trace on specific emissions, and process from an impact point of view.it identifies the key hotspots along the biofuel production chain(Somerville, 2007). According to (Nanaki & Koroneos, 2011), LCA of biofuels is a very invaluable tool that offers flexible system parameterization as well as a comprehensive integrated evaluation of the environmental impacts. The authors assert that LCA studies have the potential to impact the strategic planning and policy making of the transportation sector. This is because, it takes into considerations, aspects like technical, environmental as well as costing. LCA studies have contributed to shifting of strategic planning and in influencing operational decisions. This is because it is an effective tool in identification of improvement opportunities that may not have been captured in previous studies (Cherubini et al., 2009).

Challenges of LCA for Biofuels

Despite the advantages of LCA for biofuels, there exist challenges in the tool. Among the challenges are research gaps and uncertainty especially in areas that have not been explored or are not fully developed. Additionally, in the agricultural processes, there are uncertainties with large measurements that are not easy to obtain. For instance, biofuels use different types of feedstock that are grown in different types of soil and utilize different types of agricultural practices under different climatic conditions. These differences result in varied emission rates which cannot be generalized when determining the environment of biofuels (Van Der Voet et al., 2010). Research gaps therefore emerge due to the coverage of feedstock and the respective geographical area.

Another challenge is the incomplete coverage of impacts. This is because; most of the LCA studies are limited to greenhouse gas emissions. However, recent studies have increased the scope to include other impacts, making data and models easy to use. Even so, impacts like water use and land use have barely been considered (Van Der Voet et al., 2010). Moreover, translating the functional unit into real-life improvement poses challenges of scale and dynamics. There are factors to consider when translating the pilot projects to the real world like the rebound effects, the market mechanisms that will influence food security, and the social structure.

The current supply of biofuel is feedstock from corn and sugar cane, which are farm-grown. Other sources in the future may include feedstock from farms, rangelands, or forests. Due to the nature of feedstock and their transportation cost, they may be stored and processed at small to intermediate scale facilities (Mckone et al., 2011). Compared to oil companies and government agencies that have a hierarchical structure to make decisions, biomass production has no such hierarchy and decision making especially at the first stage of production. When implementing a life cycle assessment of biofuels to influence policies proves to be a hard task as compared the case of a formal hierarchical structure. For example, uncertain weather conditions, complex markets pricing, and tax policies influence farmers in unpredictable ways.

Additionally, due to the vast sources of feed stocks ranging from sugar cane, cane residue, rapeseed, switch grass, and other oilseeds, a wide range of information is needed because the characteristics of the feedstock are different. Therefore, it is difficult to make an overarching decision with different characteristics. This is because many alternative feed stocks will have mixed output systems. For instance, one type of crop can produce protein and at the same time produce fuel feedstock (Somerville et al., 2010).

Furthermore, it is crucial to apportion impacts biofuel feedstock from primary products, by-products co-products as well as residues. Allocation of decisions requires a clear rational basis. Mckone et al. (2011), assert that impact metrics like carbon, health, water footprint require a clear and rational basis. Biofuels feed stocks with mixed outputs are sensitive to either of the allocations like no allocation, allocation by energy, or allocation by economic approach. There is a challenge on the sensitivity of impacts to the choice of allocation (Gnansounou et al., 2009; Guinée et al., 2009). The U.S EPA's renewable fuel standard assessment has had such a challenge. For instance, when determining if rapeseed biodiesel fuels meet the EU renewable energy directive that saves on 60% of greenhouse gas emissions, researchers found out that it meets the requirements. However, only the rape cake, which is a co-product, was used along with the use of glycerol as a source of heat. Allocation by energy content was used (Thamsiriroj and Murphy, 2010). Several other studies depict this problem in the choice of the allocation method (Kendall et al., 2009; Slade et al., 2009).

Moreover, farmers and foresters are a bit reluctant in making long terms land and resource commitment to perennial feed stocks like tree crops and switch grass. Despite the land being suitable for such feed stocks, the farmers need assurance of clear markets as well as financial security. An additional issue is the indirect land use (Fargione et al., 2008). Land available for both food and biofuel feedstock is already heavily used. Biomass production will accelerate deforestation. Deforestation leads to the release of large amounts of carbon to the atmosphere causing a climate impact debt that will take decades to counter. Furthermore, the use of cropland for biofuel production has also been linked with increased emission of greenhouse gases (Searchinger et al., 2008).

CONCLUSION

Biofuels are promising alternative energy sources to fossil fuels with appeciable grenhouse gas (GHG) benefits. The increase in biofuel life-cycle assessment (LCA) studies has subtantially bridged the gap in the discrepancies reported by different researchers on the sustainability of biofuels. While the benefits of biofuels are appreciated, LCA studies also show that conversion of forest land to fuel crops has no greenhouse benefits and may even increase GHG emissions. Future research should emphasize land use, land use change and water in LCA. With paucity of data on such important parameters, discrepancies still remain which limit policy making.

REFERENCES

Cherubini, F., Bird, N. D., Cowie, A., Jungmeier, G., Schlamadinger, B., & Woess-Gallasch, S. (2009). Energy- and greenhouse gas-based LCA of biofuel and bioenergy systems: Key issues, ranges and recommendations. *Resources, Conservation and Recycling*, *53*(8), 434–447. doi:10.1016/j.resconrec.2009.03.013

Fargione, J., Hill, J., Tilman, D., Polasky, S., & Hawthorne, P. (2008). Land clearing and the biofuel carbon debt. *Science*, *2008*(319), 1235–1238. PMID:18258862

Gallego-Schmid, A., Zepon, T., & Ricardo, R. (2019). Life cycle assessment of wastewater treatment in developing countries: A review. *Water Research*, *153*, 63–79. doi:10.1016/j.watres.2019.01.010 PMID:30690219

Gnansounou, E., Dauriat, A., Villegas, J., & Panichelli, L. (2009). Life-cycle assessment of biofuels: Energy and greenhouse gas balances. *Bioresource Technology*, *100*(21), 4919–4930. PMID:19553106

Guinée, J. B., Heijungs, R., & van der Voet, E. (2009). A greenhouse gas indicator for bioenergy: Some theoretical issues with practical implications. *Int. J. LCA*, *14*, 328–339.

Hauschild, M. Z., Rosenbaum, R. K., & Olsen, S. I. (2017). Life Cycle Assessment: Theory and Practice. doi:10.1007/978-3-319-56475-3

Hill, J., Tajibaeva, L., & Polasky, S. (2016). Climate consequences of low-carbon fuels: The United States Renewable Fuel Standard. *Energy Policy*, *97*, 351–353. doi:10.1016/j.enpol.2016.07.035

Ige, E., Olanrewaju, A., Duffy, J., & Obiora, C. (2021). A review of the effectiveness of Life Cycle Assessment for gauging environmental impacts from cement production. *Journal of Cleaner Production*, *324*, 129213.

Kendall, A., & Chang, B. (2009). Estimating life cycle greenhouse gas emissions from corn-ethanol: A critical review of current US practices. *Journal of Cleaner Production*, *17*(13), 1175–1182.

Lamnatou, C., Ezcurra-Ciaurriz, X., Chemisana, D., & Plà-Aragonés, L.M. (2022). Life Cycle Assessment (LCA) of a food-production system in Spain: Iberian ham based on an extensive system. *The Science of the Total Environment*, *808*, 151900. PMID:34838553

Larson, E. D. (2006). A review of life-cycle analysis studies on liquid biofuel systems for the transport sector. *Energy for Sustainable Development*, *2*, 109–126.

Liu, W., Dongxue, H., Tie, G., Zicheng, P., Zishen, M., & Maochuan, L. (2022). Comparative life cycle assessment of cement, sintered bricks and non-sintered bricks manufacturing using water-based drilling cuttings from shale gas production in the Sichuan Basin, China. *Journal of Environmental Management, 314*, 115135. PMID:35487056

Mckone, T. E., Nazaro, W. W., Berck, P., Au, M., Lipman, T., Torn, M. S., Masanet, E., Lobscheid, A., Santero, N., Mishra, U., Barrett, A., Bomberg, M., Fingerman, K., Scown, C., Strogen, B., & Horvath, A. (2011). *Grand Challenges for Life-Cycle Assessment of Biofuels*. Academic Press.

Nanaki, E. A., & Koroneos, C. J. (2011). Comparative LCA of the use of biodiesel, diesel and gasoline for transportation. *Journal of Cleaner Production, 20*(1), 14–19.

Reijnders, L., & Huijbregts, M. A. J. (2008). Palm oil and the emission of carbon-based greenhouse gases. *Journal of Cleaner Production, 16*(4), 477–482.

Searchinger, T., Heimlich, R., Houghton, R. A., Dong, F., Elobeid, A., Fabiosa, J., Tokgoz, S., Hayes, D., & Yu, T. H. (2008). Use of U.S. croplands for biofuels increases greenhouse gases through emissions from land-use change. *Science, 319*, 1238–1240. PMID:18258860

Sebastian, K., Jinqin, Z., Zhen, X., Laner, D., & Woon, K. (2022). A systematic review of life cycle assessment of solid waste management: Methodological trends and prospects. *The Science of the Total Environment, 831*, 154903. PMID:35367543

Slade, R., Bauen, A., & Shah, N. (2009). The greenhouse gas emissions performance of cellulosic ethanol supply chains in Europe. *Biotechnology for Biofuels, 2*(15), 1–19. PMID:19682352

Somerville, C. (2007).. . *Biofuels, 17*(4), 115–119.

Somerville, C., Youngs, H., Taylor, C., Davis, S. C., & Long, S. P. (2010). Feedstocks for lignocellulosic biofuels. *Science, 329*, 790–792. PMID:20705851

Thamsiriroj, T., & Murphy, J. D. (2010). Can rape seed biodiesel meet the European Union sustainability criteria for biofuels? *Energy & Fuels, 24*, 1720–1730.

Van Der Voet, E., Lifset, R. J., & Luo, L. (2010). Life-cycle assessment of biofuels, convergence and divergence. *Biofuels, 1*(3), 435–449. doi:10.4155/bfs.10.19

Zah, R., & Gauch, M. (2007). *Life cycle assessment of energy products : Environmental impact assessment of biofuels*. Academic Press.

Chapter 3
Entrepreneurial Opportunities In Bioenergy

Prashant Kumar
https://orcid.org/0000-0001-9760-6903
Department of Bioinformatics, Kalinga University, Raipur, India

Sunil Kumar Verma
B.N. College of Engineering and Technology, Lucknow, India

ABSTRACT

Entrepreneurism is vital as entrepreneurs push technological change and innovation, consequently generating economic growth. Bio-energy entrepreneurism is the focal point of this chapter, along with entrepreneurial motivations, opportunities, and market drivers. Market drivers include government policies, trends in energy use, and progression of the energy sector into more renewables. Several factors influence entrepreneurs, which involves several factors ranging from the environment to economics, pioneering spirit, and social factors. With the help of entrepreneurial motivations and market drivers, individuals can easily visualize and plan all the bio-energy sector opportunities. Bio-energy entrepreneurs moderate the market of transformation of biomass to energy. The bio-energy field is filled with stories of failures, successes, and everything in between. There is a marginal difference between a failure story and a success story in the modern bio-energy sector, which narrow down to timing.

1. INTRODUCTION

According to an assessment by the International Energy Agency, by 2026, India is on track to become 3rd world's largest market for ethanol behind Brazil and the United States of America, having increased ethanol requirement to an anticipated three billion litres amid 2017-2021. According to the agency, by 2026, Asia is expected on the way to surpass Europe in provisions of bio-fuels output, thanks to India's expanding appetite. While by 2026, yearly global demand for biofuels is expected to increase to 186 billion litres, an increase of 28%, Asia is expected to contribute over thirty per cent of new output during the predicted era. The main driver of demand expansion is thought to be government policy. Additional

DOI: 10.4018/978-1-6684-5269-1.ch003

issues such as public transportation fuel demand, expenses, and policy design will also significantly impact. In June, Prime Minister Narendra Modi stated the government is committed to achieving a 20 per cent ethanol blend in gasoline target by 2025. Previously, the deadline had been set for 2030. According to CRISIL analysts, roughly 16-18 million tonnes of gasoline transactions are expected to be replaced due to ethanol mixing. It will be difficult for India to meet its 20 per cent blending goal in just five years, But even attaining 11 per cent blending would make it the world's third-largest ethanol market behind the United States and Brazil, the IEA study concluded, making it the world's third-largest ethanol market behind the United States and Brazil (Street, 2012). A considerable portion of India's current vehicles could be incompatible with fuel blends containing more than 10% ethanol. Retrofits are an option However, given the scale of the undertaking, this may be impractical, the IEA stated, adding that flex-fuel vehicles or vehicles otherwise compatible with 20% blends will need to be made accessible. Consumers will need to be persuaded to buy them. (Street, 2012) At present, the percentage of ethanol in gasoline is roughly 8.5 per cent. Under the Ethanol Blended Petrol (EBP) programme, the government has already reintroduced the administered price mechanism for ethanol procurement, allowing ethanol production from a variety of feedstocks such as heavy molasses, sugarcane juice, sugar, sugar syrup, damaged food grains, maise, and surplus rice stocks with the Food Corporation of India (FCI).

In Ethanol Allocation Year 2016-17 (December 2016 – November 2017), government governed oil marketing companies obtained 66.5 crore litres of ethanol, which grew to 173 crore litres in the Allocation Year 2019-20. As previously stated, the government authorised oil CPSEs to build 2nd Generation (2G) Ethanol Biorefineries using agri-residues and biomass in various parts of the country. OMCs are building 12 2-G biorefineries for a total asset of Rs 14000 crore. Ethanol purchased through the EBP programme is subject to a 5% GST rate. The EBP initiative attempts to alleviate environmental problems, lower oil imports bills, and provide farmers with a remunerative income. For Ethanol Supply Year 2O2O-21, the procurement price of ethanol was Rs 62.65 per litre for sugarcane juice, Rs 51.55 per litre for spoiled food grains, Rs 56.87 per litre for rice available with Food Corporation of India, Rs 51.55 per litre for maise, and Rs 45.69-57.61 per litre for heavy molasses. After acquiring statutory permits, sugar mills and distilleries are allowed to build ethanol facilities, and the government announced an interest endowment scheme to aid in the construction of these plants. (*International Energy Outlook 2013*, 2013)

2. ENTREPRENEURSHIP MOTIVATIONS

The primary movers or significant stakeholders in all bioenergy entrepreneurial ventures may be motivated by various factors. Lin (2008) acknowledged three key expansion motives for entrepreneurs: social welfare, environmental and entrepreneurial. Bioenergy will solve a problem or issue with the environment in a project with environmental implications. The environmental incentive can be seen in swine farms whose expansion plans are thwarted by concerns about the increased stench. An anaerobic digester that produces biogas eliminates stink dilemma whilst producing energy, used to generate electricity and operate agricultural equipment in the farm (biogas is cleaned to eliminate contaminations, subsequently compressed and consumed as compressed natural gas). (*Cultivating Entrepreneurship | University of Maryland Extension*, n.d.) There are several definitions and examples of social welfare, including delivering energy to a community at a cheap rate than traditional generating, creating employment for the community, and raising tax revenue. When public leaders sought to fund, manufacture, or recruit ethanol factories to their area, the corn ethanol boom provided multiple cases in point of social welfare motives. After it was put together, the ethanol factory delivered high-wage jobs, generated significant tax revenue, and raised maise costs.

In conclusion, prime movers display entrepreneurial instincts when they take chances with breakthrough technologies without a market. The aim to be a market leader in a new bioenergy area drives these entrepreneurs. This kind of entrepreneurial motivation may be perceived in the 1st generation cellulosic plants, which are still trying to get up and running. A further case of entrepreneurial motivation is the several minor businesses running with algae to develop a way for producing and harvesting algae for biofuels. (Han, n.d.)

Ahead of Lin's goals, Shane explained that entrepreneurship requires a human group and an innovative route in which enterprisers participate in the creative activity by relocating resources in a novel way. (Shane et al., 2003). It might be characterised as the fourth motivator of pioneering spirit, or, as Shane puts it, human motivation plays a critical role in the entrepreneurial process. This pioneering mindset will benefit bioenergy entrepreneurs to compete with fossil fuels by bringing alternative energy sources to market. An entrepreneur must be driven by a desire to respond to market conditions. The next part delves deeper into these market drivers.

3. BIOENERGY AS A BUSINESS

The acceptance and development of a dependable source of renewable bioenergy is a big challenge that most countries face. The majority of studies in renewable energy focus on the controlled models of implementation intended for this type of energy. (Järvinen, 2005) From a business standpoint, the economic benefit of bioenergy is strongly reliant on bio-mass robust usage. As a result, a grasp of bio-energy utility parameters is critical. The source of bio-mass, the structure and location of the active business, the storage and transformation centre, and the relevant supply and consumption sources are all key aspects to consider when assessing the advantages of bioenergy in a business setting.

The improvement of high-efficiency bio-mass heating systems has unlocked up fresh possibilities for the active use of wood in wooded areas with systems located close to the user. In general, the plant placement is adjusted by locating the optimum site within high-productivity wooded areas with a handy network of forest paths, a good presence of a woody business, and easy grid connectivity. The environmental impact, economic and social implications, and Encouragement of communal amenities all play a role in the economic viability of bio-energy power plants. As a result, bio-mass heating has numeral convincing benefits, both for the surrounding community and system proprietor in the case of district heating systems. (Dahiya, 2014)

Bio-energy may generally replace expensive traditional energy sources and electricity with local biomass resources, typically accessible for little or no cost as waste (*Cultivating Entrepreneurship | University of Maryland Extension*, n.d.). Bioenergy can reduce pollution and greenhouse gas emissions while simultaneously protecting customers from rising fossil fuel prices. Bioenergy production creates multiple local jobs in the feedstock assembly, planning, and transportation. It allows several other firms to expand and reduces transportation distances. (*International - US Energy Information Administration (EIA)*, n.d.)

The bioenergy industry has several difficulties. Currently, biomass heating systems are more expensive to install than traditional fossil fuel-burning systems. If their cost does not fall, the development of bio-energy usage may be slowed. It should also be noted that the quality of biomass feedstock is vastly varied compared to the more standardised commercially available fossil fuels. (Käkölä, 2003) Bioenergy production necessitates increased physical space for biomaterial supply, storage, and handling. When bioenergy production presents more logistical obstacles, it naturally necessitates greater operator engagement and diligence, which costs more money.

Where the cost of conventional energy is high, and biomass feedstock is low, biomass-heating systems appear to offer the best commercial opportunities (Ojala, 2008). If numerous bio-energy systems are linked regionally, the benefits of bio-energy can be amplified. There are no technological barriers to merging bio-mass heating, wind and solar power, and water power into a similar energy network. Societies with various energy suppliers on a single network are less vulnerable to fluctuating energy prices and climate change.

Nevertheless, the bioenergy industry's ending advantage is always quantified in the bioenergy sector, including customers, merchants, and manufacturers. As a result, it is critical to comprehend how the bioenergy business and marketplaces operate and how distinct players appear to benefit from them. When a product is needed but is inexpensive and functions well, the customer receives a value. (Järvinen, 2005; Käkölä, 2003; Ojala, 2008)

Customer value = performance / price

The above formula is founded with the premise that each product module or role has its unique expenditure, which drives up the cost of the product. In actuality, the expense of construction these functions vary by customer, as do the needs of particular consumers. As a result, each product function should meet consumer demand as outlined in the formula below. (*November & December 2008 Newsletter | Agricultural Marketing Resource Center*, n.d.; Reinis et al., 1975)

Value = Value = [(n · a/c) 1+(n. a/c)2 + (n · a/c) ?]

where:

n = the need for an object or service
a = the ability of an object or service to satisfy this need
c = the cost of the object.

The functions that should meet the customer's needs are represented by numbers in this formula, with the final function represented by a question mark.

A retailer's value of a bioenergy product may fluctuate significantly from a customer's value. This is because, for starters, the retailer is primarily concerned with the monetary gain on his investments. Retailers are looking for fast returns and products that can be sold rapidly. Furthermore, retailers aim to make the most money with the least money. As a result, the value formula for a retailer is not the same as the value formula for a customer. In its place, computing a cost-to-benefit ratio, merchants approximate the sales revenues they assume to gain by selling their products, as shown by a retailer's value formula. The formula computes the retail product value (Jordão, 2011).

Retail value = unit sales (unit price – unit cost)

Maker utilises his formula to calculate the worth of bioenergy products for a retailer or a client. This is because his interests differ slightly from those of the others. He has higher capital costs than a retailer because of the production plant and tools, which tend to compel manufacturers to retain working most of the time. Because the manufacturer is likewise concerned with the gain on his investment, he uses the following formula to determine manufacturing value. (Shane & Venkataraman, 2000)

Manufacturing value = [Customer and retailer benefits (1 to n)] + Profit/Costs

Bioenergy generation should be addressed financially in addition to the business value consideration.

In practice, this implies that all investments should pay off in the long run, and entrepreneurs and communities should also benefit from their efforts' commercial benefits and profits. As a result, the cost of invested capital might be a hurdle for the bioenergy industry. This is since, to thrive, a bioenergy firm may require public investment at first, as private funding may not be accessible. Even if this is the case, bio-energy entrepreneurs must ensure that their business strategies are appropriately constructed, taking into account a variety of business metrics such as value, profitability, and funding. With a well-crafted business strategy, an entrepreneur can also communicate his objectives to various bio-energy industry participants. (Shane & Venkataraman, 2000)

4. BIOENERGY OPPORTUNITIES: BIOFUEL VALUE CHAIN

Waste streams such as municipal solid waste or manure, plant oils, dedicated energy crops, agricultural crop leftovers, and waste streams such as municipal solid waste or manure are all examples of bio feed-stocks. (pre-consumer or post-consumer). Production, harvest, storage, and transportation costs apply to each feedstock. Examining the stages involved in producing feedstock's will provide business prospects.

The following is a breakdown of the biofuel value chain:

Breeding and genetics of feedstock's/Feedstock production/Logistics of feedstock's/Biomass processing/Conversion processes/By-product and co-product utilisation/Marketing distribution Innovative people and small businesses can make critical discoveries in the areas of biomass crop genetics and conversion biotechnology, which are both areas where innovative people and small businesses are capable of creating significant findings linked to better ways to convert biomass to fuels. Biotechnology can impact biofuels production through enzymes, catalysts, yeasts, and plant breeding. Engineering linked to equipment development and refinement is required to solve and improve the system, both at the bio-refinery and across the supply chain. Engineers and tinkerers have come up with new ways to solve equipment difficulties. These solutions have the potential to be commercialised. (Street, 2012)

Fuel has always been distributed through leading corporations and gasoline and diesel organisation. Bio-fuel companies are increasingly participating in the supply chain by growing into blenders and promoting customised pumps that provide consumers with bio-fuel options.

By-products and co-products are critical to first-generation ethanol facilities' profitability. Similarly, enhanced bioenergy facility economics will result from the capacity to market by-products and co-products. Distiller corn gluten and grains feeds are two of the essential co-products of the corn ethanol business. The maise ethanol industry also sells CO_2 and post-fermentation corn oil as by-products. A separate firm partner is frequently used to capture and market the by-product. Companies have formed in some circumstances to add value to the by-product, such as pelleting distiller's grains for sale as a denser feed, capturing CO_2 for use in the bottling business, and converting maise oil to bio-diesel. Regarding the potential for entrepreneurs to put value to a low-value product, by-products are worth investigating.

5. OPPORTUNITIES IN SMALL-SCALE BIOENERGY

Entrepreneurs who make their bio-diesel conversion systems, capture CH_4 from manure and other waste or use waste wood and other materials for heat generation have a long history in small-scale and niche industries.

5.1 Biodiesel

Biodiesel is a clean-burning, environmentally friendly fuel made from edible and non-edible oils trans-esterification. Compared to traditional fuel sources, it is renewable, cost-effective, carbon-neutral, and non-toxic. As a result, it is widely used to minimise carbon emissions in India's automotive, marine, and railway industries. Biodiesel is also mixed with petroleum diesel fuel in compression ignition engines because it improves engine lubrication and extends engine life. In India's vehicle sector, rising petroleum products and growing environmental concerns drive up demand for biodiesel.

Furthermore, the market's expansion is aided by a growing focus on sustainable development techniques. Biodiesel is also utilised to power underground mining equipment and machinery because it improves air quality, increases energy security, and provides various safety benefits. Aside from that, increased energy consumption and the growing use of biodiesel as a heating fuel in both home and commercial boilers are expected to impact the industry in the coming years positively.

5.2 Methane

This segment only analyses biogas methane produced by anaerobic digestion; added bio-gas CH_4 sources, as landfill gas not included. When high carbon materials are enclosed in an anaerobic environment, microbes digest the organic components and release bio-gas, biogas methane is produced (predominately methane). Manure, food industry wastes, human sewage, and other organic waste are all common feedstocks for biogas methane generation. Methane can generate energy in internal combustion engines. Manure is an excellent feedstock for a CH_4 digester, and it is also a feedstock for an anaerobic digester. As a result, such systems are commonly found on swine operations, dairy farms and municipal wastewater systems. The systems produce methane continually after they are deployed. Methane needs to be used or released to avert pressure from developing in the system.

- Flaring CH_4 can help the environment since it converts one carbon and one methane molecule into CO_2, which has a Twenty-four times lower greenhouse gas effect than CH_4.
- If there is a cost for reducing CO_2 equivalents greenhouse gases, such as renewable energy credits or carbon trading.
- Odors can be reduced with the help of Anaerobic digestion, and it has been effectively utilised to lessen odours at a processing facility and farms, letting the facility continue in crowded locations or grow with less opposition from fellow citizens.
- farmers can become bioenergy entrepreneurs. Instead of being the primary product.

5.3 Biomass and Wood

Residential wood heating has been a staple of human civilisation for millennia. In the US nowadays, heating with the help of wood-burning in homes is widespread in densely forested areas. In certain circumstances, technology has remained relatively unchanged. Individual homeowners cut and process their wood in most situations, although some acquire wood from small businesses or other individuals. (Verma & Hussain, n.d.) Pellet stoves, which burn wood or other biomass pellets and have extra controls than wood stoves, have become more popular. For mass transport and packing, wood pellets are homogeneous, dense, and flow freely. Wood pellets have grown in popularity in Europe due to their renewable nature and lower greenhouse gas emissions than coal or other fossil fuels. Pellets for commercial and other heating applications have developed markets. (*International Energy Outlook 2013*, 2013)

- Wood harvest
- Other than commercial timber harvest, thinning, slash, trimmings, lumber mill trash, manufacturing waste, shelterbelt renovation, old pallets, and dead tree clearance are all wood sources.
- Wood processing is converting wood into wood pellets or other high-value products.
- In recent years, the international market for wood pellets has exploded.

5.5 Cooperatives

Cooperatives consist of working collectively to address their members' mutual needs. The cooperative's ownership and governance are shared by its members. Bio-energy cooperatives have sprung up to provide bio-energy to associates or utilise feedstock provided by members. Many businesses begin as cooperatives to benefit from shared resources. A few are created as a project that is way too big for a single individual or family to handle alone. Creating a cooperative can assist in overcoming the first scale. Others are created to deliver a higher-quality product or service than an individual could provide on their alone. The structure will vary based on the cooperative's goals. However, in general, a group of like-minded people pool their capital, labour, or other resources to develop a business that benefits all members. (Ojala, 2008) A cooperative's goal could be social welfare, environmental, or entrepreneurial to return to the developmental objectives.

- Examples of bio-energy cooperatives
- Biodiesel on a small scale (oilseeds are given, bio-diesel is the end product, and members use it for environmental reasons)
- Bio-mass pellet mill
- Methane digester

6. LARGE SCALE BIO-ENERGY OPPORTUNITIES

Large scale bio-energy signifies to businesses that can process tens of millions of gallons of water/year or else millions of tonnes of biomass. Two instances are multi-million gallon per year bio-refinery or megawatt-scale biomass to heat and power facility. These amenities necessitate a substantial financial investment. Entrepreneurship can occur at this magnitude, yet we often associate entrepreneurs with small start-up businesses. We have looked at small businesses how they contribute to large and small scale bio-energy for the objectives of this chapter. Bio-energy entrepreneurs analyse the supply chain at all scales and how challenges might be resolved in creative and novel ways. For specialist expertise and equipment, bioenergy facilities continually depend on technology vendors. Heavy machinery, computers, analytical equipment, waste management and chemistry are just a few of the areas covered by these technological companies. (*Ethanol Impacts | Agricultural Economics*, n.d.)

6.1 Opportunities at the Periphery

There are other options outside of the bio-fuel or bio-energy trades. As a method to contribute, entrepreneurs may be at ease to work on the outside. These off-the-beaten-path options may not immediately benefit the bio-fuel and bio-energy industries, but they do take advantage of their presence. This peripheral activity includes businesses that pop up to satisfy the individual requirements of construction employees throughout peak construction periods. (Ojala, 2008)Following the development boom, these enterprises rely on community patronage as well. Such businesses gain an advantage by having a solid first year of operations, which allows them to pay off debt sooner.

As part of bio-energy booms, various peripheral establishments started, which includes the following:

- Motels
- Cafeterias
- Apparel stores
- Bank
- Service station
- Studio apartment

7. GOVERNMENT INITIATIVE IN SECTOR OF BIOENERGY

7.1 National Bio-gas Fertilizer Management Program

Construction of family size biogas plants of 2 to 6 cu. m. efficiency is being made by Uttar Pradesh New and Renewable Energy Agency from the additional financial year 2010-11 of Rural Development Department and KVIC as per the design and drawing approved by the Ministry under this scheme, such as Dinbndhu, KVIC or installation of permissible prefabricated biogas plants, domestic fuel supply, lighting, and for the production of organic fertiliser for construction of family-size biogas plants of 2 to 6 cu. m. efficiency is being made. The rest of the cost is borne by the beneficiary as part of the Central funds of Rs. 9000/- for General Category beneficiaries and Rs. 11000/- for Scheduled Caste beneficiaries. The cost of a two cubic metre biogas plant is around Rs. 22,000/- per plant, and the plant can cook food for five people simultaneously by providing light for 4 to 5 hours per day from 1 lamp. The plant will also produce 6 tonnes of high-quality organic fertiliser per year, and the plant will provide the equivalent energy of around 18 LPG cylinders per year.

7.2 Biogas Based Power Generation Program

In this system, a generator's gas produced from a biogas plant is used to power plants with capacities ranging from 3 to 250 kV, per a ministry-approved design and drawing. After the beneficiary completes the construction plant, the Ministry will reimburse the beneficiary with a reward ranging from Rs 30000-40000/ kV in declining order. A biogas plant with a capacity of 85 cu. m. may be built with 155 cattles. A hundred units of electricity/day and approximate 270 tonnes of high-quality organic fertiliser are generated by a 100 per cent bio-gas-powered generator with a 10 kV capacity that costs around Rs. 16 Lacks.

7.3 Bagasse Based Co-Generation Project's in Sugar Mills

There is a lot of bagasse accessible in the state's sugar mill, which has much potential for co-generation. Uttar Pradesh New and Renewable Energy Agency began its endeavours in 1994 as an architect to establish power production plants based on co-generation using bagasse from Uttar Pradesh's sugar mills.

Throughout 1994, the state government created a strategy to encourage the development of similar enterprises in the state. PPAs agreed with the power corporation were later specified as model PPAs after founding the U. P. Electricity Regulatory Commission.

7.4 Biomass-based Power Projects

There are numerous options to build power plants of various capacities using biomass other than bagasse and a range of processes (gasification, co-generation, and combustion). According to a TERI report, the state has a biomass power generation capacity of 3757 megawatts. To establish such initiatives, region-specific surveys must be conducted to estimate the projects' true potential. The following is an updated list of other biomass-based initiatives in the state besides bagasse.

7.5 Biomass-Based Gasifier Power Project

Gasifier Plants (from 40 kilowatts to 1 megawatt) based on gasification process employing primarily rice husk biomass may meet the electrical and thermal energy requirements. Entrepreneurs are putting up gasifier plants with or without grants from the union government to seek aid from rising costs and shortage of oil and electricity. By the end of 2013-14, more than 171 industrial units in the state had established gasifier plants, resulting in a total capacity of 40-44 megawatts.

7.6 Biomass-Based Co-Generation Power Project

MW-scale co-generation power plants based on combustion processes using agro-residues have been built in several sectors. These projects are put up in pulp and paper mills that demand thermal (process steam) and electrical energy. They can meet those industries' thermal and electrical energy needs. The establishment of non-bagasse-based co-generation power projects is being supported in the state's various paper mills. Until the end of 2013-14, a total of 201 megawatts of capacity had been built. (*R&D | Ministry of New and Renewable Energy, Government of India*, n.d.)

7.7 Industrial Waste Based Power Projects

The processing of waste disposal from various industrial units, including distilleries, food processing companies, dairies, pulp and paper mills, can generate more than 150 MW of energy for the state.

108 MW of capacity was generated from waste by various industrial units (mostly distilleries) until the 2013-14 fiscal year. (*R&D | Ministry of New and Renewable Energy, Government of India*, n.d.)

8. EXAMPLES OF INDIAN ENTREPRENEURISM

8.1 Bio-solutions India

The Hamburg-based company purchases agricultural waste through farmers and manufactures biodegradable packaging and tableware in Ramanagara, near Bengaluru. The patented process turns it into self-binding fibres by simply churning agro-waste with water in massive machinery. The result is biodegradable packaging and dinnerware that only takes three months to decompose. This addresses two environmental issues: crop waste disposal and plastic pollution while also giving farmers a new source of revenue. BIO-SOLUTIONS tested their packaging with Big Basket, an online grocery store, from early 2017 to mid-2018, by providing limited quantities of packaging for fruits and vegetables. (Roopa, n.d.)

8.2 Kriya Labs

It offers a way to convert fifteen to twenty million tonnes of agricultural stubble every year, primarily rice straw, which has no market matched to different types of agro-waste such as wheat straw and bagasse. They use their method to turn rice straw into pulp. Paper (and its derivatives), bioethanol (biofuels in general), textiles, and other speciality chemicals such as cellulose acetate and carboxymethyl cellulose can benefit from this pulp. The procedure is safe for the environment. Pulp quality is determined by the eventual use and can be adjusted accordingly.

8.3 Farm2Energy

Sukhbir Singh Dhaliwal's start-up in Punjab assists farmers by baling the stubble from their fields for free. The company, founded in July 2016, provides bio-mass supply solutions to various clients, including farmers, landowners, and bio-mass users. The company uses integrated supply systems to treat paddy straw, maise stover, sugar cane garbage, and wheat straw before supplying it to the bio-fuels, bio-power, and biobased industries. It also aids farmers in the management and utilisation of biomass. Biopellets, bio-char, and torrefied pallets are also available through Farm2Energy. Biopellets (produced from a mixture of paddy straw and sugarcane trash) are a sustainable alternative to traditional fossil fuels used in various commercial and industrial applications. Biochar increases soil productivity, resulting in increased output. The third product, torrefied pellets made from torrefied biomass, can be used as a coal alternative. Torrefaction can be defined as a thermal breakdown that occurs at 200 to 320 degrees Celsius without oxygen. This increases the density of the biomass and enhances the burn quality. (Roopa, n.d.)

The Delhi-based company collects oil from FBOs (food business operators) and converts it into biodiesel. It also provides a wide range of environmental services, including mixed biodiesel supply, industrial tank cleaning, agricultural and blood waste recycling, and a biogas anaerobic digestion facility, among other things. (*This Woman Entrepreneur's Start-up Collects Used Cooking Oil to Convert into Biodiesel*, 2021)

9. CONCLUSION

Biomass-based energy is still not cost-competitive with conventional energy in many situations, which is a significant barrier for bio-energy entrepreneurs that hope to offer energy in straight rivalry against fossil fuels. Several revolutions arise when the fossil-fuel business is under pressure from environmental regulations. One of the main reasons for the elevated prices of renewable fuels and energy is the high cost of feedstocks. Trees are one instance of a minimal cost bio-mass feedstock. Even if the plants are free (as in government wildfire thinning), the expense of chipping, harvesting, and shipping in January 2014 was 2 to 4 times that of Wyoming coal per million Btu. (Han, 2007). With increased competition, the primary picture of expensive feedstocks is expected to remain. Biofuels markets are still developing and unstable, which will make it difficult for businesses to obtain financial financing and a steady market for their products. Government consents for bio-fuels exist, but new proposals for revisions and repeals emerge each year, creating uncertainty for investors. The bioenergy industry's immaturity and unpredictability significantly differ, not the primary concern. For bioenergy entrepreneurs, the characteristics of a self-starter, devoted, risk-taker, unlimited energy come into play (Xiang, 2014). These characteristics are essential for the next phase of bio-energy development to succeed. An entrepreneur needs the foresight to see what lies in the future, the courage to take risks in an ambiguous environment, and the drive to keep moving forward.

REFERENCES

Cultivating Entrepreneurship | University of Maryland Extension. (n.d.). Retrieved January 20, 2022, from https://extension.umd.edu/resource/cultivating-entrepreneurship

Dahiya, A. (2014). *Bioenergy: Biomass to Biofuels*. Academic Press.

Database of State Incentives for Renewables & Efficiency®—DSIRE. (n.d.). Retrieved January 20, 2022, from https://www.dsireusa.org/

Ethanol Impacts | Agricultural Economics. (n.d.). Retrieved January 20, 2022, from https://agecon.unl.edu/ethanolimpacts

Han, H.-S. (n.d.). *Getting Biomass to Market: Harvesting Methods and Costs*. Academic Press.

International—U.S. Energy Information Administration (EIA). (n.d.). Retrieved January 20, 2022, from https://www.eia.gov/international/data/world

Järvinen, P. (2005). *Action research as an approach in design science*. Academic Press.

Jordão, T. (2011). The strategic planning for renewable energy sources deployment in the Czech Republic with the support of balanced scorecard. *International Journal of Energy and Environment*, *5*, 364–376.

Käkölä, T. (2003). *Software business models and contexts for software innovation: Key areas software business research*. doi:10.1109/HICSS.2003.1174425

November & December 2008 Newsletter | Agricultural Marketing Resource Center. (n.d.). Retrieved January 20, 2022, from https://www.agmrc.org/renewable-energy/renewable-energy-climate-change-report/renewable-energy-climate-change-report/november--december-2008-newsletter

Ojala, P. (2008). Experiences of implementing a value-based approach. *WSEAS Transactions on Information Science and Applications*, *5*(4), 385–395.

R&D | Ministry of New and Renewable Energy, Government of India. (n.d.). Retrieved January 20, 2022, from https://mnre.gov.in/research-and-development/bio-energy

Reinis, Z., Lojda, Z., Heyrovský, A., Horáková, D., & Reisenauer, R. (1975). Influence of beta-blocking agents on experimental atherosclerosis of cocks. *Casopis Lekaru Ceskych*, *114*(43), 1320–1325.

Roopa, N. (n.d.). *Startups are helping North India's farmers to dispose of crop residue cleanly (and reduce pollution)*. Retrieved January 20, 2022, from https://scroll.in/article/909699/startups-are-helping-north-indias-farmers-to-dispose-of-crop-residue-cleanly-and-reduce-pollution

Shane, S., & Venkataraman, S. (2000). The Promise of Entrepreneurship as a Field of Research. *Academy of Management Review*, *25*(1), 217–226. https://doi.org/10.2307/259271

Street, S. (2012). *Meeting U.S. Fuel Standards*. Academic Press.

This woman entrepreneur's startup collects used cooking oil to convert into biodiesel. (2021, February 15). https://yourstory.com/herstory/2021/02/woman-entrepreneur-startup-cooking-oil-biodiesel/amp

Verma, S. K., & Hussain, R. (n.d.). Comparative analysis of growth kinetics. *Concentration of Sugar and Alcohol Tolerance of Various Yeast Strains Isolated From Five Different Sources*, *2*(8), 7.

Xiang, J. (2014). Research on Teaching Methods for Communication Engineering Students in Colleges. *Open Journal of Social Sciences, 2*(5), 9–12. doi:10.4236/jss.2014.25003

Chapter 4
Biofuel Policies in India: An Assessment of Policy Barriers

Sunil Kumar Verma
B.N. College of Engineering and Technology, Lucknow, India

Prashant Kumar
https://orcid.org/0000-0001-9760-6903
Department of Bioinformatics, Kalinga University, Raipur, India

ABSTRACT

Energy is one of the most precious and demanded commodities among various industries and consumers to sustain the current lifestyle. Energy is a crucial element, which unswervingly influences the country's economic development. Numerous methods are adopted to reduce global warming, embracing clean energy from wind, solar, and biomass sources. This chapter speaks about the current situation of energy demand, the innovations in biofuel sources, and the obstacles regarding the commercialization and production of microalgal biofuel.

1. INTRODUCTION

1.1. Green Energy Projection and Foundation of Energy Demand

The term 'energy' is supposed to be a derivative of the Greek word '*Energeia*' which means activity, and as we advanced, the explanation of energy transformed according to its field. When coming to the biological system, cells store energy in biomolecules such as lipids, complex carbohydrates, and proteins. The first law of thermodynamics can conclude that energy can neither be created nor destroyed; it can only be transformed from one form to another (Figure 1).

DOI: 10.4018/978-1-6684-5269-1.ch004

Figure 1. Various types of Energy

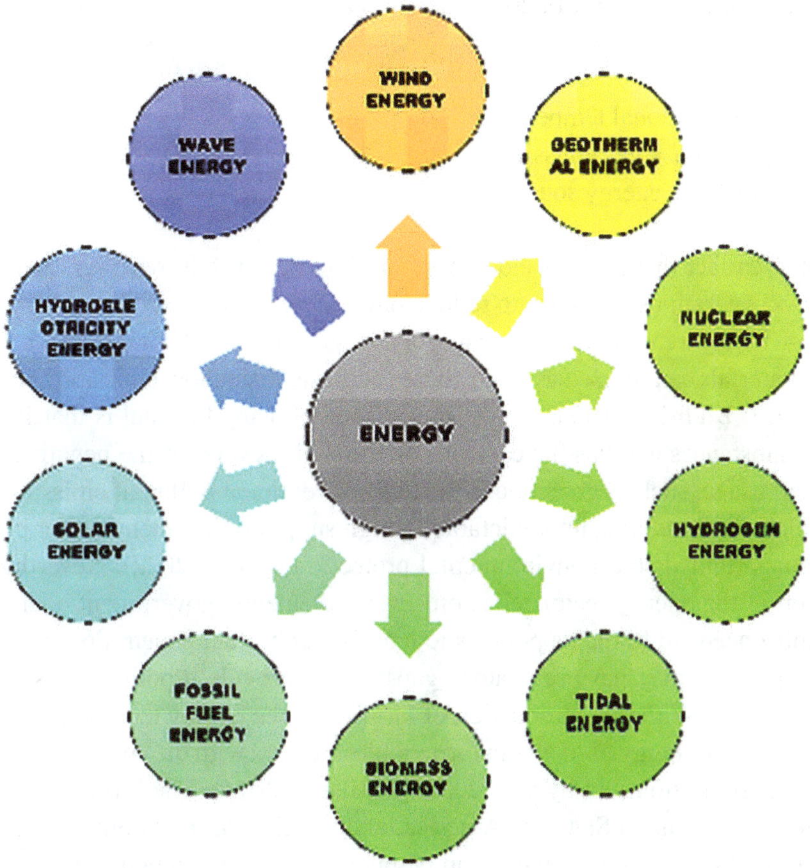

Worldwide to sustain the current lifestyle, energy is one of the most precious and demanded commodities among various industries and consumers. Energy is a crucial element, which unswervingly influences the country's economic development (Cleveland et al.,2000). It can be comprehended to the advanced energy accessibility the improved chances of nations growth. There is a solid and direct correlation between energy to per capita GNP, countries with the higher gross national product have higher energy consumption per individual (Stern, 2011). The exponential boom of energy use was originated through fossil hydrocarbon burning, with the additional support of coal in the 19^{th}century, followed by Crude oil use in the 20^{th}century, and currently, prolonged with the help of natural gas (Hall, 2016; IEA, 2016; Sayre, 2010). Currently, fossil fuels' central portion of energy demands is fulfilled. Worldwide, there is an increase in fuel price due to decrease in fossil fuel reserves (Monari et al., 2016), in addition with increased greenhouse gas emissions leading to global warming are two major concerning coercions for the transport area (Reyimu and Özçimen, 2017; Hashim et al., 2017; Lecksiwilai et al., 2017). Combustion of fossil fuel is required for power generation despite having several negative consequences such as pollution (Liu et al., 2017). Several pollutants such as SO2, NOx, CO, CO2, ground-level ozone, particulate matter, volatile organic compounds (VOC) are emitted.

There has been an increase in global temperature by 2 °C due increase in CO2 emission in the last decade (Friedling stein et al., 2014). In 2015, the Paris agreement was signed by around 194 countries whose primary aims are

1. Reduction of 2 °C in global temperature.
2. Aim to curb greenhouse gas emissions.
3. Reconnoitre substitute energy sources.

It is a well-known fact that there is a decrease in global fossil fuel reserves, and its current usage rate will last a few generations. We have to reduce our energy demands in the coming future. We need to search for eco-friendly, renewable fuel prospects (Dresselhaus and Thomas,2001). Biofuel produced from biological materials can be used as a substitute for diesel due to low emission levels when matched (Tsolcha et al., 2017; Bildirici, 2017). Major significance of using bio-fuel is that It is made from renewable organic substances and has the capacity to eliminate several of the negative features of fossil fuel extraction and usage, such as greenhouse gas and conventional pollutant emissions, finite resource depletion, and reliance on volatile, unpredictable foreign suppliers. The demand for bio-fuels also gave boost farm revenues (united states environmental protection agency, 2022). Nevertheless, sustainable biofuel promotion in the energy sector relies on several elements: government course of action. Currently, each country needs to frame its policies to have better biofuel penetration in the energy market. (Su et al., 2015) policies may have regulatory gaps which provide a political system to countersign contradictory objectives. With the introduction of bio-fuel in the energy market, by 2010 around 3% of worldwide transportation energy used was in form of bio-fuel and with the current increase it is expected to reach 27% of transportation fuel by year 2050 (Timilsina, 2014). The biofuel importance was first accredited by Brazil, the United States of America, and the European Union; now, this importance is being understood by Asian countries which can be understood by the amount of bio-fuel produced by them as mentioned in table 1 (Mohan et al., 2006).

Table 1. Biofuel Production

Country	Ethanol	Bio-diesel	Total	Reference
Brazil	19 000	227	19 227	(FAO, 2008)
Canada	1 000	97	1 097	
China	1 840	114	1 954	
India	400	45	445	
Indonesia	0.00	409	409	
Malaysia	0.00	330	330	
United States of America	26 500	1 688	28 188	
European Union	2 253	6 109	8 361	
Others	1 017	1 186	2203	

Brazils bio-fuel program was a massive success due to strong support of stringent legal directives and operative enactment (Mohan et al., 2006). In consideration of India's perspective, in 2008, A National

Policy on Bio-fuel was approved to standardise biofuel marketing in India via various firms in a transparent and operative mode following all accepted legal guidelines. However, we cannot replicate Brazil's success in India because of different climates and inconsistent rainy season patterns. Since then, in the last fifteen years, the Indian government has implemented different policies to encourage the usage of biofuels. In order to accomplish all the policy objectives, we need to have competent policy strategies to have appropriate information of the market and its dynamics for the day to day scenario (Barisa et al., 2015).

2. BIO-FUEL EVOLUTION AND ITS HOLDUPS

Bio-fuel is a non-hazardous fuel derivative of numerous biomass feedstock, and environmentally hazardous fossil fuel can be substituted (Voloshin et al., 2016). Biodiesel, bioethanol, methane and hydrogen are several biofuels produced by several countries using different types of biomass. Biofuels have turn-out to be crucial sustainable sources for restraining the production of Greenhouse gaseous (Huang et al., 2012).

Crucial benefits of bio-fuels above traditional fossil fuels are: -

1. Bio-fuels are sustainable as well as renewable resources (Razzak et al., 2013)
2. The toxic emission level via biofuel combustion is minimum (Surriya et al., 2015)
3. The Carbon dioxide emission is negated (Razzak et al., 2013; Surriya et al., 2015).

Figure 2. Different generation of biofuels

Biofuels		
First Generation	**Second Generation**	**Third Generation**
•Edible Feedstocks •Product- Bioethanol, biodiesel, vegetable oil	•Non-edible Feedstocks •Product- Bioethanol, biomethane	•Microalgae, Macroalgae & Microbes •Product- Biohydrogen

We can classify biofuel production into different generations as aforementioned in figure-2 (Hombach et al., 2016; Correa et al., 2017). First-generation feedstock biofuel is derived from cereal crops, sweet sorghum, Sugarcane, sugar beet and molasses for bio-ethanol production (Hemaiswarya et al., 2012) (Voloshin et al., 2016). Second-generation feedstock biofuel is derived from non-edible oils from *Jatropha curcas*, *Pongamiapinnata*, *Hevccabrasiliensis*, *Calophylluminophyllum,* which are usually used for biodiesel production (Hemaiswarya et al., 2012). Maity et al.(2014) stated that grass, wood and straw fall under second-generation feedstock biofuel. First-generation feedstock's biofuel and second-generation feedstock's biofuel are burdened with various concerns despite tax-credit policy and blending mandates, which permitted limited feedstock is to the energy market (Doshi et al., 2016).

First-generation feedstock biofuel and Second generation feedstock biofuel limitations: -

1. Jeopardising food safety owing to trade-off between fuel vs food.
2. Excessive land requirement.
3. High agricultural inputs
4. High investment cost
5. Hostile retail prices
6. Lower efficiency
7. Lower yield (Doshi et al., 2016; Juneja et al., 2013).

Due to these restrictions, researchers were looking for substitute feedstock and acknowledged a 3rd generation feedstock bio-fuel identified as algae, which insinuates numerous benefits as it, does not rely on edible crops, does not release elevated gaseous pollutants, and it does not require fertiliser or productive land (Mackenzie, 2013; Kim and Lee, 2015).

Several used macro-algae feedstocks are; some examples are *Enteromorphacompressa, Cladophoraglomerata, Ulva Lactuca, Laminaria*sp., *Macrocystispyrifera* (Chen et al., 2015; Xu et al., 2014). Third generation feedstock biofuel is primarily used to yield bio-diesel, which is along with chain alkyl ester, non-toxic with bio-degradable properties, carbon-neutral and nearly Sulphur-free, aromatics and its chemical, and energy properties are near to gasoline making it usable for the diesel engines without the requirement of minor to not at all adjustments (Luque, 2010).

2.1 Micro-Algae - Model for Viable Green Fuel

Micro-algae provide economic benefit, social benefit and environmental benefit using viable 3rd generation bio-fuel feedstock upon terrestrial oleaginous crops due to significant volumetric biomass, areal biomass, lipid yield, capability to adapt along with capability to grow in all sorts of water with a prospect of being grown all through year regardless of climatic changes (Ziolkowska and Simon, 2014; Ribeiro et al., 2017). Micro-algae have a high lipid content in contrast to oleaginous crops available in the market till now; that is, yearly algal oil production can be multiple folds to that of feedstock crop per ha. Additionally, compared to food crops, algae have a short doubling time and 20 to 30 times higher growth rate, providing multiple harvesting phases in a brief time (Chisti, 2007; Schenk et al., 2008; Ziolkowska and Simon, 2014). Bio-diesel produced from micro-algae can be used as a substitute for fossil diesel which can be very helpful to fulfil human's energy demand. Additional bio-diesel is carbon-neutral fuel, which means the quantity of $CO2$ released due to burning of bio-diesel stays equivalent to the amount of co2 absorbed by algae (Ziolkowska and Simon, 2014). For economic gains, carbohydrate's and lipids can be used for the production of various fuels such as gasoline, bio-ethanol, bio-diesel, bio-gas, hydrocarbon and jet fuel, while protein can be cast-off in feed, nutraceutical and food manufacturing, bio-fertilisers, surfactants, industrial enzymes (Griffiths et al., 2011; Yen et al.,2013). An additional economic value is that algae can be cultivated in sea-water, fresh-water, brackish-water, and sewage water (Cheng and He, 2014; Mata et al., 2010; Mathimani et al., 2017).

Micro-algae do not require basic cultivation needs such as portable water or fertile land. It can be easily cultivated off-shore or near-shore for biofuel, secluding land requirements. When comparing land use for algal fuel production to replace traditional fossil fuel, it would require less than > 1/7th of corn cultivated region of United States of America (Ziolkowska and Simon, 2014). Some governments provide various privileges such as cost-sharing, decreasing the total risk of the initial investment, providing subsidies on the bio-diesel production process (Su et al., 2015). Biofuel is an answer to climate change, energy insecurity and other multifarious issues (Pradhan and Ruysenaar, 2014). Hence, the bio-fuel programs success is motivated by effective and efficient guidelines following up with bio-fuel yield from various feedstock and countries as shown in Table-2. Learning from Brazil's success, developing countries counting India should prepare a roadmap to boost bio-fuel missions, blend goals and provide grants within directives, missions, and biofuel policies. The forthcoming section will have enlightened India's comprehensive breakdown of the legal charter governing bio-fuel.

Table 2. Bio-fuel yield from various feedstock and countries

Crop	Countries	Bio-fuel	Bio-fuel Yield	Reference
Sugar cane	India	Ethanol	4522	(FAO, 2008)
Oil palm	Malaysia	Biodiesel	4736	
Oil palm	Indonesia	Biodiesel	4092	
Maize	United States of America	Ethanol	3751	
Maize	China	Ethanol	1995	
Cassava	Brazil	Ethanol	1863	
Cassava	Nigeria	Ethanol	1480	

3. THE CHRONOLOGICAL ASSESSMENTS OF THE BIO-FUEL LEGISLATION IN INDIA

With the enactment of the *Indian Power Alcohol Act,* 1948 (Act No. 22 of 1948), India got its first legislation on biofuel in 1948. Its primary objective was to empower India's 'power alcohol' industry (Cleveland and Morris, 2013). Earlier, in 2000 legislature has abolished this act (Indian Power Alcohol (Repeal) Act, 2000). Finally, in 2002 the Ministry of Petroleum and Natural Gas had released a notification on the *Ethanol Blending Programme*. From January 2003, the notification required 5% ethanol blending with petrol in 9 sugarcane supplying 3 Union Territories and States (Amit Aradhay,2010). The Ethanol Blending Program remained ineffective because of inaccessibility of sugar molasses, botched-up ethanol pricing formula, bureaucratic interruptions by State organisations, late purchasing, with several other factors (Ray et al., 2011).

3.1 Bio-diesel Program

The Indian government is pining to execute its bio-diesel programme for the welfare of green energy resources and rural employment. In order to fulfil this objective, in July 2002, with the guidance of NITI Aayog, the government formed the *Committee on Development of Bio-fuel*. In July 2003, the committee submitted the report. In taking into consideration the report, in 2003, the Indian government introduced *National Bio-diesel Mission*. The Mission has a particular emphasis on the *Jatropha* farming on wasteland. It has acknowledged the Ministry of Rural Development as a nodal ministry. National Biodiesel Mission has to achieve its objectives in 2 diverse stages. Stage-I comprises demonstration and experimentation, and it has to be executed by 2006-07 (Kumar Biswas et al., 2010). Stage-II was projected for seed purchasing, oil abstraction and nursery development plant along with setting up of trans-esterification workshop lastly mixing and marketing of bio-diesel (Bandyopadhyay and Das, 2014; Kumar Biswas et al., 2010). Stage-II targets are encouraged to be complete. In 2005, the Ministry of Rural Development primed and submitted a comprehensive Project Report to the Indian government to discuss the pilot stage of biofuel feedstock crops (Mohan et al., 2006; Raju et al., 2012). Throughout the prime period, the Ministry of Petroleum and Natural Gas (in October 2005) had revealed the *Bio-diesel Purchase Policy*, which empowers oil marketing companies to purchase bio-diesel) at the cost of Rs. 26.50 per litre (inclusive of all taxes), with effect from 1 January 2006. Oil Manufacturing Companies intend to acquire bio-diesel that fulfils the criteria proposed by the Bureau of Indian Standards, and all procurement can be made only through recognised (20) centres.

Under Entry 14, List-II (State List) in Schedule 7 of the Indian Constitution, agriculture-related laws are enlisted, which empower the state government to make laws related to biofuel feedstock plantations which result in Union government-regulated petroleum-based companies with other private industries signing an MoU with state governments to promote biofuel feedstock on government-owned wasteland or unused lands through the method of contract farming with various farmers. In this initiative, significantly states governments have taken an interest in being part of the bio-diesel mission and promoting biofuel feedstock plantations with various instruments, including incentives and policies. Such states are Odisha, Uttarakhand, Rajasthan, and others few states (Venkataraman, 2012). The national bio-diesel mission has suffered setbacks due to lack of seeds, inefficient seed collection, extraction, and absence of knowledge and self-assurance in farmers and industry players. (Bandyopadhyay and Das,2014). The commercial production and sales of bio-diesel through the procurement centre is negligible.

The government conducted a study that states that farmers' failure to adopt scientific methods suffered a massive loss in lower production biofuel feedstock (Pradhan and Ruysenaar, 2014). The study also highlighted a shortage of seed and seed oil content, which led to the shutdown of various biofuel production units. Consequently, the National Bio-diesel Mission could not emerge successfully and was ended. Bio-diesel Purchase Policy faced the same fate of failure due to 'unrealistic purchase price' and subdued answers from Oil Manufacturing Companies (Kumar Biswas et al., 2010). In 2003 bio-diesel national Mission was announced; however, due to failure until 2008, the government could not implement a formal bio-fuel policy. As a result, in September 2008 'National Policy on Biofuels, was formulated and on 24 December, 2009, it was approved by the central government. Concerning the above, the next part tries to explore the National Policy on Bio-fuels, 2008, and pinpoint the weaknesses.

3.2 Key Characteristics of the National Policy on Bio-Fuels, 2008

1. The policy's primary purpose is to gratify bio-fuel demand and make certain accessibility of a minimum level of biofuels throughout the nation (Chandel et al., 2017).
2. The policy attempts to fast-track production of bio-fuel from non-edible oil feedstock's.
3. To confront the food *v fuel* debate (Murali et al., 2016; Rajagopal, 2008).
4. To draw the interest of foreign direct investment (Chandel et al., 2017).
5. Development of National Bio-fuel Coordination Committee.
6. To establish Bio-fuel Steering Committee.
7. Secretary to supervise policy implementation.

4. COMPREHENSIVE EXPLORATION OF THE NATIONAL POLICY ON BIOFUELS

4.1 Policy's Vision and Goals

- The Policy objectives are to make the biofuels market driver in the transportation and energy sectors in the future. The policy aims to accelerate development, promotion and production of biofuels and their usage to progressively replace diesel and petrol for transportation and stationary along with other applications, though backing to environmental concerns, creating employment ultimately directing towards sustainable development.
- The policy Goal is to safeguard the minimum level of biofuels at any given time easily accessible in the market to fulfil the requirement, by 2017 achieving the target of 20% biofuel blending for biodiesel and bioethanol.

4.2. Definitions and Scope

4.2.1 Characterisations of bio-fuels in the Policy

1. 'Biofuels' are liquid or gaseous fuels produced from biomass resources and used in place of, or in addition to, diesel, petrol or other fossil fuels for transport, stationary, portable and other applications.
2. 'Biomass' resources are the biodegradable fraction of products, wastes and residues from agriculture, forestry and related industries, as well as the biodegradable fraction of industrial and municipal wastes.

4.2.2 The Scope of the Policy Incorporates

1. 'Bioethanol': ethanol produced from biomass.
2. 'Bio-diesel': a methyl or ethyl ester of fatty acids.
3. Other bio-fuels: bio-methanol, bio-synthetic fuels.

4.3 Interventions and Enabling Mechanism

4.3.1 Plantations

The policy has discovered 400 kinds of trees in India that bear non-edible oilseeds, with the possibility of all of these species being used to produce biofuels. Farmers must obtain proper permission from local communities to use non-arable land. In 'Scheduled Areas,' where tribal tribes make up most of the population, the Panchayats Act, 1996 must be followed. The fifth Schedule of the Indian Constitution identifies these categories. It is worth noting that employment in non-edible oil seeds plantations will be included under the MNREGA, 2005. Policy guarantees that the Minimum Purchase Price for oilseeds is met, and it will be executed with cyclic revisions. (Paras 5.1-5.5 of the National Policy on Bio-fuels, 2008; Patel et al., 2014).

4.3.2 Processing

The policy pushes India's government to launch a 'National Registry,' which would be in charge of developing and maintaining relevant data on biodiesel and bioethanol supply and regularly reviewing blending levels based on this data. The blending levels must meet BIS specifications and certification norms. As a result, processing industries and OMCs must collaborate to create a sound system for this cause. Motor engine manufacturers are permitted to do required modifications to present engines "to ensure compatibility with biofuels." Section 52 of the Motor Vehicles Act of 1988 similarly permits similar changes to Automobiles (Paras 5.6-5.10 of the Policy; Ray et al., 2012).

4.3.3 Financial and Marketing Enticements

In India, OMCs are responsible for the storage, distribution, and sale of biofuels; current systems must regulate these actions. The Bio-fuel Steering Committee will determine the Minimum Purchase Price for bio-fuel, which will be chosen by the National Biofuel Coordination Committee, which will examine the "entire value chain" when establishing the Minimum Purchase Price for biodiesel, and "real cost of production and import price of bio-ethanol" when determining the Minimum Purchase Price for bioethanol. (Paras 5.11-5.12 of the Policy). In the aftermath of a petrol or diesel price gap below the Minimum Purchase Price for bio-fuels, the government would reimburse the OMCs appropriately. (Ray et al., 2012).

The National Bank of Agriculture and Rural Development, the Indian Renewable Energy Development Agency, the Small Industries Development Bank of India, and other lending institutions are mandatory to offer monetary support toward the planting of non-edible oil feedstock's, as well as the establishment of oil processing and extraction units for bio-fuel production. It is worth noting that under the automatic route, FDI up to 100% is authorised in biofuel technology and projects, as long as the biofuel is only used for local consumption. For the plantation of non-edible oilseeds, however, foreign direct investment is forbidden. (Paras 5.13-5.15 of the Policy). The policy pushes the Indian government to establish a "National Bio-fuel Fund" to provide financial incentives, subsidies, and grants for novel and 2nd generation feedstocks. Bio-diesel and bio-ethanol will be exempt from all central taxes and levies, except for a 16 per cent exempt excise duty on bioethanol. Other special discounts were also available to establish bio-oil extraction plants and processing facilities. It is important to emphasise that biofuel imports will be permitted to a limited degree, as determined by the NBCC, and biofuel exports would be approved only when the domestic requirements have been met (Paras 8.1-8.2 of the Policy). Unlike other countries, India's policy does not encourage biofuel blenders and merchants. (Bandyopadhyay and Das, 2014).

4.3.4 R&D and Quality Standards for Bio-Fuels

R&D and Innovation would receive a massive boost from the policy in bio-fuels. Indigenous research and development will be given priority, and intellectual property will be safe and secured as needed. Research centres, Educational establishments, and businesses are encouraged to collaborate on research projects, and the appropriate funding will be granted. In order to achieve global technology, competitiveness and competence transfer must be accelerated (Para 5.19-5.22). The BIS issued IS-15607 for biodiesel and I.S.: 2796: 2008 for bioethanol, based on ASTM D-6751 and EN-14214, respectively, American and European standards. The BIS must regularly assess existing standards to produce new standards-compliant with global benchmarks. (Paras 6.1-6.2).

4.4 State Governments Current Contribution

The policy promotes state government to develop and execute biofuel programmes actively. State governments must embrace biofuel policy and establish a nodal point to promote biofuels. (Paras 9.1-9.2 of the Policy). Many state governments have been aggressively participating in the backing of India's bio-fuel programme following the National Policy on Biofuels. Several states have previously produced or worked on biofuel policies and vision statements.

4.5 Institutional Mechanism

The Prime Minister would spearhead the formation of the National Biofuel Coordination Committee, which offers élite coordination and policy supervision. In order to properly monitor and administer biofuel programmes, the National Bio-fuel Coordination Committee must convene regularly. The Deputy Chairman of the Planning Commission, the Ministry of New and Renewable Energy (as coordinating Ministry), the Ministry of Rural Development, the Ministry of Agriculture, the Ministry of Environment and Forests, the Ministry of Petroleum and Natural Gas, the Ministry of Science and Technology, and the Secretary of the Ministry of New and Renewable Energy make up the National Bio-fuel Coordination Committee. (Part 11 of the Policy). A Bio-fuel Steering Committee will be formed to oversee the policy's implementation under the Cabinet Secretary governance. The BSC comprises eleven Secretaries from the Indian government's various Ministries and Departments. Several ministries participated in policymaking, development and promotion of biofuels at the national level, and each Ministry must coordinate with the others. Several programmes have been launched in India to encourage the development of biofuel crops. For example, the Agriculture Ministry established the Integrated Development of Tree Borne Oil Seed Scheme, which offers farmers and non-governmental organisations financial assistance to cultivate tree-borne oil seeds. (Dalemans et al., 2018; Raju et al., 2012).

4.6 Major Limitations to Accomplish Objectives Established Under India's Biofuel Policy

Even after constant decade-long efforts in developing biofuel and its promotion, India's accomplishment was not up to the desired mark. India's progress is far behind several other countries (Kumar Biswas and Pohit, 2013). As a result, the aims outlined in India's Bio-fuel Policy appear to be "unrealistic" in light of succeeding facts. (Kumar Biswas et al., 2010).

1. Land Restrictions for the Cultivation of Biofuel Crops

Non-edible oil crops must be planted exclusively on "wastelands" in non-forest and forest areas, according to paragraphs 1.5 and 5.1 of the policy. It is worth noting that the policy does not explain the term "wasteland." The writers were able to track the meaning of 'wasteland' back to the colonial authorities, who defined it as "land that did not generate any tax since it was uncultivated," which included everything from semi-jungle to forest, and marshes land to dry fields (Kumar Biswas et al., 2010; James and James, 1999). Lord Canning devised the 'wasteland rules' in 1861 to manage the profits earned by the uncultivated areas. There are several government categories for wastelands in India, including two principal classifications, ninefold classifications, and the Wasteland Atlas. (Baka, 2014). Besides, a large chunk of wasteland is being illegally encroached by landless labours and poor people; too which government has no available record of (Kumar Biswas et al., 2010). The State Governments must finalise the government wasteland and its usage and given for plantings, according to paragraph 9.2 of the policy. According to reports, India has no legislation for delineating non-cultivable wasteland appropriate for bio-fuel (mostly Jatropha) growth. Furthermore, the appropriation of arable land for the growth of bio-fuel crops has created significant socio-economic challenges that have hampered the Bio-fuel Policy's implementation (Kumar Biswas et al., 2010; Bastos Lima, 2012).

2. Rational Problems in Feedstock Cultivation

According to the Indian government, feedstock farming is now included in the MGNREGA scheme. Jatropha was the first crop that came to mind. By 2012 Indian government has resolute Jatropha plantation on 11.2 million hectares to 13.4 million hectares of land to come across a bold target of 10% bio-diesel-diesel blending mandate. Researchers and policymakers appear to misjudge the harvest of the Jatropha plant in sub-optimal growing situations (Sreenivas et al., 2018; Bastos Lima, 2012). It was also discovered that policymakers did not refer farmers when making such judgments. Overall, farmers cultivate Jatropha solely as a fence crop, and low crop yields discourage farmers (Bastos Lima, 2012).

3. Discrepancy in State Tax Structures

The policy stipulated a 16 per cent reduction on tariff on bio-ethanol and the elimination of all central taxes and duties on bio-diesel and bio-ethanol. On the other hand, state tax policies are not all the same, and they may differ from state to state. Even though each state has its excise tax, biodiesel is exempt from the federal excise tax of 4% as economic motivation. Raju et al. (2012) demonstrated that the policy's heterogeneous tax structure provides a roadblock to its timely implementation. Due to differing state legislation and administrative controls, biofuel's swift and unrestricted movement across state boundaries has been hindered.

4. Distinct Species of Non-Edible Oil Feedstock and their Suitability

The policy's paragraph 5 lists 400 kinds of non-edible seed-bearing trees in the country, although actual trials focus solely on the Jatropha. Some research is done by developing genetic improvements to plant kinds, nevertheless with minor achievement. An alternative explanation is that biodiesel production in India is based upon the Jatropha plant, proven to be commercially impractical due to lower yield, poor seed quality, and promotion issues. (Slette and radhey, 2014). According to Kumar Biswas and Pohit (2013), most research concentrates on plant-based materials. However, it neglects other elements such as soil conditions and agro-climatic factors, which stay necessary for boosted crop output.

5. Organisational limitations

National Bio-fuel Mission had planned to create the National Bio-fuel Development Board but later dissolved for unspecified reasons. The government has failed to establish any establishment proposed in the national biofuel mission to ease the transfer of benefits straight to farmers; therefore, it is essential to tackle these problems within prevailing established machinery or create innovative machinery (Kumar Biswas et al., 2010).

4.7 Recent Developments

To generate awareness and stimulate biofuel in India, on an industry level, the Ministry of Petroleum and Natural Gas, in synchronisation with the Petroleum Conservation Research Association, has launched a *Working Group* (PTI, 2015). The central government will amend the National Policy on Bio-fuels and Motor Spirit and High-Speed Diesel Control Order to ease bio-fuel sale (Business Line, 2015). Also, the Indian government is anticipated to disclose a Strategy on 'Flexible-Fuel Cars', which can operate either bioethanol or on petrol or its combination. These 'flex-fuel vehicles' will play a crucial role in fighting ever-increasing pollution and boosting ethanol consumption.

5. A WAY FORWARD

Through intensive study, it has been pointed that the Oil Manufacturing Company's solitary dependence on the ethanol produced from molasses confines enactment of the Ethanol Blending program. As recommended by Standing Committee on Petroleum & Natural Gas, we have again restated that "other than sugarcane, ethanol can be produced from corn, maise, wheat and food grains having high starch content", which has efficaciously executed in the United States of America. We can use various short generation biofuel feedstock crops that can be easily raised on monsoon-dependent agricultural lands, which could play an essential role in increasing the total productivity of biofuel feedstock and improving farmers' income. So, the Indian government must distribute the apposite amount of resources to fast-track the current Research and Development in manufacturing ethanol. The Indian government has to revive all the committees made under the policies and coordinate among other ministries to make up a comprehensive plan for effective implementation of the Bio-fuel program. The government must review its policies and make changes to have a fruitful bio-fuel program if required.

6. CONCLUSION

There are four vital types of machinery for any biofuels to succeed if taken into run-through, i.e., economics, technology, policy support, and environmental sustainability. Researchers emphasised various challenges for commercially viable biodiesel production, such as cultivation, harvesting, and production costs. However, legal issues and its solution with bio-fuel marketing in the energy sector are still not extensively considered, particularly in emerging countries like India. Unless we have policy support, we cannot convert beneficial technology into practical usage. Incidentally, India's biofuel policy accepted obligation in blending ethanol and bio-diesel programs. Still, the biofuel program did not earn significant success primarily due to biomass cultivation cost and the food vs fuel conflict. Henceforth, this chapter addresses the micro-agal feedstock potentials in biofuel production in addition to the policy concerns, which hampers its commercialisation in India. This chapter also displays important data related Indian government's national Bio-fuel Policy, which encompass legal hurdles for biofuel marketing. It focuses on the Indian government initiative to address industries problems regarding the bio-diesel program. Ethanol blending and its various policies. In order to accomplish India's aspiring biofuel targets, there needs revolution throughout the grass root levels, especially from the public and government are required for prosperous biofuel future in India.

REFERENCES

Amit Aradhey. (2010). *India-Biofuels Annual, Global Agricultural Information Network Report Number, IN1058*. USDA.

Aradhey, A. (2011). India Biofuels Annual. 2011. USDA Foreign Agricultural Service. Global Agricultural Information Network (GAIN). GAIN report number: IN 1159.

Awalgaonkar, N., Tibdewal, S., Singal, V., Mathew, J., & Karthikeyan, A. K. (2015). Biodiesel Production Status: Are the Present Policies Good Enough for the Growth of Biodiesel Sector in India? In *Engineering Asset Management - Systems, Professional Practices and Certification, Lecture Notes in Mechanical Engineering* (pp. 1199–1211). Springer.

Azbar, N., Çetinkaya Dokgöz, F. T., Keskin, T., Korkmaz, K. S., & Syed, H. M. (2008). 2009. Continuous fermentative hydrogen production from cheese whey wastewater under thermophilic anaerobic conditions. *Int. J. Hydrogen Energy, IWBT*, *34*, 7441–7447.

Baka, J. (2014). What wastelands? A critique of biofuel policy discourse in South India. *Geoforum*, *54*, 315–323.

Bandyopadhyay, K. R., & Das, K. (2014). Biofuels in South Asia: Connecting the Dots. *Renewable Energy Law and Policy Review*, *5*, 145–161.

Barisa, A., Romagnoli, F., Blumberga, A., & Blumberga, D. (2015). Future biodiesel policy designs and consumption patterns in Latvia: A system dynamics model. *Journal of Cleaner Production*, *88*, 71–82.

Basavaraj, G., Rao, P. P., Ravinder Reddy, C., Kumar, A. A., Srinivasa Rao, P., & Reddy, B. V. S. (2012). A Review of national biofuel policy in India: A critique- need for promotion of alternative feedstocks. *J Biofuels*, *3*, 65–78.

Bastos Lima, M. (2012). *An institutional analysis of biofuel policies and their social implications lessons from Brazil, India and Indonesia*. UNRISD Occasional Paper: Social Dimensions of Green Economy and Sustainable Development.

Bastos Lima, M. G., & Gupta, J. (2014). The extraterritorial dimensions of biofuel policies and the politics of scale: Live and let die? *Third World Quarterly*, *35*(3), 392–410.

Bell, M. L., Zanobetti, A., & Dominici, F. (2014). Who is more affected by ozone pollution? A systematic review and meta-analysis. *American Journal of Epidemiology*, *180*, 15–28. PMID:24872350

Benson, D., Kerry, K., & Malin, G. (2014). Algal biofuels: Impact significance and implications for EU multi-level governance. *Journal of Cleaner Production*, *72*, 4–13.

Bildirici, M. E. (2017). The effects of militarization on biofuel consumption and CO2 emission. *Journal of Cleaner Production*, *152*, 420–428.

Blanchard, R. (2015). A review of biofuels in In-dia: challenges and opportunities. In *World Energy Engineering Congress*. Association of Energy Engineers.

Business Line. (2015). *Centre mulls revising National Policy on Biofuels*. Business Line.

Chandel, A. K., Bhatia, L., Garlapati, V. K., Roy, L., & Arora, A. (2017). Biofuel Policy in Indian Perspective: Socioeconomic Indicators and Sustainable Rural Development. In *Sustainable Biofuels Development in India*. Springer.

Chen, F., Qiao, Z., Fan, Z., Zheng, Q., Wu, Y., Zhang, M., Cui, Y., Deng, Y., Luo, B., Zhang, W., Ji, K., Qiao, X., Zhao, X., & Li, X. (2017). The effects of Sulphur dioxide on acute mortality and years of life lost are modified by temperature in Chengdu, China. *The Science of the Total Environment*, *576*, 775–784. PMID:27810762

Chen, H., Zhou, D., Luo, G., Zhang, S., & Chen, J. (2015). Macroalgae for biofuels production: Progress and perspectives. *Renewable & Sustainable Energy Reviews*, *47*, 427–437.

Cheng, D., & He, Q. (2014). Assessment of Environmental Stresses for Enhanced Microalgal Biofuel Production – An Overview. *Frontiers in Energy Research*, *2*, 1–8.

Chisti, Y. (2007). Biodiesel from microalgae. *Biotechnology Advances*, *25*(3), 294–306. PMID:17350212

Cleveland, C. J., Kaufmann, R. K., & Stern, D. I. (2000). Aggregation and the role of energy in the economy. *Ecological Economics*, *32*, 301–317.

Cleveland, C. J., & Morris, C. G. (2013). *Handbook of Energy: Diagrams, Charts, and Tables: 1* (1st ed.). Elsevier Science.

Correa, D. F., Beyer, H. L., Possingham, H. P., Thomas-Hall, S. R., & Schenk, P. M. (2017). Biodiversity impacts of bioenergy production: Microalgae vs. first generation biofuels. *Renewable & Sustainable Energy Reviews*, *74*, 1131–1146.

Czyrnek-Delêtre, M. M., Smyth, B. M., & Murphy, J. D. (2017). Beyond carbon and energy: The challenge in setting guidelines for life cycle assessment of biofuel systems. *Renewable Energy*, *105*, 436–448.

Dalemans, F., Muys, B., Verwimp, A., Van den Broeck, G., Bohra, B., Sharma, N., Gowda, B., Tollens, E., & Maertens, M. (2018). Redesigning oilseed tree biofuel systems in India. *Energy Policy*, *115*, 631–643.

Demirbas, A. (2008). Production of Biodiesel from Tall Oil. *Energy Sources. Part A, Recovery, Utilization, and Environmental Effects*, *30*, 1896–1902.

Devalia, J. L., Rusznak, C., Herdman, M. J., Trigg, C. J., Tarraf, H., & Davies, R. J. (1994). Effect of nitrogen dioxide and sulphur dioxide on airway response of mild asthmatic patients to allergen inhalation. *Lancet*, *344*, 1668–1671. PMID:7996960

Doshi, A., Pascoe, S., Coglan, L., & Rainey, T. J. (2016). Economic and policy issues in the production of algae-based biofuels: A review. *Renewable & Sustainable Energy Reviews*, *64*, 329–337.

Dresselhaus, M. S., & Thomas, I. L. (2001). Alternative energy technologies. *Nature*, *414*, 332–337. PMID:11713539

Dwivedi, G., Sharma, M. P., & Kumar, M. (2014). Status and Policy of Biodiesel Development in India. *Int. J. Renew. Energy Res*, *4*, 246–254.

Economic Times. (2017). *Make ethanol from surplus food grains: Draft policy*. https://goo.gl/Af6zy7

Fernandes, B. S., Peixoto, G., Albrecht, F. R., Saavedra del Aguila, N. K., & Zaiat, M. (2010). Potential to produce biohydrogen from various wastewaters. *Energy for Sustainable Development*, *14*, 143–148.

Ferrell, J., & Sarisky-Reed, V. (2010). *National Algal Biofuels Technology Roadmap (No. DOE/EE—0332)*. EERE Publication and Product Library. doi:10.2172/1218560

Friedlingstein, P., Andrew, R. M., Rogelj, J., Peters, G. P., Canadell, J. G., Knutti, R., Luderer, G., Raupach, M. R., Schaeffer, M., van Vuuren, D. P., & Le Quéré, C. (2014). Persistent growth of CO2 emissions and implications for reaching climate targets. *Nature Geoscience*, *7*, 709–715.

Goswami, K., & Choudhury, H. K. (2015). To grow or not to grow? Factors influencing the adoption of and continuation with *Jatropha* in North East India. *Renewable Energy*, *81*, 627–638.

Griffiths, M.J., Dicks, R.G., Richardson, C., & Harrison, S.T.L. (2011). *Advantages and Challenges of Microalgae as a Source of Oil for Biodiesel*. doi:10.5772/30085

Haas, M. J., Michalski, P. J., Runyon, S., Nunez, A., & Scott, K. M. (2003). Production of FAME from acid oil, a by-product of vegetable oil refining. *Journal of the American Oil Chemists' Society*, *80*, 97–102.

Hall, C. A. S. (2016). Energy Return on Investment: A Unifying Principle for Biology, Economics, and Sustainability. Springer.

Hashim, H., Narayanasamy, M., Yunus, N. A., Shiun, L. J., Muis, Z. A., & Ho, W. S. (2017). A cleaner and greener fuel: Biofuel blend formulation and emission assessment. *Journal of Cleaner Production*, *146*, 208–217.

Hemaiswarya, S., Raja, R., Carvalho, I.S., Ravikumar, R., Zambare, V., & Barh, D. (2012). *An Indian scenario on renewable and sustainable energy sources with emphasis on algae*. Academic Press.

Hill, J., Nelson, E., Tilman, D., Polasky, S., & Tiffany, D. (2006). Environmental, economic, and energetic costs and benefits of biodiesel and ethanol biofuels. *PNAS*, *103*, 11206–11210.

Hirayama, S., Ueda, R., Ogushi, Y., Hirano, A., Samejima, Y., Hon-Nami, K., & Kunito, S. (1998). Ethanol production from carbon dioxide by fermentative microalgae. *Studies in Surface Science and Catalysis*, 657–660.

Hombach, L. E., Cambero, C., Sowlati, T., & Walther, G. (2016). Optimal design of supply chains for second generation biofuels incorporating European biofuel regulations. *Journal of Cleaner Production*, *133*, 565–575.

Hon-Nami, K. (2006). A unique feature of hydrogen recovery in endogenous starch-to-alcohol fermentation of the marine microalga, *Chlamydomonas perigranulata. Applied Biochemistry and Biotechnology, 131*, 808–828. PMID:18563656

Hu, X., Zhang, Y., Ding, Z., Wang, T., Lian, H., Sun, Y., & Wu, J. (2012). Bioaccessibility and health risk of arsenic and heavy metals (Cd, Co, Cr, Cu, Ni, Pb, Zn and Mn) in TSP and PM2.5 in Nanjing, China. *Atmospheric Environment, 57*, 146–152.

Huang, D., Zhou, H., & Lin, L. (2012). Biodiesel: An Alternative to Conventional Fuel. *Energy Procedia, 16*, 1874–1885.

Indian Power Alcohol (Repeal) Act 2000 | National Portal of India [WWW Document]. https://www.india.gov.in//indian-power-alcohol-repeal-act-2000

Issariyakul, T., Kulkarni, M. G., Dalai, A. K., & Bakhshi, N. N. (2007). Production of biodiesel from waste fryer grease using mixed methanol/ethanol system. *Fuel Processing Technology, 88*, 429–436.

James, G. A., & James, G. (1999). *Ethical Perspectives on Environmental Issues in India* (1st ed.). APH Publishing Corporation.

Juneja, A., Ceballos, R. M., & Murthy, G. S. (2013). Effects of Environmental Factors and Nutrient Availability on the Biochemical Composition of Algae for Biofuels Production: A Review. *Energies, 6*, 4607–4638.

Kampa, M., & Castanas, E. (2008). Human health effects of air pollution. *Environmental Pollution, 151*, 362–367. PMID:17646040

Khanna, M., Önal, H., Crago, C. L., & Mino, K. (2013). Can India Meet Biofuel Policy Targets? Implications for Food and Fuel Prices. *American Journal of Agricultural Economics, 95*, 296–302.

Kim, I., Seo, Y. H., Kim, G.-Y., & Han, J.-I. (2015). Co-production of bioethanol and biodiesel from corn stover pretreated with nitric acid. *Fuel, 143*, 285–289.

Kim, S.-K., & Lee, C.-G. (2015). *Marine bioenergy: Trends and developments*. CRC Press.

Koçar, G., & Civaş, N. (2013). An overview of biofuels from energy crops: Current status and future prospects. *Renewable & Sustainable Energy Reviews, 28*, 900–916.

Kumar, S., Chaube, A., & Jain, S. K. (2012). Critical review of Jatropha biodiesel promotion policies in India. *Energy Policy, 41*, 775–781.

Kumar, S., Shrestha, P., & Salam, P. A. (2013). A review of biofuel policies in the major biofuel producing countries of ASEAN: Production, targets, policy drivers and impacts. *Renewable & Sustainable Energy Reviews, 26*, 822–836.

Kumar Biswas, P., & Pohit, S. (2013). What ails India's biodiesel programme? *Energy Policy, 52*, 789–796.

Kumar Biswas, P., Pohit, S., & Kumar, R. (2010). Biodiesel from jatropha: Can India meet the 20% blending target? *Energy Policy, 38*, 1477–1484.

Kumar Tiwari, A., Kumar, A., & Raheman, H. (2007). Biodiesel production from jatropha oil (Jatropha curcas) with high free fatty acids: An optimized process. *Biomass and Bioenergy*, *31*, 569–575.

Kumari, V., Shah, S., & Gupta, M. N. (2007). Preparation of Biodiesel by Lipase-Catalyzed Transesterification of High Free Fatty Acid Containing Oil from Madhuca indica. *Energy & Fuels*, *21*, 368–372.

Lali, A. (2016). Biofuels for India: What, when and how. *Current Science*, *110*, 552–555.

Lecksiwilai, N., Gheewala, S. H., Silalertruksa, T., & Mungkalasiri, J. (2017). LCA of biofuels in Thailand using Thai Ecological Scarcity method. *Journal of Cleaner Production*, *142*, 1183–1191.

Liu, L., Cheng, S. Y., Li, J. B., & Huang, Y. F. (2007). Mitigating Environmental Pollution and Impacts from Fossil Fuels: The Role of Alternative Fuels. *Energy Sources. Part A, Recovery, Utilization, and Environmental Effects*, *29*, 1069–1080.

Luque, R. (2010). Algal biofuels: The eternal promise? *Energy & Environmental Science*, *3*, 254–257.

Mackenzie, A. (2013). Synthetic biology and the technicity of biofuels. Studies in History and Philosophy of Science Part C: Studies in History and Philosophy of Biological and Biomedical Sciences. *Philosophical Perspectives on Synthetic Biology*, *44*, 190–198.

Maity, J. P., Bundschuh, J., Chen, C. Y., & Bhattacharya, P. (2014). Microalgae for third generation biofuel production, mitigation of greenhouse gas emissions and wastewater treatment: Present and future perspectives–A mini review. *Energy*, *78*, 104–113.

Mariod, A., Klupsch, S., Hussein, I. H., & Ondruschka, B. (2006). Synthesis of Alkyl Esters from Three Unconventional Sudanese Oils for Their Use as Biodiesel. *Energy & Fuels*, *20*, 2249–2252.

Mata, T. M., Martins, A. A., & Caetano, N. S. (2010). Microalgae for biodiesel production and other applications: A review. *Renewable & Sustainable Energy Reviews*, *14*, 217–232.

Mathimani, T., Uma, L., & Prabaharan, D. (2017). Optimization of direct solvent lipid extraction kinetics on marine trebouxiophycean alga by central composite design – Bioenergy perspective. *Energy Conversion and Management*, *142*, 334–346.

Meng, X., Yang, J., Xu, X., Zhang, L., Nie, Q., & Xian, M. (2009). Biodiesel production from oleaginous microorganisms. *Renewable Energy*, *34*, 1–5.

Miao, X., & Wu, Q. (2006). Biodiesel production from heterotrophic microalgal oil. *Bioresource Technology*, *97*, 841–846. PMID:15936938

Mofijur, M., Masjuki, H., Kalam, M., Rahman, S. A., & Mahmudul, H. (2015). Energy scenario and biofuel policies and targets in ASEAN countries. *Renewable & Sustainable Energy Reviews*, *46*, 51–61.

Mohan, M. R., Phillippe, G. T., & Shiju, M. (2006). Biofuel laws in Asia: Instruments for energy access, security, environmental protection and rural empowerment. *Asian Biotechnology and Development Review*, *8*(2), 51–75.

Monari, C., Righi, S., & Olsen, S. I. (2016). Greenhouse gas emissions and energy balance of biodiesel production from microalgae cultivated in photobioreactors in Denmark: A lifecycle modeling. *Journal of Cleaner Production*, *112*, 4084–4092.

MoP & NG. (2006). *Bio-diesel Purchase Policy*. Government of India.

Moser, B. R., & Vaughn, S. F. (2012). Biodiesel from Corn Distillers Dried Grains with Solubles: Preparation, Evaluation, and Properties. *Bioenerg. Res.*, *5*, 439–449.

Murali, P., Hari, K., & Prathap, D. P. (2016). An Economic Analysis of Biofuel Production and Food Security in India. *Sugar Tech*, *18*, 447–456.

National Policy on Biofuels. (2008). Ministry of New & Renewable Energy, Government of India. https://mnre.gov.in/file-manager/UserFiles/biofuel-policy.pdf

Nebel, B. A., & Mittelbach, M. (2006). Biodiesel from extracted fat out of meat and bone meal. *European Journal of Lipid Science and Technology*, *108*, 398–403.

Ngo, H. L., Zafiropoulos, N. A., Foglia, T. A., Samulski, E. T., & Lin, W. (2008). Efficient Two- Step Synthesis of Biodiesel from Greases. *Energy & Fuels*, *22*, 626–634.

Park, J.-Y., Kim, D.-K., Wang, Z.-M., Lu, P., Park, S.-C., & Lee, J.-S. (2008). Production and characterization of biodiesel from tung oil. *Applied Biochemistry and Biotechnology*, *148*, 109–117. PMID:18418744

Pimentel, D., & Patzek, T. W. (2005). Ethanol Production Using Corn, Switchgrass, and Wood; Biodiesel Production Using Soybean and Sunflower. *Natural Resources Research*, *14*, 65–76.

Planning Commission. (2003). *Report of the Committee on Development of Biofuels*. Government of India. http://mnre.gov.in/file-manager/UserFiles/biofuelpolicy. pdf

Pohit, S., Biswas, P. K., Kumar, R., & Goswami, A. (2010). Pricing model for biodiesel feedstock: A case study of Chhattisgarh in India. *Energy Policy*, *38*, 7487–7496.

Pradhan, S., & Ruysenaar, S. (2014). Burning Desires: Untangling and Interpreting "Pro-Poor" Biofuel Policy Processes in India and South Africa. *Environment & Planning A*, *46*, 299–317.

PTI. (2015). *Steps for Implementation of Biofuel Programme*. http://pib.nic.in/newsite/PrintRelease.aspx?relid=124076

Rajagopal, D. (2008). Implications of India's biofuel policies for food, water and the poor. *Water Policy*, *10*(S1), 95–106.

Raju, S. S., Parappurathu, S., Chand, R., Joshi, P. K., Kumar, P., & Msangi, S. (2012). *Biofuels in India: potential, policy and emerging paradigms*. Policy Paper - National Centre for Agricultural Economics and Policy Research.

Raju, S. S., Shinoj, P., & Joshi, P. K. (2009). Sustainable development of biofuels: Prospects and challenges. *Economic and Political Weekly*, *26*, 65–72.

Ramadhas, A. S., Jayaraj, S., & Muraleedharan, C. (2005). Biodiesel production from high FFA rubber seed oil. *Fuel*, *84*, 335–340.

Rashid, U., & Anwar, F. (2008). Production of biodiesel through optimized alkaline-catalyzed transesterification of rapeseed oil. *Fuel*, *87*, 265–273.

Ravindranath, N. H., Sita Lakshmi, C., Manuvie, R., & Balachandra, P. (2011). Biofuel production and implications for land use, food production and environment in India. *Energy Policy, 39*, 5737–5745.

Ray, S., Goldar, A., & Miglani, S. (2012). The ethanol blending policy in India. *Economic and Political Weekly, 7*, 23–25.

Ray, S., Miglani, S., & Goldar, A. (2011). *Ethanol blending policy in India: demand and supply issues.* ICRIER Policy Series.

Razzak, S. A., Hossain, M. M., Lucky, R. A., Bassi, A. S., & de Lasa, H. (2013). Integrated CO2 capture, wastewater treatment and biofuel production by microalgae culturing—A review. *Renewable & Sustainable Energy Reviews, 27*, 622–653.

UPES Report. (2016). *UPES, CSTEP, and PLR Chambers, Fuel Blending in India: Learning and Way Forward (Report, 2016).* Author.

Reyimu, Z., & Özçimen, D. (2017). Batch cultivation of marine microalgae Nannochloropsis oculata and Tetraselmis suecica in treated municipal wastewater toward bioethanol production. *Journal of Cleaner Production, 150*, 40–46.

Ribeiro, L. A., Pereira da Silva, P., Ribeiro, L., & Dotti, F. L. (2017). Modelling the impacts of policies on advanced biofuel feedstocks diffusion. *Journal of Cleaner Production, 142*, 2471–2479.

Rodolfi, L., Chini Zittelli, G., Bassi, N., Padovani, G., Biondi, N., Bonini, G., & Tredici, M. R. (2009). Microalgae for oil: Strain selection, induction of lipid synthesis and outdoor mass cultivation in a low-cost photobioreactor. *Biotechnology and Bioengineering, 102*, 100–112. PMID:18683258

Sahoo, P. K., Das, L. M., Babu, M. K. G., & Naik, S. N. (2007). Biodiesel development from high acid value polanga seed oil and performance evaluation in a CI engine. *Fuel, 86*, 448–454.

Sayre, R. (2010). Microalgae: The Potential for Carbon Capture. *BioSci., 60*, 722–727.

Schenk, P. M., Thomas-Hall, S. R., Stephens, E., Marx, U. C., Mussgnug, J. H., Posten, C., Kruse, O., & Hankamer, B. (2008). Second Generation Biofuels: High-Efficiency Microalgae for Biodiesel Production. *Bioenerg. Res., 1*, 20–43.

Sengupta, M., & Poddar, A. (2013). National policy on biofuel under the scanner. *International Journal of Emerging Technology and Advanced Engineering, 3*, 521–526.

Singh, S. (2009). *India Biofuels Annual.* USDA Foreign Agricultural Service, GAIN Report Number IN9080.

Sivaramakrishnan, R., & Incharoensakdi, A. (2017). Microalgae as feedstock for biodiesel production under ultrasound treatment–A review. *Bioresource Technology, 250*, 877–887. PMID:29221914

Slette, J., & Aradhey, A. (2014). *India Biofuels Annual.* USDA Gain Report: IN4055.

Slocombe, S.P., Zhang, Q., Ross, M., Anderson, A., Thomas, N.J., & Lapresa, Á., Rad-Menéndez, C., Campbell, C.N., Black, K.D., Stanley, M.S., & Day, J.G. (2015). Unlocking nature's treasure-chest: Screening for oleaginous algae. *Scientific Reports, 5*, srep09844. PMID:26202369

Smyth, B., Gallachóir, B. Ó., Korres, N., & Murphy, J. (2010). Can we meet targets for biofuels and renewable energy in transport given the constraints imposed by policy in agriculture and energy? *Journal of Cleaner Production*, *18*(16-17), 1671–1685.

Sorda, G., Banse, M., & Kemfert, C. (2010). An overview of biofuel policies across the world. *Energy Policy*, *38*, 6977–6988.

Sreenivas, C., Rao, A. B., & Patwardhan, A. (2018). *Critical Evaluation of Biodiesel Production Initiatives in India, Biofuels: Greenhouse Gas Mitigation and Global Warming*. Springer.

Standing Committee Report. (2015). *Ethanol Blended Petrol & Bio Diesel Policy*. Standing Committee on Petroleum and Natural Gas, MoPNG.

Stern, D. I. (2011). The role of energy in economic growth. *Annals of the New York Academy of Sciences*, *1219*, 26–51. PMID:21332491

Su, Y., Zhang, P., & Su, Y. (2015). An overview of biofuels policies and industrialization in the major biofuel producing countries. *Renewable & Sustainable Energy Reviews*, *50*, 991–1003.

Surriya, O., Saleem, S. S., Waqar, K., Kazi, A. G., & Öztürk, M. (2015). Bio-fuels: A Blessing in Disguise. In *Phytoremediation for Green Energy* (pp. 11–54). Springer.

Tai, A. P. K., Martin, M. V., & Heald, C. L. (2014). Threat to future global food security from climate change and ozone air pollution. *Nature Climate Change*, *4*, 817–821.

The Hindu. (2016). Flex fuel policy of government to curb pollution worry automakers (Sriram Lakshman). *The Hindu*.

Trivedi, J., Aila, M., Bangwal, D. P., Kaul, S., & Garg, M. O. (2015). Algae based biorefinery—How to make sense? *Renewable & Sustainable Energy Reviews*, *47*, 295–307.

Tsolcha, O.N., Tekerlekopoulou, A.G., Akratos, C.S., Aggelis, G., & Genitsaris, S., Moustaka-Gouni, M., & Vayenas, D.V. (2017). Biotreatment of raisin and winery wastewaters and simultaneous biodiesel production using a Leptolyngbya-based microbial consortium. *Journal of Cleaner Production*, *148*, 185–193.

USDA GAIN Report. (2010). *Amit Aradhey, India-Biofuels Annual*. Global Agricultural Information Network Report Number, IN1058.

Veljković, V. B., Lakićević, S. H., Stamenković, O. S., Todorović, Z. B., & Lazić, M. L. (2006). Biodiesel production from tobacco (*Nicotiana tabacum* L.) seed oil with a high content of free fatty acids. *Fuel*, *85*, 2671–2675.

Venkataraman, N. S. (2012). Have we given a quiet burial to jatropha biofuel projects? *Deccan Herald*. https://www.deccanherald.com/content/246514/have-we-given-quiet-burial.html

Voloshin, R. A., Rodionova, M. V., Zharmukhamedov, S. K., Nejat Veziroglu, T., & Allakhverdiev, S. I. (2016). Review: Biofuel production from plant and algal biomass. *International Journal of Hydrogen Energy*, *41*, 17257–17273.

Xu, X., Kim, J. Y., Oh, Y. R., & Park, J. M. (2014). Production of biodiesel from carbon sources of macroalgae, *Laminaria japonica*. *Bioresource Technology*, *169*, 455–461. PMID:25084043

Yacobucci, B. D., & Bracmort, K. (2010). *Calculation of lifecycle greenhouse gas emissions for the Renewable Fuel Standard (RFS). R40460*. Congressional Research Service.

Yen, H.-W., Hu, I.-C., Chen, C.-Y., Ho, S.-H., Lee, D.-J., & Chang, J.-S. (2013). Microalgaebased biorefinery – From biofuels to natural products. *Bioresource Technology*, *135*, 166–174. PMID:23206809

Zhang, J., & Jiang, L. (2008). Acid-catalyzed esterification of *Zanthoxylum bungeanum* seed oil with high free fatty acids for biodiesel production. *Bioresource Technology*, *99*, 8995–8998. PMID:18562195

Zhou, A., & Thomson, E. (2009). The development of biofuels in Asia. *Applied Energy*, *86*, S11–S20.

Ziolkowska, J. R., & Simon, L. (2014). Recent developments and prospects for algae-based fuels in the US. *Renewable & Sustainable Energy Reviews*, *29*, 847–853.

Chapter 5
Green House Gases:
Challenges, Effect, and Climate Change

Kannadhasan S.
https://orcid.org/0000-0001-6443-9993
Study World College of Engineering, India

Nagarajan R.
https://orcid.org/0000-0002-4990-5869
Gnanamani College of Technology, India

ABSTRACT

Changes in weather patterns over decades or longer periods of time are referred to as climate change. Natural and human forces both contribute to climate change. Since the Industrial Revolution, humans have contributed to climate change through emitting greenhouse gases and aerosols, as well as changing land use, leading global temperatures to rise. More storms, floods, droughts, and rising sea levels may result from increasing global temperatures and the melting of ice sheets, sea ice, and glaciers. The sun's rays supply energy to the whole world. GHGs are required to trap heat and maintain a temperature that permits life to flourish on the planet. The greenhouse effect is a natural and necessary occurrence for life on Earth to survive. If the greenhouse effect did not exist, the world would be around 33°C cooler right now. Humans have contributed to the increase in atmospheric GHGs via rising fossil fuel usage and deforestation in recent decades. Increased GHG emissions have been the primary cause of global warming during the last century.

INTRODUCTION

Since 1850, three primary datasets have been used to calculate global surface temperatures. These figures indicate a warming of +0.8°C to +1.0°C since 1900. 4 Since 1950, land-only data has shown increasing trends of +1.1°C to +1.3°C, indicating that land temperatures respond to climate change quicker than ocean temperatures. While multi-decadal time periods (30 years or more) are often used to assess global warming, attribution of trends over shorter time periods may be problematic due to natural variability.

DOI: 10.4018/978-1-6684-5269-1.ch005

Natural climatic variability is caused by internal interactions between the atmosphere, ocean, land surface, and sea ice. These oscillations happen whether or not there is climate change, and are generally referred to as "noise" or natural changes around a "normal" figure. Temperature rises as GHG levels increase, which has a substantial indirect influence on water vapour. Due to the greenhouse effect of water vapour, greater global temperatures induced by GHGs increase the atmosphere's capacity to store water vapour, resulting in a positive feedback loop. The quantity of water vapour in the atmosphere increases by roughly 7% when the global temperature rises by 1°C. (Dones et al., 2016; Frischknecht et al., 2003; Inter-Governmental Panel on Climate Change (IPCC), 2001; Villigen & Swiss, 2003)

Water vapour is a potent and quick feedback that increases any initial force by two to three, despite the fact that CO2 is the most significant human control knob on climate. Water vapour is a large contribution to climate change, despite the fact that it is not a significant beginning force. By continuing to address criteria pollutant nonattainment issues, state and municipal governments may accomplish significant GHG emission reductions. The most efficient approach to accomplish this goal is to ensure that the techniques utilised to achieve emission reductions for criteria pollutant compliance also provide benefits in terms of GHG reduction, rather than ineffective or counterproductive GHG reduction efforts. Because strategies that reduce GHG emissions nearly invariably result in reduced emissions of other air pollutants, STAPPA and ALAPCO believe it is vital to focus on the link between GHG mitigation and conventional air pollution management. The most well-known harmonised procedures are used to address CO2, particle matter (PM), nitrogen oxides (NOx), sulphur dioxide (SO2), carbon monoxide (CO), and air toxics, all of which are important CO2 sources. The most harmful greenhouse gases include CO2, methane, nitrous oxide, hydrofluorocarbons, perfluorocarbons, and sulphur hexafluoride. Ozone precursors (NOx and NMVOCs) have an indirect greenhouse effect since they are also GHGs. The focus of this study is on CO2 for two reasons. To begin with, CO2 is thought to be responsible for more than half of the expected global warming consequences. In 1997, CO2 emissions accounted for almost 82 percent of total emissions in the United States. (Dones et al., n.d.; Hondo, 2003; Lee, 1997; Ozawa et al., 2003; Spath et al., 1999). GHG emissions refer to the amount of greenhouse gases released into the atmosphere. Second, the primary source of CO2 is fossil-fuel combustion, which is governed by state and local governments via emission categories. According to the Intergovernmental Panel on Climate Change (IPCC), the Earth's climate system is continually changing, both worldwide and locally, with some of the changes linked to human activity and resulting in greenhouse gas emissions. Energy delivery systems, especially fossil-fuel systems, are the principal contributors to these gas emissions. This article primarily leverages data from a recent Swiss study addressing Life-Cycle Assessment (LCA) issues based on environmental inventories of European-wide energy systems to provide GHG emission comparisons.

The ocean covers 71% of the Earth's surface and is home to 1.3 billion cubic kilometres of salty water. This enormous body of water has a huge influence on the global environment. It functions as a global thermostat, lowering heat gradients induced by differences in incoming solar radiation over time and space, controlling the composition of numerous trace chemicals in the atmosphere, and functioning as a dump for a range of manmade toxins. Climate fluctuations are often connected to changes in the atmosphere (Fraedrich, 2001; Fraedrich et al., 2005; Fraedrich & Lunkeit, 2008; Kunz et al., 2008; Lorenz, 1955). However, analysing the atmosphere on its own is challenging. The land, oceans, and ice-covered regions of the Earth's surface are all intricately tied to the atmosphere (known as the cryosphere). The biosphere, which comprises both terrestrial and marine biological systems, has a substantial relationship.

CLIMATE CHANGE

The atmosphere, land, ocean, cryosphere, and biosphere make up the basic framework of the climate system. The ocean has an impact on the rate at which greenhouse gases are emitted. This is performed by absorbing surplus greenhouse gases from the environment, as well as some greenhouse-induced heat. Both of these measures contribute to slowing global warming. The global ground-level temperature of the atmosphere would presently be one to two degrees Celsius higher if it weren't for the cooling effect of the ocean. The question is whether the ocean will continue to counteract global warming, and whether the rate at which it does so will slow or speed up over time. We need to know the precise answers to these questions before we can determine the total amount of global climate change. As a consequence, further study is required on (1) gas exchange processes at the air-sea interface, with an emphasis on CO:, (2) greenhouse gas transit routes from the surface to the ocean's deep layers and sediments, and (3) ocean circulation characteristics.

The wind is the primary driving force behind the ocean's surface circulation, which moves heat and water in horizontal planes. Excessive evaporation on the sea surface from thick water descending towards the ocean's core causes the thermohaline circulation to cool and/or increase salt content. On the one hand, this process allows deeper ocean layers to interact with the atmosphere, but on the other hand, it "opens" the huge volume of the deep ocean, allowing it to store heat and CO2. This sinking of heavy surface water may now only be seen in a few isolated areas of the ocean. Deepest convection occurs in the northern North Atlantic (Norwegian and Greenland Seas) and near Antarctica (Weddell Sea). Warm surface water in the North Atlantic is cooled, allowing more CO: to dissolve before sinking as equally salty water into the deep ocean, where it has a long history of interaction with the atmosphere. After being kept off from direct touch with the atmosphere for a long time, upwelled deep water in the Southern Hemisphere swiftly transforms into colder, denser water with a high CO2 content and re-enters the ocean's deep and bottom layers.

As these two water masses spread across the ocean, each with its own set of characteristics, the "resident" deep ocean water rises slowly but progressively. The rate and method of deep water creation, as well as oceanic heat transfer to high latitude continental areas, are all determined by surface water circulation in the North Atlantic. Excess salinity (and the associated heat flux, which is controlled by the rate at which surface water flows into the Atlantic) generates strong oceanic fronts, defines the boundaries between ice-covered and open ocean areas, and heats northern Europe. The whole system has been found to be exceedingly unstable across decadal, historical, and geologic time periods. Little changes in the sea ice limit, as well as salinity, temperature, and volume of inflowing surface waters, may have a substantial influence on Europe's climate, economy, and living conditions, according to historical records..

Scientists, engineers, industry, and regulators are all interested in the future role of hydrogen H2 as a synthetic fuel and energy carrier. In the future, hydrogen is likely to play a larger role in transportation. An internal combustion engine or a fuel cell may be used to create a hydrogen-powered vehicle. In the second method, hydrogen is burnt. Electrochemical traction is created in a fuel cell, which then supplies energy to power an electric motor. A fuel cell car's efficiency in converting chemical energy to mechanical energy might be far higher than that of a normal gasoline-powered internal combustion engine vehicle. Hydrogen's attraction as a future transportation technology derives from its capacity to boost overall economic growth while drastically reducing environmental pollution. Hydrogen generation, storage, delivery, and usage in all aspects of the energy economy, including transportation, will be major problems for any future hydrogen economy. To ensure that overall environmental damage is avoided and the full potential of hydrogen as a clean fuel is realised, thorough environmental evaluations will be necessary for all technologies employed at each step, from synthesis to usage.

We examine what we know about hydrogen's fate and behaviour in the atmosphere, as well as its major sources and sinks. Hydrogen is a greenhouse gas, contrary to common assumption, and its global warming potential is evaluated in contrast to carbon dioxide. Then we assess the global warming implications of switching from a fossil-fuel to a hydrogen-based energy source. The researchers found that substituting hydrogen-based energy systems for fossil-fuel-based energy systems would only result in a minor reduction in global warming unless hydrogen leakage from any future hydrogen economy is tightly monitored. Gases that trap heat in the atmosphere are referred to as "greenhouse gases" (GHGs). Carbon dioxide, for example, is naturally created and released into the atmosphere as a consequence of both natural and human activities. Humans are the only ones who manufacture and release other greenhouse gases, such as fluorinated gases. The most major greenhouse gases that enter the atmosphere as a consequence of human activity are carbon dioxide, methane, nitrous oxide, and fluorinated gases. The behaviours described above are frighteningly similar to those that occur in the Earth's climate system. Shortwave solar radiation from the Sun reaches the planet's surface, where it is absorbed and heated by the planet's comparably clear atmosphere. The gross radioactive flux density generated by the planet's surface (land, water) much outnumbers the input irradiative flux density received from the Sun, just as a person's thermal infrared radiation far outnumbers his or her caloric intake on a global and annual average. The atmosphere absorbs thermal infrared radiation from the planet's surface, which is subsequently heated and radiates, in part downward toward the surface.

Due to downwelling infrared radiation from the atmosphere to the surface, the net upward infrared flux from the surface is significantly decreased in contrast to the gross blackbody flux. The greenhouse effect assumes that a greenhouse maintains its heat by collecting and releasing longwave radiation via its glass surfaces. (True greenhouses maintain their warm temperatures by preventing warm air from flowing away from the surface.)

GREEN HOUSE GASES

In the atmosphere, a greenhouse gas absorbs and emits heat infrared radiation. This method is the source of the difficulties. The principal greenhouse gases in the Earth's atmosphere include carbon dioxide, chlorofluorocarbons, methane, and nitrous oxide. The following are the greenhouse gases that produce an increase in air and earth temperature and are present in the troposphere: CO2 is a substance that exists in the form of a gas. It contributes to global warming by producing around 55% of the greenhouse gases produced by human activities. The chief causes are fossil fuel combustion (67 percent), deforestation, and various kinds of land removal and burning (33 percent). CO2 levels in the atmosphere were 355 parts per million in 1990, and they are growing at a 1.5 parts per million per year pace. CFCs account approximately 24% of all greenhouse gas emissions created by humans. They also deplete the ozone layer in the stratosphere. CFCs are mostly generated through leaky air conditioners and refrigerators, among other sources. CFC levels in the atmosphere are now at 0.00225 parts per million and growing at a rate of 0.5 percent per year. Methane is responsible for 18% of the global increase in greenhouse gas emissions. When microorganisms break down dead organic material in damp, oxygen-deficient environments like swamps, natural wetlands, paddy fields, landfills, and the digestive tracts of cows, sheep, and termites, methane is produced. The current amount of methane in the atmosphere is 1.675 parts per million, and it's growing at a 1% yearly rate.

Human-caused carbon dioxide emissions account for 6% of total emissions. The breakdown of nitrogen fertilisers in soil, animal wastes, and nitrate-contaminated groundwater, as well as the burning of biomass and nitrogen-rich fuels, all contribute to its production (particularly coal). The atmospheric concentration of N20 is 0.3 parts per million, and it is increasing at a rate of 0.2 percent each year. One of the main objectives of climate research is to figure out how the statistical characteristics of the climate system vary when the value of external or internal factors changes. It was recently shown that an approach for quasi-equilibrium systems may be quantified and has formal connections to Kubo 20 response theory. The irreversibility of the ten climate systems, notably their entropy production, is the emphasis. The maximal entropy production principle (MEPP) asserts that an out-of-equilibrium non-linear system adjusts to maximise entropy formation. Despite the fact that MEPP's general validity is in question, it has proved effective as a heuristic in 15 climate research, requiring a full re-evaluation of entropy generation's function in the climate system. This has fuelled a push to adopt a new generation of diagnostic tools based on the second law of thermodynamics for auditing climate models and defining a set of parameterizations for use in conceptual and intermediate complexity models, as well as recreating historical climate conditions.

Carnot efficiency, entropy production, and the degree of climate system irreversibility are all linked to climate system irreversibility. As a consequence, it is now possible to conduct a new fruitful inquiry into the onset and decrease of snowball conditions as a function of solar constant changes. The research makes use of Planet Simulator, a lightweight climate model (PLASIM). Ignoring mixing processes and assuming local thermodynamic equilibrium, this remains true everywhere except in the upper atmosphere, where there is no mass. For estimating the steady-state climate response, slab ocean climate models are excellent. The global atmospheric energy balance is much improved over prior versions of the model by re-feeding the kinetic energy losses owing to surface friction and horizontal and vertical momentum dispersion.

The sensitivity of all three temperature indicators is positive, as expected: the sensitivity of the surface temperature is well within the range anticipated by IPCC climate models, while the sensitivity of the two bulk thermodynamic temperatures is lower. Atmospheric thermodynamics is the study of heat-to-work (and vice versa) changes in the earth's atmosphere in relation to weather and climate. In addition to the globally averaged surface temperature, the intensity of the Lorenz energy cycle, the Carnot efficiency, the generation of material entropy, and the degree of irreversibility of the system are all linear with the logarithm of CO2 concentration. The generalised sensitivity diagram below shows how when the climate system warms, it gets less efficient, more irreversible, and creates more entropy. Changes in the intensity of latent heat fluxes are the most critical components, emphasising the need of addressing climate change's influence on the hydrological cycle at a fundamental level. All variables may be effectively reparameterized with reference to just this one due to the identified variables' monotonic (and, in particular, linear) reliance on the logarithm of CO2 concentration.

Using a basic yet Earth-like climate model and a novel theoretical framework, we study the influence of global warming on climate system thermodynamics. In addition to the globally averaged surface temperature, the intensity of the Lorenz energy cycle, the Carnot efficiency, the generation of material entropy, and the degree of irreversibility of the system are all linear with the logarithm of CO2 concentration. These high sensitivity levels imply that as the climate heats, it becomes less efficient, more irreversible, and more entropy is produced. Using the essential notions of classical thermodynamics, atmospheric thermodynamics investigates processes such as moist air properties, cloud formation, atmospheric convection, boundary layer meteorology, and vertical stability in the atmosphere. Atmospheric thermodynamics is concerned with water and its changes. Among the subjects covered are the ideal gas law, specific heat capacities, adiabatic (entropy-conserving) processes, and wet adiabatic processes. The bulk of troposphere gases are made up of water vapour, which is one of the most important trace components of air.

The essential function of atmospheric thermodynamics is described in basic equations of air motion, which are either grid resolved or subgrid parameterized, in terms of adiabatic and diabatic forces acting on air parcels. Numerical weather and climate predictions are created using these equations. The sun provides heat to the Earth. The Earth and its atmosphere emit heat into space. They disperse the bulk of the solar heat, maintaining the Earth's normal temperature. Greenhouse gases warm the Earth and envelop it in a warm blanket. According to the second law of thermodynamics, heat cannot flow spontaneously from a lower-temperature substance to a higher-temperature material. "Global warming" is one of the most often used words in environmental study. The temperature of the Earth has been continuously increasing over the last five decades. Many experts believe that effective and acceptable pollution control techniques are urgently needed. Humans are exposed to a variety of harmful gases, each of which creates its own set of health problems. Carbon dioxide, sulphur dioxide, and hydrogen sulphide are among the gases that may be removed from exhaust gases using various processes. As a consequence of greenhouse gases, shortwave radiation may enter the earth's atmosphere, heating the land and oceans. Long-wave radiation from the earth's surface is unable to reach the atmosphere due to these greenhouse gases. These activities contribute to the greenhouse effect. A large number of scientists have looked into greenhouse gases, their impacts, and potential remedies. The research and studies on greenhouse gases are the topic of this review. The greenhouse effect has produced considerable global warming in recent decades, notably in the past four to five decades. Warmer temperatures extend the growth season and make frigid areas more livable in general. When utilised in a high-intensity way, natural gas produces more greenhouse gases than fossil fuels. Carbon dioxide (CO2) and methane (CH4) are two climate-changing gases (CH4). The temperature of the Earth has been continuously increasing over the last five decades. Many experts believe that effective and acceptable pollution control techniques are urgently needed. Humans are exposed to a variety of harmful gases, each of which creates its own set of health problems. The research and studies on greenhouse gases are the topic of this review.

Many of us can eat without much thought, preparation, or time thanks to contemporary culinary technology. Fast food, ready-to-eat meals, and processed snacks give quick nutrition while causing the least amount of interruption to our daily routines. The immediate impact of a satisfied appetite is obvious, but the environmental impact of food and agriculture is less apparent. Recent publications, such as Michael Pollen's The Omnivore's Dilemma, have questioned whether a more ecologically friendly food system is feasible. Furthermore, the effects of global climate change raise worries about food's carbon footprint. This short will examine the most up-to-date scientific studies of food and climate systems in order to assist us in making intuitive and environmentally sound eating decisions. Greenhouse gas emissions are produced through food production, packaging, transportation, sale, and consumption.

Food-related processes, such as soil nitrous oxide generation, produce three major greenhouse gases: carbon dioxide, methane, and nitrous oxide. Depending on their persistence (how long they stay active) and impact, various GHGs in the atmosphere have varied global warming potentials. Because methane remains in the atmosphere for a shorter period of time while absorbing more heat, it has a greater cumulative impact than carbon dioxide. Nitrous oxide is the most potent global warming gas.

Food's carbon footprint is a similar measure of the impact of global warming (also known as CO2e: carbon dioxide equivalent). The carbon footprint of a single unit of food refers to the total amount of greenhouse gas emissions produced by that unit of food. Food consumption accounts for around 15% of total GHG emissions in the United States. We may work to reduce GHG emissions in a variety of "hot spots" across the food chain.

In the United States, agriculture contributes for around 6% of total GDP, which includes all agricultural processes but excludes those beyond them. GHG emissions refer to the amount of greenhouse gases released into the atmosphere. Though we may make educated predictions, exact data on agricultural GHG emissions in the United States is difficult to come by owing to the country's diverse food supply and consumption. Agriculture is the largest source of GHG emissions, which are mostly caused by microbial activity in the soil and manure. When fossil fuels are used to power agricultural machines, carbon dioxide is created. When fertiliser emissions escape from the soil, they constitute a substantial source of nitrous oxide emissions into the atmosphere.

There are many procedures involved in getting food to markets. Food miles, or the average distance travelled by food from farm to fork, are a measurement of the food system's transportation infrastructure. According to the Leopold Center, produce supplied to the upper Midwest travels an average of 1,500 kilometres from industrial centres throughout the continental United States. Food transportation is responsible for around 11% of total GHG emissions from the food chain. Experts believe that changes in technology and infrastructure will contribute in the decrease of emissions related to transportation.

Food consumption is the second-largest source of greenhouse gas emissions in the household, behind transportation. The energy needed for refrigeration is one of the primary GHG consequences of food production. Certain commodities need continual refrigeration as soon as they leave the farm. Food waste wastes the energy needed to produce it, as well as the greenhouse gas emissions produced as a consequence. Meat waste in retail, industry, and households accounts for 20% of the meat sector's total GHG emissions (How to reduce home emissions is discussed further down.)

The proportional GHG emissions from these three food chain components vary depending on the kind of food. Livestock and meat production release more greenhouse emissions than plant-based agriculture. When cooked and stored, plant-based proteins like lentils and dry beans contribute significantly to total emissions.

Agriculture produces three major greenhouse gases: carbon dioxide (CO2), methane (CH4), and nitrous oxide (NOx) (N20). While nitrous oxide and methane are the most common greenhouse gases in the atmosphere, they have longer atmospheric lifetimes and absorb more long-wave radiation. As a consequence, even little volumes of methane and nitrous oxide might significantly contribute to global warming. Agriculture may have both good and negative consequences as a result of climate change. According to several studies, warmer temperatures lengthen growing seasons, and more carbon dioxide in the air leads to higher agricultural productivity. As a result of increasing temperatures and decreased soil moisture, production patterns may shift northward, demanding greater irrigation. Changes, on the other hand, will most likely vary significantly from one location to the next. Geographical factors will have a big impact on agriculture's potential to benefit from climate change. While some projections seem to be optimistic in certain areas, the risk of increased climatic unpredictability and extremes is not always included. Droughts, heat waves, severe thunderstorms, and tornadoes may become more common, outweighing the agricultural advantages. Farmers are finding it increasingly difficult to respond as climate change becomes more unpredictable.

The two categories of greenhouse gases generated by biomass power plants that contribute to global warming are direct and indirect greenhouse gases. Direct greenhouse gases include carbon dioxide (CO2), nitrous oxide (N2O), and methane (CH4). Indirect greenhouse gases include NMVOCs, nitrogen oxides (NOx) (nitrogen monoxide (NO) and nitrogen dioxide (NO2)), sulphur dioxide (SO2), and carbon monoxide (CO). By absorbing and storing infrared radiation released by the earth's surface and lower atmosphere, and then reflecting it back to the planet's surface, direct greenhouse gases contribute to global warming. The environment is a system of systems that is a component of a higher system of systems, the human system, from the perspective of human life and development. Because it is clear that preserving and mending the environment in the interests of human life and growth is difficult, it is also impossible to damage the environment indiscriminately because it is the medium through which humans live. As a result, we must create compromises based on our knowledge and experience that respect both human and environmental concerns. Their effect and advantages are tracked so that remedial action may be taken if necessary.

According to current research, sustainability (development) involves not just the environment, but also the whole human system as well as the fundamental assets (i.e. public assets) on which people rely. Infrastructures and technologies, particularly those that are crucial, as well as the environment, property, and public welfare, are all important assets in the human system. Sustainability assessment is a method for recognising, anticipating, and assessing the influence of arbitrary inputs, such as fluctuations, on society's long-term growth in a broad sense (e.g. legal rules, ordinances, regulations, political intent, plan, program, and project). The aforementioned evaluation, which is based on current knowledge of the human system and its assets, may be conducted at any time to ensure effective territory management. The evolution of the environment, as well as the global political, social, and economic situation, demonstrates the need to address cases and actions that, because of their gravity, result in critical situations that lead to relevant crises of the type denoted as a humanitarian catastrophe or social crisis. As a result, a safety concept linked to development principles produced and accomplished through safety management is necessary from the viewpoints of human security, human system development, quality environmental conservation, state existence, stability, and development..

Basic (or normal) management, which is connected to emergency and crisis management, aims for security and long-term growth. As a result of land and regional growth, industrial zones are becoming closer to residential neighbourhoods. As a result, the potential of harm rises, and society (community) refuses to take any chances. Risk management, and by extension risk engineering, which involves risk assessment, risk reduction, and harm explanation, is critical in this situation. Finally, risk engineering is linked to technological systems (the human factor's impact on complex process safety is only addressed in advanced versions), and it may be utilised to revitalise landscapes by applying engineering principles in a wider sense.

As a result, risk management is essential. On the one hand, it aims to provide the best possible protection for people, property, and the environment, while on the other, it aims to restore damaged landscapes utilising engineering techniques and results. One of the most important concerns confronting the globe today is environmental deterioration. Some settings may never be retrieved, depending on the extent of the loss. Flora and wildlife that formerly thrived in these areas are now extinct. City planners, industry, and resource management must address the long-term environmental implications of expansion to minimise future repercussions. Future environmental damage might be avoided with proper planning. Despite the fact that human activities are often connected to environmental deterioration, ecosystems adapt throughout time. Some ecosystems deteriorate over time, whether or not they are touched by human activity, to the point that they can no longer sustain the life that is "meant" to live there. Landslides, earthquakes, tsunamis, and wildfires all have the potential to devastate local plant and animal ecosystems, rendering them untenable. This might happen as a result of a natural catastrophe or long-term resource deterioration caused by the introduction of an invasive alien species into a new ecosystem.

When the environment is unable to deal with the new species, degradation occurs. These contaminants may build up in sediments and biomagnify in the tissues of top-of-the-food-chain animals, resulting in concentrations many times greater than those seen in water or air. Ozone is a gas that may be found on the ground as well as in the stratosphere, the Earth's upper atmosphere. Ozone is a contaminant that may impair people's health at ground level. On the other side, ozone creates a stratospheric layer that shields life on Earth from the sun's destructive ultraviolet (UV) rays. Ozone-depleting substances like chlorofluorocarbons, hydro chlorofluorocarbons, and halons are man-made chemicals that slowly deplete the "healthy" ozone layer. These chemicals have been used as coolants, foaming agents, fire extinguishers, solvents, insecticides, and aerosol propellants in the past, and they are still used in certain situations today. More UV radiation penetrates the Earth as the protective ozone layer thins, possibly causing skin cancer, cataracts, and immune system damage. UV radiation has the potential to harm delicate crops, such as soybeans, resulting in lower yields.

CONCLUSION

Environment-related ramifications Toxic contaminants in the air, as well as those deposited on soils and surface waters, may harm animals. Animals, like people, may develop health issues if they are exposed to excessive levels of air toxics over time. Air toxics have been linked to birth abnormalities, reproductive failure, and illness in animals, according to a study. In aquatic ecosystems, persistent hazardous air pollutants (those that take a long time to degrade in the environment) are of special concern. Air pollution has a range of effects on crops and plants. Ground-level ozone may diminish agricultural crop and commercial forest yields, smother tree seedling development and survival, and make plants more susceptible to disease, pests, and other environmental challenges (such as harsh weather). Acid rain and increased UV radiation due by ozone depletion may have an influence on agriculture and forestry, as previously stated. Climate change is happening on a global scale. Because of a complex mix of naturally existing gases in the Earth's atmosphere, some of the sun's heat is preserved near the surface. The "greenhouse effect" maintains a steady temperature on the earth. Unfortunately, evidence is accumulating that people have disrupted nature's equilibrium by emitting vast volumes of greenhouse gases such as carbon dioxide and methane. As a consequence, the Earth's atmosphere seems to be retaining more of the sun's heat, leading to global warming. According to many experts, climate change will have significant consequences for human health, agriculture, water resources, forests, animals, and coastal regions.

REFERENCES

Dones, Zhou, & Tian. (n.d.). Life cycle assessment. In *Integrated assessment of sustainable energy systems in China — The China Energy Technology Program*. Academic Press.

Dones, R., Bauer, C., Bolliger, R., Burger, B., Faist, M., Frischknecht, R., Heck, T. N., Jungbluth, A., & Röder. (2016). Life Cycle Inventories of Energy Systems: Results for Current Systems in Switzerland and other UCTE Countries. Academic Press.

Fraedrich, K. (2001). Simple Climate Models. In Progress in Probability 49. Birkh¨auser.

Fraedrich, K., Jansen, H., Kirk, E., Luksch, U., & Lunkeit, F. (2005). The planet simulator: Towards a user friendly model. *Met. Zeit.*, *14*, 299–304.

Fraedrich, K., & Lunkeit, F. (2008). Diagnosing the entropy budget of a climate model. *Tellus. Series A, Dynamic Meterology and Oceanography*, *60*, 921–931.

Frischknecht, R., Jungbluth, N., Althaus, H. J., Doka, G., Dones, R., Hellweg, S., Hischier, R., Nemecek, T., Rebitzer, G., & Spielmann, M. (2003). *Overview and Methodology — Data v1.01*. Swiss Centre for Life Cycle Inventories. www.ecoinvent.ch

Hondo. (2003). Life cycle GHG emission analysis of power generation systems: Japanese case. *Energy*.

Inter-Governmental Panel on Climate Change (IPCC). (2001). *The Scientific Basis. Contribution of Working Group I to the Third Assessment Report*. Author.

Kunz, T., Fraedrich, K., & Kirk, E. (2008). Optimization of simplified GCMs using circulation indices and maximum entropy production. *Climate Dynamics*, *30*, 803–813.

Lee, Y. (1997). Comparison of energy sources in terms of their full energy chain emission factors of greenhouse gases. *Energy Policy*, *25*(1), 1–6.

Lorenz, E. N. (1955). Available potential energy and the maintenance of the general circulation. *Tellus*, *7*, 157–167.

Ozawa, H., Ohmura, A., Lorenz, R. D., & Pujol, T. (2003). The second law of thermodynamics 30 and the global climate system, A review of the maximum entropy production principle. *Reviews of Geophysics*, *41*, 1018.

Spath, Mann, & Kerr. (1999). Life-cycle assessment of coal-fired power production. National Renewable Energy Laboratory, NREL/TP-570-25119.

Villigen & Swiss. (2003). *Centre for Life Cycle Inventories*. www.ecoinvent.ch

Chapter 6
Biofuels From Bio-Waste and Biomass

Kondapalli Vamsi Krishna
Christ University (Deemed), India

Sompalli Bhavana
Christ University (Deemed), India

Koushik Koujalagi
Christ University (Deemed), India

Alok Malaviya
Christ University (Deemed), India

ABSTRACT

The planet's limited natural fossil fuel reserves are anticipated to be very soon owing to massive usage. Biofuels would be a critical alternative source that may reduce global warming and CO2 emissions. The food-versus-fuel dilemma is, however, one of the key drawbacks of first-generation biofuels like corn ethanol, sugarcane ethanol, etc. Cellulose and hemicellulose, the primary constituents of lignocellulosic feedstocks, could be reduced to sugars by either thermochemical/biological processes before being fermented to generate biofuels. However, owing to structural heterogeneity, more complicated operational techniques are required before the production technology can be commercialized, and several challenges must be addressed. This chapter provided an assessment of various feedstocks, availability, various processing techniques, obstacles, and current technical developments in the generation of biofuels from biomass.

DOI: 10.4018/978-1-6684-5269-1.ch006

INTRODUCTION

The earth's finite natural fossil fuel reserves are being exhausted at a quicker pace as the world's energy demands rise as a result of advances in industrialization and modernization (Mohapatra, Ray, & Ramachandran, 2019; Ruan et al., 2019; Saini, Saini, & Tewari, 2015). Petroleum, coal, natural gas, bitumen, oil shales, and tar are examples of fossil fuels that have aided in the wealth and success of many nations throughout the world, as well as meeting energy demands for power, transportation, and heating (Dahman, Dignan, Fiayaz, & Chaudhry, 2019). Fossil fuels account for almost 80% of all primary energy use. With the transportation sector accounting for only 58% (Raud, Kikas, Sippula, & Shurpali, 2019). From a global point, the energy and transport industry is associated with heavy fossil fuel consumption (Dahman, Dignan, et al., 2019). As the world's population grows and industry and technology advance, this need grows year after year.

According to the 2011 consumption rate, the oil will last 54 years, natural gas 64 years, and coal 112 years (Khattab & Watanabe, 2019). Indeed, the varied technical breakthroughs driven by the use of heavy fossil fuels have improved human existence and made it more comfortable; yet, this comfort has arrived at a significant expense. Global warming and greenhouse gas (GHG) emissions have come from the extensive consumption of fossil fuels. Ozone (O_3), Carbon Dioxide (CO_2), methane (CH_4) and Nitrous oxide (NO_3) are the main greenhouse gases that emit and absorb infrared radiation owing to their unique characteristics and chemical makeup. The elevated level of these gases in the atmosphere had several negative consequences for the planet. Roughly 98% of the greenhouse gases released into the atmosphere are a direct consequence of the production of fossil fuels (Dahman, Syed, Begum, Roy, & Mohtasebi, 2019; Khattab & Watanabe, 2019). The *Mauna Loa* research center revealed CO_2 levels of 413 ppm in March 2019, up from 280 ppm in preindustrial Period. The proportion of atmospheric CO_2 buildup has escalated to 2.28 ppm per year between 2008 and 2017, up from 0.8 ppm per year between 1960 and 1969. The CO_2 content in the atmosphere has increased by 43% since 1750, and as a consequence of this rise, the average surface temperature of the earth has increased by 0.85°C since pre-industrial times (Raud et al., 2019). Each of these elements has shifted the phenomena of global warming from a local myth to a grave concern worldwide that concerns our own humankind's survival on our planet. Because of the continued and increasing release of greenhouse gases, the average temperature of the Earth has been steadily rising. Changes in climate, weather system disruption, glacier retreat, rising sea level, coastal flooding, associated biodiversity loss, and human health effects are the main consequences of fossil fuel consumption (Dahman, Dignan, et al., 2019; Dahman, Syed, et al., 2019). All these collectively coupled with the climb in the energy consumption and demands, depletion of fossil fuels, and severe environmental concerns have sparked to watch out for an alternative, sustainable, renewable, and green energy source that could not only meet the global demand and also helps to lessen the global greenhouse emissions (Agbor, Cicek, Sparling, Berlin, & Levin, 2011; Raud et al., 2019). In this chapter, the focus will be on research into the types, compositions, and application of various forms of biomass accessible for biofuel generation and also the numerous pretreatment and processing options that may be used to various bio-wastes to transform them into useable biofuels cost-effectively and sustainably to fulfill rising energy demand.

NECESSITY OF BIOFUELS

Changes in climate and security of energy supply should be at the forefront of the States international agenda, impacting societal, economical, and political choices. Every country's primary goal should be to find a renewable, cost-effective, biodegradable, and environmentally benign alternative energy source. Biofuels, without a doubt, would be a significant alternative source in which biomass-based biofuels might ameliorate global warming and reduce CO_2 emissions owing to the preceding consumption of this gas by the biomass during photosynthesis, resulting in carbon-neutral emission (Coyle, 2007; Sindhu et al., 2019). The term 'biofuel' refers to any liquid, gas, or solid fuel predominantly produced from a renewable biomass feedstock. Bioethanol, biomethanol, biosynthetic gas (biosyngas), biodiesel, biogas (biomethane), biochar, bio-oil, biohydrogen, and FischerTropsch produced liquids are the common biofuels (Ali, Saleem, Khan, & Watson, 2019; Ayhan Demirbas, 2008). Mankind has been consuming biofuels for an extended time, ever since humans began burning firewood for various home and domestic needs. Biofuels can play an significant role in the future by addressing the enormous fuel market and ensuring global energy security (Kumar et al., 2020). Biofuels offer several benefits over fossil fuels, including renewable, sustainable, abundant availability, biodegradability, CO_2/GHG (greenhouse gas) neutrality, minimal or zero SO_x emissions, reduced NO_x emissions, and the ability to be generated locally and in a safer manner. Biofuels also have several advantages, including the consumption of local resources, the stimulation of rural agricultural development and investment, the reduction of energy dependence on imports, the creation of jobs, the support of rural economic development, and the improvement of local livelihoods (Ruan et al., 2019). Biofuels are regarded as the most essential sustainable fuel resource due to the benefits outlined above. They also contribute to the lessening of greenhouse emissions and, as a result, to the improvement of environmental health. As a result, many governments have set various goals and regulations to encourage and expand the biofuel business and so use energy resources. It is estimated that diverse biofuels will account for 30% of world energy demand (Guo, Song, & Buhain, 2015).

BIOFUEL FACTS, FIGURES, AND POLICIES

The Center for Strategic and International Studies in its report 'International Energy Outlook 2011' stated that renewables are the fastest-growing energy source. It is rising at a rate of 2.8% per year and would arrive at roughly 15% of the world energy market by 2035. It also stated that by 2035, liquid biofuels will include 29% of world energy consumption. This translates to about 5 million barrels per day of the total production (Dahman, Syed, et al., 2019). Governments of different nations have made biofuel policies in current years for the domestic production and usage of biofuels. Brazil has been extremely doing well in biofuel policy that mandates the minimum level of blending of ethanol with petrol. As of today, it is projected that the petrol sold in Brazil contains almost 20 -25% of ethanol in terms of volume (Lamers, Hamelinck, Junginger, & Faaij, 2011). In the USA, the US Energy Independence and Security Act of 2007 have laid down a aim of production of 15 billion gallons of corn ethanol by 2015, and by this year, 21 billion gallons of cellulosic and advanced biofuels (Iens & Argione, 2011). Similarly to these countries, all other big nations, including China, Canada, India, Japan, and South Africa, have legislated biofuel policies that mandated the targets of biofuel blending requirements in fossil fuels, as well as other policies and schemes that encourage the arrangement and commercial biofuel production (Dahman, Syed, et al., 2019; Timilsina, Mevel, & Shrestha, 2011).

INDIA AMONG THE OTHERS

India, a country with over a billion people and limited space, has one of the world's fastest-expanding economies. Energy is critical to the country's socio-economic growth. India is primarily reliant on fossil fuels to fulfill its energy requirements. India has relatively limited fossil fuel supplies and must import oil, coal, and gas, with oil import dependency being particularly severe. To fulfill the rising energy demand, the Indian government is urgently seeking an alternative to fossil fuels. The 2009 'National Policy on Biofuels' envisions biofuels as a vital tool of stimulating rural development and creating local job possibilities. In actuality, India's biofuel business is still in its infancy. India's biofuel production amounts to less than 1% of world output. Within the different renewable energy sources accessible, India has a plethora of bioenergy choices that have been employed to fulfill cooking needs for a long time. Various bioenergy policies and initiatives have been implemented in India in recent years to boost bioenergy production. However, due to policy limitations, the government is unable to fully use the enormous potential of bioenergy solutions. Bioenergy policy implementation is fraught with difficulties in terms of technological, institutional, financial, environmental, and social factors. However, the objective for mixing gasoline with ethanol was around 5%.Currently, bio-ethanol from sugar cane molasses and biodiesel from non-edible oilseed crops such as Jatropha (*Jatropha curcas*) and Pongamia (*Millettia pinnata*) are being encouraged for commercial usage in India (Gopinathan & Sudhakaran, 2009; Kothari et al., 2020; Roy & Chandra, 2019).

Agriculture is the basis of India's GDP, bringing subsistence to 60% of the population, and generating a considerable amount of agro waste as a result of agricultural strength. It is also the second-largest producer of sugarcane following Brazil (Ravindranath et al., 2005). India generates 686 million tons of gross agricultural residue biomass per year, with 234 million tons (34%) accessible as an excess for biofuel energy production. Biomasses from agriculture are used as feedstock in agro-industries (Naik, Goud, Rout, & Dalai, 2010). Residues and by-products from agriculture include rice husks, cottonseed, cereal straw, corn cobs, cotton stalks, various grasses, maize, sorghum stover, sunflower seed hulls, peanut shells, , and wood sawdust and chips, to name a few (Lal, 2008).

CLASSIFICATION OF BIOFUELS

Biofuels can be classified in different ways which are based on different aspects like production process, commercialization stages, physical state, and biomass. This chapter will focus on research into the types, compositions, and applications of various forms of biomass accessible for biofuel generation. The emphasis was on the numerous pretreatment and processing options that may be used to various bio-wastes to transform them into useable biofuels cost-effectively and sustainably to fulfill rising energy demand.

Biofuels are fuels that are produced by biological processes from various sources like plants, crops, agriculture residues. Based on the production route, biofuels can be categorized as primary biofuels and secondary biofuels. Primary biofuels are made up of a variety of materials such as wood chips, pellets, fuel wood, and organic materials that are burnt straight to provide energy for food preparation, heating, and power generation. Secondary biofuels are made from a wide range of biomass and may be utilized for transportation, energy generation, and a number of industrial purposes. Ethanol, diesel, and biogas are examples of secondary biofuels (Dahman, Syed, et al., 2019; Datta, Hossain, & Roy, 2019; Ruan et al., 2019).

Biofuels are categorized into conventional and advanced biofuels based on the level of commercialization of technology. Bioethanol made from sugar and starch, biodiesel made from transesterification, and anaerobic digesting methane is all examples of conventional biofuels. Sugarcane and sugar beet, corn, wheat, vegetable oil, animal fat, recovered household fat, waste frying oil, etc., are the most used feedstocks for conventional biofuels. All of these things are sold in different countries throughout the world. Hydrotreated vegetable oil, cellulosic (lignocellulosic) bioethanol, biomass-to-liquid (BtL) biodiesel, microalgae biodiesel, and biohydrogen are examples of advanced biofuels. All these are not yet commercialized. They are in different phases (research and development, demonstration, or early commercial stages) to be commercialized (Awogbemi, Von Kallon, Onuh, & Aigbodion, 2021; Ruan et al., 2019).

Based on the physical state the biofuels can be segregated into three types - solid, liquid, and gaseous biofuels. Different biofuels, their benefits, and drawbacks have been discussed in the following section.

Solid Biofuels

Solid biofuels are raw solid feedstock used to produce biofuels such as wood, wood chips, wood pellets, and charcoal. Lignocellulosic biomass and numerous forms of solid waste are examples of solid biomass. They can be utilized directly as biofuel or transformed into other types of biofuel over various methods. Wood was utilized for numerous domestic uses such as cooking and heating in the early nineteenth century, and it is still the major fuel today. Wood chips are little fragments of wood that have been chipped from tree trunks or branches and used as fuel for heating and energy generating. Similarly, wood pellets are created by grinding wood pieces with sawdust and compressing them via a pelletizer, and they may be utilized as a fuel in household heating and energy generation (Goyal, Seal, & Saxena, 2008; Guo et al., 2015; Sindhu et al., 2019). In the lack of air, biochar, as well identified as charcoal, is a porous, carbon-enriched greyish black solid generated via torrefaction, gasification, hydrothermal carbonization, and other processes that may reach temperatures over 400°C (usually between 300°C and 900°C). It offers an advantage over other raw feedstocks like firewood, wood chips, and so on in terms of lower combustion emissions and higher energy content (Glaser, Haumaier, Guggenberger, & Zech, 2001; Kwapinski et al., 2010). Biochar can be utilized as a solid fuel in boilers, as a feedstock for subsequent gasification processes, to produce activated carbon, and so on (Kambo & Dutta, 2015).

Liquid Biofuels

Bioethanol, biodiesel, Bio-methanol, Biobutanol, and Biooil are examples of liquid biofuels. They're becoming increasingly popular because to the fact that they're highly flammable, have a high energy-to-mass ratio, and are simple to carry.

Bioethanol

The most often employed liquid biofuel in our day is ethyl alcohol, which has the chemical formula CH_3 - CH_2 - OH and is manufactured by the microbial fermentation of sugars or starch using a range of feedstocks. The United States (from maize) and Brazil (from sugar cane) are the world's leading producers of bioethanol, accounting for almost 90% of worldwide output (Cheng & Timilsina, 2011; Vohra, Manwar, Manmode, Padgilwar, & Patil, 2014). It possesses 35% oxygen and therefore can minimize automotive emissions and may be utilized in the same way as conventional fuels. However, they have a poor volume energy density and need more bioethanol per kilometer than gasoline. As a result, it is mixed with gasoline in various ratios ranging from 5% to 25% ethanol and 95% to 75% gasoline and may be used in automobiles without any modifications (Naik et al., 2010; Vohra et al., 2014).

Biobutanol

Biobutanol, C4 alcohol with molecular formula C_4H_9OH, a biofuel made from the ABE (Acetone Butanol and Ethanol) fermentation of the traditional feedstock. It was produced via Clostridium acetobutylicum fermentation of corn and molasses, with the end products being acetone, butanol, and ethanol, and therefore the procedure was designated as ABE fermentation. Byproducts of ABE fermentation include acetic, lactic, propionic, butyric acids, hydrogen, isopropanol, CO_2, and lipids. Biobutanol has greater energy density, is less corrosive than ethanol, safer to use, and has no need for engine modifications as it uses a fuel to air ratio closer to gasoline. Nevertheless, butanol becomes lethal once its concentration reaches 13g/L in the fermentation medium. As a result, much research is being done to genetically enhance butanol-producing cultures that can withstand elevated acetone butanol concentrations of 20.0 to 30.0 g/L while simultaneously recovering the products (Dürre, 2007; Kolesinska et al., 2019; Xue & Cheng, 2019).

Biodiesel

Biodiesel is a colorless to yellowish fuel that resembles conventional diesel. It is a combination of long-chain fatty acid methyl esters (FAME) synthesized in the presence of alcohol and alkaline catalysts from animal fats, edible and non-edible vegetable oil, waste or recycled oil, waste grease and also from algal. Unlike petroleum diesel, which only includes hydrocarbons, biodiesel contains carbon, hydrogen, and oxygen. It has a high flash point, making it harmless to store, handle, and transport, as well as a higher cetane number, indicating that it has greater ignition quality, a low sulfur content, and is biodegradable (Atabani et al., 2012; Hoekman, Broch, Robbins, Ceniceros, & Natarajan, 2012; Ma & Hanna, 1999; V. K. Mishra & Goswami, 2018). It does, however, have a low calorific value, a high pour point (meaning it solidifies at lower temperatures), and high viscosity. As a result, it may be mixed with normal crude oil or petroleum diesel and utilized. The most common biodiesel blends are B2 (2% biodiesel and 98% normal diesel), B5 (5% biodiesel and 95% regular gasoline), and B20 (20% biodiesel and 80% regular diesel) (Morone & Cottoni, 2016).

Biooil

Biooil is a dark, free-flowing organic liquid having a smoky odor which is a mixture of aldehydes, acids, alkenes, aromatics, alcohols, esters, ketones, phenols, oligomers, furans, sugars, guaiacols, syringols, and nitrogen compounds produced by the depolymerization or fragmentation of biomass by fast/slow pyrolysis and thermal liquefaction. Pyrolysis is a practice to heat the plant biomass at 300–900°C in the absence of air. They are also referred to as pyrolysis liquid, pyrolysis oil, wood oil, bio-crude oil, wood distillate, wood liquid, etc.,(Czernik & Bridgwater, 2004). The major disadvantages of bio-oil are high water content, which is a result of the presence of water in the feedstock and also the production of water in the dehydration reactions in the pyrolysis process, high viscosity, less oxygen, low energy density, immiscible with hydrocarbon fuels and corrosiveness which made the bio-oil cannot be used as a transportation fuel. The major challenge preventing the commercialization of bio-oil is the high production cost. Research and development efforts have to be made to reduce the water content, high acidity and further improve the quality of the fuel by pretreatment of raw feedstocks, optimization of operation conditions, adopting strategies such as hydrotreating, hydrocracking, supercritical fluids, solvent addition/esterification and steam reforming (D. Vamvuka, 2011; Guo et al., 2015; Xiu & Shahbazi, 2012).

Gas Biofuels

The third type of biofuels, the gaseous biofuels includes biogas, biohydrogen, biosyn gas, etc., generally have advantages like more reactive, the requirement of fewer oxidants, simple reactors, and low wastage in the yields.

Bio Hydrogen

It is a colorless, odorless, and lightest combustible gas produced from renewable sources that include crop residues, energy crops, agriculture residues, livestock residues, algal biomass, industrial wastes, etc., by either thermochemical conversion or biological conversion. Thermochemical transformation comprises water and biomass thermochemical decomposition, syngas water-gas shift reaction, natural gas (or methane) steam reforming, pyrolysis bio-oil steam reforming, and so on(H. Balat & Kirtay, 2010; M. Balat & Balat, 2009). Biophotolysis is the initial phase in biological conversion technology, which involves using photosynthetic bacteria to convert CO_2 and water into H_2 or photo heterophilic microbes to convert carbohydrates or organic acids into H_2 and CO_2.Following that, photofermentative bacteria photo-decompose organic compounds, followed by dark fermentation, which generates H_2 utilising anaerobic or facultative anaerobic bacteria, and ultimately, microbial electrolysis transforms electron equivalents in organic compounds to H_2 by uniting microbial metabolism with bioelectrochemical reactions in a microbial electrolysis cell. There are practical encounters that have to be addressed like light dependency in the biophotolysis and photofermentation steps, the buildup of acidogenic by-products that may constrain the dark fermentation step, and cell lysis in the electrolysis step (A. Demirbas, 2008a; Meramo-Hurtado, Puello, & Cabarcas, 2020; Vendruscolo, 2014).

Biogas

Biogas also termed biomethane is a gas biofuel that is made by anaerobic digestion of organic wastes like plant residues into smaller molecules in natural environments by the consortium of microorganisms which usually contain 60–65% of methane (CH_4), 30–35% of CO_2, and a small percentage of water vapor, H_2 andH_2S (Ge, Xu, & Li, 2016; Martínez-Gutiérrez, 2018). The purification step is necessary to remove the moisture, CO_2, H_2S, and other impurities before supplying as a natural gas substitute. Because of the presence of higher CO_2 content, it produces lower energy than natural gas. Anaerobic digestion has been employed to manage waste and generate energy by various livestock farms, wastewater treatment plants, and private residences. It is also applied to generate energy and bio-based compressed natural gas (CNG) that is applied to power natural gas cars (Guo et al., 2015; Wei, 2016).

Syngas

Syngas is a gaseous biofuel which is the result of gasification or pyrolysis of carbon-rich materials such as natural gas, petroleum, coal, and dry plant biomass and it contains CO_2,CO, N_2, H_2, and hydrocarbons (e.g., CH_4, C_2H_4, and C_2H_6) and very minor quantities of moisture, H_2S,NH_3, and tars(Yan, Luo, Hu, Xiao, & Cheng, 2010; Y. Zhang, Zhao, Gao, Li, & Huang, 2015). The main plus of the syngas is versatility as they can be used as starting material to synthesize transportation fuels like methanol, methane, ethanol, dimethylether, and extra products via the gas-liquid platform. Because of stringent feedstock requirements, biomass-based biogas gasification is currently in the early demonstration phases and additional research is needed to know the economics of biomass and attain the financial feasibility for commercial-scale production (Ruiz, Juárez, Morales, Muñoz, & Mendívil, 2013).

BASED ON FEEDSTOCK SOURCES

Based on the different types of feedstocks the biofuels can be divided into first and second-generation biofuels.

First-generation biofuels

First-generation biofuels, often stated to as conventional biofuels, are derived mostly from food crops for instance cereals, sugar cane, tuber crops, sugar beets, and oil crops. The first generation of biofuels covers biodiesel, vegetable oils, biogas, bio alcohols, biomethanol, and Syngas (Ruan et al., 2019).There are over 350 oil-containing crops classified for biodiesel production. Many countries generate first-generation biofuel from a variety of sources (Hirani, Javed, Asif, Basu, & Kumar, 2018). Brazil produces bioethanol mostly from sugarcane, whereas the United States, Canada, and Europe generate bioethanol primarily from maize, wheat, sugar beet, and sorghum, respectively. Maize accounts for approximately 67% of worldwide bioethanol output (Mohapatra et al., 2019).From 2001 to 2017, a large increase in the global production of bioethanol was observed. In 2001, the production was 4874 million gallons which reached up to 27,737 million gallons in 2017 (Mohapatra et al., 2019). Globally, 40-billion-liter bioethanol is produced by the US and 2 billion liters by Canada from corn or wheat. In the case of Brazil, they produce 25 billion liter bioethanol from sugarcane whereas China produces 3 billion liters from corn, cassava. From wheat, sugarcane, and sugar beet, France and Germany produce 1 billion liters and 750 million liter bioethanol respectively. India supplies 1 billion liters of bioethanol from sugarcane and molasses and Australia produces 500 million liters from sugarcane (Mohapatra et al., 2019).

The production of biodiesel in the United States, France, Germany, and Italy is mostly dependent on rapeseed oil and soyabean oil. Brazil mostly employs soybean oil in the manufacturing of biodiesel (Hoekman et al., 2012). It was reported that annual biodiesel production is increasing exponentially. Approximately 85% of biodiesel is produced by European countries (Atabani et al., 2012; V. K. Mishra & Goswami, 2018). The production of biodiesel is increasing by the year. They are produced from oil of soybean, sunflower, canola, or palm oil. Due to their environmental benefits, they are widely used (https://www.mdpi.com/1996-1073/11/12/3366/html). The production was 0.8 billion L in 2000, which reached 4 billion L in 2005. A four-fold increase in production was seen in 2010. By 2012, the production reached over 22.5 billion liters. European countries are the largest producer of biodiesel consisting of 80% of global production, followed by the USA (Hirani et al., 2018). In 2001, the Republic of China started the production of bioethanol using corn. By 2007, they started using grain-based feedstock producing about 1.4 million metric tons of bioethanol, which got replaced by nonfood crops like cassava and sorghum (Mussatto et al., 2010). Thailand started bioethanol production using cassava and cane molasses (Silalertruksa & Gheewala, 2009). The pros and cons of the first-generation biofuels were given in Table 1.

Table 1. Pros and Cons of First-generation biofuels (Datta et al., 2019; Ruan et al., 2019)

Pros	Cons
Sustainable **Renewable form of energy** **Feedstock production is simple using current resources and infrastructure** **Environmental and social advantages** **A huge number of feedstocks are accessible**	**Requires a huge amount of fertilizer, water, and availability of land, lowering the overall energy ratio.** **Added to increased food prices due to equal demand with food.** **Could potentially have negative influence on biodiversity** **Increased carbon footprint compared to other generations of biofuel** **Merging with conventional energy sources** **Compete for the availability of lands**

Sugar containing feedstocks

Sugar containing feedstocks for the generation of biofuels are sugarcane, sugar beet, and sweet sorghum with yields of 62–74 tons/ha, 54–111 tons/ha, and 50–62 tons/ha, respectively which are grown mainly in Brazil, Germany, France, and India. In comparison to maize and wheat, sugarcane remains the biggest provider to bioethanol production and the minimum water user in terms of crop biomass output (Mohapatra et al., 2019).

Sugarcane

Sugarcane (*Saccharum officinarum* L.) belongs to the grass family *Poaceae* (Khattab & Watanabe, 2019). Sugarcane is sugar-producing crop, signifying a substantial quantity of cultivated crops in the world which is grown typically in tropical and subtropical areas (Hiloidhari et al., 2019). It is one of the most favorable sources for bioethanol production and is the second most used raw material for the same. It is widely grown in Brazil which accounts for 27% of global production (Bertrand, Vandenberghe, Soccol, Sigoillot, & Faulds, 2016; Mączyńska et al., 2019). The yearly worldwide production of sugarcane is about 1741.3 million tons where Asian countries put in 44% of it and India is the top country in production (Hiloidhari et al., 2020). Sugarcane was preferred for ethanol production owing to its high sucrose juice composition (Bertrand et al., 2016). Bioethanol from sugarcane represents almost 1/3 of total ethanol production in 2010 (Rasool & Hemalatha, 2016). Each ton of sugarcane could yield 740 kg of juice and 260 kg of wet bagasse. Also, when compared to corn ethanol, sugarcane ethanol produces 45% more ethanol per unit of land (Crago, Khanna, Barton, Giuliani, & Amaral, 2010). It is estimated that one ton of sugarcane at the processing center contains 135 kg of sucrose which can produce almost 70L of ethanol (Naqvi & Yan, 2015). Sugarcane is primarily constituted with 12-17 % sugars by wet weight, with 68-72 % moisture (90% sucrose and 10% glucose or fructose) (Ayodele, Alsaffar, & Mustapa,

2020). Sugarcane produces directly monomeric sugars which can be directly used for fermentation. Cane juice which can be extracted from the sugarcanes is roughly 95% and the leftover solid residue is called bagasse which is also a potential biomass feedstock and used for steam and electricity generation by burning in boilers (Rasool & Hemalatha, 2016). The sugarcane bagasse and its potential as a feedstock will be studied in the forthcoming sections.

Ethanol production from sugar cane starts with cane storage and cane crushing for juice extraction. In the sugar mills, first, the sugarcane stalks are washed, chopped, and shredded with the help of blades or knives (Naqvi & Yan, 2015). The bioethanol process from sugar cane initiates with the lignin or hemicellulosic fibers pretreatment to release cellulose. Cellulose fibers are depolarized to yield free sugars. The juice is warmed to 115°C and treated with lime and sulfuric acid, resulting in the precipitation of unwanted inorganics. The sugars released are used up to make ethanol from which the ethanol is distilled (Canilha et al., 2012).In annexed plants, crystalline sugar is co-produced along with ethanol. In these plants, sucrose crystals are formed after the juice is concentrated which is removed by centrifugation and this leaves syrup called molasses. Molasses contains 31% sucrose and 15% inverted sugar (Jayus, Nurhayati, Mayzuhroh, Arindhani, & Caroenchai, 2016). Sugar molasses of 25% as a substrate could yield about 32.32 g/L bioethanol by batch fermentation using *Pichia veronae* strain HSC-22 (Hamouda, Nassar, Madian, Abu Amr, & El-Gendy, 2015). Both sugarcane and molasses have to be diluted to adjust the sugar concentration before fermentation which enables the optimal growth conditions for microorganisms with larger yields. This also contains sufficient minerals and organic nutrients making it appropriate for ethanol production by the *S. cerevisiae* (Bertrand et al., 2016). *S. cerevisiae* strain AS2.1190 immobilized on sugar cane pieces was observed in electron microscopy for its yield (77.13 g/l) of bioethanol by 10 batch fermentations (Liang et al., 2008). A typical fermentation process requires 4–12 h per batch. After the run, separation of yeast and distillation is performed to separate ethanol (Saran, Malaviya, & Chaubey, 2019). In a study *S. cerevisiae,* KP096551 was used for enhancing the production of bioethanol using sugar cane molasses as a substrate. The yield of bioethanol was raised to 8.55% (Rasmey, Hassan, Aboseidah, & Abdulwahid, 2018).

Sugar beet

Sugar beets (*Beta vulgaris* L.), the tuber crops are one of the chieffavorable sources of bioethanol production in the EU primarily in Northwest EU. In the EU, 30% of the bioethanol is produced from Sugar beets in 2011 (Bertrand et al., 2016; Marzo, Díaz, Caro, & Blandino, 2019). It is regarded as the most efficient source for ethanol as the yield of ethanol through fermentation ranges from 100 and 120 L per hectare (Panella, 2010).Countries in Asia like China, Japan, India, and Turkey are keen to use Sugar beet for biofuel production. Based on the data from SNSF in France, it was estimated that the average production of sugar beet is 70 tons/hectare, with the sugar content of 17% on the wet basis which gives atypical production of 12 tons of sugar/hectare. Sugar beetroots are made up of about 75% water, 18% sugar (mainly sucrose), and 7% insoluble and soluble materials (Bertrand et al., 2016). It has been observed that water-soluble sugars extracted from 1 kg sugar beet may generate up to 0.07 kg ethanol (Šantek, Gwehenberger, Šantek, Narodoslawsky, & Horvat, 2010). Sucrose is the chief sugar present and maltose; fructose and glucose are in lesser quantities. Conventional sugar beetroots are used to generate sugar and for biofuel production, specialized non-grade varieties bred are used for total sugar production. The Beet tops have feed value but are usually left in the field at harvest and they constitute 4.6 -7.5 tons per hectare (Panella, 2010).

Ethanol production from sugar beets begins with washing the roots, followed by cutting them into thin strips for better extraction of sugar in a root cutter. Washing is usually done using pre-washing drums, jet washers (Šantek et al., 2010).Then the cut pieces are transported to a diffusion step where sugars are extracted by hot water by a process called counter-current extraction. This is called diffusion juice, the remains are called Pulp which is further processed by pressing and dewatering which accounts for almost 22-28% of dry mass (Vučurović & Razmovski, 2012). The dehydrated pulp contains 80 to 94% fermentable components (pentosans, pectin's, and cellulose) and only 12 to 16% lignin, crude protein, and mineral substances (Marzo et al., 2019). This also can be utilized as biofuel feedstock like sugarcane bagasse with the developing technologies in biofuel production (Khattab & Watanabe, 2019).The extracted juice also called green juice is used for fermentation by *S. cerevisiae* to produce the ethanol which is recovered by a distillation process (Dodić et al., 2009).The molasses obtained after the sugar extraction process are also approximately 53% of sugar and can be used for bioethanol production (Vučurović & Razmovski, 2012). However, the major drawbacks of ethanol from sugar beetroots are storage and production costs (Panella, 2010).Utilization of molasses and pulp also for biofuel production, developing new breeds, and performing anaerobic fermentation to yield biomethane have to be evaluated to decrease the production costs. Different countries have integrated sugar beet in the bioethanol policies for its high sugar content and it does not vie with food (Rulli, Bellomi, Cazzoli, De Carolis, & D'Odorico, 2016). A flocculating yeast strain, *Saccharomyces cerevisiae* IR2 was employed for bioethanol production from sugar beet juice. The study concluded that Raw sugar beet juice was shown to be an ideal substrate for ethanol synthesis, requiring neither any pH adjustments nor nitrogen supply (Ogbonna, Mashima, & Tanaka, 2001).

Sweet sorghum

Sweet sorghum (*Sorghum bicolor* L.) Moench, a Graminaceous crop can be used as a promising feedstock for biofuel production because it has a shorter growing cycle. Sorghum is grown on roughly 35 million hectares throughout the world, with a global average grain production of 1.5 t DM/yr (van der Weijde et al., 2013). Sorghum has a superior energy balance than other competitive crops due to its low input needs, flexibility, and extraordinary biological output (Jianliang Yu, XuZhang, & Tan, 2008). Further benefits over other crops comprise a shorter growing period, tolerance towards drought, and can be grown in colder regions and temperate zones unlike sugarcane (tropical and subtropical areas) and sugar beet (temperate climate) (Wortmann et al., 2010). All the plant parts - grains, stalk juice, and plant biomass can be used for biofuel production. The average productivity is 49.7 tons per hectare which produces roughly 63 % of juice. The juices of Sweet sorghum stalks have sugars (6% fructose , 85% sucrose and 9% glucose), invertase enzyme, and other compounds and impurities (Bertrand et al., 2016). The occurrence of the reducing sugars and invertase enzyme makes the juice unsuitable to sugar production rather it makes it suitable to produce ethanol (Ratnavathi, Chakravarthy, Komala, Chavan, & Patil, 2011).

Ethanol production begins with crushing the stalks for juice which leaves out the bagasse that is usually rich in cellulose, hemicellulose, and lignin (Quintero, Montoya, Sánchez, Giraldo, & Cardona, 2008). Fresh juice is sterilized by heating up to 100°C and then purified by adding lime milk and precipitation by CO_2 (Bertrand et al., 2016). The settled solids are filtered from the juice and purified juice is acquired which is now called thin juice. This can be sent for fermentation after regulating the sugar concentration to 16-18 % which usually contains glucose, fructose, and sucrose (Ratnavathi CV, 2017). *S. cerevisiae* then converts the juice or syrup into ethanol. Sugar is converted to ethanol and CO_2, with minor end products such as glycerol, fuel oils, aldehydes, and ketones being produced in lower amounts.

After that, it is distilled and concentrated by dehydration to produce ethanol (Ratnavathi et al., 2011). Distillation purifies the ethanol generated in the fermented mash, yielding 95 percent pure ethanol, which is then dehydrated to obtain 99% purity (Ramesh C. Ray & Chandrasekaran Trilokesh, 2019). Enzymatic hydrolysis of the sorghum grains has been reported, as *S. cerevisiae* cannot break down starch which is the primary component in sorghum grains, a combination of enzymes like α-amylase and amyloglucosidase are utilized (Ramesh C. Ray & Chandrasekaran Trilokesh, 2019). The use of sorghum stalks for producing bioethanol was carried out using *S. cerevisiae*. The stalks treated with dilute acids and a total of 20.5% of Bioethanol were revived by this method (Chauhan, Hajare, Mamo, & Madebo, 2021). A comparative study using six different yeast strains, Vin7, SBg, N96, and GSL for production of bioethanol from sweet stem sorghum juice, was carried out in a YEPS medium that consisted of yeast extract, phosphate, and sucrose. Among all the strains Vin7 and SB9 showed higher sugar consuming strain and thereby higher bioethanol yield (87.4%) (Bulawayo, Bvochora, Muzondo, & Zvauya, 1996).

Starch-based feedstocks

The most popular starch-based feedstocks used in bioethanol production include corn, wheat, and tuber crops including cassava, yam, and sweet potato. Starch polysaccharides are broken down into sugars during bioethanol production. Yeasts then turn these carbohydrates into ethanol.

Corn

Corn (*Zea mays*) is the most employed feedstock for first-generation bioethanol production in the United States, because of its high starch content and the conversion of starch to alcohol (ethanol) is relatively easy (Woiciechowski et al., 2016). The corn grows well in temperature 10-30°C. (https://farm-energy.extension.org/corn-for-biofuel-production/). Based on corn variety and its quality, ethanol production varies. The yield of bioethanol is high when kernels are used as it has high starch content (Singh, 2012). The starch in corn kernel has alpha-linked glucose polymer which can be easily broken down to monomers (glucose) and then these glucose molecules are fermented to alcohol using *S. cerevisiae*. It was reported that about 72% of the dry mass of corn kernel contains starch. The amount of hemicellulose/cellulose, Protein, Oil, sugars, and ash present in corn kernels is about 10.5, 9.5, 4.5, 2, and 1.5 % of the dry mass respectively. The moisture content in the corn kernel was found to be 15% (Woiciechowski et al., 2016). In the US mostly 40% of the corn cultivated is used for bioethanol production (Naqvi & Yan, 2015). The ethanol production from corn consists of majorly 5 steps which are raw material pretreatment, hydrolysis, fermentation, separation, and dehydration (Quintero et al., 2008). It was reported that about 500 gallons of ethanol are produced from corn cultivated on oneacre of land. Around 4 tons of corn grains are harvested from one acre of land (Sexton & Zilberman, 2011). Around 3460 - 4020 L per hectare ethanol is produced from corn in Brazil (Kim & Dale, 2004). It was reported that about 2.7 gallons of ethanol can be produced from one bushel of corn and 17 pounds of distillers dried grains (DDG) can be obtained as a co-product. This DDG can be fed to livestock. There are two ways by which the corn is processed, wet milling and dry milling. Wet milling is mainly used for the production of high fructose corn syrups which are used in the food industry. They are used as sweeteners in non-alcoholic beverages and baked products. In dry milling, the corn is mashed, and slurry is prepared by adding a heat-stable enzyme and water. The slurry prepared is then converted into corn mash by adding other enzymes. This corn mash is further used for the fermentation process. After the fermentation, the distillation of the liquid part is done to get pure ethanol and the remaining part is centrifuged. The solid material is DDG and is further used (Martin, 2010).

Cassava

Cassava (*Manihot esculenta*) is a shrub and a tuber crop. 25-35% starch is present in the roots and tubers of cassava, sweet potato. Bioethanol production from tubers is more cost-effective than that from grains. About 281 million tons of cassava is produced globally. Out of Asian countries, Thailand and Indonesia produce about 26 million tons and 28 million tons of cassava. Around 77-94% of starch and 59-70% of the moisture is present in cassava (Woiciechowski et al., 2016). Mainly China and Thailand use cassava for bioethanol production due to their obtainability all through the year (Mohapatra et al., 2019). Cassava roots have a 30% of starch content by which 180 L of 96% pure ethanol per ton of cassava can be produced (Bertrand et al., 2016). Bioethanol production using cassava is led by Thailand (50%), followed by China (44%). Around 3400-3600L per hectare ethanol is produced from cassava in Brazil (Mohapatra et al., 2019).

Production process from cereal crops

Unlike the sugar crops, the production of bioethanol from the cereal crops is different as the key component in these crops is starch, a macromolecule made of a lengthy polymer of glucose units that can't be turned into ethanol immediately (Mohapatra et al., 2019). Hydrolysis is a crucial step that breaks down the starch molecule into individual glucose monomers. The dry-milling process and wet-milling process are the two major processes used for bio-ethanol production (Naqvi & Yan, 2015). In the dry milling, the corn is washed, crushed, milled, ground into flour, and made into slurry by adding water (Bertrand et al., 2016). The starchy material is gelatinized for the amylose and amylopectin susceptible for enzymes to hydrolyze in the next step (Woiciechowski et al., 2016). The enzymatic hydrolysis of starch has two steps - liquefaction and saccharification (Bertrand et al., 2016) and in the liquefaction step, a heat-stable α-amylase, which hydrolyzes α-1, 4 bonds between glucose units in the stanch molecule randomly, yielding dextrins and gluco-oligosaccharides with reduced starch viscosity. Dextrins are further hydrolyzed by dextrins/pullulanase that hydrolyzes both α-1, 4 and α-1, 6 glycosidic linkages and finally convert to glucose by glucoamylase (Baras, Gacesa, & Pejin, 2002). Gluco-amylases also known as amyloglucanases, attacks starch from the non-reducing end to produce glucose (Nikolov, Meagher, & Reilly, 1989).Wet milling process, consists of an additional step is performed where dilute sulfuric acid is added to the corn to make a slurry and left for 24–48 hrs where the corn grains decompose into different components. The corn starch is then fermented into ethanol (Naqvi & Yan, 2015). In the fermentation step, *S. cerveciae* converts the glucose into ethanol and CO_2 where in, bioethanol is separated from the culture that has 8–11% (w/w) ethanol in the distillation section (Quintero et al., 2008). Few studies were provided in the Table 2 and the general process is mentioned in the Fig 1.

Table 2. Various studies used first-generation biomass for the production of biofuels

Substrate type	Organism	Fermentation type	Product and yield	References
Sugarcane molasses	*Saccharomyces cerevisiae*	Submerged fermentation	Bioethanol - 0.669 g/g	(Jayus et al., 2016)
Sugarcane molasses	*Kluyveromycesmarxianus* K21	Submerged fermentation	Bioethanol - 0.65 g/L/h	(Lin et al., 2020)
Sugar beet pulp	*S. cerevisiae*	Simultaneous saccharification and fermentation (SSF)	Bioethanol - 26.9 ± 1.2 g/L	(Berłowska et al., 2016)
Sugar beet pulp	*Pichia stipitis*	Simultaneous saccharification and fermentation (SSF)	Bioethanol - 86.5 ± 2.1%	
Sugar Beet Pulp	*S. cerevisiae (strain DTN)*	Submerged fermentation	Bioethanol - 0.446 ± 0.017 g/g	(Vučurović & Razmovski, 2012)
Sugar beet raw juice	*S. cerevisiae*	Submerged fermentation	Bioethanol-90 g L1 h1	(Dodić et al., 2009)
Sweet sorghum juice	*S. cerevisiae* BY4741	Submerged fermentation	Bioethanol-113.7 ± 3.1 g L1	(Sasaki et al., 2015)
Cane molasses	***Zymomonas mobilis*** MTCC 92	Submerged fermentation	Bioethanol-58.7 ± 0.09 and 59.1 ± 0.08 g/l	(Behera, Mohanty, & Ray, 2012)

Figure 1. First-generation biofuel production process

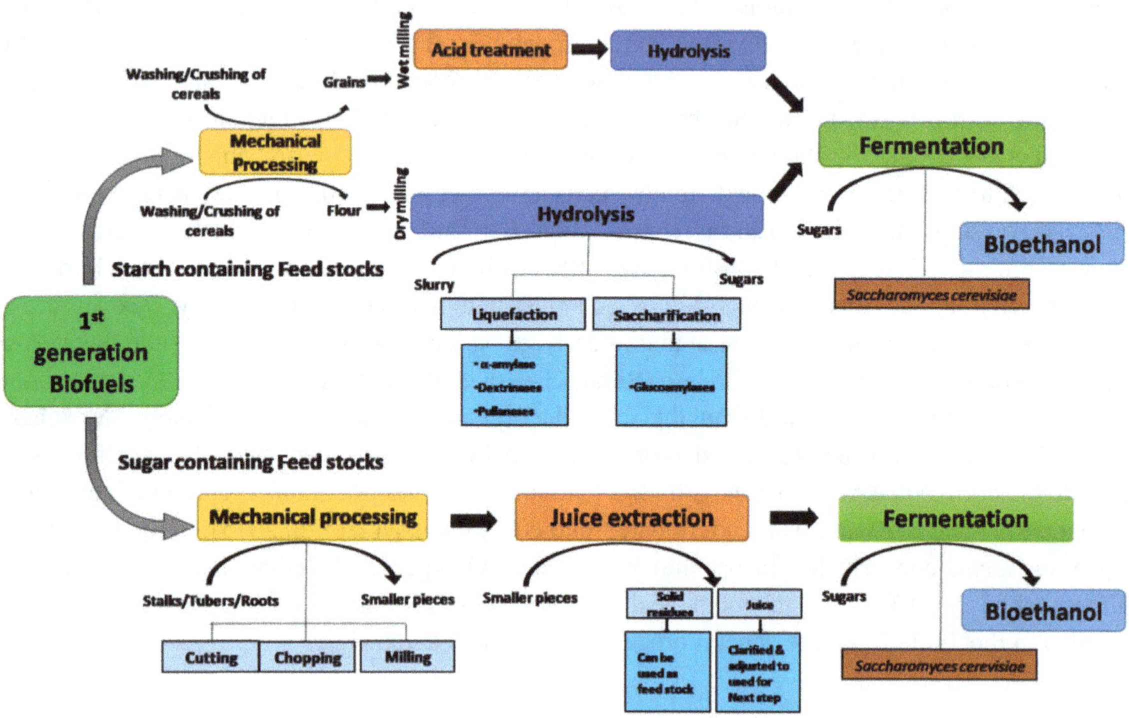

Second-generation biofuels

Second-generation biofuel is also acknowledged as lignocellulosic biofuels are produced from a broad diversity of feedstock which is majorly crop, agricultural, and forest leftovers and are examples of non-edible biomass. These feedstocks are inexpensive, plentiful, and renewable (Lee & Lavoie, 2013; Raud et al., 2019). They are produced in a sustainable way which doesn't impact that much on global CO_2 levels (Naik et al., 2010). Since they are taken from non-food material, a crisis for food is not seen (Hirani et al., 2018). The non-food materials including bagasse, clover, de-oiled seed residues, straw, leaves, grass, and stem biomass are widely used for 2G biofuels. The 2G biofuel mainly includes bioethanol, biodiesel, dimethyl-ether, biosyngas FT (Fischer-Tropsch) biodiesel (Ruan et al., 2019). The production of 2G biofuel can be done by thermochemical route or biochemical route. Unlike first-generation biofuel, 2G biofuel does not need additional water, land, and fertilizers (Dahman, Dignan, et al., 2019). This fuel is combined with other conventional fuels like petroleum-based fuels for combustion. This is applied in internal combustion engines (Datta et al., 2019). Second-generation bioethanol is produced by utilizing cellulose, a glucose polymer feedstock. During the production process, the raw material will undergo pretreatment by physical, chemical, or biological methods (Prachand Shrestha, Anthony L Pometto III, 2012). A wide variety of lignocellulosic biomass is used for the 2G biofuel production which includes agricultural and timber industry residues and co-products, food crop residues, perennial lignocellulosic plants like reed canary grass, *Miscanthus, Eucalyptus* (Ho, Ngo, & Guo, 2014). It was reported that by 2016, about 67 second-generation bioethanol facilities were started throughout the world. The United States is the leading in second-generation biofuel production (Mohapatra et al., 2019). The lignocellulosic biomass contains cellulose, hemicellulose, and lignin. Hemicellulose gets decomposed fast at temperature 200-380°C, cellulose at 320-400°C and lignin is the most stable one and decomposes at above 400°C (J. Zhang & Zhang, 2019). The content of cellulose, hemicellulose, and lignin vary from different types of feedstocks. The biomass that can be used for the production of biofuels can be divided into non-woody and woody biomass. The pros and cons of the second-generation biofuels were given in the Table 3.

Table 3. Pros and Cons of second-generation biofuels (Carriquiry, Du, & Timilsina, 2011; Saini et al., 2015)

Pros	Cons
Minimal feedstock costs No rivalry with food Using the entire plant rather, as well as leftovers, thereby higher energy generation Better yield with lesser area The percentage of usable feedstocks is higher than is easily available presently Environmentally feasible	Overuse of agricultural lands eventually may lead to soil erosions The process of production is complex. Lack of commercially viable technologies Advanced research technologies are lesser-known

Woody based feedstock

Hardwoods and softwoods make up the woody biomass, which is used to make biofuels. The chemical compositions and the physical properties of these materials are different from each other. Based on these ethanol production varies (Wijeyekoon & Vaidya, 2021). Reportedly, the use of wood-based feedstock for bioethanol production can also lower carbon emissions more effectively. Also, the higher the area of production less is the production costs (Perez-Verdin et al., 2009). The utilization of novel non-edible lignocellulosic materials in the creation of chemicals and fuels is a strategic goal. These raw minerals are an intriguing and potential alternative to fossil fuels as a source of energy (García, González Alriols, & Labidi, 2014). Short-rotation woody crops have been marketed as a long-term biofuel feedstock for heat and/or electricity generation. Short rotation woody crops (SRWC) are perennial plants with consistent production and well-developed technique for cultivation (Stolarski et al., 2015). Some of the plants that are used for woody feedstock are Hybrid poplar (*Populus* spp.), Willows (*Salix* spp), the American sycamore (*Plantanus occidentalis*), Sweetgum (*Liquidambar* spp.), Black locust (*Robina pseudocacia*), Silver maple (*Acer saccharinum*), Loblolly Pine (*Pinus taeda*) (García et al., 2014; Stolarski et al., 2015). Mainly the wood material in the plants is made of cellulose microfibrils which are together gripped by hemicellulose and lignin. The vessels and tracheid are present in the center which is surrounded by the microfibril cellulose. Based on this layer the toughness of the wood increases and will undergo various pre-treatment. Combining multiple pretreatments has also been investigated and might be beneficial in achieving appropriate segregation of the various components and better yields (Alvira, Tomás-Pejó, Ballesteros, & Negro, 2010). To not only improve the efficiency of subsequent bio refinery processing stages, but also to avoid high-contaminated hemicellulosic or lignin-containing streams, the correct mixture of pretreatment steps (first, for hemicelluloses solubilization, and then, for effective de-lignification) must be used (Wijeyekoon & Vaidya, 2021). The production of biofuel from woody biomass depends on the feedstocks from both forests and plantations. Lignocellulosic biomass can create 442 billion liters of second-generation ethanol, while total agricultural wastes and discarded crops can yield 491 billion liters of ethanol every year (Kim & Dale, 2004). About 30% of the global forest biomass is produced from the US. For biochemical conversion of sugar to fuels, woody biomass is not preferred because of their high lignin content and recalcitrance to digestion. It was reported that 1.2 million liters of ethanol was produced from 4 million dry tons of wood every year in Mississippi (Perez-Verdin et al., 2009). Owing to the high energy requirement for the conversion, bioethanol production is difficult (Mohapatra et al., 2019).

Non-woody biomass

The other sort of lignocellulosic feedstock that can be well-thought-out for biofuel production is non-woody biomasses which are cheaper, simple to process, and need less energy for the production of bioethanol. They can be sorted out broadly into agriculture residues, process residues, native plants and non-wood plant fibers, solid wastes, etc. which were discussed in the coming topics (García et al., 2014; Mohapatra et al., 2019).

Agriculture residues

Corn stover, Cereal straws, rice husks, etc., are the various agriculture residues that can be utilized as a potential feedstock for biofuel production.

Corn stover

The residues available in the field after the corn grains harvest from the corn plant are usually the cobs, stalks, leaves, husks and tassel are called corn stovers (Liew, Hassim, & Ng, 2014). Globally approximately 377 million tons of corn stover is produced per year whereas the USA is the major source as the USA mainly uses corn ethanol (Hiloidhari et al., 2019; Thompson & Tyner, 2014). It has 37- 61% cellulose (Axelsson et al., 2012), 19.3- 31% of hemicellulose, and 6.9 - 18% of lignin (Ruan et al., 2019). Xylose and arabinose are the predominant five-carbon sugars. Stover is a byproduct of maize production and not a food-based product. In hemicellulose and they too can be fermented to ethanol with the suitable microorganism (Woiciechowski et al., 2016). Development for cost-effective technologies for the biomass conversion into sugars is a need and diverse trials are in progress to extract the sugars for the fermentation process. Some of the studies using corn stover as the substrate to produce biofuel were given in Table 4.

Cereal Straws

A variety of cereal grains have been used as food for humans containing 60-70% starch. They generate enormous quantities of straws as agricultural residues like dry stalks and leaves in the fields following the removal of nutrient grains and seeds. Barley, wheat, rice, oats, and sorghum are all considered essential cereal crops. Different cereal crops such as barley and oat from Europe, sorghum and corn from the United States, and rice and wheat from India and China are predicted to produce 1580.2 million tones every year (Kim & Dale, 2004). All these different cereal straws have 33%–47% cellulose which can be utilized in biofuel production (Tye, Lee, Wan Abdullah, & Leh, 2016). Pyrolysis can be applied to these crop residues which produce biochar. Rice straw was found to be a good potential for biochar as the yield was around 43.8%. Similarly, comparative studies of anaerobic digestion were conducted with goat manure and three crop residues, namely rice straw, corn stalks and wheat straw found better for biogas production (Bian et al., 2016).Table 4 has provided with few studies using various cereal straws for biofuel generation.

Rice husk/hulls

Oryza sativa, commonly called rice is an herbaceous plant that belongs to the Poaceae family. Rice husk (RH) is a major byproduct of the paddy process, accounting for 20–25% of the weight. India is the world's second-biggest rice grower, producing 102.0 MT of rice and around 20.4–33.6 MT of RH per year (Potumarthi, Baadhe, Nayak, & Jetty, 2013).It is a lignocellulosic substance with a low bulk density that contains around 20–25% lignin, 35–40% cellulose, and 15–20% hemicellulose. It has the potential to act as a feedstock for the production of gasoline-ethanol as of its availableness and affordable cost (Jun Yu, Zhang, He, Liu, & Yu, 2009). Various studies using rice husk for biofuel production were given in Table 4.

Process residues

Sugar cane bagasse, Cassava bagasse, Sweet potato residues are the various process residues that can be used as potential feedstock for biofuel production.

Sugarcane bagasse

Sugarcane wastes, particularly bagasse (SB) and leaves (SL) have been investigated for biotechnological and non-biotechnological applications. For decades, SB and SL have been studied for use in lignocellulosic bioconversion, allowing for the cost-effective use of residual substrates in the manufacture of bioethanol and value-added commercial goods like as xylitol, specialty enzymes, organic acids, single-cell protein (Chandel, da Silva, Carvalho, & Singh, 2012). The leftover dry matter residue after extraction of juice from sugar cane is termed sugar cane bagasse. The major advantage is its accessibility at the plant site where the juice is extracted. It is composed of 40–45% of cellulose, 27–37.5% of hemicellulose, and 10–20% of lignin (Canilha et al., 2011; Zhao, Peng, Cheng, & Liu, 2009). From a ton of bagasse, 300 L of ethanol can be produced (Woiciechowski et al., 2016). However, it has a higher amount of complex inbuilt structure and where individual sugars aren't accessible for the yeast to produce ethanol and therefore a cost-effective pretreatment has to be done before enzyme hydrolysis to break the polysaccharides into monosaccharides. In the following chapters, we go through the techniques in greater depth. Studies for biofuel production using sugarcane bagasse were provided in Table 4.

Cassava Bagasse

Cassava bagasse is a waste by-product of cassava starch manufacture that is both affordable and widely available. For every 250–300 tons of cassava root processed, roughly 280 tons of bagasse with an 85% moisture content are generated (Pandey et al., 2000). Cassava is a starch based tuber crop produced in many Asian, Africa and Latin American. India also has many cassava processing plants. Cassava bagasse contains 30-35% of carbohydrates which can be employed for biofuel production (Mohapatra et al., 2019). Few studies for biofuel production using cassava bagasse were given in Table 4.

Sweet Potato Residues (SPRs)

The remaining residue after the removal of starch from sweet potatoes is called Sweet Potato Residues (SPRs) contributing for 10% of the total dry matter of sweet potatoes. The annual production of sweet potatoes is about 71 million tons globally which harvests 2 million tons of SPRs. Viscosity was one of the SPRs main drawbacks to utilize as feedstock (Duvernay, Chinn, & Yencho, 2013). Requirements of heavier instruments of corrosion resistance are required which is a major bottleneck. However recent enzymatic methods for utilization of SRPs opened up the opportunities to utilize the same as feedstock (Izmirlioglu & Demirci, 2012). Various studies employing the SRPs were provided in Table 4.

Native Plants

Grasses are indigenous plants that grow naturally and do not require special care, making them a cost-effective biomass source. Carbohydrates and lignin make up the majority of them. These polysaccharides may be broken into sugars and then fermented to produce bioethanol. The most commonly used grass plants are Napier grass, Giant Miscanthus (*Miscanthus* spp.), Switch grass, and Bermuda grass. *Miscanthus* spp. is a perennial sterile grass hybrid that proliferates through rhizomes. It is a type of saw grass that can yield 5 to 8 times as much ethanol per acre as compared to corn. It has high lignocellulosic content, 40%–60% of cellulose, 20%–40% of hemicellulose, and 10%–30% lignin. Numerous species of

Miscanthus have diverse carbohydrate and lignin content which can be utilized for biofuel production. They intern help in protecting soil erosion and fixing atmospheric carbon as contrast to other annual crops (Axelsson et al., 2012). *Panicum virgatum* is generally called Switchgrass, a perennial herbaceous grass that is native to North America and requires low requirements like *Miscanthus* (Adler, Sanderson, Boateng, Weimer, & Jung, 2006). It has 37% to 40% of cellulose, 25% to 29% of hemicellulose, and 18% to 25% lignin. This grass has a high net energy benefit and effective soil carbon absorption, making it a good crop for soil conservation on productive land (Mohapatra et al., 2019). The next grass feedstock in the list is *Pennisetum purpureum* commonly called Napier grass a native to eastern and central Africa and has been introduced in tropical and subtropical countries. The plant's significant properties include high cellulosic fiber content, minimal nitrogen fertilizer use, and the ability to grow rapidly, making it an ideal feedstock for biofuel synthesis (Zahran, 1999). The grass has cellulose in the range of 40% to 50%, hemicellulose in the range of 20% to 40%, and lignin in the range of 10% to 25% (Takara & Khanal, 2015). *Conodont dactylon*, termed Bermuda grass, has high carbohydrate content similar to the different grass feedstocks. It has 40%–55% of carbohydrates (cellulose and hemicellulose) and 20%–25% of lignin (Takara & Khanal, 2015). Cocksfoot grass, reed canary grass, big bluestem grass, and alfalfa are in addition other potential grass feedstock that can be considered for biofuel production (Njoku, Uellendahl, & Ahring, 2013). Various studies were provided in Table 4 which provides information regarding their use as substrates for biofuel production.

Natural Non-woody Plant Fibers - Bast crops

Bast crops are highly resourceful mop crops that can grow on waste or even brackish water. Bast fibers are the fibrous plant fiber's outer layer, accounting for one-third of its weight. Some of the Bast fibre crops with high cellulose content and potential as biofuel feedstocks are flax, ramie, kenaf, sun hemp, and industrial hemp (Cherney & Small, 2016). Industrial hemp is considered a vital sustainable potential option for biofuel production as said by European Energy Agency. The global yearly production of hemp is around 0.1 million tons. Industrial hemp has 70.0–90.0% of cellulose and other fibers such as flax, jute, and ramie has 60.0%–80.0%, 51.0%–84.0%, and 68.0%–76.0% of cellulose respectively (SaifulAzry, Amel B. Ahmed, 2011). The biofuel output from industrial hemp and other bast fibre crops, on the other hand, is mostly unknown.

Food processing wastes and Solid wastes

Enormous amounts of different fruits and vegetables are used and processed every day, because of increasing market demand which generates huge quantities of industrial wastes, which include peels, seeds, shells, pomace, skin, etc. from various fruits and vegetables (Hiloidhari et al., 2019). Fruit leftovers such as tomato, asparagus, kiwi, and potato residues may also be employed to produce biofuels by anaerobic digestion (Siciliano, Limonti, Mehariya, Molino, & Calabrò, 2019). Different solid wastes that include a municipal solid waste (paper, plastics, etc.), wastewater sludge, food waste, animal manure, etc., in addition, can be used as a source for biofuel as they have cellulose (Ho et al., 2014). Fruit & vegetable processing sectors have developed massively leading huge amount of industrial wastes generation like leaves or straw, seeds, pulp, or pomace. Up to 60% of citrus peels are generated in citrus fruit processing industries which are a possible substrate for multiple fermentation processes (Wilkins, Suryawati, Maness, & Chrz, 2007). Some of the studies using various wastes were provided in the Table 4.

BIOMASS STRUCTURE AND COMPONENTS

The many varieties of lignocellulosic biomass that are accessible have previously been mentioned in previous topics. Cellulose (30–50%), hemicellulose (25–30%), and lignin (10–35%) make up the majority of lignocellulosic biomass, which is structured into a complex structure. Subject to the biomass type, source, growing circumstances, maturity, and harvest time, these percentages vary (Agbor et al., 2011; McKendry, 2002). In the upcoming sections the different components of lignocellulosic biomass processing procedures, challenges, and recent technological advances in the production of biofuels from these biomasses will be discussed.

Cellulose $(C_6H_{10}O_5)_x$

This is the most abundant polysaccharide in lignocellulosic biomass and the primary component. The essential building component is cellobiose, which is made up of a linear chain of D-glucose monomers connected by β-(1,4)-glycosidic linkages (Himmel et al., 2007). In wood pulp, the number of glucose molecules in a chain can range from 800 to 10,000 units until 17000 units. This is called the degree of polymerization. These linear cellulose chains unite to form tightly packed cellulose fibers via intramolecular and intermolecular hydrogen bonds and van der Waals forces. Because of their ordered packing, these cellulose fibers sequentially form crystalline structures. They are water-insoluble and extremely resistant to depolymerization. The regions of cellulose with less crystalline packing are referred to as amorphous structures (Xuebing Zhao, 2012). Because these regions are less resistant to biodegradation, they may be properly broken to produce sugars via enzymatic hydrolysis and alcoholic fuel (Khattab & Watanabe, 2019; Robak & Balcerek, 2018).

Hemicellulose

Hemicellulose, which is made up of several sugars, is the next most common polymer. It's a polymer made up of short, branched-chain sugars that's heterogeneous. It is composed of two distinct monomeric sugars - Pentoses (C5 sugar) and hexoses (C6 sugars). D-xylose and L-arabinose are pentoses, while d-mannose, d-galactose, and d-glucose are hexoses. Xylan is the most plentiful hemicellulose component of all (Peng, Peng, Xu, & Sun, 2012). All of the pentose and hexose combine to form short, branching structures. They are amorphous and provide hemicellulose the ability to be partially soluble in water. Sugar acids such as uronic acids, as well as monosaccharides, are found in hemicellulose, which contains d-glucuronic, 4-O-methylglucuronic, and d-galacturonic acids. Because of its low molecular weight and simple structure with short lateral chains, hemicellulose is easier to hydrolyze than cellulose. Hemicelluloses in secondary cell walls interfere with cellulose and lignin, which create a network by hydrogen-bonding to cellulose microfibers, covering them, and providing the structural backbone for plant cell walls (Koshijima & Watanabe, 2003; J. X. Sun, Sun, Sun, & Su, 2004). As a result, enzymes have difficulty accessing the sugars in cellulose that can be turned into biofuel. To make the cellulose more digestible, at least half of the hemicellulose must be eliminated by pretreatment procedures (Saha, 2003).

Lignin

Lignin is a 4-hydroxyphenylpropanoid-based aromatic compound found in lignocellulosic biomass. They are known as monolignols or lignin precursors, and they are linked by ether (C-O-C) and carbon-carbon (C-C) bonds with varying degrees of methoxylation. There are three varieties of monolignols: p-coumaroyl, conifineryl alcohol, and sinapyl alcohol, which are derived from the precursor's p-hydrophenyl (H), guaiacyl (G), and syringyl (S). They are connected by various links to form a complex three-dimensional polyphenolic network. Lignin is hydrophobic and defiant to chemical and biological destruction. It is located in the center of the lamella and serves as a binding substance for the cell wall and plant cells. Lignin forms an amorphous mixture with hemicellulose and cellulose fibrils embedded within it and is thus referred to as a 'glue' or binding substance that holds the various components of lignocellulosic biomass together (Boerjan, Ralph, & Baucher, 2003; Ghaffar & Fan, 2013). The successful fragmentation of lignin is a critical step in gaining entry to the polysaccharide's cellulose and hemicellulose for the enzymes to break down and create sugars for bioethanol conversion. Pretreatment procedures in the processing of lignocellulosic biomass can remove lignin, allowing value-added goods such as dyes, insecticides, and industrial plastics to be converted (Dahman, Dignan, et al., 2019).

Conversion technology

Given the variety of biomasses available, it is possible to generate a wide range of biofuel products from the same feedstock. Using thermochemical and biochemical processes, lignocellulosic biomass may be converted into a variety of biofuels as mentioned in Fig 2.

Figure 2. Second-generation biofuel production process

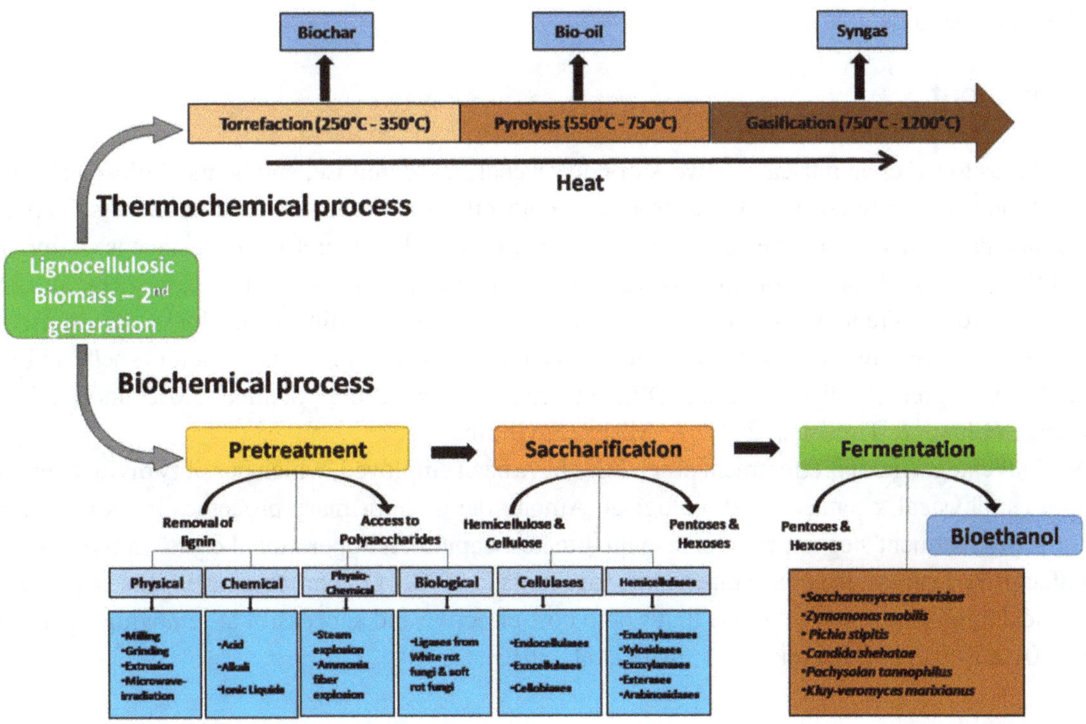

Thermochemical Process

Thermochemical processing, as the name implies, entails the conversion of biomass into a variety of products by thermal decomposition and chemical reformation, and primarily includes heating of the biomass in the presence of varying oxygen concentrations. Combustion, gasification, liquefaction, hydrogenation, and pyrolysis are examples of thermochemical conversion processes (Goyal et al., 2008). The biomass is roasted at low temperatures between 250 to 350°C in the absence of oxygen in this process to form biochar (Glaser et al., 2001; Kwapinski et al., 2010). Pyrolysis is a procedure that changes biomass to biooil by heating it to temperatures ranging from 350–550°C and reaching 750°C in the lack of oxygen (Mohan, Pittman, & Philip, 2017). The biomass undergoes full depolymerization and is transformed largely into syngas, with charcoal and biooils as by-products, in the gasification process, which happens at even higher temperatures ranging from 750 to 1,200°C and with insufficient oxygen supply (Lee & Lavoie, 2013; McKendry, 2002)(J. Zhang & Zhang, 2019).

Biochemical Process

Biochemical pretreatment of biomass employing biological, physical (heat), or chemical catalysts are used in the biochemical process. For the degradation of polysaccharides into simple sugars, biocatalysts such as enzymes are used, and fermentative microorganisms (yeast or bacteria) are used for fermentation, where the sugars are transformed into biofuel. Each method of biomass conversion has its own set of obstacles and opportunities. The biochemical route is thought to be more environmentally friendly, practical, and cost-effective. The biochemical method is limited to the conversion of polysaccharides, but the thermochemical approach is more widely used nowadays since it can convert all of the organic constituents of biomass into various biofuels (Robak & Balcerek, 2018; Vohra et al., 2014). Further to this section onwards, the focus will be on the conversion processes for lignocellulosic biomass via biochemical methods.

Pretreatment

Lignin holds together an intricate network of polysaccharides, cellulose, and hemicellulose in lignocellulosic biomass. The removal of lignin from the lignocellulosic biomass would be the first step in the conversion process, allowing the polysaccharides to be obtained. The pretreatment stage is an important part of the biothermal conversion process. The pretreatment phase should be able to create highly digestible solids, avoid degradation of cellulose and hemicellulose, avoid inhibitor development in subsequent processes, and easily recover lignin. The pretreatment process is influenced by factors such as biomass crystallinity, degree of polymerization (DP), surface area accessibility, hemicellulose, and lignin concentration (Cheng & Timilsina, 2011; A. Mishra & Ghosh, 2017).

Various categories of pretreatment procedures have been employed, which possibly divided into three categories: physical, chemical, and biological. Among the three primary processes in biofuel production, the pretreatment step appears to have the greatest capital and operational costs. In the bioethanol manufacturing process, the pretreatment stage accounts for 40% of the processing expenses. As a result, it's critical to pick the best technique that's both cost-effective and successful at removing lignin (Naik et al., 2010; Robak & Balcerek, 2018; Saini et al., 2015).

Physical methods like mechanical pretreatment enable us to accomplish higher hydrolysis efficiency. Chipping, grinding, and milling are examples of techniques that may be used to enhance surface area, minimize biomass chip size, and lower the degree of polymerization and crystallinity of cellulose. These methods are regularly combined with other methods. Following chipping, milling, or grinding, the particle size is condensed to 10–30 mm and 0.2–2 mm, respectively, depending on the kind of physical pretreatment used. Two types of milling are being practiced -ball milling and wet disc milling. The most major disadvantage of milling is a large amount of energy requirement (Hideno et al., 2009; Inoue, Yano, Endo, Sakaki, & Sawayama, 2008). Wet disc milling has been shown to reduce energy consumption when compared to other milling pretreatment methods, but ball milling has been shown to produce elevated glucose and xylose yields after enzymatic hydrolysis (Da Silva, Inoue, Endo, Yano, & Bon, 2010). Extrusion pretreatment is a type of thermophysical method where the materials are subjected to different processes like mixing, heating, and shearing which disrupts the (X. Chen et al., 2014; Wahid, Hjorth, Kristensen, & Møller, 2015). Microwave irradiation is a physical pretreatment method that degrades and partially eliminates lignin and hemicellulose and modifies the cellulose structure and thus increases the enzymatic susceptibility. It is based on the direct application of electromagnetic radiation to a substance (for example, an organic solvent or plant tissue) that can absorb and convert electromagnetic energy (microwaves) into heat (Ganzler, K., Salgó, A., & Valkó, 2012; Hu & Wen, 2008).

In chemical procedures, acids, alkaline agents, organic solvents, ionic liquids, and other chemicals can be employed to break apart the lignocellulosic structures. However, this approach has a number of drawbacks, including the creation of hazardous substances, carbohydrate loss, and the high cost of chemicals employed in the process (Haghighi Mood et al., 2013). Acid hydrolysis is the most popular chemical approach for biomass pretreatment, in which the acid decomposes lignin and hydrolyzes hemicellulose into monosaccharides, allowing for easier access to cellulose for enzymatic hydrolysis. Concentrated acid (72% H_2SO_4) used at low temperature or dilute acid (2–5% H_2SO_4) applied at high temperature can do this. Other than sulphuric acid, other acids also were in use like H_3PO_4 (Hong, Xue, Weng, & Guo, 2012). H_2SO_4 and H_3PO_4 are generally inexpensive and effective at hydrolyzing lignocellulose. Similarly, Hydrochloric acid (HCl) is more volatile and simpler to recover than sulfuric acid (H_2SO_4), and it destroys biomass more effectively (A. Demirbas, 2008b). Nitric acid (HNO_3) has a high rate of cellulose to sugar conversion (Tutt, Kikas, & Olt, 2012). Both acids, however, are more costly than sulphuric acid. However, the primary disadvantages of using concentrated acid are its toxicity and corrosiveness, as well as the cost of acid recovery and the production of fermentation inhibitors, whereas the main disadvantage of using dilute acid is the need for more energy (Arora, Sharma, Kumar, & Sani, 2019). Similarly, alkali treatment includes the use of sodium (NaOH) (Wang, Keshwani, Redding, & Cheng, 2010), potassium (KOH) (Hendriks & Zeeman, 2009), calcium ($CaOH_2$) (Xu & Cheng, 2011), that can break the bonds between lignin and carbohydrates and increases the accessibility of the cellulose for enzyme hydrolysis. It is more beneficial as the lignin and hemicellulose are removed efficiently and also increase the surface area for the enzymes to hydrolyze better. Alkaline pretreatment, alternatively, is carried out at lower temperatures and does not necessitate the use of sophisticated reactors, making it more suitable for use on-farm. Long residence times (from hours to days) and the necessity to neutralize the pretreated slurry are the key disadvantages (Haghighi Mood et al., 2013). A new approach for the pretreatment is the use of chemicals as solvents like methanol, ethanol, or acetone in grouping with acid or base catalyst which can help in the disruption of the lignocellulose structure and increase the accessibility for cellulose fibers. This approach is termed Organosolv pretreatment. The catalysts used consist of organic or inorganic acids (hydrochloric and sulfuric acids), bases (ammonia, sodium hydroxide, and lime) (Araque et al., 2008; Mesa et al., 2011; F. Sun & Chen, 2008). Though lignin and hemicellulose are removed efficiently, the high cost of the solvents, flammability, and volatility are the drawbacks (Araque et al., 2008).

A different latest physio-chemical pretreatment technique is the use of ionic liquids (IL) which are the organic salts composed of large organic cations and small inorganic anions which can notably break the lignocellulosic structure by breaking the bonds between cellulose, hemicellulose, and lignin. IL's are a novel category of solvents that are completely made up of ions and have low melting points (b100°C), high polarities, great thermal stabilities, and minimal vapor pressure (cations and anions) (Peter Wasserscheid; Wilhelm Keim, 2010). The imidazolium cation, which may couple with anions such as acetate, bromide, chloride, sulfate, nitrate, methanoate, and triflate, is present in the most prevalent forms. A suggested approach of dissolution implies that these ionic liquids fight for hydrogen bonding with lignocellulosic components, destroying their three-dimensional network (Moulthrop, Swatloski, Moyna, & Rogers, 2005). They have the benefit of simultaneously dissolving carbohydrates and lignin, which might open up new processing options for lignocellulosic feedstocks. Though its significant potential for improving cellulosic biomass digestibility, ILs are still in the research stage, and issues such as the high cost of the stage of chemicals must be examined for commercial usage (Miyafuji, 2015). Steam explosion or also called Steam pretreatment (SP) is a physio-chemical pretreatment process in which the biomass's lignin and hemicellulose are divided up by steam heating (Thermo), shearing owing to abrupt decompression and extension of moisture (mechano), and glycosidic bond degradation (chemical). SP may be efficiently increased by using H_2SO_4 (Wen Hua Chen, Pen, Yu, & Hwang, 2011) or SO_2 (Mackie, Brownell, West, & Saddler, 1985) as a catalyst. The use of an acid catalyst improves hemicellulose sugar recovery, lowers inhibitory compound generation, and improves enzymatic hydrolysis on solid residue. It is an appealing procedure since it uses few chemicals. It uses less energy and has no recycling or environmental costs. However, one of the drawbacks of SP is that it does not destroy the lignin-carbohydrate matrix, which enhances the likelihood of condensation and precipitation of soluble lignin components, making the biomass less digestible (Agbor et al., 2011; Haghighi Mood et al., 2013). Ammonia fiber explosion (AFEX) is a method based on the steam explosion where the ammonia is added with the biomass along with steam. For a limited period (< 30 min), high pressure (1.72–2.06 Mpa) and moderate temperatures (60–120°C) are applied, followed by a quick pressure release. This increases the ease of access of cellulose for enzymatic hydrolysis by partially solubilizing hemicelluloses. The rapid expansion of ammonia gas causes the breakdown of the lignin-carbohydrate complex and disruption of biomass fibers, leading to improved digestibility of biomass (Shishir P.S. Chundawat, Balan Venkatesh, 2007; Teymouri, Laureano-Perez, Alizadeh, & Dale, 2005).

Biological strategies entail the employment of microbes such as white and soft rot fungi, which may breakdown the components of lignocellulosic biomass. A wide range of naturally occurring microorganisms (single or in co-culture) or enzyme systems are used in biological pretreatment (pure or mixture). As a result, microbial pretreatment (fungal, bacterial, microbial consortia, and ensiling) and enzymatic pretreatment may be distinguished. These Lignin degrading enzymes include laccases and lignin peroxidases. Two of the most significant enzymes in the ligninolytic enzyme system are lignin peroxidase (EC 1.11.1.7) and manganese peroxidase (EC 1.11.1.7). Laccases (EC 1.10.3.2.) are copper-based enzymes that help break down lignin. Laccases work along with lignin peroxidase and manganese peroxidase to degrade lignin completely. Several white-rot fungi have been studied, including *Cyathus stercolerus, Ceriporiopsis subvermispora*, *Ceriporia lacerate, Pleurotus ostreaus*, *Pycnoporus cinnarbarinus*, and *Phanerochaete chrysosporium*, with excellent delignification efficiency and enhanced hydrolysis efficiency (Chiaramonti et al., 2012; Sindhu, Binod, & Pandey, 2016). Because no chemicals are used in these procedures, they are both ecologically benign and inexpensive to run. However, the hydrolysis rate is lower; therefore the process would take longer. The inadequate hydrolysis rate is the most notable drawback of using fungus to pretreat woody materials. Enzymatic hydrolysis can be utilised to convert leftover lignocellulose into sugars after pretreatment. Apart from the kind of feedstock, the pretreatment method must be selected depending on the biofuels cost economics (Arora et al., 2019; Dahman, Syed, et al., 2019; Raud et al., 2019).

Saccharification

The saccharification of the cellulose fibrils exposed in the pretreatment step into fermentable sugars via a hydrolysis process is the next step in the biochemical process. The activity of hydrolytic enzymes breaks down cellulose and hemicellulose into glucose and xylose. Cellulases, hemicellulases, and auxiliary enzymes are required for full biomass decomposition. Cellulases are the primary enzymes, which are very specialized and may be split into three classes depending on where they attack. Endocellulases (EG) (EC.3.2.1.4) target cellulose fiber areas with poor crystallinity and produce cello-oligosaccharides with free-chain endings. Exocellulases (CBH) (EC.3.2.1.91) further breakdown the molecule by cleaving cellobiose units from free chain ends. Endo and exocellulases hydrolyze cellulose to form cello-oligosaccharides and cellobioses, which are then hydrolyzed into individual glucose molecules by Cellobioses or β-glucosidases (BG) (EC.3.2.1.21) (Horn, Vaaje-Kolstad, Westereng, & Eijsink, 2012). Aerobic microorganisms such as *Actinomycetales* and anaerobic microbes such as *Clostridiales* may both create cellulases. *Trichoderma reesei* is an aerobic fungus that can synthesize all three cellulases (Arora et al., 2019). Cellulase blends provide an advantage over individual enzymes since they operate synergistically to increase the hydrolysis rate. Commercial grade cellulases are available from Novozymes (Enzyme cocktails - Cellic CTec2, Cellic CTec3) and Genencor (Multifect CL, Multifect GC from *T. reesi*) (Saini et al., 2015). The hydrolysis process is influenced by elements such as substrate content, cellulase activity and loading, temperature, pH, hydrolysis duration, cellulose molecular structure and crystallinity, cellulose fiber surface area, and so on. The use of thermostable cellulases from fungi such as *Chaetomium thermophilum, Myceliophthora thermophila, Talaromyces emersonii, Thermoascus aurantiacus*, and *Tribulus terrestris* in bioethanol production is being researched (Guo et al., 2015). Endo-1, 4-β-xylanase (EC 3.2.1.8), and β-xylosidase are the primary enzymes involved (EC 3.2.1.37). Endoxylanases cleave the key chain of xylan at random, resulting in a variety of xylooligosaccharides, whereas β-xylosidase liberates xylose by eliminating terminal monosaccharides from the non-reducing end of short oligosaccharides. The L-arabinose residues are hydrolyzed by α-L-arabinofuranosidase (EC 3.2.1.55). α-D-glucuronidase (EC 3.2.1.139) cleaves α -1, 2 bonds between glucuronic acid residues and β-D-xylopyranosyl, and acetylxylan esterase (EC 3.1.1.72) eliminate the O-acetyl of acetyl xylan. In reality, hemicellulose biodegradation is a complicated process that necessitates the coordinated activity of many hemicellulases (Chávez, Bull, & Eyzaguirre, 2006; Peng et al., 2012; Sunna & Antranikian, 1997). The cost of enzymes is the most significant constraint for commercial biochemical conversion. With various research studies conducted throughout the world in recent years, the cost has been significantly decreased (A. Demirbas, 2008b; Mohapatra et al., 2019).

Fermentation

Hexoses and pentoses are end products of the biomass saccharification phase that can be fermented to produce biofuels. Microbes employed in commercial bioethanol production should be able to use a wide range of substrates, produce a large amount of ethanol, tolerate high temperatures, produce low pH and inhibitors throughout the pretreatment and saccharification processes, and tolerate high sugar and ethanol concentrations (Saini et al., 2015). *Saccharomyces cerevisiae*, *Zymomonas mobilis*, and certain genetically modified microorganisms can ferment hydrolyzed biomass containing various sugars to produce ethanol. They cannot, however, ferment pentoses and have a lower tolerance for high sugar concentrations (Robak & Balcerek, 2018). *Pichia stipites* (Agbogbo, Haagensen, Milam, & Wenger, 2008), *Candida shehatae*, *Pachysolan tannophilus* (Perego, Converti, Palazzi, Del Borghi, & Ferraiolo, 1990), and *Kluyveromyces marixianus* (Hasunuma & Kondo, 2012) may ferment pentoses as well as hexoses, but their efficiency is lower, and they are also sensitive to process variables such as temperature, pH, inhibitors, and so on. For biomass conversion into biofuel, a variety of techniques have been investigated. The first way is Separate hydrolysis and fermentation (SHF), in which cellulose saccharification (hydrolysis) and fermentation are performed in separate units, ensuring that the hydrolysis and ethanol production processes are carried out under optimal circumstances. The buildup of carbohydrates, on the other hand, may limit enzyme function. Simultaneous saccharification and fermentation, or SSF, is the next strategy, in which saccharification and fermentation are done in the same unit. The sugars generated by cellulose hydrolysis would be quickly fermented into ethanol in this method. This prevents the sugars from accumulating, resulting in higher ethanol production with fewer enzymes. The operating temperature is the primary drawback of this process, since the optimal temperature for hydrolysis is generally 50°C, while yeast viability is lower (28-37°C) (Olofsson, Bertilsson, & Lidén, 2008; Wingren, Galbe, & Zacchi, 2003). The third method on the list is a variation of SSF in which both hexoses and pentoses can be fermented simultaneously, thus the term simultaneous saccharification and co-fermentation (SSCF). The utilization of mixed cultures can help achieve this. Because bacteria that can utilize hexoses grow faster than those that can use pentoses, the conversion of hexoses to ethanol is more common. A second technique, on the other hand, is to utilize a single microbe that can take in together hexoses and pentoses, letting for high conversion and ethanol yield. To further understand the commercialization aspect of this process, more study is needed (W. C. Chen et al., 2017; McMillan, Newman, Templeton, & Mohagheghi, 1999). Consolidated bioprocessing (CPB) technology is a newer technique that goes a step further. CBP method for biofuel generation from lignocelluloses entails combining all three biological activities in a single bioreactor: synthesis of saccharolytic enzymes, hydrolysis of polysaccharides in pretreated biomass, and fermentation of hexose and pentose sugars. Production of enzymes that can successfully hydrolyze lignocellulosic biomass to fermentable sugars, utilization of hexoses and pentoses, efficient ethanol production (titer, yield, and productivity), fermentation inhibitors, resistance to ethanol, and stressful conditions such as low pH, high osmotic pressure, high temperature, low nutrition capacity, and fluctuating processes are the basic requirements of a microbe for this technology. No natural microbe, however, possesses all of the characteristics necessary for this technique (Hasunuma & Kondo, 2012; Olguin-Maciel, Singh, Chable-Villacis, Tapia-Tussell, & Ruiz, 2020). *S. cerevisiae*, a usual good ethanol producer with high inhibitor tolerance, can be transformed to produce the enzyme cellulases and hemicellulases, or *Clostridium thermocellum*, *C. phytofermentans*, that produces cellulases and hemicellulases, can be modified to be an proficient ethanol producer using genetic engineering technology (M. Balat, Balat, & Öz, 2008; Mohapatra et al., 2019). Because just one reactor is needed, charges for enzyme manufacturing may be avoided; the complete substrate is utilised for sugar release and no part of it is used for cellulase production. However, due to factors such as a long fermentation duration, low ethanol production, limited tolerance to ethanol, and the creation of fermentation inhibitors, this configuration has not yet been fully developed at a commercial level. Significant research has to be done to commercialize this process (Mohapatra et al., 2019; Raud et al., 2019).

Table 4. Various studies used lignocellulosic biomass for the production of biofuels

Substrate type	Organism	Fermentation type	Product and yield	References
Cassava bagasse	*S. cerevisiae*	Submerged fermentation	Bioethanol-0.25 m3/ton	(Martinez, Feiden, Bariccatti, & Zara, 2018)
Cassava bagasse	*Rhizopus oligosporus*	Solid-state Fermentation	Bioethanol-89.2%	(Escaramboni, Fernández Núñez, Carvalho, & de Oliva Neto, 2018)
Cassava cellulosic waste	*S. cerevisiae*	Submerged fermentation	Bioethanol-32.4% alcohol	(Elemike, Oseghale, & Okoye, 2015)
Sugarcane straw	*S. cerevisiae* Y 904	Solid-state Fermentation	Bioethanol-290 L/ton	(Pratto, dos Santos-Rocha, Longati, de Sousa Júnior, & Cruz, 2020)
Sugarcane-Leaf Waste	*S. cerevisiae* BY4743	Steam-salt alkali unfiltered (SSA-U)	Bioethanol-31.06 g/L	(Moodley & Gueguim Kana, 2019)
Sugar beet cossettes	*S. cerevisiae*	Submerged fermentation	Bioethanol-0.618g/L	(Pavlečić, Rezić, Šantek, Horvat, & Šantek, 2017)
Sugar Beet Pomace	*S. cerevisiae* PM-16	Solid-state Fermentation	Bioethanol-358.458 mg/g	(Rodríguez et al., 2010)
Sorghum pioneer	*S. cerevisiae*	Submerged fermentation	Bioethanol-0.49 g/l	(Koradiya, Duggirala, Tipre, & Dave, 2016)
Corn stover	*S. cerevisiae*	Submerged fermentation	Bioethanol-40.0 g/L	(Lau & Dale, 2009)
Sweet potato (hydrolysate)	*S. cerevisiae*	Submerged fermentation	Bioethanol-67.8 g/L	(Duvernay et al., 2013)
Switchgrass hydrolysate	*C. beijerinckii* P260	Submerged fermentation	Acetone -0.41 g/L; Butanol 0.97 g/L; Ethanol 0.10 g/L	(Qureshi et al., 2010)
King grass	*S. cerevisiae*	Submerged fermentation	Bioethanol - 30.8 g/L	(Cardona, Rios, Peña, Peñuela, & Rios, 2016)
Cocksfoot grass	*S. cerevisiae*	Simultaneous saccharification and fermentation (SSF)	Bioethanol - 31.49 g/L	(Njoku et al., 2013)
Apple & pear peel extracts	*C. acetobutylicum*	Submerged fermentation	Butanol-14 g/L	(Raganati, Procentese, Olivieri, Russo, & Marzocchella, 2016)
Rice straw	*C. acetobutylicum*	Submerged fermentation (Anaerobic)	Butanol-80 g/kg ,Acetone-21 g/kg	(Amiri, Karimi, & Zilouei, 2014)
Rice straw	*C. acetobutylicum*	Submerged fermentation (Anaerobic)	Acetone-18.2 g ; Butanol - 44.2 g; ; Ethanol -0.6 g	(Moradi, Amiri, Soleimanian-Zad, Ehsani, & Karimi, 2013)
Rice straw	*Pleurotus ostreatus*	Solid-state Fermentation	Methane -263 and 214 L/kg	(Mustafa, Poulsen, & Sheng, 2016)
Rice straw hydrolysate	*C. saccharoperbutylacetonicum*	Submerged fermentation (Anaerobic)	Acetone-5.1 g/L ; D Butanol-1.26 g/L/; Ethanol-8.1 g/L	(Wen Hsing Chen, Chen, & Lin, 2013)
Sago waste residues	*Bacillus coagulans*, *C. bifermentans*and *Clostridium* spp.	Submerged fermentation	Butanol-1.53 g/L; 3.36 g/L; 1.56 g/L; 2.5 g/L	(Johnravindar, Murugesan, Wong, & Elangovan, 2017)
Orange peel waste	*S. cerevisiae* & *C. acetobutylicum*	Submerged fermentation	Bio-butanol-19.5 g/L	(Joshi, Waghmare, Sonawane, & Waghmare, 2015)
Cauliflower waste	*C. acetobutylicum* NRRL B 527	Submerged fermentation	Acetone-Butanol-Ethanol-5.35 g/L	(Khedkar, Nimbalkar, Chavan, Chendake, & Bankar, 2017)
Lettuce residues	*C. acetobutylicum* DSMZ 792	Submerged fermentation	Butanol-1.1 g/L	(Procentese, Raganati, Olivieri, Elena Russo, & Marzocchella, 2017)
Sweet potato peel	*S. cerevisiae*	Submerged fermentation	Bioethanol-6.39g/L	(Ojewumi et al., 2018)

FUTURE RESEARCH DIRECTIONS

To address the extraordinary speed of climate change effected by the amassing of greenhouse gases (GHGs) in the atmosphere, there is a clear urgent need to convert energy dependency from fossil fuels to renewable energy sources. There are possibly two ways existed to reduce GHG emissions, one by reducing the usage of fuels and later being the replacement of harmful fuel with a less harmful one. Changes in climate and energy policy should be at the center of the country's global agenda, seeing as they have an impact on social, economic, and political decisions. To address these issues, employing environmentally acceptable, biodegradable, and cost-effective alternatives like biofuel might be an excellent choice. First-generation biofuels include corn ethanol, sugarcane ethanol, pure plant oils, and starch-based biodiesel. Because their technologies and implementation are easy, biofuel production technologies of the first generation have been successfully created and are widely used. However, the major drawback of first-generation biofuels is the food-versus-fuel conundrum; growing production of these fuels is one of the factors driving up food prices. As a reason, the lignocellulosic feedstock can develop alternative biofuels, which are frequently referred to as "second generation" biofuels. Under laboratory conditions; bioethanol synthesis from lignocellulosic biomass offers a good fuel output. However, various obstacles must be overcome before the manufacturing method can be commercialized. Various efforts to commercialize second-generation biofuels are underway across the world. In the case of biochemical fuels, breakthroughs in research and engineering of microorganisms meant to process specific feedstocks are required, followed by large-scale demonstrations to establish commercial practicality. The high capital and processing expenses of bioethanol from lignocellulosic biomass are a major drawback since specialized technology and infrastructure are required to facilitate the conversion process. When the production is scaled up, the price will fall. A number of public and private sources, including venture capital funds and internal resources from big organizations, have financed notable work in critical areas including safety, research, and development.

Development of biomass feedstocks with lesser lignin content and higher cellulose content.

Development of enzymes like cellulases and hemicellulases that depolymerize the biomass to achieve higher activity thus increasing the content of fermentable sugars and also reducing the content of inhibitors.

Development of novel microorganisms through metabolic engineering that can show resistance towards high ethanol, high temperatures, and can ferment pentoses along with hexoses.

This assistance has aided in moving it forward; nevertheless, additional assistance is required to make it a reality.

CONCLUSION

Technological innovations driven by the use of fossil fuels have enhanced human civilization and made it more comfortable; yet, this comfort has come at significant expenses. Biofuels have evolved to represent one of the main crucial sustainable fuel sources, and they are viewed as vital in our attempts to reduce greenhouse gas emissions, enhance environmental wellbeing, and create innovative energy sources. Biofuels, unlike fossil fuels, are generated from organic sources like biomass, which includes plants, crops, and agricultural waste. Furthermore, because of the abundance of agricultural feedstock such as forest residues, bagasse, native plants & fibers, corn stover, SPRs, bast crops & so on, developing countries, such as India, can be targeted for socioeconomic development using a biorefinery approach. Some of the obstacles that are identified that could hamper the biofuel economy are, the lack of refueling structure; costly zero-emission biofuel extraction; some forms of biofuel-based automobiles exhibit technology gaps, fuel cell lifespan. However, owing to a lack of knowledge and inadequate valuation, these useful wastes are often destroyed in large quantities for waste management rather than being used constructively. These wastes can be employed as a raw material in numerous investigations and industries, as well as an alternative source for the manufacture of various products such as biogas and biofuel. Adapted biogas processes are paving a new path to mitigate these swings by adapting operating and storage principles, whether that is through feedstock, biogas, or power, in response to changing net energy requirements.

When spoken about the future of Biofuels, the Country's developmental aspects for biofuel usage are influenced by factors like environmental worries on net carbon emissions, as well as security risks on the reliability and expense of fuel imports. Next few decades, the development of more efficient biofuel technologies might double or triple yields, as well as raise the energy output/input ratio in some of the less efficient bioenergy crops like maize or wheat. The biofuel sector is notably upsetting food production and prices in several parts of the world, a measured evidence-based assessment of biofuels that addresses a wide range of socio-economic and technological gaps in the biofuel sector. It has to be noted that biofuels are essentially economic, as biofuels long-term viability is highly reliant on their profitability. Elements like environmental and economic stability, as well as how technology permits these items to evolve must be taken into account. Biofuel production is now most viable in ecologically limited tropical locations due to the increased biofuel yields per acre and cheaper input prices. Also, Biofuel production has a primary profit element, which is based on resources that would substantially undercut the rise in oil prices. If nations throughout the world can use biofuels in the coming years, it will have a tremendous influence on the global economic structure and environment.

ACKNOWLEDGMENT

Kondapalli Vamsi Krishna and Sompalli Bhavana would like to acknowledge the fellowship from KSTePS, Department of Science and Technology (DST), Govt. of Karnataka for fellowship to pursue their Ph.D. work. Koushik Koujalagi would like to thank Christ (Deemed to be University) for a Ph.D. fellowship.

REFERENCES

Adler, P. R., Sanderson, M. A., Boateng, A. A., Weimer, P. J., & Jung, H. J. G. (2006). Biomass yield and biofuel quality of switchgrass harvested in fall or spring. *Agronomy Journal*, *98*(6), 1518–1525. doi:10.2134/agronj2005.0351

Agbogbo, F. K., Haagensen, F. D., Milam, D., & Wenger, K. S. (2008). Fermentation of acid-pretreated corn stover to ethanol without detoxification using Pichia stipitis. *Applied Biochemistry and Biotechnology*, *145*(1–3), 53–58. doi:10.100712010-007-8056-4 PMID:18425611

Agbor, V. B., Cicek, N., Sparling, R., Berlin, A., & Levin, D. B. (2011). Biomass pretreatment: Fundamentals toward application. *Biotechnology Advances*, *29*(6), 675–685. doi:10.1016/j.biotechadv.2011.05.005 PMID:21624451

Ali, M., Saleem, M., Khan, Z., & Watson, I. A. (2019). The use of crop residues for biofuel production. In Biomass, Biopolymer-Based Materials, and Bioenergy: Construction, Biomedical, and other Industrial Applications. doi:10.1016/B978-0-08-102426-3.00016-3

Alvira, P., Tomás-Pejó, E., Ballesteros, M., & Negro, M. J. (2010). Pretreatment technologies for an efficient bioethanol production process based on enzymatic hydrolysis: A review. *Bioresource Technology*, *101*(13), 4851–4861. doi:10.1016/j.biortech.2009.11.093 PMID:20042329

Amiri, H., Karimi, K., & Zilouei, H. (2014). Organosolv pretreatment of rice straw for efficient acetone, butanol, and ethanol production. *Bioresource Technology*, *152*, 450–456. doi:10.1016/j.biortech.2013.11.038 PMID:24321608

Araque, E., Parra, C., Freer, J., Contreras, D., Rodríguez, J., Mendonça, R., & Baeza, J. (2008). Evaluation of organosolv pretreatment for the conversion of Pinus radiata D. Don to ethanol. *Enzyme and Microbial Technology*, *43*(2), 214–219. doi:10.1016/j.enzmictec.2007.08.006

Arora, R., Sharma, N. K., Kumar, S., & Sani, R. K. (2019). Lignocellulosic Ethanol: Feedstocks and Bioprocessing. Bioethanol Production from Food Crops. doi:10.1016/B978-0-12-813766-6.00009-6

Atabani, A. E., Silitonga, A. S., Badruddin, I. A., Mahlia, T. M. I., Masjuki, H. H., & Mekhilef, S. (2012). A comprehensive review on biodiesel as an alternative energy resource and its characteristics. *Renewable & Sustainable Energy Reviews*, *16*(4), 2070–2093. doi:10.1016/j.rser.2012.01.003

Awogbemi, O., Von Kallon, D. V., Onuh, E. I., & Aigbodion, V. S. (2021). An overview of the classification, production and utilization of biofuels for internal combustion engine applications. *Energies*, *14*(18), 1–42. doi:10.3390/en14185687

Axelsson, L., Franzén, M., Ostwald, M., Berndes, G., Lakshmi, G., & Ravindranath, N. H. (2012). Perspective: Jatropha cultivation in southern India: Assessing farmers' experiences. *Biofuels, Bioproducts & Biorefining*, *6*(3), 246–256. doi:10.1002/bbb.1324

Ayodele, B. V., Alsaffar, M. A., & Mustapa, S. I. (2020). An overview of integration opportunities for sustainable bioethanol production from first- and second-generation sugar-based feedstocks. *Journal of Cleaner Production*, *245*, 118857. Advance online publication. doi:10.1016/j.jclepro.2019.118857

Balat, H., & Kirtay, E. (2010). Hydrogen from biomass - Present scenario and future prospects. *International Journal of Hydrogen Energy*, *35*(14), 7416–7426. doi:10.1016/j.ijhydene.2010.04.137

Balat, M., Balat, H., & Öz, C. (2008). Progress in bioethanol processing. *Progress in Energy and Combustion Science*, *34*(5), 551–573. doi:10.1016/j.pecs.2007.11.001

Balat, M., & Balat, M. (2009). Political, economic and environmental impacts of biomass-based hydrogen. *International Journal of Hydrogen Energy*, *34*(9), 3589–3603. doi:10.1016/j.ijhydene.2009.02.067

Baras, J., Gacesa, S., & Pejin, D. (2002). Ethanol is a strategic raw material. *Hemijska Industrija*, *56*(3), 89–104. doi:10.2298/HEMIND0203089B

Behera, S., Mohanty, R. C., & Ray, R. C. (2012). Ethanol fermentation of sugarcane molasses by Zymomonas mobilis MTCC 92 immobilized in Luffa cylindrica L. sponge discs and ca-alginate matrices. *Brazilian Journal of Microbiology, 43*(4), 1499–1507. doi:10.1590/S1517-83822012000400034

Berłowska, J., Pielech-Przybylska, K., Balcerek, M., Dziekońska-Kubczak, U., Patelski, P., Dziugan, P., & Kręgiel, D. (2016). Simultaneous Saccharification and Fermentation of Sugar Beet Pulp for Efficient Bioethanol Production. *BioMed Research International, 2016*. Advance online publication. doi:10.1155/2016/3154929

Bertrand, E., Vandenberghe, L. P. S., Soccol, C. R., Sigoillot, J. C., & Faulds, C. (2016). First generation bioethanol. In Green Energy and Technology (Vol. 0). https://doi.org/ doi:10.1007/978-3-319-30205-8_8

Bian, R., Ma, B., Zhu, X., Wang, W., Li, L., Joseph, S., ... Pan, G. (2016). Pyrolysis of crop residues in a mobile bench-scale pyrolyser: Product characterization and environmental performance. *Journal of Analytical and Applied Pyrolysis, 119*, 52–59. doi:10.1016/j.jaap.2016.03.018

Boerjan, W., Ralph, J., & Baucher, M. (2003). Lignin Biosynthesis. *Annual Review of Plant Biology, 54*, 519–546. doi:10.1146/annurev.arplant.54.031902.134938

Bulawayo, B., Bvochora, J. M., Muzondo, M. I., & Zvauya, R. (1996). Ethanol production by fermentation of sweet-stem sorghum juice using various yeast strains. *World Journal of Microbiology & Biotechnology, 12*(4), 357–360. doi:10.1007/BF00340211 PMID:24415310

Canilha, L., Chandel, A. K., Suzane Dos Santos Milessi, T., Antunes, F. A. F., Luiz Da Costa Freitas, W., Das Graças Almeida Felipe, M., & Da Silva, S. S. (2012). Bioconversion of sugarcane biomass into ethanol: An overview about composition, pretreatment methods, detoxification of hydrolysates, enzymatic saccharification, and ethanol fermentation. *Journal of Biomedicine & Biotechnology, 2012*, 1–15. Advance online publication. doi:10.1155/2012/989572 PMID:23251086

Canilha, L., Santos, V. T. O., Rocha, G. J. M., Almeida, E., Silva, J. B., Giulietti, M., Silva, S. S., ... Carvalho, W. (2011). A study on the pretreatment of a sugarcane bagasse sample with dilute sulfuric acid. *Journal of Industrial Microbiology & Biotechnology, 38*(9), 1467–1475. doi:10.100710295-010-0931-2 PMID:21210180

Cardona, E., Rios, J., Peña, J., Peñuela, M., & Rios, L. (2016). King Grass: A very promising material for the production of second generation ethanol in tropical countries. *Biomass and Bioenergy, 95*, 206–213. doi:10.1016/j.biombioe.2016.10.008

Carriquiry, M. A., Du, X., & Timilsina, G. R. (2011). Second generation biofuels: Economics and policies. *Energy Policy, 39*(7), 4222–4234. doi:10.1016/j.enpol.2011.04.036

Chandel, A. K., da Silva, S. S., Carvalho, W., & Singh, O. V. (2012). Sugarcane bagasse and leaves: Foreseeable biomass of biofuel and bio-products. *Journal of Chemical Technology and Biotechnology (Oxford, Oxfordshire), 87*(1), 11–20. doi:10.1002/jctb.2742

Chauhan, N. M., Hajare, S. T., Mamo, B., & Madebo, A. A. (2021). Bioethanol Production from Stalk Residues of Chiquere and Gebabe Varieties of Sweet Sorghum. *International Journal of Microbiology, 2021*, 1–16. Advance online publication. doi:10.1155/2021/6696254 PMID:33679985

Chávez, R., Bull, P., & Eyzaguirre, J. (2006). The xylanolytic enzyme system from the genus Penicillium. *Journal of Biotechnology*, *123*(4), 413–433. doi:10.1016/j.jbiotec.2005.12.036 PMID:16569456

Chen, W. C., Lin, Y. C., Ciou, Y. L., Chu, I. M., Tsai, S. L., Lan, J. C. W., Chang, Y.-K., & Wei, Y. H. (2017). Producing bioethanol from pretreated-wood dust by simultaneous saccharification and co-fermentation process. *Journal of the Taiwan Institute of Chemical Engineers*, *79*, 43–48. doi:10.1016/j.jtice.2017.04.025

Chen, W. H., Chen, Y.-C., & Lin, J.-G. (2013). Evaluation of biobutanol production from non-pretreated rice straw hydrolysate under non-sterile environmental conditions. *Bioresource Technology*, *135*, 262–268. doi:10.1016/j.biortech.2012.10.140 PMID:23186674

Coyle, W. (2007). The Future of Biofuels: A Global Perspective. *Amber Waves : The Economics of Food, Farming, Natural Resources, and Rural America*, *5*(5), 365–372. doi:10.22004/ag.econ.125366

Crago, C. L., Khanna, M., Barton, J., Giuliani, E., & Amaral, W. (2010). Competitiveness of Brazilian sugarcane ethanol compared to US corn ethanol. *Energy Policy*, *38*(11), 7404–7415. doi:10.1016/j.enpol.2010.08.016

Czernik, S., & Bridgwater, A. V. (2004). Overview of applications of biomass fast pyrolysis oil. *Energy & Fuels*, *18*(2), 590–598. doi:10.1021/ef034067u

Da Silva, A. S. A., Inoue, H., Endo, T., Yano, S., & Bon, E. P. S. (2010). Milling pretreatment of sugarcane bagasse and straw for enzymatic hydrolysis and ethanol fermentation. *Bioresource Technology*, *101*(19), 7402–7409. doi:10.1016/j.biortech.2010.05.008 PMID:20576565

Dahman, Y., Dignan, C., Fiayaz, A., & Chaudhry, A. (2019). An introduction to biofuels, foods, livestock, and the environment. Biomass, Biopolymer-Based Materials, and Bioenergy: Construction, Biomedical, and other Industrial Applications. doi:10.1016/B978-0-08-102426-3.00013-8

Dahman, Y., Syed, K., Begum, S., Roy, P., & Mohtasebi, B. (2019). Biofuels: Their characteristics and analysis. Biomass, Biopolymer-Based Materials, and Bioenergy: Construction, Biomedical, and other Industrial Applications. doi:10.1016/B978-0-08-102426-3.00014-X

Datta, A., Hossain, A., & Roy, S. (2019). An overview on biofuels and their advantages and disadvantages. *Asian Journal of Chemistry*, *31*(8), 1851–1858. doi:10.14233/ajchem.2019.22098

Demirbas, A. (2008). Biofuels sources, biofuel policy, biofuel economy and global biofuel projections. *Energy Conversion and Management*, *49*(8), 2106–2116. doi:10.1016/j.enconman.2008.02.020

Demirbas, A. (2008a). Biohydrogen generation from organic waste. *Energy Sources. Part A, Recovery, Utilization, and Environmental Effects*, *30*(5), 475–482. doi:10.1080/15567030600828909

Demirbas, A. (2008b). Products from lignocellulosic materials via degradation processes. *Energy Sources. Part A, Recovery, Utilization, and Environmental Effects*, *30*(1), 27–37. doi:10.1080/00908310600626705

Dodić, S., Popov, S., Dodić, J., Ranković, J., Zavargo, Z., & Jevtić Mučibabić, R. (2009). Bioethanol production from thick juice as intermediate of sugar beet processing. *Biomass and Bioenergy*, *33*(5), 822–827. doi:10.1016/j.biombioe.2009.01.002

Dürre, P. (2007). Biobutanol: An attractive biofuel. *Biotechnology Journal, 2*(12), 1525–1534. doi:10.1002/biot.200700168

Duvernay, W. H., Chinn, M. S., & Yencho, G. C. (2013). Hydrolysis and fermentation of sweetpotatoes for production of fermentable sugars and ethanol. *Industrial Crops and Products, 42*(1), 527–537. doi:10.1016/j.indcrop.2012.06.028

Elemike, E. E., Oseghale, O. C., & Okoye, A. C. (2015). Utilization of cellulosic cassava waste for bio-ethanol production. *Journal of Environmental Chemical Engineering, 3*(4), 2797–2800. doi:10.1016/j.jece.2015.10.021

Escaramboni, B., Fernández Núñez, E. G., Carvalho, A. F. A., & de Oliva Neto, P. (2018). Ethanol biosynthesis by fast hydrolysis of cassava bagasse using fungal amylases produced in optimized conditions. *Industrial Crops and Products, 112*(December), 368–377. doi:10.1016/j.indcrop.2017.12.004

Ganzler, K., Salgó, A., & Valkó, K. (2012). Microwave extraction A novel sample preparation method for chromatography. *Comprehensive Sampling and Sample Preparation, 2*, 135–149. doi:10.1016/B978-0-12-381373-2.00048-X

García, A., González Alriols, M., & Labidi, J. (2014). Evaluation of different lignocellulosic raw materials as potential alternative feedstocks in biorefinery processes. *Industrial Crops and Products, 53*, 102–110. doi:10.1016/j.indcrop.2013.12.019

Ge, X., Xu, F., & Li, Y. (2016). Solid-state anaerobic digestion of lignocellulosic biomass: Recent progress and perspectives. *Bioresource Technology, 205*(January), 239–249. doi:10.1016/j.biortech.2016.01.050 PMID:26832395

Ghaffar, S. H., & Fan, M. (2013). Structural analysis for lignin characteristics in biomass straw. *Biomass and Bioenergy, 57*, 264–279. doi:10.1016/j.biombioe.2013.07.015

Glaser, B., Haumaier, L., Guggenberger, G., & Zech, W. (2001). The "Terra Preta" phenomenon: A model for sustainable agriculture in the humid tropics. *Naturwissenschaften, 88*(1), 37–41. doi:10.1007001140000193

Glaser, B., Haumaier, L., Guggenberger, G., & Zech, W. (2001). The "Terra Preta" phenomenon: A model for sustainable agriculture in the humid tropics. *Naturwissenschaften, 88*(1), 37–41. doi:10.1007001140000193
PMID:11302125

Gopinathan, M. C., & Sudhakaran, R. (2009). Biofuels: Opportunities and challenges in India. *In Vitro Cellular & Developmental Biology. Plant, 45*(3), 350–371. doi:10.100711627-009-9217-7

Goyal, H. B., Seal, D., & Saxena, R. C. (2008). Bio-fuels from thermochemical conversion of renewable resources: A review. *Renewable & Sustainable Energy Reviews, 12*(2), 504–517. doi:10.1016/j.rser.2006.07.014

Guo, M., Song, W., & Buhain, J. (2015). Bioenergy and biofuels: History, status, and perspective. *Renewable & Sustainable Energy Reviews, 42*, 712–725. doi:10.1016/j.rser.2014.10.013

Haghighi Mood, S., Hossein Golfeshan, A., Tabatabaei, M., Salehi Jouzani, G., Najafi, G. H., Gholami, M., & Ardjmand, M. (2013). Lignocellulosic biomass to bioethanol, a comprehensive review with a focus on pretreatment. *Renewable & Sustainable Energy Reviews, 27*, 77–93. doi:10.1016/j.rser.2013.06.033

Hamouda, H. I., Nassar, H. N., Madian, H. R., Abu Amr, S. S., & El-Gendy, N. S. (2015). Response Surface Optimization of Bioethanol Production from Sugarcane Molasses by Pichia veronae Strain HSC-22. *Biotechnology Research International*, *2015*, 1–10. doi:10.1155/2015/905792

Hasunuma, T., & Kondo, A. (2012). Consolidated bioprocessing and simultaneous saccharification and fermentation of lignocellulose to ethanol with thermotolerant yeast strains. *Process Biochemistry*, *47*(9), 1287–1294. doi:10.1016/j.procbio.2012.05.004

Hendriks, A. T. W. M., & Zeeman, G. (2009). Pretreatments to enhance the digestibility of lignocellulosic biomass. *Bioresource Technology*, *100*(1), 10–18. doi:10.1016/j.biortech.2008.05.027

Hideno, A., Inoue, H., Tsukahara, K., Fujimoto, S., Minowa, T., Inoue, S., ... Sawayama, S. (2009). Wet disk milling pretreatment without sulfuric acid for enzymatic hydrolysis of rice straw. *Bioresource Technology*, *100*(10), 2706–2711. doi:10.1016/j.biortech.2008.12.057

Hiloidhari, M., Bhuyan, N., Gogoi, N., Seth, D., Garg, A., & Singh, A. … Kataki, R. (2019). Agroindustry wastes: Biofuels and biomaterials feedstocks for sustainable rural development. In Refining Biomass Residues for Sustainable Energy and Bioproducts: Technology, Advances, Life Cycle Assessment, and Economics. https://doi.org/ doi:10.1016/B978-0-12-818996-2.00016-8

Himmel, M. E., Ding, S. Y., Johnson, D. K., Adney, W. S., Nimlos, M. R., Brady, J. W., & Foust, T. D. (2007). Biomass recalcitrance: Engineering plants and enzymes for biofuels production. *Science*, *315*(5813), 804–807. doi:10.1126cience.1137016 PMID:17289988

Hirani, A. H., Javed, N., Asif, M., Basu, S. K., & Kumar, A. (2018). A review on first- and second- generation biofuel productions. *Biofuels: Greenhouse Gas Mitigation and Global Warming: Next Generation Biofuels and Role of Biotechnology*, 141–154. doi:10.1007/978-81-322-3763-1_8

Ho, D. P., Ngo, H. H., & Guo, W. (2014). A mini review on renewable sources for biofuel. *Bioresource Technology*, *169*(July), 742–749. doi:10.1016/j.biortech.2014.07.022 PMID:25115598

Hoekman, S. K., Broch, A., Robbins, C., Ceniceros, E., & Natarajan, M. (2012). Review of biodiesel composition, properties, and specifications. *Renewable & Sustainable Energy Reviews*, *16*(1), 143–169. doi:10.1016/j.rser.2011.07.143

Hong, B., Xue, G., Weng, L., & Guo, X. (2012). Pretreatment of moso bamboo with dilute phosphoric acid. *BioResources*, *7*(4), 4902–4913. doi:10.15376/biores.7.4.4902-4913

Horn, S. J., Vaaje-Kolstad, G., Westereng, B., & Eijsink, V. G. (2012). Novel enzymes for the degradation of cellulose. Biotechnol Biofuels. *Biotechnology for Biofuels*, *5*(1), 45. doi:10.1186/1754-6834-5-45 PMID:22747961

Hu, Z., & Wen, Z. (2008). Enhancing enzymatic digestibility of switchgrass by microwave-assisted alkali pretreatment. *Biochemical Engineering Journal*, *38*(3), 369–378. doi:10.1016/j.bej.2007.08.001

Iens, J. O. H. N. W., & Argione, J. O. F. (2011). Biofuels and biodiversity notifications. *Environmental Impact of Biofuels*, *21*(June), 1085–1095.

Inoue, H., Yano, S., Endo, T., Sakaki, T., & Sawayama, S. (2008). Combining hot-compressed water and ball milling pretreatments to improve the efficiency of the enzymatic hydrolysis of eucalyptus. *Biotechnology for Biofuels*, *1*(1), 1–9. doi:10.1186/1754-6834-1-2 PMID:18471309

Izmirlioglu, G., & Demirci, A. (2012). Ethanol production from waste potato mash by using Saccharomyces cerevisiae. *Applied Sciences (Switzerland)*, *2*(4), 738–753. doi:10.3390/app2040738

Jayus, Nurhayati, Mayzuhroh, A., Arindhani, S., & Caroenchai, C. (2016). Studies on Bioethanol Production of Commercial Baker's and Alcohol Yeast under Aerated Culture Using Sugarcane Molasses as the Media. *Agriculture and Agricultural Science Procedia*, *9*, 493–499. doi:10.1016/j.aaspro.2016.02.168

Johnravindar, D., Murugesan, K., Wong, J. W. C., & Elangovan, N. (2017). Waste-to-biofuel: Production of biobutanol from sago waste residues. *Environmental Technology (United Kingdom)*, *38*(13–14), 1725–1734. doi:10.1080/09593330.2017.1283362

Joshi, S. M., Waghmare, J. S., Sonawane, K. D., & Waghmare, S. R. (2015). Bio-ethanol and bio-butanol production from orange peel waste. *Biofuels*, *6*(1–2), 55–61. doi:10.1080/17597269.2015.1045276

Kambo, H. S., & Dutta, A. (2015). A comparative review of biochar and hydrochar in terms of production, physico-chemical properties and applications. *Renewable & Sustainable Energy Reviews*, *45*, 359–378. doi:10.1016/j.rser.2015.01.050

Khattab, S. M. R., & Watanabe, T. (2019). Bioethanol From Sugarcane Bagasse: Status and Perspectives. In Bioethanol Production from Food Crops. https://doi.org/ doi:10.1016/B978-0-12-813766-6/00010-2

Khedkar, M. A., Nimbalkar, P. R., Chavan, P. V., Chendake, Y. J., & Bankar, S. B. (2017). Cauliflower waste utilization for sustainable biobutanol production: Revelation of drying kinetics and bioprocess development. *Bioprocess and Biosystems Engineering*, *40*(10), 1493–1506. doi:10.100700449-017-1806-y

Kim, S., & Dale, B. E. (2004). Global potential bioethanol production from wasted crops and crop residues. *Biomass and Bioenergy*, *26*(4), 361–375. doi:10.1016/j.biombioe.2003.08.002

Kolesinska, B., Fraczyk, J., Binczarski, M., Modelska, M., Berlowska, J., Dziugan, P., ... Kregiel, D. (2019). Butanol synthesis routes for biofuel production: Trends and perspectives. *Materials (Basel)*, *12*(3). Advance online publication. doi:10.3390/ma12030350

Koradiya, M., Duggirala, S., Tipre, D., & Dave, S. (2016). Pretreatment optimization of Sorghum pioneer biomass for bioethanol production and its scale-up. *Bioresource Technology*, *199*, 142–147. doi:10.1016/j.biortech.2015.08.156 PMID:26384087

Koshijima, T., & Watanabe, T. (2003). *Preparation and Characterization of Lignin-Carbohydrate Complexes*. doi:10.1007/978-3-662-05191-7_1

Kothari, R., Vashishtha, A., Singh, H. M., Pathak, V. V., Tyagi, V. V., Yadav, B. C., Ashokkumar, V., & Singh, D. P. (2020). Assessment of Indian bioenergy policy for sustainable environment and its impact for rural India: Strategic implementation and challenges. *Environmental Technology and Innovation*, *20*, 101078. doi:10.1016/j.eti.2020.101078

Kumar, M., Sun, Y., Rathour, R., Pandey, A., Thakur, I. S., & Tsang, D. C. W. (2020). Algae as potential feedstock for the production of biofuels and value-added products: Opportunities and challenges. *The Science of the Total Environment*, *716*, 137116. doi:10.1016/j.scitotenv.2020.137116 PMID:32059310

Kwapinski, W., Byrne, C. M. P., Kryachko, E., Wolfram, P., Adley, C., Leahy, J. J., Novotny, E. H., & Hayes, M. H. B. (2010). Biochar from biomass and waste. *Waste and Biomass Valorization*, *1*(2), 177–189. doi:10.100712649-010-9024-8

Lal, R. (2008). Crop residues as soil amendments and feedstock for bioethanol production. *Waste Management (New York, N.Y.)*, *28*(4), 747–758. doi:10.1016/j.wasman.2007.09.023 PMID:18053700

Lamers, P., Hamelinck, C., Junginger, M., & Faaij, A. (2011). International bioenergy trade - A review of past developments in the liquid biofuel market. *Renewable & Sustainable Energy Reviews*, *15*(6), 2655–2676. doi:10.1016/j.rser.2011.01.022

Lau, M. W., & Dale, B. E. (2009). Cellulosic ethanol production from AFEX-treated corn stover using Saccharomyces cerevisiae 424A(LNH-ST). *Proceedings of the National Academy of Sciences of the United States of America*, *106*(5), 1368–1373. doi:10.1073/pnas.0812364106 PMID:19164763

Lee, R. A., & Lavoie, J. M. (2013). From first- to third-generation biofuels: Challenges of producing a commodity from a biomass of increasing complexity. *Animal Frontiers*, *3*(2), 6–11. doi:10.2527/af.2013-0010

Liang, L., Zhang, Y. P., Zhang, L., Zhu, M. J., Liang, S. Z., & Huang, Y. N. (2008). Study of sugarcane pieces as yeast supports for ethanol production from sugarcane juice and molasses. *Journal of Industrial Microbiology & Biotechnology*, *35*(12), 1605–1613. doi:10.100710295-008-0404-z PMID:18685877

Liew, W. H., Hassim, M. H., & Ng, D. K. S. (2014). Review of evolution, technology and sustainability assessments of biofuel production. *Journal of Cleaner Production*, *71*, 11–29. doi:10.1016/j.jclepro.2014.01.006

Lin, S. P., Kuo, T. C., Wang, H. T., Ting, Y., Hsieh, C. W., Chen, Y. K., Hsu, H.-Y., & Cheng, K. C. (2020). Enhanced bioethanol production using atmospheric cold plasma-assisted detoxification of sugarcane bagasse hydrolysate. *Bioresource Technology*, *313*, 123704. doi:10.1016/j.biortech.2020.123704 PMID:32590306

Ma, F., & Hanna, M. A. (1999). Biodiesel production: A review. *Bioresource Technology*, *70*(1), 1–15. doi:10.1016/S0960-8524(99)00025-5

Mackie, K. L., Brownell, H. H., West, K. L., & Saddler, J. N. (1985). Effect of sulphur dioxide and sulphuric acid on steam explosion of aspenwood. *Journal of Wood Chemistry and Technology*, *5*(3), 405–425. doi:10.1080/02773818508085202

Mączyńska, J., Krzywonos, M., Kupczyk, A., Tucki, K., Sikora, M., Pińkowska, H., … Wielewska, I. (2019). Production and use of biofuels for transport in Poland and Brazil – The case of bioethanol. *Fuel*, *241*(September), 989–996. doi:10.1016/j.fuel.2018.12.116

Martin, M. A. (2010). First generation biofuels compete. *New Biotechnology*, *27*(5), 596–608. doi:10.1016/j.nbt.2010.06.010 PMID:20601265

Martinez, D. G., Feiden, A., Bariccatti, R., & Zara, K. R. de F. (2018). Ethanol production from waste of cassava processing. *Applied Sciences (Switzerland)*, *8*(11), 1–8. doi:10.3390/app8112158

Martínez-Gutiérrez, E. (2018). Biogas production from different lignocellulosic biomass sources: advances and perspectives. *3 Biotech*, *8*(5), 1–18. doi:10.1007/s13205-018-1257-4

Marzo, C., Díaz, A. B., Caro, I., & Blandino, A. (2019). Status and Perspectives in Bioethanol Production From Sugar Beet. Bioethanol Production from Food Crops. doi:10.1016/B978-0-12-813766-6.00004-7

McKendry, P. (2002). Energy production from biomass (part 1): Overview of biomass. *Bioresource Technology*, *83*(1), 37–46. doi:10.1016/S0960-8524(01)00118-3 PMID:12058829

McMillan, J. D., Newman, M. M., Templeton, D. W., & Mohagheghi, A. (1999). Simultaneous saccharification and cofermentation of dilute-acid pretreated yellow poplar hardwood to ethanol using xylose-fermenting Zymomonas mobilis. *Applied Biochemistry and Biotechnology - Part A Enzyme Engineering and Biotechnology*, *77–79*(1-3), 649–665. doi:10.1385/ABAB:79:1-3:649 PMID:15304685

Meramo-Hurtado, S. I., Puello, P., & Cabarcas, A. (2020). Process analysis of hydrogen production via biomass gasification under computer-aided safety and environmental assessments. *ACS Omega*, *5*(31), 19667–19681. doi:10.1021/acsomega.0c02344 PMID:32803062

Mesa, L., González, E., Cara, C., González, M., Castro, E., & Mussatto, S. I. (2011). The effect of organosolv pretreatment variables on enzymatic hydrolysis of sugarcane bagasse. *Chemical Engineering Journal*, *168*(3), 1157–1162. doi:10.1016/j.cej.2011.02.003

Mishra, A., & Ghosh, S. (2017). Key Pretreatment Technologies for An Efficient Bioethanol Production from Lignocellulosics. *Advances in Biofeedstocks and Biofuels*, *1*, 55–83. doi:10.1002/9781119117322.ch3

Mishra, V. K., & Goswami, R. (2018). A review of production, properties and advantages of biodiesel. *Biofuels*, *9*(2), 273–289. doi:10.1080/17597269.2017.1336350

Miyafuji, H. (2015). Application of ionic liquids for effective use of woody biomass. *Journal of Wood Science*, *61*(4), 343–350. doi:10.100710086-015-1489-4

Mohan, D., Pittman, C. U., & Philip, S. (2017). Pyrolysis of Wood/Biomass for Bio-oil: A Critical Review Dinesh. *Progress in Energy and Combustion Science*, *62*(4), 848–889. doi:10.1016/j.pecs.2017.05.004

Mohapatra, S., Ray, R. C., & Ramachandran, S. (2019). Bioethanol From Biorenewable Feedstocks: Technology, Economics, and Challenges. Bioethanol Production from Food Crops. doi:10.1016/B978-0-12-813766-6.00001-1

Moodley, P., & Gueguim Kana, E. B. (2019). Bioethanol production from sugarcane leaf waste: Effect of various optimized pretreatments and fermentation conditions on process kinetics. *Biotechnology Reports (Amsterdam, Netherlands)*, *22*, e00329. doi:10.1016/j.btre.2019.e00329 PMID:31008065

Moradi, F., Amiri, H., Soleimanian-Zad, S., Ehsani, M. R., & Karimi, K. (2013). Improvement of acetone, butanol and ethanol production from rice straw by acid and alkaline pretreatments. *Fuel*, *112*, 8–13. doi:10.1016/j.fuel.2013.05.011

Morone, P., & Cottoni, L. (2016). Biofuels: Technology, economics, and policy issues. Technology, economics, and policy issues. In Handbook of Biofuels Production: Processes and Technologies: Second Edition. doi:10.1016/B978-0-08-100455-5.00004-7

Moulthrop, J. S., Swatloski, R. P., Moyna, G., & Rogers, R. D. (2005). High-resolution 13C NMR studies of cellulose and cellulose oligomers in ionic liquid solutions. *Chemical Communications*, *40*(12), 1557–1559. doi:10.1039/b417745b PMID:15770258

Mussatto, S. I., Dragone, G., Guimarães, P. M. R., Silva, J. P. A., Carneiro, L. M., Roberto, I. C., Vicente, A., Domingues, L., & Teixeira, J. A. (2010). Technological trends, global market, and challenges of bio-ethanol production. *Biotechnology Advances*, *28*(6), 817–830. doi:10.1016/j.biotechadv.2010.07.001 PMID:20630488

Mustafa, A. M., Poulsen, T. G., & Sheng, K. (2016). Fungal pretreatment of rice straw with Pleurotus ostreatus and Trichoderma reesei to enhance methane production under solid-state anaerobic digestion. *Applied Energy*, *180*, 661–671. doi:10.1016/j.apenergy.2016.07.135

Naik, S. N., Goud, V. V., Rout, P. K., & Dalai, A. K. (2010). Production of first and second generation biofuels: A comprehensive review. *Renewable & Sustainable Energy Reviews*, *14*(2), 578–597. doi:10.1016/j.rser.2009.10.003

Naqvi, M., & Yan, J. (2015). First-Generation Biofuels. Handbook of Clean Energy Systems, 1–18. doi:10.1002/9781118991978.hces207

Nikolov, Z. L., Meagher, M. M., & Reilly, P. J. (1989). Kinetics, equilibria, and modeling of the formation of oligosaccharides from D-glucose with Aspergillus niger glucoamylases I and II. *Biotechnology and Bioengineering*, *34*(5), 694–704. doi:10.1002/bit.260340514 PMID:18588154

Njoku, S. I., Uellendahl, H., & Ahring, B. K. (2013). Comparing oxidative and dilute acid wet explosion pretreatment of cocksfoot grass at high dry matter concentration for cellulosic ethanol production. *Energy Science & Engineering*, *1*(2), 89–98. doi:10.1002/ese3.11

Ogbonna, J. C., Mashima, H., & Tanaka, H. (2001). Scale up of fuel ethanol production from sugar beet juice using loofa sponge immobilized bioreactor. *Bioresource Technology*, *76*(1), 1–8. doi:10.1016/S0960-8524(00)00084-5 PMID:11315804

Ojewumi, M. E., Job, A. I., Samson Taiwo, O., Mopelola Obanla, O., Ayoola, A. A., Ojewumi, E. O., & Oyeniyi, E. A. (2018). Bio-Conversion of Sweet Potato Peel Waste to Bio-Ethanol Using Saccharomyces Cerevisiae. *International Journal of Pharmaceutical and Phytopharmacological Research*, *8*(3), 46–54. www.eijppr.com

Olguin-Maciel, E., Singh, A., Chable-Villacis, R., Tapia-Tussell, R., & Ruiz, H. A. (2020). Consolidated bioprocessing, an innovative strategy towards sustainability for biofuels production from crop residues: An overview. *Agronomy (Basel)*, *10*(11), 1834. Advance online publication. doi:10.3390/agronomy10111834

Olofsson, K., Bertilsson, M., & Lidén, G. (2008). A short review on SSF - An interesting process option for ethanol production from lignocellulosic feedstocks. *Biotechnology for Biofuels*, *1*(1), 1–14. doi:10.1186/1754-6834-1-7 PMID:18471273

Pandey, A., Soccol, C. R., Nigam, P., Soccol, V. T., Vandenberghe, L. P. S., & Mohan, R. (2000). Biotechnological potential of agro-industrial residues. II: Cassava bagasse. *Bioresource Technology*, *74*(1), 81–87. doi:10.1016/S0960-8524(99)00143-1

Panella, L. (2010). Sugar Beet as an Energy Crop. *Sugar Tech*, *12*(3–4), 288–293. doi:10.100712355-010-0041-5

Pavlečić, M., Rezić, T., Šantek, M. I., Horvat, P., & Šantek, B. (2017). Bioethanol production from raw sugar beet cossettes in horizontal rotating tubular bioreactor. *Bioprocess and Biosystems Engineering*, *40*(11), 1679–1688. doi:10.100700449-017-1823-x PMID:28770369

Peng, F., Peng, P., Xu, F., & Sun, R. C. (2012). Fractional purification and bioconversion of hemicelluloses. *Biotechnology Advances*, *30*(4), 879–903. doi:10.1016/j.biotechadv.2012.01.018 PMID:22306329

Perego, P., Converti, A., Palazzi, E., Del Borghi, M., & Ferraiolo, G. (1990). Fermentation of hardwood hemicellulose hydrolysate by Pachysolen tannophilus, candida shehatae and Pichia stipitis. *Journal of Industrial Microbiology*, *6*(3), 157–164. doi:10.1007/BF01577690

Perez-Verdin, G., Grebner, D. L., Sun, C., Munn, I. A., Schultz, E. B., & Matney, T. G. (2009). Woody biomass availability for bioethanol conversion in Mississippi. *Biomass and Bioenergy*, *33*(3), 492–503. doi:10.1016/j.biombioe.2008.08.021

Potumarthi, R., Baadhe, R. R., Nayak, P., & Jetty, A. (2013). Simultaneous pretreatment and saccharification of rice husk by Phanerochete chrysosporium for improved production of reducing sugars. *Bioresource Technology*, *128*, 113–117. doi:10.1016/j.biortech.2012.10.030 PMID:23196230

Pratto, B., dos Santos-Rocha, M. S. R., Longati, A. A., de Sousa Júnior, R., & Cruz, A. J. G. (2020). Experimental optimization and techno-economic analysis of bioethanol production by simultaneous saccharification and fermentation process using sugarcane straw. *Bioresource Technology*, *297*, 122494. doi:10.1016/j.biortech.2019.122494 PMID:31813817

Procentese, A., Raganati, F., Olivieri, G., Elena Russo, M., & Marzocchella, A. (2017). Pre-treatment and enzymatic hydrolysis of lettuce residues as feedstock for bio-butanol production. *Biomass and Bioenergy*, *96*, 172–179. doi:10.1016/j.biombioe.2016.11.015

Quintero, J. A., Montoya, M. I., Sánchez, O. J., Giraldo, O. H., & Cardona, C. A. (2008). Fuel ethanol production from sugarcane and corn: Comparative analysis for a Colombian case. *Energy*, *33*(3), 385–399. doi:10.1016/j.energy.2007.10.001

Qureshi, N., Saha, B. C., Hector, R. E., Dien, B., Hughes, S., Liu, S., Iten, L., Bowman, M. J., Sarath, G., & Cotta, M. A. (2010). Production of butanol (a biofuel) from agricultural residues: Part II - Use of corn stover and switchgrass hydrolysates. *Biomass and Bioenergy*, *34*(4), 566–571. doi:10.1016/j.biombioe.2009.12.023

Raganati, F., Procentese, A., Olivieri, G., Russo, M. E., & Marzocchella, A. (2016). Butanol production by fermentation of fruit residues. *Chemical Engineering Transactions*, *49*, 229–234. doi:10.3303/CET1649039

Ramesh, C., Ray, K. B. U., & Chandrasekaran Trilokesh, C. L. (2019). Sweet Sorghum for Bioethanol Production: Scope, Technology, and Economics. In Bioethanol Production from Food Crops (pp. 81–100). doi:10.1016/B978-0-12-813766-6/00005-9

Rasmey, A.-H., Hassan, H., Aboseidah, A., & Abdulwahid, O. (2018). Enhancing Bioethanol Productivity from Sugarcane Molasses by Saccharomyces cerevisiae Y17 KP096551. *Egyptian Journal of Botany*. doi:10.21608/ejbo.2018.1820.1126

Rasool, U., & Hemalatha, S. (2016). A review on bioenergy and biofuels: Sources and their production. *Brazilian Journal of Biological Sciences*, *3*(5), 3. doi:10.21472/bjbs.030501

Ratnavathi, C. V. (2017). Prospects of Sweet Sorghum as a Raw Material for Ethanol Production. *Recent Advances in Petrochemical Science*, *3*(3), 51–53. doi:10.19080/RAPSCI.2017.03.555613

Ratnavathi, C. V., Chakravarthy, S. K., Komala, V. V., Chavan, U. D., & Patil, J. V. (2011). Sweet Sorghum as Feedstock for Biofuel Production: A Review. *Sugar Tech*, *13*(4), 399–407. doi:10.100712355-011-0112-2

Raud, M., Kikas, T., Sippula, O., & Shurpali, N. J. (2019). Potentials and challenges in lignocellulosic biofuel production technology. *Renewable & Sustainable Energy Reviews*, *111*(May), 44–56. doi:10.1016/j.rser.2019.05.020

Ravindranath, N. H., Somashekar, H. I., Nagaraja, M. S., Sudha, P., Sangeetha, G., Bhattacharya, S. C., & Abdul Salam, P. (2005). Assessment of sustainable non-plantation biomass resources potential for energy in India. *Biomass and Bioenergy*, *29*(3), 178–190. doi:10.1016/j.biombioe.2005.03.005

Robak, K., & Balcerek, M. (2018). Review of second generation bioethanol production from residual biomass. *Food Technology and Biotechnology*, *56*(2), 174–187. doi:10.17113/ftb.56.02.18.5428 PMID:30228792

Rodríguez, L. A., Toro, M. E., Vazquez, F., Correa-Daneri, M. L., Gouiric, S. C., & Vallejo, M. D. (2010). Bioethanol production from grape and sugar beet pomaces by solid-state fermentation. *International Journal of Hydrogen Energy*, *35*(11), 5914–5917. doi:10.1016/j.ijhydene.2009.12.112

Roy, M. M., & Chandra, A. (2019). Promoting biofuels: The case of ethanol blending initiative in India. *Clean Technologies and Environmental Policy*, *21*(5), 953–965. doi:10.100710098-019-01687-z

Ruan, R., Zhang, Y., Chen, P., Liu, S., Fan, L., Zhou, N., . . . Li, B. (2019). Biofuels: Introduction. In Biomass, Biofuels, Biochemicals: Biofuels: Alternative Feedstocks and Conversion Processes for the Production of Liquid and Gaseous Biofuels. doi:10.1016/B978-0-12-816856-1.00001-4

Ruiz, J. A., Juárez, M. C., Morales, M. P., Muñoz, P., & Mendívil, M. A. (2013). Biomass gasification for electricity generation: Review of current technology barriers. *Renewable & Sustainable Energy Reviews*, *18*, 174–183. doi:10.1016/j.rser.2012.10.021

Rulli, M. C., Bellomi, D., Cazzoli, A., De Carolis, G., & D'Odorico, P. (2016). The water-land-food nexus of first-generation biofuels. *Scientific Reports*, *6*(March), 1–10. doi:10.1038/srep22521

Saha, B. C. (2003). Hemicellulose bioconversion. *Journal of Industrial Microbiology & Biotechnology*, *30*(5), 279–291. doi:10.100710295-003-0049-x PMID:12698321

Saini, J. K., Saini, R., & Tewari, L. (2015). Lignocellulosic agriculture wastes as biomass feedstocks for second-generation bioethanol production: concepts and recent developments. *3 Biotech, 5*(4), 337–353. doi:10.1007/s13205-014-0246-5

Šantek, B., Gwehenberger, G., Šantek, M. I., Narodoslawsky, M., & Horvat, P. (2010). Evaluation of energy demand and the sustainability of different bioethanol production processes from sugar beet. *Resources, Conservation and Recycling*, *54*(11), 872–877. doi:10.1016/j.resconrec.2010.01.006

Saran, S., Malaviya, A., & Chaubey, A. (2019). Introduction, Scope and Significance of Fermentation Technology. In High Value Fermentation Products (Vol. 1, pp. 1–25). Academic Press.

Sasaki, K., Tsuge, Y., Sasaki, D., Kawaguchi, H., Sazuka, T., Ogino, C., & Kondo, A. (2015). Repeated ethanol production from sweet sorghum juice concentrated by membrane separation. *Bioresource Technology*, *186*, 351–355. doi:10.1016/j.biortech.2015.03.127 PMID:25857769

Sexton, S. E., & Zilberman, D. (2011). Biotechnology and biofuel. In Frontiers of Economics and Globalization (Vol. 10). doi:10.1108/S1574-8715(2011)0000010014

Shishir, P. S. (2007). Effect of Particle Size Based Separation of Milled Corn Stover on AFEX Pretreatment and Enzymatic Digestibility. *Biotechnology and Bioengineering*, *96*(2), 219–231. doi:10.1002/bit.21132 PMID:16903002

Shrestha, P., Pometto, A. L. III, Khanal, S. K., & van Leeuwen, J. (2012). Second-Generation Biofuel Production from Corn-Ethanol Industry Residues. *Green Energy and Technology*, *62*, 71–87. doi:10.1007/978-1-4471-2324-8_5

Siciliano, A., Limonti, C., Mehariya, S., Molino, A., & Calabrò, V. (2019). Biofuel production and phosphorus recovery through an integrated treatment of agro-industrial waste. *Sustainability (Switzerland)*, *11*(1), 1–17. doi:10.3390u11010052

Silalertruksa, T., & Gheewala, S. H. (2009). Environmental sustainability assessment of bio-ethanol production in Thailand. *Energy*, *34*(11), 1933–1946. doi:10.1016/j.energy.2009.08.002

Sindhu, R., Binod, P., & Pandey, A. (2016). Biological pretreatment of lignocellulosic biomass - An overview. *Bioresource Technology*, *199*, 76–82. doi:10.1016/j.biortech.2015.08.030 PMID:26320388

Sindhu, R., Binod, P., Pandey, A., Ankaram, S., Duan, Y., & Awasthi, M. K. (2019). Biofuel production from biomass: Toward sustainable development. In *Current Developments in Biotechnology and Bioengineering*. Waste Treatment Processes for Energy Generation. doi:10.1016/B978-0-444-64083-3.00005-1

Singh, V. (2012). Effect of corn quality on bioethanol production. *Biocatalysis and Agricultural Biotechnology*, *1*(4), 353–355. doi:10.1016/j.bcab.2012.06.001

Stolarski, M. J., Krzyzaniak, M., Łuczyński, M., Załuski, D., Szczukowski, S., Tworkowski, J., & Gołaszewski, J. (2015). Lignocellulosic biomass from short rotation woody crops as a feedstock for second-generation bioethanol production. *Industrial Crops and Products*, *75*, 66–75. doi:10.1016/j.indcrop.2015.04.025

Stolarski, M. J., Szczukowski, S., Tworkowski, J., & Klasa, A. (2013). Yield, energy parameters and chemical composition of short-rotation willow biomass. *Industrial Crops and Products*, *46*, 60–65. doi:10.1016/j.indcrop.2013.01.012

Sun, F., & Chen, H. (2008). Organosolv pretreatment by crude glycerol from oleochemicals industry for enzymatic hydrolysis of wheat straw. *Bioresource Technology*, *99*(13), 5474–5479. doi:10.1016/j.biortech.2007.11.001 PMID:18077155

Sun, J. X., Sun, X. F., Sun, R. C., & Su, Y. Q. (2004). Fractional extraction and structural characterization of sugarcane bagasse hemicelluloses. *Carbohydrate Polymers*, *56*(2), 195–204. doi:10.1016/j.carbpol.2004.02.002

Sunna, A., & Antranikian, G. (1997). Xylanolytic enzymes from fungi and bacteria. *Critical Reviews in Biotechnology*, *17*(1), 39–67. doi:10.3109/07388559709146606 PMID:9118232

Takara, D., & Khanal, S. K. (2015). Characterizing compositional changes of Napier grass at different stages of growth for biofuel and biobased products potential. *Bioresource Technology*, *188*, 103–108. doi:10.1016/j.biortech.2015.01.114 PMID:25727997

Teymouri, F., Laureano-Perez, L., Alizadeh, H., & Dale, B. E. (2005). Optimization of the ammonia fiber explosion (AFEX) treatment parameters for enzymatic hydrolysis of corn stover. *Bioresource Technology, 96*(18), 2014–2018. doi:10.1016/j.biortech.2005.01.016

Thompson, J. L., & Tyner, W. E. (2014). Corn stover for bioenergy production: Cost estimates and farmer supply response. *Biomass and Bioenergy*, *62*, 166–173. doi:10.1016/j.biombioe.2013.12.020

Timilsina, G. R., Mevel, S., & Shrestha, A. (2011). Oil price, biofuels and food supply. *Energy Policy*, *39*(12), 8098–8105. doi:10.1016/j.enpol.2011.10.004

Tsubota, Y., & Ichiryu, K. (2005). Development of riding simulator. *Proceedings of the 6th International Conference on Fluid Power Transmission and Control, ICFP 2005, 6*, 686–689.

Tutt, M., Kikas, T., & Olt, J. (2012). Influence of different pretreatment methods on bioethanol production from wheat straw. *Agronomy Research, 10*(1), 269–276.

Tye, Y. Y., Lee, K. T., Wan Abdullah, W. N., & Leh, C. P. (2016). The world availability of non-wood lignocellulosic biomass for the production of cellulosic ethanol and potential pretreatments for the enhancement of enzymatic saccharification. *Renewable & Sustainable Energy Reviews*, *60*, 155–172. doi:10.1016/j.rser.2016.01.072

Vamvuka, D. (2011). Bio-oil, solid and gaseous biofuels from biomass pyrolysis processes—An overview. *International Journal of Energy Research*, *35*, 835–862. doi:10.1002/er

Vamvuka, D. (2011). Bio-oil, solid and gaseous biofuels from biomass pyrolysis processes—An overview. *International Journal of Energy Research*, *35*(10), 835–862. doi:10.1002/er.1804

van der Weijde, T., Alvim Kamei, C. L., Torres, A. F., Vermerris, W., Dolstra, O., Visser, R. G. F., & Trindade, L. M. (2013). The potential of C4 grasses for cellulosic biofuel production. *Frontiers in Plant Science*, *4*(MAY), 1–18. doi:10.3389/fpls.2013.00107 PMID:23653628

Vendruscolo, F. (2014). Biohydrogen Production from Starch Residues. *International Journal of Bioengineering and Life Sciences, 8*(12), 1399–1406. Retrieved from https://waset.org/publications/10000017

Vohra, M., Manwar, J., Manmode, R., Padgilwar, S., & Patil, S. (2014). Bioethanol production: Feedstock and current technologies. *Journal of Environmental Chemical Engineering*, *2*(1), 573–584. doi:10.1016/j.jece.2013.10.013

Vučurović, V. M., & Razmovski, R. N. (2012). Sugar beet pulp as support for Saccharomyces cerivisiae immobilization in bioethanol production. *Industrial Crops and Products*, *39*(1), 128–134. doi:10.1016/j.indcrop.2012.02.002

Wahid, R., Hjorth, M., Kristensen, S., & Møller, H. B. (2015). Extrusion as pretreatment for boosting methane production: Effect of screw configurations. *Energy & Fuels*, *29*(7), 4030–4037. doi:10.1021/acs.energyfuels.5b00191

Wang, Z., Keshwani, D. R., Redding, A. P., & Cheng, J. J. (2010). Sodium hydroxide pretreatment and enzymatic hydrolysis of coastal Bermuda grass. *Bioresource Technology*, *101*(10), 3583–3585. doi:10.1016/j.biortech.2009.12.097 PMID:20089396

Wasserscheid, P., & Keim, W. (2010). Ionic Liquids - New Solutions for Transition Metal Catalysis. *Angewandte Chemie International Edition*, *39*(21), 3772–3789. doi:10.1002/1521-3773(20001103)39:21<3772::AID-ANIE3772>3.0.CO;2-5 PMID:11091453

Wei, S. (2016). The application of biotechnology on the enhancing of biogas production from lignocellulosic waste. *Applied Microbiology and Biotechnology*, *100*(23), 9821–9836. doi:10.100700253-016-7926-5 PMID:27761635

Wijeyekoon, S. L. J., & Vaidya, A. A. (2021). Woody biomass as a potential feedstock for fermentative gaseous biofuel production. *World Journal of Microbiology & Biotechnology*, *37*(8), 1–15. doi:10.100711274-021-03102-6 PMID:34258684

Wilkins, M. R., Suryawati, L., Maness, N. O., & Chrz, D. (2007). Ethanol production by Saccharomyces cerevisiae and Kluyveromyces marxianus in the presence of orange-peel oil. *World Journal of Microbiology & Biotechnology*, *23*(8), 1161–1168. doi:10.100711274-007-9346-2

Wingren, A., Galbe, M., & Zacchi, G. (2003). Techno-economic evaluation of producing ethanol from softwood: Comparison of SSF and SHF and identification of bottlenecks. *Biotechnology Progress*, *19*(4), 1109–1117. doi:10.1021/bp0340180 PMID:12892470

Woiciechowski, A. L., Medeiros, A. B. P., Rodrigues, C., de Souza Vandenberghe, L. P., de Andrade Tanobe, V. O., Dall'Agnol, A., Gazzoni, D. L., & Soccol, C. R. (2016). Feedstocks for biofuels. *Green Energy and Technology*, *0*, 15–39. doi:10.1007/978-3-319-30205-8_2

Wortmann, C. S., Liska, A. J., Ferguson, R. B., Lyon, D. J., Klein, R. N., & Dweikat, I. (2010). Dryland performance of sweet sorghum and grain crops for biofuel in nebraska. *Agronomy Journal*, *102*(1), 319–326. doi:10.2134/agronj2009.0271

Xiu, S., & Shahbazi, A. (2012). Bio-oil production and upgrading research: A review. *Renewable & Sustainable Energy Reviews*, *16*(7), 4406–4414. doi:10.1016/j.rser.2012.04.028

Xu, J., & Cheng, J. J. (2011). Pretreatment of switchgrass for sugar production with the combination of sodium hydroxide and lime. *Bioresource Technology*, *102*(4), 3861–3868. doi:10.1016/j.biortech.2010.12.038 PMID:21194931

Xue, C., & Cheng, C. (2019). Butanol production by Clostridium. In Advances in Bioenergy (1st ed., Vol. 4). doi:10.1016/bs.aibe.2018.12.001

Yan, F., Luo, S., Hu, Z., Xiao, B., & Cheng, G. (2010). Hydrogen-rich gas production by steam gasification of char from biomass fast pyrolysis in a fixed-bed reactor: Influence of temperature and steam on hydrogen yield and syngas composition. *Bioresource Technology*, *101*(14), 5633–5637. doi:10.1016/j.biortech.2010.02.025 PMID:20194019

Yu, J., XuZhang, & Tan, T. (2008). Ethanol production by solid state fermentation of sweet sorghum using thermotolerant yeast strain. *Fuel Processing Technology*, *89*(11), 1056–1059. doi:10.1016/j.fuproc.2008.04.008

Yu, J., Zhang, J., He, J., Liu, Z., & Yu, Z. (2009). Combinations of mild physical or chemical pretreatment with biological pretreatment for enzymatic hydrolysis of rice hull. *Bioresource Technology*, *100*(2), 903–908. doi:10.1016/j.biortech.2008.07.025 PMID:18760920

Zahran, H. H. (1999). Rhizobium -Legume Symbiosis and Nitrogen Fixation under Severe Conditions and in an Arid Climate. *Microbiology and Molecular Biology Reviews*, *63*(4), 968–989. doi:10.1128/MMBR.63.4.968-989.1999 PMID:10585971

Zhang, J., & Zhang, X. (2019). The thermochemical conversion of biomass into biofuels. Biomass, Biopolymer-Based Materials, and Bioenergy: Construction, Biomedical, and other Industrial Applications. doi:10.1016/B978-0-08-102426-3.00015-1

Zhang, Y., Zhao, Y., Gao, X., Li, B., & Huang, J. (2015). Energy and exergy analyses of syngas produced from rice husk gasification in an entrained flow reactor. *Journal of Cleaner Production*, *95*, 273–280. doi:10.1016/j.jclepro.2015.02.053

Zhao, X., Peng, F., Cheng, K., & Liu, D. (2009). Enhancement of the enzymatic digestibility of sugarcane bagasse by alkali-peracetic acid pretreatment. *Enzyme and Microbial Technology*, *44*(1), 17–23. doi:10.1016/j.enzmictec.2008.07.011

Zhao, X., Zhang, L., & Liu, D. (2012). Biomass recalcitrance. Part I: The chemical compositions and physical structures affecting the enzymatic hydrolysis of lignocellulose. *Biofuels, Bioproducts & Biorefining*, *6*(3), 246–256. doi:10.1002/bbb.1331

Chapter 7
Biogas:
Renewable Natural Gas

Bela Khiratkar
Ajeenkya D.Y. Patil University, India

Shankar Mukundrao Khade
https://orcid.org/0000-0002-1974-0709
Ajeenkya D.Y. Patil University, India

Abhishek Dutt Tripathi
Banaras Hindu University, India

ABSTRACT

Biogas is a renewable natural gas used in production of energy which is generated by breakdown of organic matter by anaerobic digestion. It is a composition of methane, carbon dioxide, and some other gases in small quantities. The most common natural resources for production of biogas are aquatic sediments, animal waste, crop residues, wastewater sludge, and many others. Some other sources are landfills, water lagoons, etc., which are a result of human activity. There are several technologies used for biogas production, mainly biodigesters, wastewater treatment plants, and landfill gas recovery systems. Production of biogas is highly dependent on the feedstock availability and the policy support by the government.

INTRODUCTION

Fossil fuels are among the most highly used sources of non-renewable energy which lead to 80% of the global needs. To attain sustainable economic growth, antagonistic and fast-paced deployment of renewable resources is needed. One of the most common sources of energy is fossil fuels. To sustain the non-renewable sources, we need accelerated and antagonistic deployment of efficient energy sources by the second half of the century (IRENA, 2017a). The use of efficient bioenergy conversion and modern technologies are gaining high importance recently which are used to fulfill the demand of global energy

DOI: 10.4018/978-1-6684-5269-1.ch007

needs. Biogas is one of the most cost-effective and ecologically beneficial forms of renewable energy (Deublein & Steinhauser, 2011). It could be a viable option for the world's insatiable energy demand while also reducing waste and greenhouse gas (GHG) emissions. Anaerobic digestion may transform sewage sludge, agricultural and crop leftovers, animal dung and industrial organic wastes, and wastewaters into biogas. The type of the raw materials used for digestion, as well as the pH of the medium, influence the composition of biogas (Muoz et al. 2015). Under anaerobic conditions, bacteria decompose complex organic matter into simpler components, resulting in biogas and fertilizer (Gemmeke, 2009 & Piwowar, 2020). The technique of using anaerobic respiration for the production of biogas has numerous environmental benefits. It helps to treat organic wastes along with introducing green energy, protecting the environment, and also helps to reduce greenhouse gas emissions (Falk, 2011; Theuerl, 2019; Daniel, 2017; Gemmeke, 2009; Stolze, 2017]. There are four phases to the biodegradation of complex organic compounds. Hydrolysis, acidogenesis, acetogenesis, and methanogenesis, are the respective steps (Rohstoffe, 2012). The composition of biogas has a higher concentration of methane which generally makes up to 50% to 75%. This methane content makes the biogas flammable and also generates a blue flame which can be used as an energy source.

The calorific value is fixed on the basis of the concentration of CH_4 as the other substances do not play a major role in the energy content. The calorific value of raw biogas is $21 MJ/m^3$ considering that the density of $1.22\ kg/Nm^3$ CH_4 present in the raw biogas is up to 50% which is also similar to that of air ($1.29\ kg/Nm^3$) (Rutz and Janssen 2008). So as to know more about the various applications of biogas, it is a must that the concentration of CH_4 is appreciably higher (i.e. >90%). This also states that the utilization capacity of any biogas is determined by the composition of the gas. Upgrading biogas is a widely used, well-developed, and easily obtainable technique. It is becoming more widely used around the world, although it is still in its infancy when compared to the biogas producing industry. After treatment, biogas can be used for a variety of purposes, including electricity and heat generation, natural gas grid connection, and automotive biofuel (Rosen T, 2019).

BIOLOGICAL REACTIONS

Hydrolysis, acidogenesis, acetogenesis, and methanogenesis, are the procedures followed after then numerous complex biochemical reactions and as a product of these, biogas is formed (Ghodrat et al., 2018). The complete reaction of conversion of food wastes into biogas is as follows (Deublein & Steinhauser, 2011):

$$C_cH_nO_oN_nS_s + wH_2O ® mCH_4 + nNH_3 + sH_2S + (c\text{-}m)\ CO_2$$

where, $m = ⅛\ (4c+h-2o-3n-2s)$

$w = ¼\ (4c-h-2o+3n+3s)$

Carbohydrates ($C_6H_{12}O_6$), proteins ($C_{13}H_{25}O_7N_3S$), and lipids ($C_{12}H_{24}O_6$) come under the degradable fraction of the food waste.

Individual stages are controlled by various microorganism consortia, which have syntrophic interactions with the microbial consortia participating in the other stages. On hydrolyzing bacteria excrete exo-enzymes such as cellulase, protease, lipase, amylase, and xylanase, initiation of hydrolysis is done. These hydrolytic exo-enzymes are absorbed in the surface of the substrate which leads to the progressive conversion of the polymers to monomers and then further to oligomers (e.g., glucose, fatty acids, glycerol, and amino acids). The enzymes used in all these conversions are water-soluble. Hydrolysis is the stage that is considered to be the restricting stage in the biogas production from the wastewaters which contains a substantial number of recalcitrant compounds such as lignocelluloses.

The launched monomers and oligomers are then degraded to quick-chain fatty acids (propionate, acetate, butyrate, and lactate), alcohols, and gaseous byproducts (NH_3, H_2, CO_2, and H_2S) via the acidogenesis stage. All the unnecessary oxygen present can be broken down and digested with the help of facultative anaerobic microorganisms which are present inside the first 2 degrees. The anaerobic environment present inside proves to be obligatorily for the microorganisms. On the 0.33 level. All the organic materials which were formed in the last stages are then transformed into hydrogen, carbon dioxide, and acetic acid. Methanogens present then produce methane from either CO_2, acetate, or methyl under strictly anaerobic conditions.

PROCESS OF BIOMETHANATION

The anaerobic biological conversion of organics substances occurs in 4 steps. The four levels of methane fermentation are shown in Fig.1.

Figure 1. Steps of anaerobic digestion process- anaerobic treatment of industrial effluents

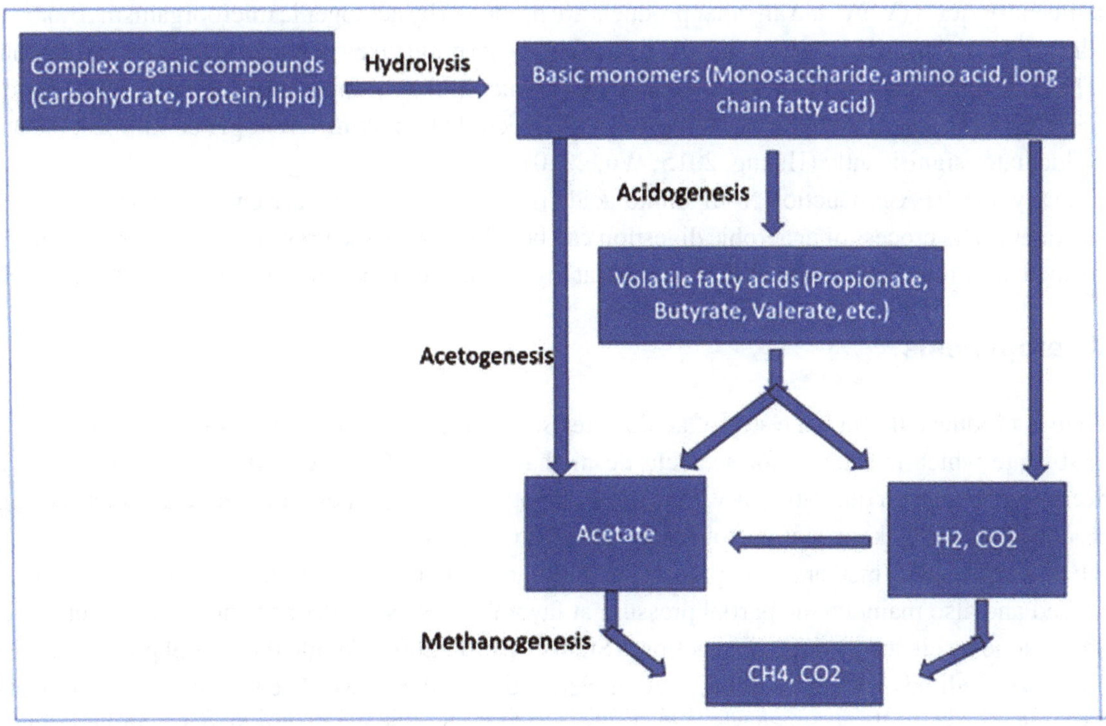

1. Hydrolysis

The anaerobic digesters usually encounter the organic biomass which consists of many compound polymers which the microorganisms have to break down. As they are not pretreated or hydrolyzed, it is expected that the microorganisms break these complex compounds into simpler forms (Gujer W, 1983). In the stage of hydrolysis, the microorganisms break down the organic macromolecules into much smaller compounds which are then used up by the acidogenic bacteria for anaerobic digestion.

In hydrolysis, the hydrolytic bacteria secrete extracellular enzymes which are capable of converting carbohydrates, lipids, and proteins to sugars, long-chain fatty acids, and amino acids respectively (Daniel, 2017). Once the enzymatic cleavage is formed, the products formed in the step of hydrolysis are diffused through the cell membrane of the acidogenic microorganisms (Gemmeke, 2009). For complex substances (such as cellulose, lignin, hemicellulose, etc.) enzymes are added to the process of hydrolysis of carbohydrates as these cannot be hydrolyzed by the microorganisms alone (Stolze, 2017).

Prior research has suggested that hydrolysis is a rate-determining step. It was found in the previous research that methanogenesis can be said to be the rate-determining stage which is dependent on the ratio of the rate of hydrolysis of methanogenic microorganisms (Lin, 2010; Luo, 2012).

For all the steps which take place in hydrolysis, there is an optimum temperature and pH at which the hydrolytic activity is at its peak. At 30-50°C and pH 5-7 are the optimal temperatures and pH respectively for hydrolysis. It has not been found yet whether or not the process of hydrolysis is improved below a pH of 7 (Huang, 2015).

2. Acidogenesis

Volatile Fatty Acids (VFA) and all other products are produced by acidogenic microorganisms by absorbing the products during the process of hydrolysis. This absorption takes place from their cell membranes.

The concentrations of the intermediates of the product in the acidogenesis stage majorly depend on the condition present in the digester. It has been observed that under different pH conditions, the VFA also fluctuates significantly (Huang, 2015; Wu, 2010).

Usually, the VFA production from amino acids is examined in protein-rich wastes such as sewage wastewaters. The process of anaerobic digestion can be inhibited by the presence of ammonia which can be formed after the deamination of amino acids at high concentrations (Kovacs E., 2013; Park J, 2014).

3. Acetogenesis

Acetate is produced through the step of acidogenesis, the original substrate is likely to be rendered into the substrate which is suitable for acetoclastic methanogenesis (Fournier G.P, 2008). The process of conversion of the larger quantities of VFAs along with other intermediates into acetate in which hydrogen is produced as well is known as acetogenesis (Hansen C.L., 2013).

Hydrogenotrophic methanogens present in the digester help to consume the additional hydrogen produced and also maintain the partial pressure at favorable levels for the continuation of acetogenesis which is done by creating exergonic reactions (Stams, A.J.M., 2009). While the partial pressure is being maintained, the lipids present go through a completely different process of acetogenesis via β-oxidation and acetogenesis. While acetogenesis leads to the production of acetate from glycerol and β-oxidation helps in the production of acetate from LCFAs (Cirne D.G., 2007).

4. Methanogenesis

Methanogenesis is the final step of anaerobic digestion. So as to produce methane gas, the intermediates accessible are all consumed by methanogenic organisms present in the batch (Ferry, J.G., 2010) [27]. The group of obligate anaerobic archea represents the methanogenic microorganisms which are the testers for detecting acute sensitivity to oxygen levels. *Echinococcus vannielli* and *Methanococcus voltae* were the microorganisms that were found. 99% of these were killed within 10 hours of them being exposed to oxygen (Kiener A., 1983).

Roughly around ⅔rd of the methane production from acetate is accounted for by acetoclastic methanogenesis, while hydrogenotrophic methanogenesis accounts for the rest ⅓rd methane production. It has also been observed that methanol, methylamines, and formate also undergo methanogenesis (Belay N., 1986 & Lovley D.R., 1983).

When the biogas production stops, which can even take up to 40 days it is considered the end of the methanogenesis in the batch reactors (Theuerl S., 2019). The volatile solid content and the ability to dewater the sludge can help to evaluate the sludge's extent of digestion.

PROCESS PARAMETERS

Along with all the biological processes, the consistency of the conditions provided is important. There are several factors such as temperature changes, changes in the substrate concentration, etc. which can affect the production of biogas. The optimum condition for the microorganisms which affect the degradation of the microbes can depend on two different stages: the first stage is for hydrolysis/ acidification and the second stage is for acetogenesis/ methanation. This is stated, given that the entire process of degradation takes place in the same reaction system and that the appropriate conditions for methanogenesis must be fulfilled, because if they don't receive the optimal conditions then there is no chance of their survival within the mixed culture as their growth rate is also very less along with the high sensitivity to environmental factors.

1. Hydrogen Partial Pressure

So as to get an undisturbed process between the acetogenic bacteria producing H_2 and the H_2 consuming bacteria, a restricted spatial symbiosis is required. The concentration of hydrogen should be properly balanced: for the production of methane, methanogenesis requires a sufficient amount of hydrogen. However, it is proved that the partial pressure of hydrogen surrounding the acetogenic bacteria should be less, causing them to stop the production of hydrogen. The substrate and the bacteria present determine the maximum permissible hydrogen pressure.

2. Concentration of Microorganisms

Since the step of methanogenesis has a lower growth rate, biogas plants usually are considered to have a longer start phase. This can also take around 3 months in which the necessary amount of sludge which is inoculated, will start the plant at its full capacity as fast as it can, as the buildup is not available at the starting phase.

3. Type of Substrate

The rate of anaerobic degradation is measured by the substrate, and it must be taken into consideration for the process of operation and technology. The microorganism's metabolism tends to run out once the substrate component runs short. Hence, it is important to feed the batch lacking substances (carbohydrates, fat, proteins, mineral substances, and trace elements) and also the substrate. For example, sugar hydrolyzes and acidifies in a really short span of time. Whereas cellulose is substantially slow and can take a pretty long time to degrade. Even medium- and heavy-weight degradable compounds hydrolyze and are eventually transformed to methane after 20 days or longer of residence.

4. Light

Methanogenesis is not killed by light, although it is substantially hampered by it. As a result, the creation of methane should take place in complete darkness.

5. Temperature

Mesophilic (35°C) and thermophilic (55°C) are the two different temperatures that are used for the process of anaerobic digestion. The mesophilic digesters are preferred over the thermophilic digesters even though the mesophilic digestion takes place at a much lower temperature and hence is slower than thermophilic digestion and yields less biomass. Mesophilic digesters are preferred due to their lower heater energy costs (Moset V., 2015). Thermophilic digestion takes place at a higher temperature compared to mesophilic digestion. Consequently, the rate of reaction is increased which leads to higher loading rates and hence an increase in the production of biogas (Moset V., 2015 & Shi, X. S., 2017). This thermophilic digestion is believed to have higher levels of pathogen clearance, which could be advantageous in some jurisdictions where pathogenic activity in effluents is regulated (Smith, 2005).

6. pH

There are no pH buffer ions which are released which degradation of the proteins and the hydrocarbons are also easier to acidify. Hence the value of pH is observed to decrease easily. As the degradation of carbohydrates and other substances takes place, the partial pressure of hydrogen increases. With the increase in the hydrogen pressure, the formation of acidic intermediate products is reduced. pH is kept in the neutral range by some natural processes in the fermenter.

Disturbance in the fermentation process can be checked by the drop in the pH and increase in levels of CO_2. The very first sign of acidification occurring is the increase of the propionic acid concentration.

7. Carbon and Nitrogen Ratio

During the process of anaerobic digestion, not much quantity of biomass is produced hence the need for nutrients is also comparatively low. The substrates which have a lesser C/N ratio will lead to an increase in the production of ammonia and inhibition of methane production. The high C/N ratio indicates the lack of nitrogen which leads to negative consequences on the protein formation mechanics and hence changes are observed in the structural, energy, and material of the microorganisms. Therefore, it is necessary to maintain the balance. For example: In China, a mixture of latrine waste and rice straw is used and in Nepal, co-fermentation occurs between human waste and elephant dung.

8. Biogas Removal

The degassing nature of the substrate is of utmost importance whenever the produced gases lead to metabolism inhibition. The transfer of materials from liquid to gaseous phase can be affected by any technical means. The K_La value of the produced biogas can be increased by 50% by this transfer from liquid to gaseous phase and inhibition of the microorganisms by H_2S can be reduced.

INHIBITORS

Composition of the substrate, concentration of inhibitors, and adaptation of bacteria to inhibitors help to determine the inhibition. Less quantity of inhibitors is needed for anaerobic bacteria as they degrade if high amounts of inhibitors are present. It has been observed that usually the anaerobic processes are mostly adaptable and hence relatively insensitive to inhibitors and also to toxic concentrations under other conditions.

REACTOR DESIGNS AND OPERATIONS

Below mentioned are the criterion which needs to be considered before choosing the optimum biogas reactor's design:

- Dry matter (DM) content of the substrate: Wet digestion (DM < 12%) and dry digestion (DM > 12%)
- Material feeding mode: semi-continuous (i.e. addition once per working day), intermittent (i.e. no addition of substrate during dwell period).
- Number of process phases: Single-phase (all steps take place in the same reactor) and two-phase (hydrolysis and methanogenesis take place in separate reactors)
- Process temperature: Psychrophilic (<25 °C), mesophilic (37 to 42 °C) and thermophilic (50 to (50 to 60°C) (Rohstoffe, 2012).

CONCLUSION

Various organic wastes can be easily treated by the process of anaerobic digestion and the production of biogas. There are several environmental and economic benefits of the biogas production. Production of biogas is affected by factors such as the type of organic waste, type of bioreactor, pH, temperature, levels of oxygen etc. The aim of this book chapter is to present a summary of the process of biomethanation, process parameters, and the factors for reactor designs and operations. A review of the literature suggests that there is a considerable prospect of increasing biogas generation in the field. The use of certain inorganic and organic additives to boost biogas production appears to be promising.

REFERENCES

Azman, S. (2016). *Anaerobic Digestion of Cellulose and Hemiceullose in the Presence of Humic Acids*. Wageningen University.

Belay, N., Sparling, R., & Daniels, L. (1986). Relationship of formate to growth and methanogenesis by Methanococcus thermolithotrophicus. *Applied and Environmental Microbiology*, *52*(5), 1080–1085. doi:10.1128/aem.52.5.1080-1085.1986 PMID:3098165

Cirne, D. G., Paloumet, X., Björnsson, L., Alves, M. M., & Mattiasson, B. (2007). Anaerobic digestion of lipid-rich waste—Effects of lipid concentration. *Renewable Energy*, *32*(6), 965–975. doi:10.1016/j.renene.2006.04.003

Daniel-Gromke, J., Rensberg, N., Denysenko, V., Stinner, W., Schmalfuß, T., Scheftelowitz, M., Nelles, M., & Liebetrau, J. (2017). Current Developments in Production and Utilization of Biogas and Biomethane in Germany. *Chemieingenieurtechnik (Weinheim)*, *90*(1-2), 17–35. doi:10.1002/cite.201700077

Dinopoulou, G., Rudd, T., & Lester, J. N. (1988). Anaerobic acidogenesis of complex wastewater: I. The influence of operational parameters on reactor performance. *Biotechnology and Bioengineering*, *31*(9), 958–968. doi:10.1002/bit.260310908 PMID:18584703

Falk, H. M., & Benz, H. C. (2011). Monitoring the Anaerobic Digestion Process. Information Resource Center der Jacobs University.

Ferry, J. G. (2010). The chemical biology of methanogenesis. *Planetary and Space Science*, *58*(14-15), 1775–1783. doi:10.1016/j.pss.2010.08.014

Fournier, G. P., & Gogarten, J. P. (2008). Evolution of Acetoclastic Methanogenesis in Methanosarcina via Horizontal Gene Transfer from Cellulolytic Clostridia. *Journal of Bacteriology*, *190*(3), 1124–1127. doi:10.1128/JB.01382-07 PMID:18055595

Gemmeke, B., Rieger, C., Weiland, P., & Schröder, J. (2009). *Biogasanlagen im Vergleich*. Fachagentur Nachwachsende Rohstoffe E.V.

Gómez, D., Ramos-Suárez, J. L., Fernández, B., Muñoz, E., Tey, L., Romero-Güiza, M., & Hansen, F. (2019). Development of a Modified Plug-Flow Anaerobic Digester for Biogas Production from Animal Manures. *Energies*, *12*(13), 2628. doi:10.3390/en12132628

Gujer, W., & Zehnder, A. J. B. (1983). Conversion processes in anaerobic digestion. *Water Science and Technology*, *15*(8-9), 127–167. doi:10.2166/wst.1983.0164

Hansen, C. L., & Cheong, D. Y. (2013). Agricultural Waste Management in Food Processing. In *Handbook of Farm, Dairy, and Food Machinery Engineering*. Academic Press. doi:10.1016/B978-0-12-385881-8.00023-9

Huang, W., Wang, Z., Zhou, Y., & Ng, W. J. (2015). The role of hydrogenotrophic methanogens in an acidogenic reactor. *Chemosphere*, *140*, 40–46. doi:10.1016/j.chemosphere.2014.10.047 PMID:25466820

Kiener, A., & Leisinger, T. (1983). Oxygen sensitivity of methanogenic bacteria. *Systematic and Applied Microbiology*, *4*(3), 305–312. doi:10.1016/S0723-2020(83)80017-4 PMID:23194731

Kovács, E., Wirth, R., Maróti, G., Bagi, Z., Rákhely, G., & Kovács, K. L. (2013). Biogas production from protein-rich biomass: Fed-batch anaerobic fermentation of casein and of pig blood and associated changes in microbial community composition. *PLoS One*, *8*(10), 8. doi:10.1371/journal.pone.0077265 PMID:24146974

Li, Y., Park, S. Y., & Zhu, J. (2011). Solid-state anaerobic digestion for methane production from organic waste. *Renewable & Sustainable Energy Reviews*, *15*(1), 821–826. doi:10.1016/j.rser.2010.07.042

Lin, L., Yan, R., Liu, Y., & Jiang, W. (2010). In-depth investigation of enzymatic hydrolysis of biomass wastes based on three major components: Cellulose, hemicellulose and lignin. *Bioresource Technology*, *101*(21), 8217–8223. doi:10.1016/j.biortech.2010.05.084 PMID:20639116

Lovley, D. R., & Klug, M. J. (1983). Methanogenesis from Methanol and Methylamines and Acetogenesis from Hydrogen and Carbon Dioxide in the Sediments of a Eutrophic Lake. *Applied and Environmental Microbiology*, *45*(4), 1310–1315. doi:10.1128/aem.45.4.1310-1315.1983 PMID:16346271

Luo, K., Yang, Q., Li, X., Yang, G., Liu, Y., Wang, D., Zheng, W., & Zeng, G. (2012). Hydrolysis kinetics in anaerobic digestion of waste activated sludge enhanced by α-amylase. *Biochemical Engineering Journal*, *62*, 17–21. doi:10.1016/j.bej.2011.12.009

Ma, J., Frear, C., Wang, Z., Yu, L., Zhao, Q., Li, X., & Chen, S. (2013). A simple methodology for rate-limiting step determination for anaerobic digestion of complex substrates and effect of microbial community ratio. *Bioresource Technology*, *134*, 391–395. doi:10.1016/j.biortech.2013.02.014 PMID:23489573

Moset, V., Poulsen, M., Wahid, R., Højberg, O., & Møller, H. B. (2015). Mesophilic versus thermophilic anaerobic digestion of cattle manure: Methane productivity and microbial ecology. *Microbial Biotechnology*, *8*(5), 787–800. doi:10.1111/1751-7915.12271 PMID:25737010

Park, J., Park, S., & Kim, M. (2014). Anaerobic degradation of amino acids generated from the hydrolysis of sewage sludge. *Environmental Technology*, *35*(9), 1133–1139. doi:10.1080/09593330.2013.863951 PMID:24701908

Piwowar, A. (2020). Agricultural Biogas—An Important Element in the Circular and Low-Carbon Development in Poland. *Energies*, *13*(7), 1733. doi:10.3390/en13071733

Rohstoffe, F. N. (2012). Guide to Biogas from Production to Use. Federal Ministry of Food, Agriculture and Consumer Protection, Fachagentur Nachwachsende Rohstoffe E.V.

Rosén, T., & Ödlund, L. (2019). System Perspective on Biogas Use for Transport and Electricity Production. *Energies*, *12*(21), 4159. doi:10.3390/en12214159

Sarker, S., Lamb, J. J., Hjelme, D. R., & Lien, K. M. (2019). A Review of the Role of Critical Parameters in the Design and Operation of Biogas Production Plants. *Applied Sciences (Basel, Switzerland)*, *9*(9), 1915. doi:10.3390/app9091915

Shi, X.-S., Dong, J.-J., Yu, J.-H., Yin, H., Hu, S.-M., Huang, S.-X., & Yuan, X.-Z. (2017). Effect of hydraulic retention time on anaerobic digestion of wheat straw in the semicontinuous continuous stirred-tank reactors. *BioMed Research International*, *2017*, 1–6. doi:10.1155/2017/2457805 PMID:28589134

Smith, S. R., Lang, N. L., Cheung, K. H. M., & Spanoudaki, K. (2005). Factors controlling pathogen destruction during anaerobic digestion of biowastes. *Waste Management (New York, N.Y.)*, *25*(4), 417–425. doi:10.1016/j.wasman.2005.02.010 PMID:15869985

Stams, A. J. M., & Plugge, C. M. (2009). Electron transfer in syntrophic communities of anaerobic bacteria and archaea. *Nature Reviews. Microbiology*, *7*(8), 568–577. doi:10.1038/nrmicro2166 PMID:19609258

Stolze, Y., Bremges, A., Maus, I., Pühler, A., Sczyrba, A., & Schlüter, A. (2017). Targeted in situ metatranscriptomics for selected taxa from mesophilic and thermophilic biogas plants. *Microbial Biotechnology*, *11*(4), 667–679. doi:10.1111/1751-7915.12982 PMID:29205917

Theuerl, S., Herrmann, C., Heiermann, M., Grundmann, P., Landwehr, N., Kreidenweis, U., & Prochnow, A. (2019). The Future Agricultural Biogas Plant in Germany: A Vision. *Energies*, *12*(3), 396. doi:10.3390/en12030396

van Lier, J. B. (1995). *Thermophilic Anaerobic Wastewater Treatment: Temperature Aspects and Process Stability.* Available online: https://edepot.wur.nl/200091

Van Lier, J. B., Mahmoud, N., & Zeeman, G. (2008). Anaerobic Wastewater Treatment. In *Biological Wastewater Treatment: Principles, Modeling and Design* (pp. 401–442). International Water Association.

Verma, S. (2002). *Anaerobic Digestion of Biodegradable Organics in Municipal Solid Wastes*. Columbia University.

Wu, X., Yao, W., & Zhu, J. (2010). Effect of pH on continuous biohydrogen production from liquid swine manure with glucose supplement using an anaerobic sequencing batch reactor. *International Journal of Hydrogen Energy*, *35*(13), 6592–6599. doi:10.1016/j.ijhydene.2010.03.097

Chapter 8
Algae as Superfood

Shital Uddhav Giri
Ajeenkya D.Y. Patil University, India

Namdev Gopal Krishna Hadapad
Ajeenkya D.Y. Patil University, India

Aditya Akhade
Ajeenkya D.Y. Patil University, India

Parth Bhilare
Ajeenkya D.Y. Patil University, India

ABSTRACT

With the increasing population, there are problems like the shrinking of arable land, pollution, and feeding the entire growing population high-quality food and reaching consumer demand, and the list goes on. Novel superfoods are rich in nutrition and produced with the most sustainable methods, including algae, specifically microalgae. Microalgae have been studied for decades. Microalgae are one of the largest and most poorly understood organisms on planet earth, but in recent years, the research has been increased for the search of renewable and sustainable energy sources. Microalgae don't require much land for cultivation and thus do not compete with conventional agricultural land. Microalgae can double their size in 24 hours. CO2, solar rays, inorganic nutrients, and water are the basic requirements for them to grow. There are microalgae and macroalgae (e.g., seaweed) that grow in saline and hypersaline water. This chapter presents the promising potential of algae (microalgae and macroalgae) as a superfood.

BACKGROUND

In the year 1753 Carolus Linnaeus a Swedish botanist coined the term algae. Algae come under a division of the plant kingdom called Thallophyte which means, simple plants without roots stems, or leaves known as thallus. Algae are eukaryotic in nature but blue-green algae (cyanobacteria) are prokaryotic in nature. The composition of the eukaryotic algal cell consists of the cell wall, nucleus, endoplasmic reticulum, mitochondria, Golgi bodies, stigma or eyespot, vacuoles, flagella, and plastids.

DOI: 10.4018/978-1-6684-5269-1.ch008

COMPOSITION OF ALGAE

The cell surface plays a very vital role in both the external and internal sides of the cell. Various functions of cell surface include cell interaction, cell signaling, excretion and secretion of various compounds, protection, and servers as an osmatic barrier. The cyanobacteria are enveloped by two parts of the gelatinous sheath and cell wall. The sheath is very slimy in nature which acts as a protective layer. Due to their slimness blue-green algae are also called my xophyceae. Algae also consist of a cell wall outside the plasma membrane, it presents a granular structure and lacks microfibrils. The cell wall is composed of Glycosamine, muramic acid and diaminopimelic acid. It has a structure of gram-negative type.

Protoplast

Cyanobacteria lack chloroplast but have Chromoplasm. Protoplasm is totally differentiated into two regions.

1. Centro Plasm (Central Area)

It is a clear and transparent region of the protoplast that contains chromatin or genetic material.

2. Chromoplasm (Peripheral Denser Region)

In this region, thylakoids are present which consists of chlorophyll A responsible for providing green color in blue-green algae, and several types of carotenoids are embedded into it. thylakoids are not restricted to peripheral reagion they are found scattered all over the cell (*Gloeotrichia* sp). phycobilins are on the surface of thylakoids. The best-known phycobilins are phycoerythrin and phycocyanin. Phycobilins are water-soluble whereas chloroplast and carotenoids are fat-soluble. phycocyanin C is the characteristic of blue-green algae and is responsible for the blue color in blue-green algae.

Gas Vacuoles

Tiny, transparent structures which are commonly found in bacteria and arches. Gas vacuoles are responsible for the floating of algae. Nitrogen and oxygen are majorly found gases but also have a minute amount of CO and CO2 which is liberated from metabolic activity. Gases are produced when there is not enough sunlight reaching the algae, especially in deep-underwater algae (*Trichodesmium sp*), they have low photosynthesis, low osmotic pressure, low production of carbohydrates and therefore increasing the density of the cell. Under these circumstances, gases are produced at a high rate and gain buoyancy.

Eukaryotic Cell Composition

The eukaryotic algal cell is simple, unicellular, motile, freshwater algae. It consists of membrane-bound organelles.

Cell Wall

Most of the algae have a cell wall made up of multiple microfibrils like cellulose, callose, pectin, chitin, and hemicellulose. In a few marine forms of algae, lime and silica are found as cell wall components because of the environment they have adapted to grow.

Plastids

Plastids are pigments that contain photosynthetic pigments. They are the site for the manufacture and storage of chemical compounds. Plastids in algae are slightly different from plant plastids that is the presence of pyrenoids in algae and absent in plants. Plastids are differentiated based on the pigment they contain, and they are chloroplast and chromophore. Chloroplast consists of chlorophyll A and chlorophyll B (e.g., *chara* and *volvox*). Chlorophyll A is common for all the algae present for photosynthetic reaction. Chromophores have chlorophyll A and other carotenoids (*sargassum, laminaria*), and these carotenoids are responsible for a different color. Carotenoids are light-harvesting pigments, they capture the light and pass it on to the light reaction centre (chlorophyll A).

Flagella

It provides motility to the algae. These are emerging from the cytoplasm. Different variations of flagella can be found. Whiplash flagella (acronematic) have a smooth surface. While tinsel flagella (pantonematic) have a rough surface with many rows of minute hair-like structures grown to the main flagellum strand. Algae can be either of this flagellum are they may have both. isokont flagella are when flagella are of similar length and heterokont flagella are when flagella are not equal in length.

Eyespot

Photoreceptive organs are found associated with flagella and chloroplast (e.g., *Chlamydomonas*) and help the organism to sense the light and move in the direction of light.

Figure 1. Cellular structure of algae

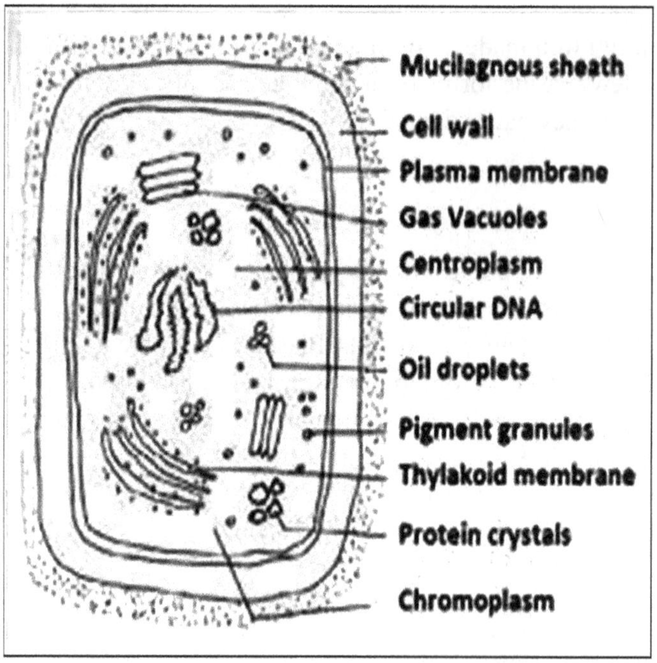

COMPOSITION OF SPIRULINA

The composition of spirulina is unique and extraordinary that it is called a superfood. Spirulina is composed of carbohydrates, protein, fats, vitamins, pigments and minerals. They also help in combating various kinds of diseases like anemia, vascular diseases, obesity, hepatotoxicity and certain cancers.

Protein

The protein content in spirulina is high which is 50 to 70 g/100 g of dry matter. Spirulina protein id totally different from other organisms which provide protein because spirulina doesn't contain any hard envelope like cellulose but has a very fragile envelope and has higher digestibility up to 83%-90% which makes it easier for people to consume and digest. It also contains Isoleucine, leucine, lysine, methionine, phenylalanine, threonine, tryptophan, and valine are necessary amino acids that cannot be synthesized by the human body. People with a vegetarian diet don't usually have sufficient lysine and methionine. It is known a complete protein because of its amino acid composition as it has every amino acid (essential and nonessential). The protein content of various foods: Eggs 13%, Soybean 42%, Grain 29%, Peanuts 25.3%, Spirulina 60%-70%.

Carbohydrate

Carbohydrates are considered a great source of quick energy for our body. Spirulina consists of 10%-15% in the form of polysaccharides. Glucose, fructose, and sucrose - are found in little quantity. Carbohydrates offer 50-70 percent of the body's total energy requirements. Meso inositol phosphate is a carbohydrate present in larger and it consists of c phosphorus and inositol.

Lipids

Water insoluble lipids are soluble in organic solvents such as ether and chloroform. Dairy, cheeses, olives, walnuts, canola, soybeans, and flaxseed are the most prevalent sources of lipids. Spirulina has a lipid content of 5-6% of its dry mass. Many of the food sources which have fatty acids content are majorly meat products like beef, chicken, fish, egg, etc. spirulina is a vegetarian source that has the highest amount of fatty acid in it. Spirulina has gamma- linolenic, linoleic, and oleic acids in them. Gamma-linolenic acid is known to have the ability to suppress cancer growth.

Vitamins

Vitamins are organic compounds required by the body in the trace amount to perform a specific cellular metabolic activity to keep the body in functional condition. They cannot be synthesized by the human body, and therefore they must be supplied through the diet. Spirulina contains a vitamin known as vitamin B12 (COBALMINE) it is found in coordination complex with cobalt. Spirulina has the highest content of vitamin B12 compared to other seaweeds found in the water. Usually, vitamin b12 is prominently found in meat-based food, and people who consume a vegetarian diet lack this vitamin. Beta-carotene is abundant in spirulina., it contains700-1700 mg/kg. Recommended daily allowance of vitamin A per day is 1 mg per day. Consumption of 1-2g of spirulina is enough to gain enough beta-carotene.

Minerals

Minerals are essential components for the body to function properly and are required in trace amounts and are found in all kinds of food we consume. Particularly in iron, calcium, phosphorus and potassium are found in high amounts. Food with high iron content is hard to find, Cereals are considered one of the good sources of food with high iron content in them (150-250 mg/kg) but they also contain phytates, and oxalates are known to suppress the bioavailability of iron.

Figure 2. Spirulina cultivated at lab scale level

Table 1. Amino acid profile of spirulina

Amino Acids	% of Total Amino Acids
Alanine	7.9
Arginine	7.3
Aspartic Acid	10.1
Cystine	1.2
Glutamic Acid	14.7
Glycine	5.4
Histidine*	1.9
Isoleucine*	5.6
Leucine*	8.7
Lysine*	5.3
Methionine*	2
Phenylalanine*	4.9
Proline	4.2
Serine	5.2
Threonine*	3.4
Tryptophan*	1.6
Tyrosine	4.5
Valine*	6.1

Algae are main producers that provide a rich source of several nutrients. While the high protein content of diverse species is one of the primary reasons for considering them as an unusual source protein, microalgae oils rich in several PUFAs appear to be especially good for youngsters, pregnant women, vegetarians, and people with seafood allergies. Because algae are a rich source of vitamins, minerals, antioxidants, and natural colorants, incorporating the entire biomass into food and feed might give color, increase nutritional value, and enhance texture and oxidation resistance.

Combining microalgae biomass into traditional foods is a means to create novel items that are both appealing and healthful. Even when used in small quantities in the nutrition of various animals, algae have been given credit with immune health, rising weight, the amount of eggs, reproductive performance, or reducing cholesterol, indicating the potential of future farming methods to improve the quality of meat and eggs. It's also not unexpected that they're important in aquaculture because they're food supply for these creatures. Finally, microalga Spirulina, humanity's greatest and most perfect resource of organic nutrition, has become a nutritious food globally due to its high protein content and many bioactive components.

heterotrophic. In a scientific sense, the term algae refer to a group of reduced plants that contain chlorophyll in their cells and are common in oceanic biotopes, since they are also found in other environments (Blazencic, 2007). They are divided into macroalgae (macroscopic algae) and microalgae (microscopic algae) based on their size.

Blue-green algae, often known as cyanobacteria due to their prokaryotic cell type, are a form of microalgae that has been identified as one of the most promising living organisms for the separation of new and biochemically active bioactive substances (Singh et al., 2005). Algae are grown and utilised as food all around the world. They are rich in vitamins and minerals, enzymes, protein, unsaturated fats, antioxidants, and other nutrients. (Pulz and Gross, 2004; Svircev, 2005; Blazencic, 2007; Gouveia et al., 2008b).

Chemical Composition of Algae

One of the key reasons for considering microalgae as an unusual source of proteins is their high protein content, and because they can synthesize all the amino acids, they could also be a supply of essential amino acids. Furthermore, in comparison to normal plant proteins, the general standard of most studied algae is comparable, if not greater (Becker, 2007). Dried Spirulina biomass, for example, provides all necessary amino acids and has roughly 68 percent protein, which is three times more than beef. Chlorella, another microalga, has 50-60% protein, which is equivalent to yeast, soy flour, and milk powder proteins in terms of quality (Blazencic, 2007). Milovanovic et al. are a group of researchers that came up with a novel way to solve Milovanovic et al. (2012) looked examined the protein content of dry biomass from different cyanobacterial strains from Serbia, and found that they all have very high levels, ranging from 42.8 percent to 76.5 percent. Algae also contain a lot of fibre. Polysaccharides such as agar, alginates, and carrageenan, which are widely utilized in the food sector as gelling or thickener agents in marmalade, ice - cream, and jellies, are the most commercially important products from algae. Because they boost the human immune system, some algal polysaccharides are also beneficial in pharmacology.

ALGAE IS A SOURCE OF VITAMINS AND MINERALS

Algae is abundant in vitamins and minerals, making it a good choice for supplements. Most cultivated plants have lower vitamin content than chlorella species (Blazencic, 2007). Furthermore, the Spirulina genus has roughly ten times more -carotene than any other plant or animal nutritional source, including carrot (Mohammed et al., 2011), as well as much more vitamin B12. In comparison to green algae, leafy greens, and liver, this genus has the greatest vitamin E, thiamine, cobalamin, biotin, and inositol (Gantar and Svircev, 2008). Tocopherol is produced in large quantities by a number of microalgal species (T, the most physiologically active form of vitamin E). Rodriguez Zavala et al. (2010) reported that after 120 hours, the synthesis of -tocopherol in the heterotrophically cultivated microalga Euglena gracilis reached 3.70.2 mg/g, which is roughly 13, 18, 95, and 56 times more than sunflower, soybean, olive, and maize, respectively (some of the most common natural sources of vitamin E).

Algae as a Component of Functional Food

In industrialized nations, the food is rich in calories, which, when combined with a contemporary lifestyle, contributes to health issues such as obesity, heart disease, diabetes, and other ailments. As a result, there is a need for food items that can improve health by supplementing the diet with vitamins, minerals, PUFAs, and other nutrients, and consumers are increasingly preferring natural components over artificial ones. Algae are a unique natural supply of biologically active chemicals, yet they are under-explored. Beta1,3-glucan, a powerful immunological stimulant, free radical scavenger, and blood lipid reduction, appears to be the most essential ingredient in Chlorella (Spolaore et al., 2006). The antibacterial, antiviral, anticancer, and food additive properties of cyanobacteria have long been known (Singh et al., 2005), and drug development is the most potential element of microalgal biotechnology (Spolaore et al., 2006).

Spirulina is a freshwater and saltwater organism. It's a cyanobacterium, a single-celled microorganism that's also known as blue-green algae. Photosynthesis is a process through which cyanobacteria, like plants, may make energy from sunlight. Spirulina was first ingested by the Aztecs, but it gained popularity again after NASA claimed that it might be cultivated in space for astronauts to consume (1). Spirulina is usually taken in dosages of 1–3 grammes per day, however, amounts as high as 10 grammes have been shown to be helpful. The nutrients in this small alga are plentiful. A spoonful of dry spirulina powder (7 grammes) includes:

- Protein: 4 grams
- Vitamin B1 (thiamine): 11% of the RDA
- Vitamin B2 (riboflavin): 15% of the RDA
- Vitamin B3 (niacin): 4% of the RDA
- Copper: 21% of the RDA
- Iron: 11% of the RDA

It also has enough levels of magnesium, potassium, and manganese, as well as trace amounts of nearly all other nutrient you require.

Furthermore, the same quantity has just 20 calorie and 1.7 grammes of digestible carbohydrates. Spirulina may be the most nutrient-dense food on the world, gramme for gramme. Spirulina protein is believed to be of great quality, equivalent to that of eggs. It contains all the necessary amino acids that your body need. Spirulina is frequently said to contain vitamin B12, however this is untrue. It contains pseudo vitamin B12, which hasn't been proven to work in people.

Properties

Antioxidant and anti-powerhouse Chronic inflammation, which leads to cancer and other disorders, might be triggered by this damage. Spirulina is high in antioxidants, which defends the body from oxidative damage. Phycocyanin is the principal active ingredient. Spirulina's unusual blue-green hue is due to this antioxidant component. It has the ability to lower bad Low density lipoprotein and triglyceride levels. Cardiovascular disease is the biggest cause of mortality worldwide. A variety of risk factors have been linked to an increased risk of heart disease. Spirulina, it turns out, has a positive impact on a number of these traits. Total cholesterol, bad LDL cholesterol, and triglycerides can all be reduced, while good HDL cholesterol can be increased, for example. Two grams of spirulina each day significantly improved these indicators in a trial of 25 persons with type 2 diabetes. Several additional research have demonstrated that larger dosages of 4.5– 8 grammes per day had beneficial benefits. Oxidation of Bad LDL Cholesterol is prevented. Your body's fatty structures are vulnerable to oxidative damage. This is called as lipid peroxidation, and it is a major cause of a number of severe disorders. The oxidation of bad LDL cholesterol, for example, is a crucial phase in the development of cardiovascular disease. In humans and animals, the antioxidants in microalgae appear to be particularly efficient at reducing lipid peroxidation. 8 grammes of spirulina each day lowered oxidative damage indicators in 37 persons with type 2 diabetes in research. It also raised blood levels of antioxidant enzymes.

The Impact of COVID-19 on the Spirulina Market

Many economies throughout the world have been negatively impacted by the COVID-19 outbreak. Many businesses were badly impacted by government attempts to battle the COVID-19 epidemic, such as statewide lockdowns and quarantines. COVID-19 continues to have an influence on the majority of food manufacturers in the F&B industry. Many businesses have been forced to close across Asia, the United States, Europe, and the entire world due to social alienation, self-isolation, and town lockdowns. Consumer food and beverage producers are facing issues such as decreased demand and disturbed supply networks. The food sector has suffered a modest setback as a result of the trade embargoes implemented by a number of nations. This situation has had a detrimental influence on the confectionery sector since customers are only purchasing the most basic items. Furthermore, government agencies enacted a slew of rules and regulations governing the acquisition of basic items. The majority of customers are just purchasing necessities, which has resulted in a considerable reduction in sweets consumption. According to trade organizations in the nutraceutical industry, there has been a major effect on many locations in the global supply chain for functional foods and supplements, which has lowered spirulina biomass consumption among end-users and impacted the spirulina market's growth.

Humans have long ingested microalgae since Spirulina was gathered from Lake Texcoco and sold in Tenochtitlan markets as early as the sixteenth century (today Mexico City). Microalgae is now being used in a wide range of culinary products. Microalgae is used by the majority of these companies as a marketing technique or as a coloring agent. Spirulina (and chemicals produced from it) has the potential to be employed as components in the production of new functional meals, which is one of the most popular culinary trends. Numerous human intervention studies have shown that Spirulina has the potential to be employed in the prevention and treatment of metabolic syndrome illnesses. T. Spirulina consumption has been linked to a variety of health advantages, as well as some of one of most significant substances obtained from Spirulina.

Figure 3. Microscopic photo of cultivated spirulina

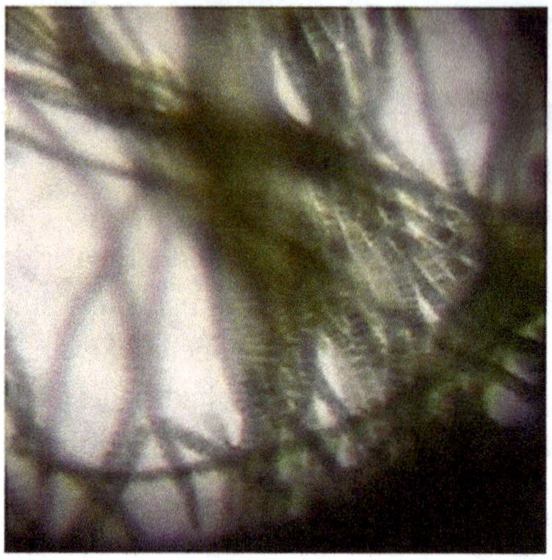

Table 3. The main fatty acids in two spirulina strains, Spirulina maxima and Spirulina platensis

Fatty Acids	Spirulina Platensis (% of Total Fatty Acids)
Alpha-linolenic (18:3 omega-3)	traces
Gamma-linolenic (18:3 omega-6)	40.1
Linoleic (18:2 omega-6)	12
Oleic (18:1 omega-6)	16.6
Stearic (18:0)	1.7
Palmitoleic (16:1 omega-6)	3.8
Palmitic (16:0)	25.8

Table 4. Spirulina's vitamin content and adult daily vitamin needs

Vitamins	Content (mg/kg)	Daily Requirement (mg) (Adult, 24-25 Years)
C	traces	15-30
Biotin	0.05	0.1-0.3
Pantothenate	4.6-25	6-10
Folate	0.50	0.40
Niacin	130.00	20.00
B12	1.5-2.0	0.003
B6	30-46	1.80
B1	34-50	1.50

Figure 4. Chocolates made from spirulina cultivated at lab

Table 5. Typical dry spirulina analysis

Minerals	Content in Spirulina (mg/kg)	Required Adult Daily Dose (47) (mg/day)
Potassium*	6400-15400	3500
Sodium	4500	500
Manganese	25-37	5
Chrome	2.8	0.5-2
Copper	8-10	1.5-3
Zinc	21-40	15
Iron*	580-1800	18
Magnesium	2000-2900	250-350
Phosphorus*	6700-9000	1000
Calcium*	1300-14000	1200

Table 6. Composition of Algae

Algae	Protein	Carbohydrates	Lipids
Aphanizomenon flos-aqua	62	23	3
Chlorella pyrenoidosa	57	26	2
Chlorella vulgaris	51-58	12-17	14-22
Porphyridium creuentum	28-39	40-57	9-14
Schizochytrium sp.	-	-	50-77
Arthrospira maxima	60-71	13-16	6-7

BIOMASS

Biomass is indeed the plant material produced by the photosynthetic process of CO2 in the air, water, and sunshine to make carbohydrates, which are the basic building block of biomass. In most cases, photosynthesis transforms less than 1% of available sunlight into chemical energy. The chemical bonds of the building elements of biomass store the sun energy that drives photosynthesis. Carbon is oxidized to generate CO2 and water if biomass is managed effectively, either chemically or biologically, by retrieving the energy contained in chemical bonds and the following 'energy' product mixed with oxygen. Because the co2 is again accessible to generate fresh biomass, the process is circular.

The physicochemical features of the big molecules that make up a specific form of biomass determine its value. For millennia, man has taken use of the energy held in these chemical bonds by igniting biomass as a biofuel and consuming plants for their sugar and starch nutritional content. Fossilized biomass has lately been mined as coal and oil. However, because the conversion of biomass to fossil fuels takes millions of years, they are not renewable on a timescale that humans can use.

Burning fossil fuels turns old biomass into new CO2, contributing to the greenhouse effect and depleting a nonrenewable resource. Because replanting harvested biomass assures that Carbon dioxide is collected and regenerated for a loop of new growth, there is no additional carbon dioxide released into the atmosphere when fresh biomass is burned. The time lag between the initial release of Carbon dioxide from fossil fuel burning and its eventual absorption as biomass, which can take several years, is a significant element that is often disregarded when contemplating the use of biomass to help mitigate global warming. One of the difficulties that the industrialized world faces is the necessity to notice this time lag and take necessary measures to reduce the lag period. A similar issue confronts the developing world, which uses its biomass resources for fuel, but it does not conduct a replacement planting program. For commercial energy farming, a variety of plants have indeed been proposed or are being studied. Woody crops, dwarf varieties crops (all perennial crops), starch and sugar crops, and oilseeds are all potential energy crops.

Biomass: Impact on the Carbon Cycle and Greenhouse Gas Emission

Biomass has the longest history of any source of energy, yet it still delivers enough energy to supply 11% of the world's fundamental energy demands after 44 years. The majority of these are utilized in underdeveloped nations, where low-quality wood is used for cooking and heating, accounting for 38 EJ. The remaining 6 EJ are utilized in industrialized nations for current technology heating and power generating.

Heating was formerly the most popular way of extracting energy from biomass. However, technological advancements have made more efficient and sustainable energy conversion processes possible. Mineral oils may be converted into practically any use using solids, liquids, or gaseous biofuels.

The amount and frequency with which biofuels are utilized varies substantially. In example, biomass provides just 3% of total basic energy demands in industrialized regions, compared to more than 40% in undeveloped countries. Table II demonstrates the range of biomass contributions to energy demands in various nations, ranging from 1 percent in Japan, Ireland, and the Netherlands to 94 percent in Tanzania and Ethiopia. Finland, Sweden, and Austria are among the industrialized nations that get a lot of energy from bioenergy, thanks to the use of wood, wood waste, and grass as fuel.

Biomass Production by Using Microalgae

Microalgae may thrive in a variety of conditions, including clean, saline, and polluted water, and they can even be used to clean. Agricultural, municipal, and industrial wastewater discharge a variety of living and non-living organisms into the environment. These liquid waste streams can be repurposed as fertilizer sources for tiny algae growth systems. The symbiotic interaction between autotrophic growth and heterotrophic/mixotrophic microalgae can aid in improving water treatment efficiency.

This method minimizes carbon dioxide emissions from microalgae biofuels and biogas production technologies while saving power. The extraction of phosphorus and nitrogen from polluted water may be linked with microalgae extraction for the manufacture of biofuels, nutritional supplements, or other bio-based compounds. As a food supply for tiny algae culture while generating bioenergy and high-value bioproducts

Biomass From Forests

Forests cover 38% of EU territory and provide a variety of ecosystem services, including carbon storage and depletion, accommodation, water management (quality, quantity, flow), air quality standards, soil depletion control, and recreation, wood, and non-timber goods. Forest growth differs greatly from that of the rest of the EU in several aspects, including site circumstances, soil, age formation, and species formation (70 percent of forests are made up of two or more tree species).

Forest management functions are driven by forest management principles, EU forestry variety, and the services they may give, impacting the potential supply of woody biomass to fulfil demand for wood-based goods and biofuels. Other elements have an impact on wood's integration inside the EU, which is crucial. Forest ownership is linked to data changes (60 percent of forests are owned by private individuals who are progressively moving to rural regions), forest area division (16 million forestland owners), and the economic advantages of forest management. This chapter focuses on biomass wood from European forests, as well as forest produce and markets. It's worth noting that forest biomass, which includes non-wood forest resources, plays a major role in the bioeconomy of several EU areas region in EU applied in this chapter to summarize forest-based sector data.

Biomass Properties

1. Moisture content
2. Calorific value
3. Proportion of fixed carbon and volatile matter
4. Ash / residue content
5. Alkali metal content
6. Cellulose / lignin ratio
7. Bulk density

Yields

The quantity of dry matter generated by each form of biomass in each production facility affects the available land area's potential manufacturing capacity, or yield. When the crop's production is quantified in dmt/ha and paired with HHV biomass, the crop's energy yield may be estimated. The highest energy output for biomass types.

The use of hybrid plants to increase biomass yields has been the subject of extensive study and development. Psychological tests on mixed popular species yielded 43 dmt / ha / an in in the US Pacific Northwest, and a reported eucalyptus harvest of 39 dmt / ha / an in Brazil (Hislop and Hall, 1996). High dmnt / ha varieties, which are correctly placed in a high-quality agricultural region, e.g., exception, and variety capable of generating high yields of dmt / ha, planted in a separate area, are anticipated to be the two primary types of variations developed for the cultivation of biomass power fields.

Biomass, a traditional third-world source of energy, has the potential to help the developed world reduce the environmental impact of burning fossil fuels to create power, but only if large-scale reforestation projects are implemented quickly. Biomass is a renowned source of renewable energy that is seen as a solution to help reduce global warming by replacing the use of fossil fuels: by 2010, renewable energy sources will provide up to 10% of the UK's electricity consumption.

The principal forms of biomass of interest for generating power are woody plants, herbaceous plants, and grasses, with the C. plant species garnering the most attention. Plants store chemical energy in the form of cellulose, hemicellulose, and lignin, with the amounts varying depending on the kind of plant.

TYPE OF BIOMASS

Wood, sawdust, straw, seed debris, compost, paper waste, household trash, polluted water, and other materials are all examples of biomass. The majority of study is currently focused on unfavorable energy and the resulting power. Common fuels will be supplemented by biomass. The primary goal of this study was to examine the research background and biomass styles of renewable energy over the last 40 years (from 1978 to 2018) in order to aid the research community in understanding future trends, as well as the international context and context, all of which provided basic information that helps those in charge of scientific policy make better decisions. The world's population is growing at an alarming rate, with today's population tripling that of 1960 and anticipated to reach 9 billion by 2050. As a result of this condition, the global proportion of energy utilized has increased. Cities have grown significantly. Wood, sawdust, straw, seed debris, compost, paper trash, household waste, polluted water, and other items are all sources of biomass. Biomass resources have been used and are becoming more significant for their economic potential, since there are large yearly quantities of agricultural output, and their products may be used as a source of power and created as energy plants, specifically for this purpose. The main goal of this study was to examine the research environment and practices in biomass renewable energy from 1978 to 2018, in order to aid the research community in understanding the current situation and future trends, as well as the international situation, and to provide basic information to aid decision- making.

The USA, India, China, Germany, and Italy are among the major countries researching on biomass as a renewable energy source, as assessed by scientific output. The Chinese Academy of Sciences is the most productive institution in the sector, followed by the National Renewable Energy Laboratory, Danmarks Tekniske University, and the Chinese Department of Education. In addition, the study identified groups based on published terms discovered through book searches. There were six communities or groups discovered. The two most important ones are concerned with converting biomass to liquid fuel. Finally, eight clusters were established based on international collaboration and biomass research. All of this is focused on three countries that fall into separate categories: the United States, India, and the United Kingdom.

CLASSIFICATION OF BIOMASS

Based on their scope and functions, biomasses can be classified into many classes. There are no specific ways to the classification of biomasses; consequently, we can classify biomasses into several groupings by relying on their magnitude, composition, software, and natural existence. Typically, biomasses can be classified into distinct methods based on the goods, starting location, and characteristic of the biomass. Type can be determined by the types of biomass present in the environment as well as its application as a substrate. Woody biomass, herbaceous biomass, animal and human waste, and aquatic biomass are some of the most well- known classifications of biomass into distinct types.

APPLICATION OF BIOMASS

Every quarter must recognize the three precepts that describe how to reduce, reuse, and recycle environmental waste rather than burning or discarding it. These methods not only aid in the control of all types of pollutants, but they also aid in the reduction of costs, as well as their availability and renewability. Biomass can be used in specialized sectors to meet the waste management hierarchy's requirements for energy, food, fertilizers, and a variety of other items. As the production of biomass increases every day, more appropriate and practical waste control measures are needed to help the ecosystem remain sustainable. Biomass can be used in the energy sector, for bio fertilizer production, for water treatment, In cement manufacturing, for the creation of thermal insulators, and for construction purposes, to name a few applications.

Application of Biomass for Biofuel Production

Biomass is one of the most effective and adaptable substrates for power generation. The provision of natural material, which includes Wood shavings, rice husk, sugarcane, timbers, bushes, leaves, and peanut shells, among other things, is the most important aspect in biomass electricity generation. Biomass is used for power generation in a wide range of applications, from small-scale use in homes for cooking to large-scale use in unique industries, such as boilers and powerhouses. Biofuel is a renewable energy source made from biomass that has been in great demand globally in recent years. One of the advantageous assets is biofuel. Biofuel is a sustainable energy source that helps to reduce air pollution, greenhouse gas emissions, and the reliance on carbon-based energy sources.

Ethanol and biodiesel are the two most common types of biofuels utilized today. Developed countries such as the United Kingdom and the United States rely on the production of bioethanol using marine yeast and seawater as a media. In the year 2019, America generated the most ethanol, Brazil placed second with 8.6 billion gallons, and the United States came in third with 15.8 billion gallons. In the year 2019, 530 million gallons of manufacturing will be produced., India is ranked fifth on this list. Biodiesel is commonly used in conventional diesel engines, either alone or in combination with petroleum diesel. Various feedstocks are used by different countries depending on their availability. The main distinction between ethanol and biodiesel Biodiesel is a type of oil, while ethanol is a type of alcohol. Ethanol is an alcohol made through the fermentation process that can be used instead of or in addition to gasoline, whereas biodiesel is made by extracting naturally existing oils from various flowers and seeds through a process called transesterification. The production of biofuel is targeted not only to meet the needs of decentralized energy generation, but also to meet the needs of shipping. This piques the interest of local organizations and entails the use of local groups' lands for the production of biofuels.

Bioprocessing Involved in Biofuel Production

Bioprocessing is the use of living things and their components, usually after various industrial processes, to produce a product that requires less energy and water and, as an outcome, produces a less affluent issue. Biofuels are a viable alternative to fossil fuels because they are non-poisonous, totally combustible, and environmentally sustainable. The production of biofuels involves a wide range of activities that are generally classified into two categories.: upstream and downstream approaches. Storage of liquid components, removal of inhibitory chemical compounds and particulates from the product, purification and sterilization, and many more procedures are included in the upstream procedures. Storage of liquid components, removal of inhibitory chemical compounds and particulates from the product, purification and sterilization, and many more procedures are included in the upstream procedures. The establishment of a microbial strain with the ability to synthesis the defined business value of a particular product is part of the upstream method. The pressure is then put through an enhanced routine in order to maximize its capacity to create the product at a low cost. Mobile isolation from the fermentation broth, purification and extraction of the desired product, and waste elimination or recycling are all examples of downstream processing. Several steps in fermentation procedure, including appropriate techniques for extracting, purifying, and characterizing the fermentation component in question. In addition to chromatography, a wide range of downstream processing options are available, including centrifugation and filtration. The chemical and physical qualities of the final product, as well as the intended grade, all influence these operations.

CONCLUSION

This study looked at the biomass's structure, composition, and classification, as well as its potential for bioenergy applications. Biomass is a mixture of biological and chemical substances that can be successfully converted into a value-added product. Biomasses can be divided into different divisions based on their applications and future possibilities. However, there are no specific criteria or methods for classifying biomasses, so we can classify biomasses into exceptional companies based on their fine, composition, application, and kind of lifestyles. Biomass can be divided into four distinct categories: woody biomass, herbaceous biomass, animal and human waste, and aquatic biomass. Exploring the relevance of lignocellulosic biomass and its utility over other current biomasses may have the capability in terms of environmental effect, bioprocessing involved in biofuel manufacture, liquid and gaseous biofuels derived from lignocellulosic biomass. The United States leads the way in researching biomass as a renewable energy source, as measured by scientific production, followed by China, India, Germany, and Italy. As a result, large nations with large populations are interested in adopting current renewable energy sources in conjunction with biomass.

REFERENCES

Adarme, C., Lim, D. K. Y., Timmins, M., Vernen, F., Li, Y., & Schenk, P. M. (2012). Microalgal biofactories: A promising approach towards sustainable omega-3 fatty acid production. *Microbial Cell Factories*, *11*(1), 96. doi:10.1186/1475-2859-11-96 PMID:22830315

Amano, H., Kakinuma, M., Coury, D. A., Ohno, H., & Hara, T. (2005). Effect of a seaweed mixture on serum lipid level and platelet aggregation in rats. *Fisheries Science*, *71*(5), 1160–1166. doi:10.1111/j.1444-2906.2005.01076.x

Batista, A. P., Raymundo, A., Sousa, I., & Empis, J. (2006). Rheological characterization of coloured oil in-water food emulsions with lutein and fycocyanin added to the oil and aqueous phases. *Food Hydrocolloids*, *20*(1), 44–52. doi:10.1016/j.foodhyd.2005.02.009

Capelli, B., & Cysewski, G. (2010). Potential health benefits of spirulina microalgae. *NUTRAfoods: International Journal of Science and Marketing for Nutraceutical Actives, Raw Materials, Finish Products, 9*(2), 19–26. Advance online publication. doi:10.1007/BF03223332

Chen, H. (2014). *Chemical composition and structure of natural lignocellulose. Biotechnology of Lignocellulose*. Springer. doi:10.1007/978-94-007-6898-7

Chisti, Y. (2008). Biodiesel from microalgae beats bioethanol. *Trends in Biotechnology, 26*(3), 126–131. doi:10.1016/j.tibtech.2007.12.002 PMID:18221809

Cuishus. F., & Butler, C. (1992). *A Review of the Potential of Bio-Diesel as a Trampon Fut ETSU-R-71*. Academic Press.

Dewey, L. (1916). *Hemp Hards as Papermaking Material*. USDA Bulletin No. 404.

Evets, L. (1994). Means to normalise the levels of immunoglobulin E, using the food supplement Spirulina. Grodenski State Medical University Russian Federation Committee of Patents and Trade. Patent (19) RU (11) 2005486.

Garrido-Cardenas, J. A., Manzano-Agugliaro, F., Acien-Fernandez, F. G., & Molina-Grima, E. (2018). Microalgae research worldwide. *Algal Research, 35*, 50–60. doi:10.1016/j.algal.2018.08.005

Grippi, D., Clemente, R., & Bernal, M. P. (2020). Chemical and bioenergetic characterization of biofuels from plant biomass: Perspectives for southern Europe. *Applied Sciences (Basel, Switzerland), 10*(10), 3571. doi:10.3390/app10103571

Johnson, P., & Shubert, E. (1986). Availability of iron to rats from spirulina, a Bluegreen algae. *Nutrition Research (New York, N.Y.), 6*(1), 85–94. doi:10.1016/S0271-5317(86)80202-0

Katinas, V., Marčiukaitis, M., Perednis, E., Dzenajavičienė, E. F. (2019). Analysis of biodegradable waste use for energy generation in Lithuania. *Renewable and Sustainable Energy Reviews, 101*, 559– 567. 5.

Khan, Z., Bhadouria, P., & Bisen, P. S. (2005). Nutritional and therapeutic potential of Spirulina. *Current Pharmaceutical Biotechnology, 6*(5), 373–379. doi:10.2174/138920105774370607 PMID:16248810

Naito, S., Maeda, N., & Fujimoto, A. (2020). *U.S. patent application No. 16/627,888. 9*. US Patent Office.

Ove Arup and Partners. (1989). *Monitoring of a Commercial Demon stration of Harvesting and Combudion of Forestry Wastes*. ETSU.

Padam, B. S., Tin, H. S., Chye, F. Y., & Abdullah, M. I. (2014). Banana by–products: An under–utilized renewable food biomass with great potential. *Journal of Food Science and Technology, 51*(12), 3527–3545. doi:10.100713197-012-0861-2 PMID:25477622

Padam, B. S., Tin, H. S., Chye, F. Y., Abdullah, M. I. (2014). Banana by–products: An under– utilized renewable food biomass with great potential. *Journal of Food Science and Technology, 51*(12), 3527–3545.

Padam, B. S., Tin, H. S., Chye, F. Y., & Abdullah, M. I. (2014). Banana by–products: An under– utilized renewable food biomass with great potential. *Journal of Food Science and Technology, 51*(12), 3527–3545. doi:10.100713197-012-0861-2 PMID:25477622

Pascaud M. (1993), The essential polyunsaturated fatty acids of spirulina and our immune response. *Bull. Inst. Océano, 12*, 49-57.

Pham Quoc, K., Dubacq, J. P., & Demandre, C. (1994). Comparative effects of exogenous fatty acid supplementations on the lipids from the cyanobacterium Spirulina platensis. *Plant Physiology and Biochemistry*, *32*(4), 501–509.

Prier, B. (1998). *Electricity from Bonon*. Financial Times Business Lid.

Raud, M., Olt, J., & Kikas, T. (2016). N2 explosive decompression pretreatment of biomass for lignocellulosic ethanol production. *Biomass and Bioenergy*, *90*, 1–6.

Raud, M., Olt, J., & Kikas, T. (2016). N2 explosive decompression pretreatment of biomass for lignocellulosic ethanol production. *Biomass and Bioenergy*, *90*, 1–6. doi:10.1016/j.biombioe.2016.03.034

Salapa, I., Katsimpouras, C., Topakas, E., & Sidiras, D. (2017). *Organosolv pretreatment of wheat straw for efficient ethanol production using various solvents*. Academic Press.

Shah, M. A., Khan, M. N. S., & Kumar, V. (2018). Biomass residue characterization for their potential application as biofuels. *Journal of Thermal Analysis and Calorimetry*, *134*(3), 2137–2145. doi:10.100710973-018-7560-9

Shklar, G., & Swartz, J. (1988). Tumor necrosis factor in experimental cancer regression with alphatocopherol, beta-carotene, canthaxanthin and algae extract. *Eur Cancer Clin Oncol*, *24*(5), 839–850. doi:10.1016/0277-5379(88)90192-7 PMID:3139418

Ubando, A. T., Del Rosario, A. J. R., Chen, W. H., & Culaba, A. B. (2021, January 15). A state-of-the-art review of biowaste biorefinery. *Environ Pollut.*, *269*, 116149. doi:10.1016/j.envpol.2020.116149 PMID:33280912

Wang, N., Zhang, C., Zhu, W., & Weng, Y. (2020). Improving interfacial adhesion of PLA/Lignin composites byone–step solvent-free modification method. *Journal of Renewable Materials*, *8*(9), 1139–1147. doi:10.32604/jrm.2020.09961

Warren, T. J., Poulter, R., & Partin, R. (1995). Converting biomass to electricity on a farm-sized scale using downdraft gasification and a spark ignon engine. *Bioresource Technology*, *52*(1), 95–98. doi:10.1016/0960-8524(95)00022-7

Watanabe, F. (2007). Vitamin B12 sources and bioavailability. *Experimental Biology and Medicine (Maywood, N.J.)*, *232*(10), 1266–1274. doi:10.3181/0703-MR-67 PMID:17959839

Chapter 9
Biofuels From Macroalgae:
A Sustainable Alternative to Conventional Energy Resources

Debraj Biswal
https://orcid.org/0000-0002-7317-6774
Government General Degree College at Mangalkote, India

Dipanwita Sarkar (Paria)
Chandernagore College, India

ABSTRACT

The steep rise in human population together with the simultaneous growth of urbanization and industrialization has triggered overuse of fossil fuels in the recent years. However, they are non-renewable and release carbon dioxide causing global warming. In this context, biofuels come as excellent substitutes because they are renewable and eco-friendly. They are obtained from biological products and include biodiesel, biomethane, bioethanol, and biohydrogen. They are classified into first, second, and third generation fuels depending on the biological feedstocks being used. Of these, the third-generation fuels, obtained from marine resources like algae, have attracted special research interests because they do not rival the food crop production or depend on land and freshwater for their cultivation. However, the potentialities of macroalgae as a biofuel resource have not been investigated thoroughly. The chapter tries to outline all the aspects of biofuel production from macroalgae pointing out the scopes for future research.

INTRODUCTION

Overexploitation of the fossil fuels and the non-renewable energy resources like coal, petroleum and natural gas in the last few years has not only led to the depletion of resources but has also added carbon dioxide (CO_2) to the atmosphere. Needless to say, CO_2 is a major greenhouse gas (GHG) and causes global warming (Naik et al., 2010). The problem has been projected to rise in the years to come. This necessitates the development of alternative energy sources to meet the growing demands of the burgeoning popula-

DOI: 10.4018/978-1-6684-5269-1.ch009

tion as well as reduce the adverse environmental effects (Mabee et al., 2005). According to Naik et al. (2010), eco-friendly and sustainable energy resources are the need of the hour for sustainable growth of the economy and industry of a nation. In this regard, several attempts have been taken by researchers to utilize plant biomass and/or organic wastes for extraction of energy. The lipid and carbohydrate contents of these substances have mainly been targeted by hydrolytic degradation, fermentation, pyrolysis and various other techniques to procure the fuels (Gao et al., 2020). They have been termed as 'biofuels' and have emerged as promising alternatives of conventional energy sources. Because of their renewability and sustainability they have been proposed as the 'future energy resources' (Reid et al., 2020). Their extraction from plant derivatives gives an added advantage of maintaining carbon balance in the atmosphere. According to Gao et al. (2020), plants help to sequester atmospheric CO_2 by photosynthesis. As such, burning of the fuels obtained from them does not become an additional source of CO_2 emission. Therefore, biofuels can combat global warming to a large extent (Kumar et al., 2016). Another advantage of biofuel production is its ability to generate new employment opportunities (Stevens & Verhe, 2004).

The biofuels sector is, however, in its stage of infancy because of several issues and challenges associated with their production that need to be addressed before they can be manufactured in large-scales (V. G. S. et al., 2021). The common challenge lies in the development of technological sophistication to extract the biofuels with minimum loses such that the final products do not exceed the price of conventional fuels. Otherwise, it will reduce their marketability (Gao et al., 2020). Currently, this field is being explored by various researchers in different parts of the world. The other problem is the direct confrontation of biofuel production with food resources because they have often been attempted to be derived from crop plants accelerating two pertinent issues – global water crisis and food scarcity (Naik et al., 2010). Land availability for cultivation of these crops is another problem under the current conditions of explosive population growth because of increasing land requirements for human habitation (Naik et al., 2010). Land is also required for augmenting crop production to meet the demands of the expanding population. Adequate freshwater supply is equally important to maintain agricultural productivity. Using freshwater for cultivation of terrestrial plants to be used as feedstock for biofuel manufacture will trigger competition between the biofuel and agricultural industries. These problems have encouraged the researchers to explore non-consumable raw materials (that are not terrestrial derivatives) for biofuel production (V. G. S. et al., 2021). Consequently, the marine resources primarily consisting of the micro and macroalgae have emerged as the preferred candidates for biofuel extraction. Some workers like Liao et al. (2021) have used the term 'blue bioeconomy' to describe the economy dependent on such biofuel resources. They offer several advantages over terrestrial plants the most important being their rich carbohydrate and lipid contents together with low (to almost zero) proportions of lignin facilitating the extraction of biofuels by employing cheaper technologies (Gao et al., 2020). Furthermore, they can be cultivated in unused, rejected or barren lands with saline water and wastewater and can withstand adverse environmental conditions (Gao et al., 2020). Despite these advantages, they have certain drawbacks that have prevented them from gaining popularity. They need further developments both in terms of technology and overcoming certain serious environmental issues associated with their production (Hughes et al., 2012; V. G. S. et al., 2021).

The current chapter tries to give a comprehensive review of the advantages of using macroalgae as the source of biofuels by going through some recent publications in this field. It discusses about the advantages and limitations of microalgae while referring to the potentialities of macroalgae as a biofuel resource. It also highlights about the various challenges and environmental issues of biofuel production from macroalgae that limits their acceptability. Last but not the least, it suggests some mitigation strategies by pointing out the research gaps in this field.

BIOFUEL GENERATIONS

The growing human population and its increasing demands puts severe stress on the fossil fuels that form the primary source of energy. According to V. G. S. et al. (2021), the fossil fuel requirements from 2010 to 2040 will augment by as much as 40%. However, they are non-renewable and tend to get exhausted if proper alternatives are not searched for. Additionally, they are also a cause of global warming, pollution and the emerging issue of global climate change (Biswal, 2021; 2022). These pressing conditions compelled the researchers to develop sustainable, renewable and eco-friendly energy resources time and again. Many workers have used the term 'green fuel' to describe them (Naik et al., 2010). They are primarily formulated from plant sources and depending on the raw materials used they have been classified into three generations.

First Generation Biofuels

Bioethanol, biodiesel and biogas form the three primary types of first generation biofuels. Bioethanol is a good alternative of gasoline and can be used in vehicles that are flexible with fuel use. It can be procured from fermentation of starch or sugars present in the plant biomass (Naik et al., 2010). Bioethanol can also be a source of ethyl tertiary butyl ether (ETBE) that blends well with gasoline. Additionally, vegetable oils together with residual oils and fat obtained from plant sources or wastes can be subjected to transesterification to obtain biodiesel. Another sustainable alternative of gasoline is biogas or biomethane that can be obtained by anaerobic breakdown of liquid manure and other plant sources or waste materials (Naik et al., 2010).

One major drawback of first generation biofuels is their dependency on food crops (e.g. wheat, rice, sugarcane and corn), edible oils and other food stuffs as raw materials. Since the growing population puts increasing demands on food production, utilizing food resources for biofuel production directly threatens the future food security (Naik et al., 2010). Many researchers have also claimed that utilization of food crops for biofuel production may inflate food prices (Gao et al., 2020). This has encouraged the cultivation of crops specially dedicated for biofuel production that can be grown on barren lands. In this context, oilseed crops have gained recognition because they can directly yield biodiesel while their lingnocellulosic biomass can be processed to manufacture second generation biofuels (Naik et al., 2010). However, their production has sparked controversy because they directly involve land use and also exerts negative impacts on biodiversity. Increased land use for fuel crops production has serious issues because increasing human population requires increased land supply for human habitation. Above all, according to many scientists, the production and utilization of first generation biofuels offsets carbon balance with negative impacts on the environment (Naik et al., 2010). Keeping these disadvantages in mind the second generation biofuels were developed.

Second Generation Biofuels

The second generation biofuels use non-food substances, mainly the lingnocellulose materials of plants, as the raw material for fuel production (Zabaniotou et al., 2008). This places them ahead of the first generation biofuels by not competing directly with the food materials. Moreover, the raw materials are abundantly available from plants and are cheap. These fuels are also beneficial in terms of maintenance of CO_2 balance because the plant biomass used as the feedstocks fixed atmospheric CO_2 (by photosynthesis) during their lifetime. Thus, the CO_2 emitted from the ignition of the biofuels derived from them balances, to some extent, the removal and addition of atmospheric CO_2 (Naik et al., 2010). However, their potentialities have not been explored thoroughly despite the plant biomass being an immense source of liquid biofuels. Here, it is worth mentioning that the plant-derived raw material is primarily composed of 75% polysaccharides that make up the cell walls. Degradation of the polysaccharides followed by fermentation and/or gasification can give some important biofuels like bioethanol and biogas (Naik et al., 2010).

Residues from the wood industry (e.g. wood shavings), agriculture (e.g. straw), forestry (e.g. short rotation forestry crops like eucalyptus and willow) and perennial grasses (e.g. *Miscantus* sp.) are the most common sources of feedstocks for second generation biofuel production. It has been proposed that they have the ability to supply a large extent of the increasing demands for liquid fuels (Naik et al., 2010). Apart from the substitute of gasoline (i.e. bioethanol) they can generate FT-diesel (Fischer-Tropsch diesel) and BTL (Biomass-to-Liquids) which can completely replace diesel (Naik et al., 2010). Additionally, Bio-SNG (Synthetic Natural Gas) obtained by gasification of the plant biomass can be an alternative to gasoline but it requires slight modifications. Similarly, Bio-DME (Dimethyl Ether) [obtained from syngas] can be used as a used as a diesel-substitute with certain alterations (Balat, 2007). The syngas can also be transformed into bio-kerosene (Balat, 2007).

One of the greatest disadvantage of second generation biofuels is the cost involved in the technology required for the conversion of the lignocellulosic raw-materials into fuels. As such, the biofuels obtained become pricey compared to their conventional alternatives (V. G. S., et al., 2021). This lack of cost-effectiveness proves a barrier to their acceptability. The drawbacks prompted the development of third generation biofuels with inexpensive, non-consumable and renewable feedstocks that can be easily processed to give cost-effective fuels.

Third Generation Biofuels

Utilization of marine resources in the form of seaweeds and cyanobacteria made the way for greater production of biofuels with lesser investments (V. G. S., et al., 2021). Their non-dependence on agricultural lands and freshwater supplies make them the feedstock of choice for biofuel production (Gajaraj et al., 2018). Marine water, brackish water, saline water and even wastewater, which are unfit for cultivation of agricultural crops, may be used for cultivating most of the non-consumable algae. Animal, industrial and municipal runoffs can also be utilized for growing algae. Additionally, they can withstand adverse climatic conditions and can also be grown in dark using suitable sugar sources in the medium (Gajaraj et al., 2018). This makes their cultivation flexible. Furthermore, the biofuels derived from macroalgae have been found to be secure, recyclable and sulphur-free making them extremely useful alternatives of conventional fuels (Mat Aron et al., 2020).

The algae are characterized by rapid rates of growth and high efficiency of fixing atmospheric CO_2 by photosynthetic processes (Peter et al., 2021). The macroalgae mainly consist of approximately 25% to 60% of carbohydrates, 5% to 47% of proteins, 18% to 55% of ash and lipid contents less than 5% (V. G. S., et al., 2021). The biochemical composition, however, varies with the biotic and abiotic factors of the environment where these algae are cultivated (Peter et al., 2021). The seaweeds have high productivity facilitating quicker generation of surplus raw materials for extraction of biofuels. The annual harvest of wet seaweeds across the globe has been recorded at 7.5 to 8 million tons (Subba Rao & Mantri, 2006). According to Jung et al. (2013) the brown algae, kelp (*Saccharina japonica*) and sea mustard (*Undaria pinnatifida*) have the highest productivity per unit area making them one of the best feedstocks for biofuel production. Eventually, the algal biofuels have emerged as promising alternatives to the earlier two generations and have gained popularity as the 'third generation biofuel'.

MACROALGAE AS A SOURCE OF BIOFUELS

The use of algae in biofuel production is recently gaining grounds because of their ability to fix carbon dioxide (CO_2) by photosynthesis. According to Aresta et al. (2005) their photosynthetic efficiency is high and approximately ranges from 6% to 8% whereas it is about 1.8% to 2.2% in biomass derived from terrestrial plants. Additionally, they do not compete with food production as was the case with first generation biofuels (Singh & Olsen, 2011). They are beneficial over second generation biofuels because of their low lignin content saving the cost incurred in its removal (Singh & Olsen, 2011). Poor quality lands, not suitable for crop cultivation, can be easily used for production of algae (Levine et al., 2010). They have rapid growth rates and the ability to utilize the nutrients (like nitrogen and phosphorus) present in wastewater, thereby, contributing to wastewater recycling as well (Singh et al., 2011).

Though both the microalgae and macroalgae have been recognized as prospective sources for biofuels, macroalgae is comparatively new and unexplored giving ample opportunities for research (Mata et al., 2009). The higher lipid contents and faster growth rates of microalgae have made them preferable over macroalgae for biodiesel production but their lower carbohydrate contents do not make them a good source for production of biomethane (Jin et al., 2013). The energy yield of macroalgae is also higher than the same amount of microalgae (Chen et al., 2015). It has been observed that the process of harvesting of macroalgae is simpler than microalgae because of their multicellular nature and plant-like features (Maceiras et al., 2011). The higher and good quality carbohydrates of macroalgae make them a good source of biofuels like bio-oils, bioethanol, and biogas.

There are three major groups of macroalgae depending on their photosynthetic pigments. They are green algae (Chlorophyta), red algae (Rhodophyta) and brown algae (Phaeophyta) (John, 2011; Kraan, 2013). The red algae is the most abundant group in terms of species followed by green algae (Jung et al., 2013). Brown algae has the least number of species compared to the other two and prefers cold water (Jung et al., 2013). Red algae flourish in the inter-tropical zones while the green algae can grow in all water types (Bucholc et al., 2014). Currently, macroalgae is cultivated extensively in Asia mainly as a food source, for production of fertilizers and for extraction of hydrocolloids (Roesijadi, 2010). Japan, Korea, China and Philippines account for approximately 72% of the world's share of macroalgae production (Chen et al., 2015).

Macroalgae has not received much attention for biofuel production but they have garnered industrial interests for extraction of hydrocolloids (Bucholc et al., 2014). Microalgae dominates the biofuel sector but their poor carbohydrate contents limit their versatility (Bucholc et al., 2014). The dry matter content of macroalgae amounts to about 10% to 15% and contains as high as 60% of carbohydrates (Roesijadi, 2010). Their cell wall is rich in carbohydrate sources like cellulose and/or starch (Roesijadi, 2010). Interestingly, their biochemical composition shows seasonal variation with the stored carbohydrates reaching its peak value in the autumn season (Kraan, 2013). Some of the major biochemical constituents of macroalgae have been represented in Table 1.

Table 1. Some major biochemical constituents of macroalgae (based on previous reports)

Macroalgae	Carbohydrates	Lipid	Protein	Reference(s)
Chlorophyta				
Ulva lactuca	11.6 to 13.2 (a)	9.6 to 11.4 (a)	11.4 to 12.6 (a)	Becker, 1994; Kaliaperumal et al., 2002; Suganya et al., 2016
Ulva intestinalis	367 ± 31 (b)	Not studied	90 (b)	Olsson et al., 2020
Enteromorpha compressa	24.75 (a)	11.45 (a)	7.26 (a)	Becker, 1994; Kaliaperumal et al., 2002; Suganya et al., 2016
Chaetomorpha linoides	27.00 (a)	12.00 (a)	9.45 (a)	Becker, 1994; Kaliaperumal et al., 2002; Suganya et al., 2016
Valoniopsis pachynema	31.50 (a)	9.09 (a)	8.78 (a)	Becker, 1994; Kaliaperumal et al., 2002; Suganya et al., 2016
Cladophora rupestris	399 ± 8 (b)	Not studied	184 (b)	Olsson et al., 2020
Phaeophyta				
Laminaria digitata	519 ± 42 (b)	Not studied	66 (b)	Olsson et al., 2020
Fucus serratus	287 ± 21 (b)	Not studied	71 (b)	Olsson et al., 2020
Chorda filum	292 ± 33 (b)	Not studied	63 (b)	Olsson et al., 2020
Saccharina latissima	557 ± 14 (b)	Not studied	69 (b)	Olsson et al., 2020
Ascophyllum nodosum	317 ± 17 (b)	Not studied	59 (b)	Olsson et al., 2020
Desmarestia aculeata	301 ± 101 (b)	Not studied	115 (b)	Olsson et al., 2020
Rhodophyta				
Laurencia papillosa	12.0 to 13.3 (a)	8.9 to 10.8 (a)	11.8 to 12.9 (a)	Becker, 1994; Kaliaperumal et al., 2002; Suganya et al., 2016
Hypnea valentiae	11.8 to 13.0 (a)	9.6 to 11.6 (a)	11.8 to 12.6 (a)	Becker, 1994; Kaliaperumal et al., 2002; Suganya et al., 2016
Acanthophora spicifera	11.6 to 13.2 (a)	10.0 to 12.0 (a)	12.0 to 13.2 (a)	Becker, 1994; Kaliaperumal et al., 2002; Suganya et al., 2016
Chondrus crispus	526 ± 35 (b)	Not studied	103 (b)	Olsson et al., 2020
Dilsea carnosa	477 ± 28 (b)	Not studied	152 (b)	Olsson et al., 2020
Brogniartella byssoides	240 ± 16 (b)	Not studied	158 (b)	Olsson et al., 2020

[(a) = biochemical constituent represented as % of dry weight; (b) = biochemical constituent represented as g/kg dry weight. For carbohydrates = Mean value ± S.D. (Standard Deviation)]

The major polysaccharide content of green algae includes starch, cellulose, mannan and ulvan. Red algae are a good source of agar, carrageenan, lignin and cellulose. Alginate, laminarin, mannitol, cellulose and glucan form the primary polysaccharides of brown algae. These polysaccharides differentiate macroalgae from microalgae (Jung et al., 2013). Moreover, the poor lignin content (about 0.03 kg per kg of dry matter) of macroalgae facilitates their fermentation or anaerobic digestion for extraction of biofuels (Horn, 2000). The process becomes much cheaper because additional investments on pretreatments for lignin removal, which forms an integral step for extraction of second generation biofuels, can be avoided (Chen et al., 2015). Apart from the polysaccharides, the macroalgae have high proportions of alkali metals and halogen contents (0.5% to 11%) compared to the biomass obtained from terrestrial plants (1% to 1.5%) (Mautner, 1954). The nutrient contents of the macroalgae [mainly nitrogen (N) and phosphorus (P)] have been found to vary with their morphology as well as the nutrient enrichment of the aquatic medium (McHugh, 2003). The effects of these constituents on biofuel production have been discussed later in this chapter.

BIOFUELS FROM MACROALGAE

The following biofuels can be obtained from macroalgae by application of thermochemical and biochemical technologies.

Biogas

Sea weeds are excellent candidates for biogas production because of their high carbohydrate content with low proportions of cellulose and lignin. As such, they can easily be fermented to obtain biogas (Saqib et al., 2013). The process mainly involves anaerobic breakdown of the organic matters in sea weeds to liberate gases like carbon dioxide, methane and small quantities of hydrogen sulphide. The first step of biogas production involves microbial enzyme mediated hydrolysis of the organic constituents of macroalgae (i.e. proteins, lipids and carbohydrates) into a soluble organic extract which is subsequently treated by acidogenic bacteria. In the second step, the enzymes liberated by acidogenic bacteria trigger the process of acidogenesis resulting in the liberation of alcohols and volatile fatty acids (VFAs). Subsequently, in the third step, the acetogenic bacteria converts the alcohols and VFAs into acetic acid and hydrogen which are finally acted upon by methanogenic bacteria to yield methane and carbon dioxide (Romagnoli et al., 2011).

It is important to note that the high water content (80 to 85%) of macroalgae qualifies them for microbial degradation over the thermochemical processes. However, the production of biogas from macroalgae is not free from challenges. High nitrogen content of macroalgae inhibits the action of the microbes and generates excessive ammonia which is toxic to the methanogenic bacteria (Demirbas, 2010). The presence of alkaline metals also hinders the anaerobic digestion processes. Additionally their cell walls need to be broken down by physical or chemical pretreatments to access the organic matter during biogas production (Demirbas, 2010). This problem is usually overcome by co-digesting the algal biomass with a substrate having high C:N (Carbon:Nitrogen) ratios such as straw or waste paper (Yen & Brune, 2007; Vivekanand et al., 2012). Some common macroalgae utilized for biogas production and their yields have been shown in Table 2.

Table 2. Some important biofuels and their macroalgal sources together with their respective yields (based on previous reports)

Biofuel	Macroalgal Source	Biofuel Yield	Reference(s)
Biogas	*Pilayella + Ectocarpus + Enteromarpha*	0.040 dm3/g - 0.054 dm^3/g of substrate	Grala et al., 2012
	Chaetomorpha linum	0.26 ± 0.045 L CH_4/g VS	Yahmed et al., 2016
	Fucus vesiculosus	144 ± 28 mL CH_4/g VS	Romagnoli et al., 2017
	Fucus vesiculosus + Fucus serratus	260 mL/g VS	Wu et al., 2019
	Ulva lactuca	498.75 mL/g VS	Hassaan et al., 2021
	Saccorhiza polyschides	227 ± 4 mL/g VS	Pardilhó et al., 2022
Bioethanol	*Sargassum* spp.	2.27 ± 0.04 g/L	Borines et al., 2013
	Gracilaria verrucosa	0.49 g/g	Shukla et al., 2016
	Sargassum latifolium	0.29 g/g of TRS	Soliman et al., 2018
	Ecklonia kurome	2.1 g/L	Sasaki et al., 2018
	Ulva linza	48.24 g/L	Greetham et al., 2020
	Ulva lactuca	0.41 g/g of Glucose	Allouache et al., 2021
Biodiesel	*Enteromorpha compressa*	90.6%	Suganya et al., 2013
	Caulerpa peltata	98.11%	Suganya & Renganathan, 2014
	Chara vulgaris	0.3 mL/g dry algae	Siddiqua et al., 2015
	Chaetomorpha linum	0.96 g; 4.85% dry wt.	Ameka et al., 2019
	Chaetomorpha antennina	0.48 g; 2.40% dry wt.	Ameka et al., 2019
	Cladophora sp.	4.58 g/200 g algal wt.	Naik & Naik, 2021
Biohydrogen	*Laminaria japonica*	67.0 mL H_2/g TS	Jung et al., 2011
	Padina tetrastromatica	78 ± 2.9 mL/0.05 g VS	Radha & Murugesan, 2017
	Ulva reticulata	63 mL H_2/g COD	Kumar et al., 2018
	Chaetomorpha antennina	74.5 mL H_2/g COD	Kumar et al., 2019
	Saccharina japonica	20.25 mLH_2/g VS_{added}	Yin & Wang, 2019
	Ulva sp.	1.52 mol H_2/mol Hexose	Margareta et al., 2020

[TRS - Total Reducing Sugars, VS – Volatile Solids, TS – Total Solids, COD – Chemical Oxygen Demand]

Bioethanol

Bioethanol is an example of liquid algal transportation fuel. Its production from algal biomass is facilitated by the presence of low proportions of hemicelluloses and lignin (Mutripah et al., 2014). Fermentation of algal biomass (e.g. with the yeast, *Brettanomyces custersii*) has been extensively explored for bioethanol production (Park et al., 2012). However, the algal biomass requires pre-treatments to enable the enzymes to act on the algal carbohydrates like carrageenan, laminarin and alginate (Wei et al., 2013; Chen et al., 2015). Pretreatment with acids like sulphuric acid (H_2SO_4) or hydrochloric acid (HCl) is most commonly employed for this purpose which facilitates the hydrolysis of structural polysaccharides (Han et al., 2007). Even though the process of acid hydrolysis is cheap, it generates by-products like levulinic acid and 5-hydroxy-methyl-furfural which impedes ethanol production (Klinke et al., 2004; Jeong et al., 2012). Enzymatic hydrolysis is a suitable alternative and at times essential. For example, digestion of laminarin (present in brown algae) requires enzymatic treatment with laminarinases (Lee et al., 2014). Similarly, cellulase enzyme is used to digest cellulose present in algal biomass (Singh & Trivedi, 2013). However, enzymatic hydrolysis is not a preferred method because it is very slow (Tan et al., 2013). Instead, a combined pretreatment method utilizing acids and enzymes has been found to be very productive (Daroch et al., 2013). Pretreatment using gamma radiation has also been found to be a sustainable method (Yoon et al., 2012). Interestingly, sea weeds like *Ulva pertusa* and *Gelidium elegans* are free from lignin and do not require any pretreatment (Yanagisawa et al., 2011).

Currently, the bacterium, *Zymomonas mobilis* and the yeast, *Saccharomyces cerevisiae* have gained popularity as the fermenting microorganisms for production of bioethanol (Horn et al., 2000). However, fermentative degradation of certain polysaccharides like mannitol requires the presence of oxygen. In these cases, the bacterium, *Zymobacter palmae* have been found to be very effective (Horn et al., 2000). Thus, the basic process of bioethanol production involves hydrolytic degradation of the algal polysaccharides into simple sugars followed by their fermentation to ethanol and separation by distillation in the final step. Examples of macroalgae commonly used for the production of bioethanol and their yields have been represented in Table 2.

Biodiesel

Biodiesel consists of a blend of monoalkyl esters of long chain fatty acids and can be used directly for running diesel engines (Demirbas, 2009). They can be derived from commercially obtained triglycerides from plant oils, animal fats and vegetables by subjecting them to transesterification reactions (Chen et al., 2015). However, the macroalgae have poorer lipid contents compared to microalgae resulting in lower yields of biodiesel. This makes microalgae, with oil productivity higher than the high yielding terrestrial oil crops, a raw material of choice for biodiesel production (Afify et al., 2010; Maceiras et al., 2011). Nevertheless, the macroalgal biomass can be converted into bio-oil consisting of a mixture of free fatty acids and lipids. The lipids can then be isolated and utilized in biodiesel production. Some common macroalgae used for biodiesel production and their yields have been enlisted in Table 2. The existence of large amounts of free fatty acids in bio-oil is, however, reported to hamper biodiesel production (Chen et al., 2015). Conversion of free fatty acids from algal oil into biodiesel has been attempted by Suganya and Renganathan (2014). Alternatively, glucose obtained from macroalgae (e.g. *Cladophora glomerata*) can be converted into free fatty acids followed by biodiesel production (Chen et al., 2015). In yet another method, the macroalgal biomass can be acted upon by oleaginous yeast to produce biodiesel while the free fatty acids obtained as the by-product could serve to nourish the yeast itself (Xu et al., 2014).

Proper dehydration of the algal biomass is essential prior to biodiesel production because presence of water lowers its yield by as much as ten times (Daroch et al., 2013). The algal biomass can then be mechanically compressed to extract the bio-oil though it is a time-consuming process and not much effective (Popoola & Yangomodou, 2006). Using organic solvents like hexane, acetone, benzene, chloroform, etc. to extract oils from algal biomass have been found to be more productive than the mechanical process (Topare et al., 2011). The extracted oil is then subjected to transesterification reaction using alcohols like butanol, propanol, ethanol, methanol, etc. However, use of methanol and ethanol is more preferred because they are cheaper and offer certain chemical and physical advantages over others (Surendhiran & Vijay, 2012). Presence of lipase enzyme or inorganic catalytic agents like acids and alkalis are equally essential for biodiesel production (Meher et al., 2006). Glycerol is an important by-product in these reactions and has to be removed from the reactor on a regular basis to prevent engine failure (Munir et al., 2013).

Biohydrogen

Electricity may be generated from the biohydrogen procured from algal biomass. It may also be used as a gaseous fuel (Behera et al., 2015). Photo fermentation and biophotolysis are two common methods for biohydrogen production (Shaishav et al., 2013). The red alga, *Gelidium amansii* can be anaerobically fermented to obtain biohydrogen (Park et al., 2011). However, the yield depends on adopting suitable pretreatment methods because acid hydrolysis of the algal biomass can produce 5-hydroxymethylfurfural which inhibits the production process reducing the output by almost 50% (Park et al., 2011). Dark fermentation of the fermentable sugars obtained from macroalgae by using mesophilic and thermophilic microbes has also been suggested as a suitable technique of biohydrogen production (Kumar et al., 2016). Furthermore, the carbohydrate content of the algae can be maximised by exposing them to stressed conditions during their cultivation. This can increase their biohydrogen yields (Kumar et al., 2016).

Heat treatment of the brown seaweed, *Laminaria japonica* at 170°C can produce hydrogen at the rate of 109.6 mL/g COD (Chemical Oxygen Demand) (Jung et al., 2011). Microwave pretreatment of *Laminaria japonica* at 160°C for about half an hour also gives good result (Yin et al., 2019). Macroalgae, on being treated with hydrogen peroxide, gives a good supply of hydrogen as well (Roy et al., 2014). Additionally, water can be subjected to biophytolysis using the photosynthetic properties of microalgae (Kumar et al., 2016). Table 2 represents some common macroalgal species from which biohydrogen can be obtained and their yields.

CHALLENGES OF BIOFUEL PRODUCTION

One of the greatest challenges of using terrestrial plants and/or their derivatives for biofuel production is the consumption of freshwater because many parts of the world are already facing challenges of potable water supply (Shilton & Guieysse, 2010). Under the growing negative impacts of climate change and non-sustainable anthropogenic activities, future availability of freshwater has already raised several questions (Biswal, 2021; 2022). These were the circumstances that prompted researchers to exploit the marine resources for generation of biofuels. Of the marine resources, the microalgae has been extensively experimented as biofuel producers (Behera et al., 2015). Their faster proliferating capabilities, high proportion of polyunsaturated fatty acids (often exceeding by almost 80% of their dry weight) and almost no lignin content make them highly efficient raw materials for generation of biodiesel (Mata et al., 2009; Jin et al., 2013). However, they are poorer in carbohydrates (compared to the macroalgae) restricting their versatility in biofuel manufacture (Chen et al., 2015). Their cultivation also has several constraints like availability of long periods of sunlight encouraging the use of photo bioreactors as an alternative. However, they did not gain acceptance because of the costs involved inflating the prices of the final product (Wilcox, 1977). The macroalgae, on the other hand, gives a good yield of biogas when their biomass is subjected to gasification (Wargacki et al., 2012).

At present, more than one hundred species of macroalgae are being used for commercial purposes (like pharmaceutical products, food and fertilisers) all over the world (Santelices, 2007). Knowledge regarding the life cycle of sea weeds and their cultivation in large-scales gained momentum in the twentieth century. It was China who first began to selectively breed sea weeds (beginning with *Laminaria* sp.) in the 1960s (Hughes et al., 2012). However, research shows that not even 1% of the available macroalgae have been explored properly (Budarin et al., 2011). They are good sources of bioethanol too (Hughes et al., 2012). Despite their potentialities, large-scale manufacture has not gained popularity because of the consequences on marine ecosystem due to their mass harvest (Springer et al., 2010; Kraan, 2013). It is to be remembered that the macroalgae, though non-consumable for humans, forms a support system for the marine environment. For instance, they have been found to grow profusely in the regions near the corals sheltering a variety of organisms together with the fishes and birds (Lorentsen et al., 2010). Massive harvesting of the seaweeds is also not eco-friendly because they take time to bounce back to their original conditions (Christie et al., 1998).

Cultivation of macroalgae can be carried out both offshore and onshore by several techniques. In the offshore methods, the lagoons are preferred for their growth and harvest while their onshore cultivation is operated along the coastal belts (V. G. S. et al., 2021). Comparison between both the cultivation techniques shows that offshore cultivation is more difficult and not sustainable (Hughes et al., 2012; V. G. S. et al., 2021). Hughes et al. (2012) have mentioned several environmental threats associated with large-scale cultivation of macroalgae in the marine environment (i.e. onshore cultivation). Since they grow in the water column just above the sea floor (also referred to as the seabed) they may severely impact the benthic flora and fauna. They may block the sunlight from penetrating deeper waters affecting the autotrophic organisms as well as competing with them for nutrients (Hughes et al., 2012). Sometimes inorganic fertilizers [e.g. nitrogen (N) and phosphorus (P)] are used to boost algae production which may lead to nutrient enrichment of marine waters and subsequent algal blooms. As such, the ecosystem along the coastline may be disrupted due to conditions of oxygen stress caused by algal overgrowth (Herbert, 1999; Springer et al., 2010). It may also result in terrestrial nutrient runoffs to the seabed beyond the photic zone resulting in nutrient enrichment of deep sea waters. This may make their habitats complicated. Accidental tearing or shredding of the macroalgae during their cultivation can create zones of organic enrichment (Cromey et al., 2009). The outcomes of these effects on marine organisms (e.g. cetaceans, fishes, etc.) have not been studied extensively raising doubts on the environmental impacts of macroalgae cultivation in the long run (Hughes et al., 2012). Despite the ability of the macroalgae to remove inorganic carbon from sea waters, a minor fraction may get converted to particulate organic matters and get settled on the sea floor with environmental consequences (Lüning & Pang, 2003). Several workers have pointed out that onshore cultivation often lowers the quality of the algae rendering them inappropriate for the chemical, pharmaceutical and the cosmetics industry (Hafting et al., 2015). This is due to lack of standardization and adoption of cost-effective sustainable strategies for macroalgal cultivation on large-scales (Hafting et al., 2015). Large-scale cultivation of seaweeds can affect the sedimentation patterns causing alterations in the local hydrodynamics (Hughes et al., 2012). However, according to some workers macroalgae cultivation in large-scales can enhance local biodiversity (Frank et al., 2012). Another important point that needs to be considered for biofuel extraction from macroalgae is the cost involved in their harvest and transportation to the factories. Since the seaweeds have a scattered distribution along the coasts the processes are quite expensive inflating the production costs of biofuels (Hughes et al., 2012).

Manufacture of biofuels require a steady supply of algal biomass which is difficult to manage because of the challenges involved in growing and harvesting of macroalgae. Since the bio-products obtained vary with the algal species (due to differences in their polysaccharide contents) the sorting process during biofuel manufacture becomes all the more complicated (V. G. S., et al., 2021). Further, the choice of the microbes for fermentation of the plant biomass is also an important and challenging task because the efficiency of the microbes differ with the raw materials and the conditions of reaction (V. G. S., et al., 2021). Pretreatment of the plant biomass is essential for maximum utilization of the polysaccharides in plant tissue. Therefore, this part also requires sufficient development to maximize the product output since many pre-treatment procedures may cause reduced outputs. Another major obstacle is the freshwater requirement in the bioprocessing stages which may augment the shortage of freshwater supplies in the future (Liao et al., 2021). Though seawater can be used in the hydrolysis and fractionation stages of macroalgae as well bioethanol production, their applicability on large-scales is still being worked upon (Joniver et al., 2021).

Often, the biochemical constituents of the macroalgae themselves hamper biofuel production. For instance, mixed sugar composition of the macroalgae decreases their efficiency as a biofuel source (Park et al., 2012). Additionally, the presence of large quantities minerals like calcium (Ca), magnesium (Mg), potassium (K) and sodium (Na) often inhibit the production of biofuels (Repérez, 2002). Apart from these, their pyrolysis for production of hydrocarbons, phenols, ketones, furanes, etc. involves huge energy expenditures adding to the production cost (Yanik et al., 2013). Keeping these constraints in mind, researchers from around the world are trying to develop this field by their continuous efforts (Bošnjaković & Sinaga, 2020; Fabris et al., 2020).

From the foregoing discussion it appears that besides technical challenges there are several environmental issues which contradict their claim of being eco-friendly. One more issue that often escapes observation is the role of geo-climatic conditions on the growth of macroalgae. It is well known that the geographical distribution of seaweeds largely depends on the prevalent climatic conditions. For instance, the red alga, *Porphyra* sp. grows commonly in Japan while the brown alga, *Laminaria japonica* is popular in China (Giugni & Giugni, 2010; Lam & Lee, 2019). This makes their distribution vulnerable under the effects of global climate change and will severely impact the location of the biorefineries (Zhou et al., 2010). The sustainability of bioeconomy also becomes a matter of doubt in the future (Zhou et al., 2010). Extinction of algal species or emergence of new invasive ones under the effect of climatic stress factors will require continuous research and developments in this field which will add to their production costs (Khan et al., 2018). According to experts, this augments the unpredictability of seaweeds which will affect the employment opportunities of the biofuels sector and availability of skilled labour because it will no longer remain a stable source of income. Indirectly, this will have strong impacts on biofuel production (V. G. S., et al., 2021). Furthermore, the environmental impacts of biorefinery construction along the ocean shores need to be thoroughly investigated to know their impacts in the future (V. G. S., et al., 2021).

FUTURE PERSPECTIVES

The biofuel sector is naïve and requires sufficient improvisation to increase their yields with lower investments. This provides several research opportunities in this area of which the environmental issues need to be properly addressed. Some current publications in this field show that the environmental issues that are discussed are mainly the ones involved with their direct combustion. Those related to macroalgae cultivation have, however, been poorly explored. This needs extensive research because of the proposed negative impacts on the marine ecosystem in the long run. Suitable mitigation strategies should be undertaken to resolve the issues failing which the sector will not only lose its sustainability but also its affirmation of being eco-friendly substitutes of conventional fuels. Regarding the technological advancements, importance should be given to the development of cost-effective techniques to make the biofuels cheap and readily available. This should not, however, compromise with their efficiency since that may restrict their applicability. The biofuel industry has a huge scope for recycling the wastewater by utilizing them in the cultivation of algae. Considering the growing amounts of wastewater production in the recent years, with increased urbanization and industrialization, developments in this sector can mitigate the problem of wastewater disposal as well.

CONCLUSION

Biofuels have tremendous potentialities to replace the conventional non-renewable energy resources and combat the production of greenhouse gases. This can not only solve the problem of future threats of energy availability but can also address the problem of global climate change. In this regard, the high carbohydrate contents and low proportions of lignin in macroalgae make them efficient sources of biofuels like bioethanol, biogas and biohydrogen. Biodiesel can also be extracted from macroalgae but microalgae gives better yield because of their higher content of unsaturated fatty acids compared to macroalgae. It is the versatility of macroalgae in terms of biofuel production that they are being considered to be more prospective biofuel source in the future. However, there are certain challenges, both technical and environmental, concerning their cultivation and biofuel extraction which have hindered their production in large-scales. Moreover, they have not been as extensively studied as the microalgae. Since, this sector is still in its developmental stages it gives ample opportunity for research. Resolution of the issues in this field can make macroalgae-derived biofuels one of the most promising alternative energy source in the future.

REFERENCES

Afify, A. E. M. R., Shalaby, E. A., & Shanab, S. M. M. (2010). Enhancement of biodiesel production from different species of algae. *Grasas y Aceites*, *61*(4), 416–422. doi:10.3989/gya.021610

Allouache, A., Majda, A., Toudert, A. Z., Amrane, A., & Ballesteros, M. (2021). Cellulosic bioethanol production from *Ulva lactuca* macroalgae. *Cellulose Chemistry and Technology*, *55*(5-6), 629–635. doi:10.35812/CelluloseChemTechnol.2021.55.51

Ameka, G. K., Doamekpor, L. K., Amadu, A. A., & Amamoo, A. P. (2019). Production of biodiesel from marine macroalgae occurring in the Gulf of Guinea, off the coast of Ghana. *Ghana Journal of Science*, *60*(1), 50–58. doi:10.4314/gjs.v60i1.5

Aresta, M., Dibenedetto, A., & Barberio, G. (2005). Utilization of macro-algae for enhanced CO_2 fixation and biofuels production: Development of a computing software for an LCA study. *Fuel Processing Technology*, *86*(14-15), 1679–1693. doi:10.1016/j.fuproc.2005.01.016

Balat, M. (2007). Global Bio-Fuel Processing and Production Trends. *Energy Exploration & Exploitation*, *25*(3), 195–218. doi:10.1260/014459807782009204

Becker, E. W. (1994). *Microalgae: Biotechnology and microbiology*. Cambridge University Press.

Behera, S., Singh, R., Arora, R., Sharma, N. K., Shukla, M., & Kumar, S. (2015). Scope of algae as third generation biofuels. *Frontiers in Bioengineering and Biotechnology*, *2*, 90. doi:10.3389/fbioe.2014.00090 PMID:25717470

Biswal, D. (2021). Impact of climate change on the livestock sector: An overview. *Uttar Pradesh Journal of Zoology*, *42*(22), 117–132.

Biswal, D. (2022). Soil nematodes as the silent sufferers of climate-induced toxicity: Analysing the outcomes of their interactions with climatic stress factors on land cover and agricultural production. *Applied Biochemistry and Biotechnology*. Advance online publication. doi:10.100712010-022-03965-x PMID:35593954

Borines, M. G., de Leon, R. L., & Cuello, J. L. (2013). Bioethanol production from the macroalgae *Sargassum* spp. *Bioresource Technology*, *138*, 22–29. doi:10.1016/j.biortech.2013.03.108 PMID:23612158

Bošnjaković, M., & Sinaga, N. (2020). The perspective of large-scale production of algae biodiesel. *Applied Sciences (Basel, Switzerland)*, *10*(22), 8181. doi:10.3390/app10228181

Bucholc, K., Szymczak-Żyła, M., Lubecki, L., Zamojska, A., Hapter, P., Tjernström, E., & Kowalewska, G. (2014). Nutrient content in macrophyta collected from southern Baltic Sea beaches in relation to eutrophication and biogas production. *The Science of the Total Environment*, *473-474*, 298–307. doi:10.1016/j.scitotenv.2013.12.044 PMID:24374591

Budarin, V. L., Zhao, Y., Gronnow, M. J., Shuttleworth, P. S., Breeden, S. W., Macquarriea, D. J., & Clark, J. H. (2011). Microwave-mediated pyrolysis of macro-algae. *Green Chemistry*, *13*(9), 2330–2333. doi:10.1039/c1gc15560a

Chen, H., Zhou, D., Luo, G., Zhang, S., & Chen, J. (2015). Macroalgae for biofuels production: Progress and perspectives. *Renewable & Sustainable Energy Reviews*, *47*, 427–437. doi:10.1016/j.rser.2015.03.086

Christie, H., Fredriksen, S., & Rinde, E. (1998). Regrowth of kelp and colonization of epiphyte and fauna community after kelp trawling at the coast of Norway. *Hydrobiologia*, *375*, 49–58. doi:10.1023/A:1017021325189

Cromey, C. J., Nickell, T. D., Treasurer, J., Black, K. D., & Inall, M. (2009). Modelling the impact of cod (*Gadus morhua* L) farming in the marine environment-CODMOD. *Aquaculture (Amsterdam, Netherlands)*, *289*(1-2), 42–53. doi:10.1016/j.aquaculture.2008.12.020

Daroch, M., Geng, S., & Wang, G. (2013). Recent advances in liquid biofuel production from algal feedstocks. *Applied Energy*, *102*, 1371–1381. doi:10.1016/j.apenergy.2012.07.031

Demirbas, A. (2009). Progress and recent trends in biodiesel fuels. *Energy Conversion and Management*, *50*(1), 14–34. doi:10.1016/j.enconman.2008.09.001

Demirbas, A. (2010). Use of algae as biofuel sources. *Energy Conversion and Management*, *51*(12), 2738–2749. doi:10.1016/j.enconman.2010.06.010

Fabris, M., Abbriano, R. M., Pernice, M., Sutherland, D. L., Commault, A. S., Hall, C. C., Labeeuw, L., McCauley, J. I., Kuzhiuparambil, U., Ray, P., Kahlke, T., & Ralph, P. J. (2020). Emerging Technologies in Algal Biotechnology: Toward the Establishment of a Sustainable, Algae-Based Bioeconomy. *Frontiers in Plant Science*, *11*, 279. doi:10.3389/fpls.2020.00279 PMID:32256509

Frank, E. D., Han, J., Palou-Rivera, I., Elgowainy, A., & Wang, M. Q. (2012). Methane and nitrous oxide emissions affect the life-cycle analysis of algal biofuels. *Environmental Research Letters*, *7*(1), 014030. doi:10.1088/1748-9326/7/1/014030

Gajraj, R. S., Singh, G. P., & Kumar, A. (2018). Third-Generation Biofuel: Algal Biofuels as a Sustainable Energy Source. In A. Kumar, S. Ogita & Y.-Y. Yau (Eds.), Biofuels: Greenhouse Gas Mitigation and Global Warming (pp. 307-325). Springer (India) Pvt. Ltd.

Gao, G., Burgess, J., Wu, M., Wang, S., & Gao, K. (2020). Using macroalgae as biofuel: Current opportunities and challenges. *Botanica Marina*, *63*(4), 355–370. doi:10.1515/bot-2019-0065

Giugni, D., & Giugni, V. (2010). Intellectual Property: A powerful tool to develop biotech research. *Microbial Biotechnology*, *3*(5), 493–506. doi:10.1111/j.1751-7915.2010.00172.x PMID:21255349

Grala, A., Zieliński, M., Dębowski, M., & Dudek, M. (2012). Effects of hydrothermal depolymerisation and enzymatic hydrolysis of algae biomass on yield of methane fermentation process. *Polish Journal of Environmental Studies*, *21*(2), 363–368.

Greetham, D., Adams, J. M., & Du, C. (2020). The utilization of seawater for the hydrolysis of macroalgae and subsequent bioethanol fermentation. *Scientific Reports*, *10*(1), 9728. doi:10.103841598-020-66610-9 PMID:32546695

Hafting, J. T., Craigie, J. S., Stengel, D. B., Loureiro, R. R., Buschmann, A. H., Yarish, C., Edwards, M. D., & Critchley, A. T. (2015). Prospects and challenges for industrial production of seaweed bioactives. *Journal of Phycology*, *51*(5), 821–837. doi:10.1111/jpy.12326 PMID:26986880

Han, K.-H., Ko, J.-H., & Yang, S. H. (2007). Optimizing lignocellulosic feedstock for improved biofuel productivity and processing. *Biofuels, Bioproducts & Biorefining*, *1*(2), 135–146. doi:10.1002/bbb.14

Hassaan, M. A., El Nemr, A., Elkatory, M. R., Eleryan, A., Ragab, S., El Sikaily, A., & Pantaleo, A. (2021). Enhancement of biogas production from macroalgae *Ulva latuca* via ozonation pretreatment. *Energies*, *14*(6), 1703. doi:10.3390/en14061703

Herbert, R. A. (1999). Nitrogen cycling in coastal marine ecosystems. *FEMS Microbiology Reviews*, *23*(5), 563–590. doi:10.1111/j.1574-6976.1999.tb00414.x PMID:10525167

Horn, H. J., Aasen, I. M., & Østgaard, M. (2000). Production of ethanol from mannitol by *Zymobacter palmae*. *Journal of Industrial Microbiology & Biotechnology*, *24*(1), 51–57. doi:10.1038j.jim.2900771

Horn, S. J. (2000). *Bioenergy from brown seaweeds* [Ph.D. thesis]. NTNU.

Hughes, A. D., Kelly, M. S., Black, K. D., & Stanley, M. S. (2012). Biogas from Macroalgae: Is it time to revisit the idea? *Biotechnology for Biofuels and Bioproducts*, *5*(1), 86. doi:10.1186/1754-6834-5-86 PMID:23186536

Jeong, T. S., Choi, C. H., Lee, J. Y., & Oh, K. K. (2012). Behaviors of glucose decomposition during acid-catalyzed hydrothermal hydrolysis of pretreated *Gelidium amansii*. *Bioresource Technology*, *116*, 435–440. doi:10.1016/j.biortech.2012.03.104 PMID:22522017

Jin, B. B., Duan, P. G., Xu, Y. P., Wang, F., & Fan, Y. C. (2013). Co-liquefaction of micro- and macroalgae in subcritical water. *Bioresource Technology*, *149*, 103–110. doi:10.1016/j.biortech.2013.09.045 PMID:24096026

John, R. P., Anisha, G. S., Nampoothiri, K. M., & Pandey, A. (2011). Micro and macro algal biomass: A renewable source for bioethanol. *Bioresource Technology*, *102*(1), 186–193. doi:10.1016/j.biortech.2010.06.139 PMID:20663661

Joniver, C. F. H., Photiades, A., Moore, P. J., Winters, A. L., Woolmer, A., & Adams, J. M. M. (2021). The global problem of nuisance macroalgal blooms and pathways to its use in the circular economy. *Algal Research*, *58*, 102407. doi:10.1016/j.algal.2021.102407

Jung, A. K., Lim, S. R., Kim, Y., & Park, J. M. (2013). Potentials of macroalgae as feedstocks for biorefinery. *Bioresource Technology*, *135*, 182–190. doi:10.1016/j.biortech.2012.10.025 PMID:23186669

Jung, K. W., Kim, D. H., & Shin, H. S. (2011). Fermentative hydrogen production from *Laminaria japonica* and optimization of thermal pretreatment conditions. *Bioresource Technology*, *102*(3), 2745–2750. doi:10.1016/j.biortech.2010.11.042 PMID:21123054

Kaliaperumal, N., Ramalingam, J. R., Kalimuthu, S., & Ezhil Valavan, R. (2002). Seasonal changes in growth, biochemical constituents and phycocolloid of some marine algae of mandapam coast. *Seaweed Research and Utilisation*, *24*(1), 73–77.

Khan, M. I., Shin, J. H., & Kim, J. D. (2018). The promising future of microalgae: Current status, challenges, and optimization of a sustainable and renewable industry for biofuels, feed, and other products. *Microbial Cell Factories*, *17*(1), 36. doi:10.118612934-018-0879-x PMID:29506528

Klinke, H. B., Thomsen, A. B., & Ahring, B. K. (2004). Inhibition of ethanol-producing yeast and bacteria by degradation products produced during pre-treatment of biomass. *Applied Microbiology and Biotechnology*, *66*(1), 10–26. doi:10.100700253-004-1642-2 PMID:15300416

Kraan, S. (2013). Mass-cultivation of carbohydrate rich macroalgae, a possible solution for sustainable biofuel production. *Mitigation and Adaptation Strategies for Global Change*, *18*(1), 27–46. doi:10.100711027-010-9275-5

Kumar, D., Eswari, A. P., Park, J.-H., Adishkumar, S., & Banu, J. R. (2019). Biohydrogen generation from macroalgal biomass, *Chaetomorpha antennina* through surfactant aided microwave disintegration. *Frontiers in Energy Research*, *7*, 78. doi:10.3389/fenrg.2019.00078

Kumar, K., Ghosh, S., Angelidaki, I., Holdt, S. L., Karakashev, D. B., Morales, M. A., & Das, D. (2016). Recent developments on biofuels production from microalgae and macroalgae. *Renewable & Sustainable Energy Reviews*, *65*, 235–249. doi:10.1016/j.rser.2016.06.055

Kumar, M. D., Tamilarasan, K., Kaliappan, S., Banu, J. R., Rajkumar, M., & Kim, S. H. (2018). Surfactant assisted disperser pretreatment on the liquefaction of *Ulva reticulata* and evaluation of biodegradability for energy efficient biofuel production through nonlinear regression modelling. *Bioresource Technology, 255*, 116–122. doi:10.1016/j.biortech.2018.01.116 PMID:29414156

Lam, M. K., & Lee, K. T. (2019). Scale-Up and Commercialization of Algal Cultivation and Biofuel Production. In A. Pandey, D.-J. Lee, Y. Chisti, & C. R. Soccol (Eds.), *Biofuels from Algae* (pp. 261–286). doi:10.1016/B978-0-444-64192-2.00019-6

Lee, K. C., Arai, T., Ibrahim, D., Kosugi, A., Prawitwong, P., Lan, D., Murata, Y., & Mori, Y. (2014). Purification and characterization of a thermostable laminarinase from *Penicillium rolfsii* c3-2(1) IBRL. *Bioresource, 9*(1), 1072–1084. doi:10.15376/biores.9.1.1072-1084

Levine, R. B., Pinnarat, T., & Savage, P. E. (2010). Biodiesel production from wet algal biomass through in situ lipid hydrolysis and supercritical transesterification. *Energy & Fuels, 24*(9), 5235–5243. doi:10.1021/ef1008314

Liao, Y. C., Chang, C. C., Nagarajan, D., Chen, C. Y., & Chang, J. S. (2021). Algae-derived hydrocolloids in foods: Applications and health-related issues. *Bioengineered, 12*(1), 3787–3801. doi:10.1080/21655979.2021.1946359 PMID:34281484

Lorentsen, S.-H., Sjøtun, K., & Grémillet, D. (2010). Multi-trophic consequences of kelp harvest. *Biological Conservation, 143*(9), 2054–2062. doi:10.1016/j.biocon.2010.05.013

Lüning, K., & Pang, S. (2003). Mass cultivation of seaweeds: Current aspects and approaches. *Journal of Applied Phycology, 15*(2/3), 115–119. doi:10.1023/A:1023807503255

Mabee, W. E., Gregg, D. J., & Saddler, J. N. (2005). Assessing the emerging biorefinery sector in Canada. *Applied Biochemistry and Biotechnology, 123*(1-3), 765–778. doi:10.1385/ABAB:123:1-3:0765 PMID:15930556

Maceiras, R., Rodríguez, M., Cancela, A., Urréjola, S., & Sánchez, A. (2011). Macroalgae: Raw material for biodiesel production. *Applied Energy, 88*(10), 3318–3323. doi:10.1016/j.apenergy.2010.11.027

Margareta, W., Nagarajan, D., Chang, J.-S., & Lee, D.-J. (2020). Dark fermentative hydrogen production using macroalgae (*Ulva* sp.) as the renewable feedstock. *Applied Energy, 262*, 114574. doi:10.1016/j.apenergy.2020.114574

Mat Aron, N. S., Khoo, K. S., Chew, K. W., Show, P. L., Chen, W.-H., & Nguyen, T. H. P. (2020). Sustainability of the four generations of biofuels – A review. *International Journal of Energy Research, 44*(12), 9266–9282. doi:10.1002/er.5557

Mata, T. M., Martins, A. A., & Caetano, N. S. (2009). Microalgae for biodiesel production and other applications: A review. *Renewable & Sustainable Energy Reviews, 14*(1), 217–232. doi:10.1016/j.rser.2009.07.020

Mautner, H. G. (1954). The chemistry of brown algae. *Economic Botany, 8*(2), 174–192. doi:10.1007/BF02984737

McHugh, D. J. (2003). *A guide to the seaweed industry*. Food and Agriculture Organization of the United Nations.

Meher, L. C., Vidya Sagar, D., & Naik, S. N. (2006). Technical aspects of biodiesel production by transesterification – a review. *Renewable & Sustainable Energy Reviews*, *10*(3), 248–268. doi:10.1016/j.rser.2004.09.002

Munir, N., Sharif, N., Naz, S., Saleem, F., & Manzoor, F. (2013). Harvesting and processing of microalgae biomass fractions for biodiesel production (a review). *Science Technology and Development*, *32*(3), 235–243.

Mutripah, S., Meinita, M. D. N., Kang, J.-Y., Jeong, G.-T., Susanto, A. B., Prabowo, R. E., & Hong, Y.-K. (2014). Bioethanol production from the hydrolysate of *Palmaria palmata* using sulfuric acid and fermentation with brewer's yeast. *Journal of Applied Phycology*, *26*(1), 687–693. doi:10.100710811-013-0068-6

Naik, K. K., & Naik, T. P. (2021). Synthesis of biodiesel by using marine Macro-algae of Karwar Region, Karnataka, India. *Gorteria Journal*, *34*(1), 26–34.

Naik, S. N., Goud, V. V., Rout, P. K., & Dalai, A. K. (2010). Production of first and second generation biofuels: A comprehensive review. *Renewable & Sustainable Energy Reviews*, *14*(2), 578–597. doi:10.1016/j.rser.2009.10.003

Olsson, J., Toth, G. B., & Albers, E. (2020). Biochemical composition of red, green and brown seaweeds on the Swedish west coast. *Journal of Applied Phycology*, *32*(5), 3305–3317. doi:10.100710811-020-02145-w

Pardilhó, S., Boaventura, R., Almeida, M., & Dias, J. M. (2022). Marine macroalgae waste: A potential feedstock for biogas production. *Journal of Environmental Management*, *304*, 114309. doi:10.1016/j.jenvman.2021.114309 PMID:34933268

Park, J. H., Hong, J. Y., Jang, H. C., Oh, S. G., Kim, S. H., Yoon, J. J., & Kim, Y. J. (2012). Use of *Gelidium amansii* as a promising resource for bioethanol: A practical approach for continuous dilute-acid hydrolysis and fermentation. *Bioresource Technology*, *108*, 83–88. doi:10.1016/j.biortech.2011.12.065 PMID:22261657

Park, J. H., Yoon, J. J., Park, H. D., Kim, Y. J., Lim, D. J., & Kim, S. H. (2011). Feasibility of biohydrogen production from *Gelidium amansii*. *International Journal of Hydrogen Energy*, *36*(21), 13997–14003. doi:10.1016/j.ijhydene.2011.04.003

Peter, A. P., Khoo, K. S., Chew, K. W., Ling, T. C., Ho, S.-H., Chang, J.-S., & Show, P. L. (2021). Microalgae for biofuels, wastewater treatment and environmental monitoring. *Environmental Chemistry Letters*, *19*(4), 2891–2904. doi:10.100710311-021-01219-6

Popoola, T. O. S., & Yangomodou, O. D. (2006). Extraction, properties and utilization potentials of cassava seed oil. *Biotechnology*, *5*(1), 38–41. doi:10.3923/biotech.2006.38.41

Radha, M., & Murugesan, A. G. (2017). Enhanced dark fermentative biohydrogen production from marine macroalgae *Padina tetrastromatica* by different pretreatment processes. *Biofuel Research Journal, 13*(1), 551–558. doi:10.18331/BRJ2017.4.1.5

Reid, W. V., Ali, M. K., & Field, C. B. (2020). The future of bioenergy. *Global Change Biology, 26*(1), 274–286. doi:10.1111/gcb.14883 PMID:31642554

Repérez, P. (2002). Mineral content of edible marine seaweeds. *Food Chemistry, 79*(1), 23–26. doi:10.1016/S0308-8146(02)00171-1

Roesijadi, G. (2010). Macroalgae as a biomass feedstock: a preliminary analysis. U.S. Department of Energy under contract, Pacific Northwest National Laboratory. doi:10.2172/1006310

Romagnoli, F., Blumberga, D., & Gigli, E. (2011). Biogas from marine macroalgae: A new environmental technology-life cycle inventory for a further LCA. *Environmental and Climate Technologies, 4*(1), 97–108.

Romagnoli, F., Pastare, L., Sabūnas, A., Bāliņa, K., & Blumberga, D. (2017). Effects of pre-treatment on Biochemical Methane Potential (BMP) testing using Baltic Sea *Fucus vesiculosus* feedstock. *Biomass and Bioenergy, 105*, 23–31. doi:10.1016/j.biombioe.2017.06.013

Roy, S., Kumar, K., Ghosh, S., & Das, D. (2014). Thermophilic biohydrogen production using pre-treated algal biomass as substrate. *Biomass and Bioenergy, 61*, 157–166. doi:10.1016/j.biombioe.2013.12.006

Santelices, B. (2007). The discovery of kelp forests in deep-water habitats of tropical regions. *Proceedings of the National Academy of Sciences of the United States of America, 104*(49), 19163–19164. doi:10.1073/pnas.0708963104 PMID:18042707

Saqib, A., Tabbssum, M. R., Rashid, U., Ibrahim, M., Gill, S. S., & Mehmood, M. A. (2013). Marine macroalgae *Ulva*: A potential feed-stock for bioethanol and biogas production. *Asian Journal of Agriculture and Biology, 1*, 155–163.

Sasaki, Y., Takagi, T., Motone, K., Shibata, T., Kuroda, K., & Ueda, M. (2018). Direct bioethanol production from brown macroalgae by co-culture of two engineered *Saccharomyces cerevisiae* strains. *Bioscience, Biotechnology, and Biochemistry, 82*(8), 1459–1462. doi:10.1080/09168451.2018.1467262 PMID:29708475

Shaishav, S., Singh, R. N., & Satyendra, T. (2013). Biohydrogen from algae: Fuel of the future. *International Research Journal of Environmental Sciences, 2*(4), 44–47.

Shilton, A., & Guieysse, B. (2010). Sustainable sunlight to biogas is via marginal organics. *Current Opinion in Biotechnology, 21*(3), 287–291. doi:10.1016/j.copbio.2010.03.008 PMID:20378331

Shukla, R., Kumar, M., Chakraborty, S., Gupta, R., Kumar, S., Sahoo, D., & Kuhad, R. C. (2016). Process development for the production of bioethanol from waste algal biomass of *Gracilaria verrucosa. Bioresource Technology, 220*, 584–589. doi:10.1016/j.biortech.2016.08.096 PMID:27619709

Siddiqua, S., Mamun, A. A., & Enayetul Babar, S. M. (2015). Production of biodiesel from coastal macroalgae (*Chara vulgaris*) and optimization of process parameters using Box-Behnken design. *SpringerPlus, 4*(1), 720. doi:10.118640064-015-1518-1 PMID:26636008

Singh, A., Nigam, P. S., & Murphy, J. D. (2011). Renewable fuels from algae: An answer to debatable land based fuels. *Bioresource Technology, 102*(1), 10–16. doi:10.1016/j.biortech.2010.06.032 PMID:20615690

Singh, A., & Olsen, S. I. (2011). A critical review of biochemical conversion, sustainability and life cycle assessment of algal biofuels. *Applied Energy, 88*(10), 3548–3555. doi:10.1016/j.apenergy.2010.12.012

Singh, D. P., & Trivedi, R. K. (2013). Production of biofuel from algae: An economic and eco-friendly resource. *International Journal of Scientific Research, 2*(11), 352–357.

Soliman, R. M., Younis, S. A., El-Gendy, N. S., Mostafa, S., El-Temtamy, S. A., & Hashim, A. I. (2018). Batch bioethanol production via the biological and chemical saccharification of some Egyptian marine macroalgae. *Journal of Applied Microbiology, 125*(2), 422–440. doi:10.1111/jam.13886 PMID:29675837

Springer, Y. P., Hays, C. G., Carr, M. H., & Mackey, M. R. (2010). Toward ecosystem-based management of marine macroalgae-the bull kelp, *Nereocystis luetkeana. Oceanography and Marine Biology, 48*, 1–41.

Stevens, C. V., & Verhe, R. (2004). *Renewable bioresources scope and modification for nonfood application*. John Wiley and Sons Ltd.

Subba Rao, P. V., & Mantri, V. A. (2006). Indian seaweed resources and sustainable utilization: Scenario at the dawn of a new century. *Current Science, 91*(2), 164–174.

Suganya, T., Nagendra Gandhi, N., & Renganathan, S. (2013). Production of algal biodiesel from marine macroalgae Enteromorpha compressa by two step process: Optimization and kinetic study. *Bioresource Technology, 128*, 392–400. doi:10.1016/j.biortech.2012.10.068 PMID:23201520

Suganya, T., & Renganathan, S. (2014). Ultrasonic assisted acid base transesterification of algal oil from marine macroalgae *Caulerpa peltata*: Optimization and characterization studies. *Fuel, 128*, 347–355. doi:10.1016/j.fuel.2014.03.037

Suganya, T., Varman, M., Masjuki, H. H., & Renganathan, S. (2016). Macroalgae and microalgae as a potential source for commercial applications along with biofuels production: A biorefinery approach. *Renewable & Sustainable Energy Reviews, 55*, 909–941. doi:10.1016/j.rser.2015.11.026

Surendhiran, D., & Vijay, M. (2012). Microalgal biodiesel-a comprehensive review on the potential and alternative biofuel. *Research Journal of Chemical Sciences, 2*(11), 71–82.

Tan, I. S., Lam, M. K., & Lee, K. T. (2013). Hydrolysis of macroalgae using heterogeneous catalyst for bioethanol production. *Carbohydrate Polymers, 94*(1), 561–566. doi:10.1016/j.carbpol.2013.01.042 PMID:23544575

Topare, N., Rout, S. J., Renge, V. C., Khedkar, S. V., Chavan, Y. P., & Bhagat, S. L. (2011). Extraction of oil from algae by solvent extraction and oil expeller method. *International Journal of Chemical Sciences, 9*(4), 1746–1750.

V, G. S., M, D. K., Pugazhendi, A., Bajhaiya, A. K., Gugulothu, P., & J, R. B. (2021). Biofuel production from Macroalgae: Present scenario and future scope. *Bioengineered, 12*(2), 9216–9238.

Vivekanand, V., Eijsink, V. G. H., & Horn, S. J. (2012). Biogas production from the brown seaweed *Saccharina latissima*: Thermal pretreatment and co-digestion with wheat straw. *Journal of Applied Phycology*, *24*(5), 1295–1301. doi:10.100710811-011-9779-8

Wargacki, A. J., Leonard, E., Win, M. N., Regitsky, D. D., Santos, C. N., Kim, P. B., Cooper, S. R., Raisner, R. M., Herman, A., Sivitz, A. B., Lakshmanaswamy, A., Kashiyama, Y., Baker, D., & Yoshikuni, Y. (2012). An engineered microbial platform for direct biofuel production from brown macroalgae. *Science*, *335*(6066), 308–313. doi:10.1126cience.1214547 PMID:22267807

Wei, N., Quarterman, J., & Jin, Y. S. (2013). Marine macroalgae: An untapped resource for producing fuels and chemicals. *Trends in Biotechnology*, *31*(2), 70–77. doi:10.1016/j.tibtech.2012.10.009 PMID:23245657

Wilcox, H. A. (1977). The U.S. Navy's Ocean food and energy farm project. In *Ocean energy resources; Proceedings of the Energy Technology Conference, Houston, Texas* (pp. 83-104). American Society of Mechanical Engineers.

Wu, Y. N., Mattsson, M., Ding, M. W., Wu, M. T., Mei, J., & Shen, Y. L. (2019). Effects of different pretreatments on improving biogas production of macroalgae *Fucus vesiculosus* and *Fucus serratus* in Baltic Sea. *Energy & Fuels*, *33*(3), 2278–2284. doi:10.1021/acs.energyfuels.8b04224

Xu, X., Kim, J. Y., Oh, Y. R., & Park, J. M. (2014). Production of biodiesel from carbon sources of macroalgae, *Laminaria japonica*. *Bioresource Technology*, *169*, 455–461. doi:10.1016/j.biortech.2014.07.015 PMID:25084043

Yahmed, N. B., Jmel, M. A., Alaya, M. B., Bouallagui, H., Marzouki, M. N., & Smaali, I. (2016). A biorefinery concept using the green macroalgae *Chaetomorpha linum* for the coproduction of bioethanol and biogas. *Energy Conversion and Management*, *119*, 257–265. doi:10.1016/j.enconman.2016.04.046

Yanagisawa, M., Nakamura, K., Ariga, O., & Nakasaki, K. (2011). Production of high concentrations of bioethanol from seaweeds that contain easily hydrolyzable polysaccharides. *Process Biochemistry*, *46*(11), 2111–2116. doi:10.1016/j.procbio.2011.08.001

Yanik, J., Stahl, R., Troeger, N., & Sinag, A. (2013). Pyrolysis of algal biomass. *Journal of Analytical and Applied Pyrolysis*, *103*, 134–141. doi:10.1016/j.jaap.2012.08.016

Yen, H. W., & Brune, D. E. (2007). Anaerobic co-digestion of algal sludge and waste paper to produce methane. *Bioresource Technology*, *98*(1), 130–134. doi:10.1016/j.biortech.2005.11.010 PMID:16386894

Yin, Y., Hu, J., & Wang, J. (2019). Fermentative hydrogen production from macroalgae *Laminaria japonica* pretreated by microwave irradiation. *International Journal of Hydrogen Energy*, *44*(21), 10398–10406. doi:10.1016/j.ijhydene.2019.03.034

Yin, Y., & Wang, J. (2019). Enhanced biohydrogen production from macroalgae by zero-valent iron nanoparticles: Insights into microbial and metabolites distribution. *Bioresource Technology*, *282*, 110–117. doi:10.1016/j.biortech.2019.02.128 PMID:30852330

Yoon, M., Choi, J. I., Lee, J. W., & Park, D. H. (2012). Improvement of saccharification process for bioethanol production from *Undaria* sp. by gamma irradiation. *Radiation Physics and Chemistry*, *81*(8), 999–1002. doi:10.1016/j.radphyschem.2011.11.035

Zabaniotou, A., Ioannidou, O., & Skoulou, V. (2008). Rapeseed residues utilization for energy and 2nd generation biofuels. *Fuel*, *87*(8-9), 1492–1502. doi:10.1016/j.fuel.2007.09.003

Zhou, D., Zhang, L., Zhang, S., Fu, H., & Chen, J. (2010). Hydrothermal liquefaction of macroalgae *Enteromorpha prolifera* to bio-oil. *Energy & Fuels*, *24*(7), 4054–4061. doi:10.1021/ef100151h

Chapter 10
A Source of Future Energy: Gas Hydrates

Anupama Kumari
https://orcid.org/0000-0003-2318-6090
Indian Institute of Technology, Roorkee, India

Mukund Madhaw
Independent Researcher, India

C. B. Majumder
Indian Institute of Technology, Roorkee, India

ABSTRACT

Natural gas hydrate (NGH) has emerged as a future source of energy for the world that can fulfill the future energy demand. The natural gas hydrates hold twice the amount of energy than all available fossil fuels. At present, the energy demand is growing, but these are not available commercially. Various field applications on the dissociation of the natural gas hydrates have been performed till now, and the output of these field applications confirms the possibility of dissociation of the natural gas hydrates. At distinct locations in the world, the huge deposits of hydrates have been identified, but the safe and economic dissociation of gas hydrates is not known yet. Hence, the evolution of new and inexpensive, suitable methods and technologies are required for the dissociation of gas hydrates. In this chapter, the authors reviewed several articles to compare the different dissociation techniques of gas hydrates and their field applications. The advantages and disadvantages of these production methods have also been discussed.

INTRODUCTION

Gas hydrates are non-stoichiometric solid formed naturally by the entrapment of gas within the crystal structure of water molecules. When water molecules entrap gas molecules at high pressure and low temperature, then create the cage-like structure of gas hydrates. It is also called combustible gas because it can be burned at ambient temperature and pressure. Gas hydrates can be formed by methane, ethane,

DOI: 10.4018/978-1-6684-5269-1.ch010

hydrogen sulfide, carbon dioxide, nitrogen, propane, isobutane, n-butane, etc (Gabitto and Tsouris 2010). At equilibrium conditions, 1m^3 of natural gas hydrates contain 160-180 m^3 of natural gas (Li et al. 2011). Naturally, the methane gas hydrates are made by combining the water and the methane gas released either by organic decomposition or by thermogenic gas leaking beneath the sea level or permafrost region (Zhou et al. 2009). The natural deposits of gas hydrates are classified as Class 1, 2, and 3 types of reservoirs. The Class 1 type of hydrate deposits are consisting of a layer of hydrate and an underlying free gas zone. The Class 2 deposits are consisting of a layer of hydrate and an underlying mobile water zone. The Class 3 type hydrate deposits are created with an isolated layer of hydrate with no underlying zone of free gas or mobile water. In Class 3 deposits, the whole layer of hydrates is in the stability zone and enclosed by the impermeable lower and upper burden layer (Huang et al. 2016).

Sir Humphrey Davy discovered the gas hydrates for the first time in 1810, but there is also a possibility that Priestley found it in 1778 (Sloan and Koh 1998). Natural gas hydrates can be found in the high altitudes/latitudes of permafrost or in the sediment layers, which occurs in deep water on the continental margins (Kvenvolden 1998; Kvenvolden and Rogers 2005). Small parts of red lakes, i.e., the continental land, seas of the Caspian Sea, Lake Baikal, and the Black Sea, also contain natural gas hydrates. These reserves were discovered in mountain permafrost regions or deep water, such as South China Sea, the Gulf of Mexico, Indian Ocean, Siberia, Arctic Alaska, multiple areas of Qilian and Mainland China (Ginsburg et al.1992; Ginsburg and Soloviev 1994; Lu et al. 2011). The curiosity for scientific research in gas hydrates increased exponentially when it became responsible for the blockage in oil and gas pipelines. Hence various research studies are in progress for the dissociation of the gas hydrates because it can solve the crisis of energy in the future and then become a potential source of energy in the world (Koh et al. 2009). Currently, the available methods for decomposing gas hydrates are chemical injection, depressurization and thermal stimulation (Kumari et al. 2020).

There is a need to find the cost-effective technologies for the economic dissociation of the gas hydrates, and the rate of dissociation will control this economic production of gas. Figure 1 depicts the available dissociation techniques of gas hydrates. Table 1 listed the list of some patents filed on new dissociation techniques of gas hydrates.

Figure 1. Dissociation techniques of gas hydrates
(Arora et al. 2015a)

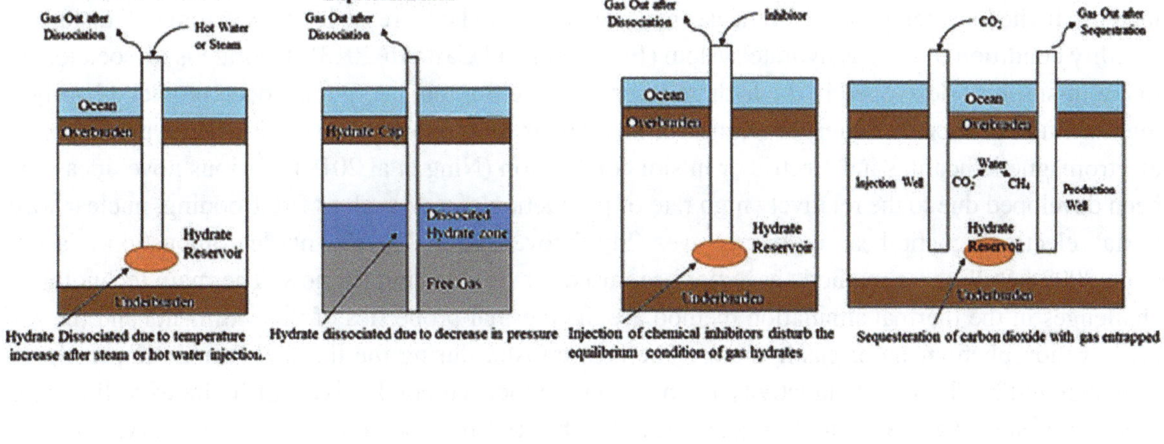

Table 1. List of some Patents filed on New Techniques for the dissociation of gas hydrate.

Assignee	Description of Patent	References
Mobil Oil Corporation, Fairfax, VA (US)	Electromagnetic radiation can be applied for the recovery of gas from the gas hydrates.	(Heinemann et al. 2001)
ConocoPhilips Company, Houston, TX(US)	The substitution of the gas has formed a more stable hydrate structure with the releasing agent without melting the gas hydrates.	(Graue et al. 2007)
New Mexico Tech Research Foundation, Socorr, N.M.	Apparatus and methods for the transportation, generation, and the dissociation of hydrates	(Guo et al. 1995)
ConocoPhilips Company, Houston, TX(US)	By the application of electromagnetic radiation, methane can be recovered.	(Singhe 2013)
The University of Tulsa, Tulsa, OK (US)	The electric heater was used in the opening of hydrate rich subset sediment, and then the released gas is collected in an overhead receiver.	(Zhang et al. 2009)
Chevron U.S.A. Inc., San Ramon, CA (US)	A two-stage method has been employed to produce fluids from a Class 3 reservoir.	(Gullapalli et al. 2009)
Halliburton Energy Services, Inc., s s Duncan, Okla.	The method or the decomposition of hydrates is to combine the acidic liquid with basic liquid to form an exothermic chemical reaction and form a salt solution.	(Chatterji and Griffith 1998)

The main objective of this chapter is to review the various dissociation methods available for gas hydrates and compare them based on different parameters such as cost, suitability for the environment, and stability conditions. We also compare these methods by focusing on their importance and problems that arise by using them and investigated the stability parameters, i.e., temperature and pressure on which gas is dissociated. We will highlight the drawbacks of various dissociation techniques and introduce some new methods, which can be used. Also studied the implementation conditions for these methods. Finally, a section of conclusions, final remarks, comparisons, issues, and directions for future research has been discussed for all dissociation techniques of gas hydrates.

THERMAL STIMULATION

In the thermal stimulation dissociation method, the gas hydrate is dissociated into water and gas after an increase in the local temperature of the gas hydrates system. The increase in temperature will disturb the stability condition of the gas hydrate system (Fitzgerald and Castaldi 2013). Hydrate is dissociated due to the in-situ heat developed in the hydrate reservoir. The thermal energy has been provided by injecting hot fluids like brine, water, or steam from the upper surface or maybe by downhole processes like electromagnetic heating and electric or in situ combustion (Ning et al.2010). Various novel ideas have been developed due to the relatively high rate of production of gas, such as fire flooding, nuclear waste burial, electromagnetic heating, microwaves (Shashkov et al. 2012), downhole combustion (Castaldi et al. 2007). Still, these methods have not been used for field testing till now. The main technological challenges in the thermal simulation method are the thermal properties of the sediments and the self-preservation phenomena of gas hydrates. The heat transfer during the thermal stimulation method is characterized by thermal conductivity mainly, and the thermal conductivity of hydrates will increase by the presence of minerals in sediments. The drawback of this method is the self-preservation of ice around the hydrate after the dissociation below the freezing point of ice. The water extracted after the dissociation of gas hydrate can again form the ice because of the endothermic heat of dissociation and prevent further hydrate dissociation (Handa 1986; Shimada et al. 2005; Chong et al. 2016). Su et al. 2012 examined the production potential of the Huff and puff technique for the production of gas in the Shenu area on the northern continental slope of south China sea. They dissociated hydrate by producing fluids and injecting hot water alternatively. They suggested that the huff and puff technique through a vertical wall is not favorable and has not industrial potential (Su et al. 2012). Ning et al. 2010 proposed a technique to use renewable solar energy to provide energy for the dissociation of marine gas hydrates with the thermal stimulation method. In the first step, the conversion of solar energy to electric energy has proceeded. Then heat energy has been created by operating the downhole heaters such as electromagnetic waves and the microwave heaters (Ning et al. 2010). Yang et al. 2010 investigated the efficiency of the single well hot water cyclic injection method to recover the gas from methane hydrate-bearing sand. In this method, the overall temperature enhances with the injection of hot water and reduces with the production of gas. This technique is favorable for the hydrate with low permeability and high saturation (Yang et al. 2010). Phirani and Mohanty 2009 suggested that the injection temperature (323 K), pressure (2 & 4 MPa), and production pressure take an important part for the hydrate reservoirs of confined class 2 type. They concluded that when the pressure and the injection temperature of the production well are high, then warm water flooding is more suitable than depressurization (Phirani and Mohanty 2009). Castaldi et al. 2007 suggested that the downhole combustion technique is very energy efficient for gas production. This method uses local source of heat in the hydrate reservoir (Castaldi et al. 2007). Li et al. 2013 designed a three-dimensional pressure vessel to predict the dissociation of methane gas hydrate by steam-assisted gravity drainage method with different rates of injection of steam (Li et al. 2013). Kawamura et.al 2007 experimented with the steam injection to examine the thermal control system on the artificial sediment with and without methane hydrate. Approximately 44% of the hydrate origin gas was dissociated after the injection of steam (Kawamura et al. 2007). Sakamoto et.al. 2004 developed a special type of apparatus and experimented with showing the physical phenomena and production behavior of gas from hydrates by the injection of hot water. The production of gas from hydrate increases with time (Sakamoto et al. 2004). Sasaki, et al. 2007 proposed a novel system for the recovery of gas from methane gas hydrates layers by injecting hot water by manufacturing dual horizontal wells. The production rate of gas increases with the rate of hot water injection and the absolute permeability (Sasaki et al. 2007). Cranganu 2009 proposed that instead of using hot fluid injection from the surface, an air or gas fuel mixture can be injected into the combustion vessel from the upper surface and then into the gas hydrate zone directly (Cranganu 2009). Li et al. 2008 examined the dissociation behavior of gas hydrate using hot brine stimulation. With the doses of brine in the vessel, the dissociation of hydrate accompanied the decrease in temperature (Li et al. 2008).

The interaction characteristics between electromagnetic waves and disperse systems, the high-frequency energy, and microwave electromagnetic fields can be used to decompose gas hydrates (Fatykhov and Bagautdinov 2005). A heat source has been generated because of the interaction between high-frequency electromagnetic radiation and medium, and the energy of propagating electromagnetic wave converges to heat. This heat source distributed over the gas hydrate volume, and then phase transition occurs in the absence of temperature gradient. In the case of large zones of phase transition can occur, due to which the decomposition temperature of gas hydrate is attained, and the energy of the electromagnetic field has been extended to the dissociation of gas hydrates (Nasyrov et al. 1997). Islam 1994 proposed a new production technique for producing natural gas from Alaskan north slope reservoirs by using electromagnetic heating with the vertical and horizontal wells. They also concluded that the porosity and the thickness of the gas hydrate reservoir are essential factors for the determination of the total amount of recovered gas (Islam 1994). Rahim et al. 2015 suggested that the plasma decomposition rate of methane to hydrogen production is faster than the methane hydrate decomposition by the microwave plasma method (Rahim et al. 2015). Liang et al. 2009 discovered that the decomposition rate is higher for the high power of microwave. However, a heat efficiency problem is raised as the temperature of the system increases. The temperature difference in the system resulted from the difference of dielectric properties of components (Liang et al. 2009). He et al. 2011 observed the decomposition behavior of gas hydrate in the unconsolidated sediment. They concluded that the sediment could supply an adequate amount of heat for decomposition, then the rate of hydrate decomposition is controlled by the intrinsic kinetics (He et al. 2011). Zhao et al. 2016 examined the dissociation of methane gas hydrates formed in the sediment by applying microwave stimulation (Zhao et al. 2016). Amit Arora 2015 proposed dissociation techniques of the gas hydrate by using microwaves (Amit Arora 2015).

Depressurization

This method is the first production method that has been applied outside the laboratory. The stability of gas hydrates is disturbed by exposing the hydrates to a low-pressure domain where the hydrates become unstable and then dissociated into water and gas. The released gas can be recovered by applying conventional technology (Ji et al. 2001; Moridis et al. 2005; Tsypkin 2005; Tang et al. 2007; Demirbas 2010a). The depressurization method is most favorable for those natural deposits where inescapable gas is found in a closure that forms the hydrate cap above it (Max and Cruickshank 1999). A well is extended and drilled into the hydrate layer and formed in the zone of free gas. The dissociated gas from this hydrate layer causes a reduction in pressure and then dissociation of the hydrate. The dissociation of hydrates in the Messoyakha fields has been done by the depressurization method (Makogon et al. 2007; Makogon 2010). There is no need for expensive continuous stimulation for the depressurization method. Hence it is an effective method for future gas hydrate dissociation et al. 2010).

Different methods have been suggested for the economic development of hydrates as an unconventional source of gas (Sawyer et al. 2000; Goel et al. 2001). The dissociation of free gas using traditional techniques generates a pressure drop. The hydrate becomes unstable because of this pressure drop and dissociates gas hydrates into the water and free gas. Three mechanisms involved in this method are dissociation, heat transfer with convection and conduction, and flow of water and gas (Goel et al. 2001; Kawamura et al. 2003). The most economical method for the dissociation of hydrates is depressurization because an external source of energy is not needed in this method as compared to other methods. But the drawback is that this method has a lower production rate due to slow reaction for the dissociation (Lee et al. 2010). In the depressurization method, the in-situ permeability or free gas fraction will affect the gas production rate. The depressurization method induced dissociation is limited by heat transfer and mass transfer, and free gas saturation is the critical factor for establishing the feasibility of gas production from marine deposits. The gas production at low pressure is fast compared to gas production at high pressure (Chong et al. 2016). Commercially this technique was used by the USSR to dissociate gas from gas hydrates in the Messoyakha field (Makogon 1997b). Kono et al. 2002 analyzed the rate of dissociation of methane gas hydrate using depressurization and also concluded that the properties of sediments could control the rate (Kono et al. 2002). Haligva et al. 2010 analyzed the decomposition of silica sand particle hydrate by depressurization method (Haligva et al. 2010). Ji et al. 2001 found out that the production rate of natural gas from a reservoir depends on well pressure, the reservoir's temperature, and the permeability of the reservoir zone (Ji et al. 2001). Kim et.al 2013 experimented on pressure cores, which were recovered from holes of UBGH (Ulleung Basin Gas Hydrate) (Kim et.al 2013). Moridis and Sloan 2007 evaluated gas recovery's potential from low saturation and dispersed hydrate formed in oceanic sediments. They also estimated the rate of gas and water production (Moridis and Sloan 2007). Lee et al. 2010 developed an experimental equipment to observe the dissociation behavior of gas hydrate in the porous rock (Lee et al. 2010). The depressurization technique was applied for field testing in Nankai Trough of Japan using two producer wells (Yu et al. 2019).

CHEMICAL INJECTION

In the chemical injection method, the equilibrium condition of hydrates must be displaced by the injection of liquid inhibitor or chemical towards the high pressure and low temperatures. This technique is also expensive, but it is less costly than thermal stimulation (Demirbas 2010a). The main factors to be considered in the chemical injection method are the concentration and temperature of chemicals, rate of injection of chemicals, pressure, and the interfacial area between hydrate and chemical. The heat of dissociation will be decreased after the chemical injection (Fan et al. 2006; Chong et al. 2016).

Table 2. List of Studies on Gas Production from Gas Hydrates by Chemical Injection Method

S.No	Description	Chemical	Advantages	Disadvantages	References
1	Hydrate saturation affects the onset time, but it has less effect on the decomposition rate. The factors, which were considered in this study, are injection rate and saturation.	N2	An optimum N2 injection rate enhances the economic benefits. The N2 injection method is advisable for the exploitation of low hydrate saturation, and it is more advantageous than depressurization.	They did not consider the effects of deposit conditions of hydrates on dissociation by N2 injection.	(Zhang et al. 2017)
2	They analyzed the effect of brine salinity and recovery characteristics of gas from the injection of various concentrations of brine.	Brine	Permeability is an essential factor in determining the production of gas from hydrates. At a low concentration of brine, gas production is excessively high.	There is a significant reduction in the rate of production of gas because of the high concentration of brine.	(Lee 2010)
3	Experimented with the gas recovery from methane gas hydrate in the porous sediment by applying the different rate of injection of ethylene glycol and concentration in a one-dimensional experimental machine	Ethylene Glycol	The efficiency of gas production is disturbed by the concentration of chemicals and the rate of injection, and at last, it attains a maximum value.		(Li et al. 2007)
4	Conducted the dissociation experiment to synthesize the gas hydrate by adding ethylene glycol as an inhibitor in the reaction vessel	Ethylene Glycol	This study has importance not only for natural gas hydrate development but also for transportation and natural gas production.		(Fan et al. 2006)
5	Developed the model to observe the production behavior from the injection of methanol and combined the depressurization mechanism.	Methanol	They prove that the methanol content is the most sensitive parameter for the time taken for the complete dissociation of the well block.	When the injection is done after high production time, the injection effect is not expected to be much during the production.	(Sung et al. 2002)
6	Examined the characteristics of the dissociation process of the hydrate.	Ethylene Glycol and Methanol and	Include the effect of inhibitors on hydrate formation in the natural gas processing, transportation, and natural gas production.		(Sira et al. 1990)
7	Examined the dissociation behavior of artificially created hydrate core samples using an aqueous inhibitor solution.	Methanol	The advantage of an inhibitor was higher at low temperatures.	Gas production behavior at a temperature lower than 19°C has not been investigated.	(Kawamura et al. 2006)
8	Investigated the dissociation behavior of methane hydrates in ethylene glycol solution and silicone oil.	Ethylene Glycol and Silicone oil	Ethylene glycol solution helps motivate the dissociation process when required during the process, and silicone oil is favorable for the transportation of slurry and storage systems.	Silicone oil is not helpful for the dissociation of methane gas hydrate.	(Kawamura et al. 2003)

In this method, various chemicals, for example, glycerin, methanol, glycol, and ethanol, are injected, which shifted the thermodynamic equilibrium stability curve of hydrates. The above techniques have been applied only in permafrost areas and till now have not been used in the dissociation of gas hydrates of marine areas (Williams et al. 2005; Hancock et al. 2005; Makogon 2010; Chong et al. 2016). These verifications are based upon several external factors such as marine and environmental, geological hazards. The advantages and disadvantages of some research work on chemical injection are listed in Table 2.

CO_2 SEQUESTRATION

The shortcomings of the chemical injection method thermal stimulation method, and depressurization method, is the low efficiency of heat transfer in a reservoir, which controls the decomposition rate of hydrates. The thermal conductivity of formation is relatively poor; hence the efficiency of the decomposition process cannot be compensated. The temperature in the hydrate decomposition region declines rapidly due to the huge absorption of heat. The above methods will decrease the reservoir's strength and further cause slope instability, seabed destruction because these methods depend on the principle of decomposition of hydrate (Khlebnikov et al. 2016). The CO_2 replacement method appears to be more profitable as it permits both recovery of gas from hydrates and sequestration of CO_2 as hydrates in the region of deep-sea sediments. This method is applicable where the ocean floor remains stable during the process and after the injection of CO_2 (Ors and Sinayuc 2014). The CO_2 sequestration process is the complex and immature technique for the dissociation of hydrates. The main drawback of CO_2 sequestration is the requirement of a high amount of CO_2, which can increase the process's cost (Ors and Sinayuc 2014). Several researchers have done work on this method for the decomposition of hydrates before and after the experiment. Studies were done by Yoon et al. 2004 revealed that the surface covering of CO_2 gas hydrate in the outermost layer is taken as an important part for deaccelerating the dissociation of methane gas hydrate within the core. But the water phase generated during the sequestration process allows the process to move forward rapidly by increasing the diffusion of carbon dioxide and methane, which results in absolute and comparatively fast production of methane gas (Yoon et al. 2004). Seo et al. 2013 suggested that various factors such as reaction kinetics, the environment of sediment, size of hydrate particle, and mass transfer should have to be considered for the replacement of CO_2 in natural gas hydrates deposits (Seo et al. 2013). Xu et al. 2017 worked for the replacement of different gas mixtures to produce methane gas from hydrates by the replacement method. Small molecules sized gases enhanced the rate of replacement by the stronger diffusivities in gas hydrates (Xu et al. 2018). Ota et al. 2005 found out that the mole fraction of CO_2 and CH_4 in the hydrate zone increases and decreases with time, respectively. They also provided a kinetic model for the exchange process by including the tendency of CH_4 re-generation in the S-Cage (Ota et al. 2005). Liu et al. 2016 investigated the thermodynamic and kinetic feasibility of CH_4 from hydrates by N_2/CO_2 mixture using the ab initio calculations and ab initio molecular dynamic simulations (Liu et al. 2016). Xu et al. 2015 examined the production of methane from a natural gas hydrate by CO_2 replacement by the estimation of similarity between various spectra of C-H, O-H, and C-O stretching vibrations in the different phases and gas hydrates with the help of Raman and NMR (Xu et al. 2015). Lee et al. 2014 developed an experimental system to obtain quantitative information for the exchange process of CO_2 and CH_4 (Lee et al. 2014). Espinoza and Santamarina 2011 designed an experiment to find the advantage of the contact- dependent sand stiffness on the formation of hydrate, CO_2 flooding and gas replacement, and dissociation (Espinoza and Santamarina 2011). Jung et al. 2010 evaluated the reaction rates, analyzed underlying phenomena, explored the region for the pressure-temperature for the optimal change, and anticipated the potential geomechanical implications the replacement in hydrates bearing sediments (Jung et al. 2010). Serrano et.al 2009 tested a continuous jet hydrate reactor in the seafloor process simulator for the CO2 replacement (Serrano et al. 2009). Pohlman et al. 2009 suggested that the decreased sedimentation rate increases the disclosure time of organic matter in sediment to anaerobic and aerobic degradation (Pohlman et al. 2009). Seo et.al 2016 experimented with investigating the replacement mechanism in the C_3H_8+CH_4 hydrate (sII) by injecting a gas mixture CO_2/N_2 (Seo et al. 2016). Yang et.al 2017 explored the potential for the production of methane from gas hydrate reservoirs by the injection of flue gas and form a carbon dioxide hydrate by the sequestration of CO_2 from flue gas (Yang et al. 2017). Wang et al. 2017 proposed a new dissociation technique for natural gas hydrate which improves the effect of replacement and reduces the cost of gas separation, but it decreases the partial pressure of methane gas in the gas phase. H_2 has an important role in disturbing the stability of methane hydrate (Wang et al. 2017b). Seo et al. 2013 suggested through the thermodynamic equilibrium studies that the composition of CO_2 in the hydrate zone and the chemical formula of mixed gas hydrate after the sequestration process can be determined at certain conditions (Seo et al. 2013). Lee et al. 2015 showed that the structural and thermodynamic properties of the mixed gas hydrates are sensible factors to determine the flow of heat in the sediments, for the required heat for the dissociation of hydrate and for the evaluation of the storage capacity of CO_2 in hydrate reservoir (Lee et al. 2015). Zhang et.al 2017 proposed a combined method using CO_2 replacement and the thermal stimulation to increase the recovery of methane gas from hydrates and CO_2 storage (Zhang et al. 2017). Ors and Sinayuc 2014 concluded the swap process of CH_4-CO_2 happens at the surface of the system mostly and the center of the system of unconsolidated sand pack accommodate CH_4 while surface contains CO_2 (Ors and Sinayuc 2014). Deusner et al. 2012 observed the production of methane from the methane gas hydrates by the injection of hot and supercritical CO_2. They concluded that the dissociation of methane gas hydrates is triggered by the heat addition, while thermally equilibrated CO_2 will react with pore water and reserves in the reservoir as immobile. They also studied the effect of temperature on the production process of methane gas from sediment by studying the effect of permeability (Deusner et al. 2014).

The above methods are not favorable for dissociating gas hydrates in marine sediments due to their limitations and disadvantages (Ning et al. 2010). Table 3 listed the advantages and disadvantages of available dissociation techniques of gas hydrates.

Table 3. Advantages and Disadvantages of different dissociation techniques of gas hydrates.

Technique	Advantages	Disadvantages	References
Depressurization	can reduce energy consumption	Advantageous for the widespread gas reservoirs in a closure	(Max and Cruickshank 1999)
Thermal Stimulation	Can overcome the problems associated with low gas production by depressurization	High loss of heat Low efficiency High consumption of energy	(Song et al. 2016; Zhang et al. 2017)
Chemical Injection	Less expensive than other methods	Not advantageous for the oceanic area Very expensive Environmental Pollution	(Demirbas 2010b; Ning et al. 2010)
CO2 replacement	----	The cost of CO2 supply is high. Efficiency and replacement speed is low.	(Lee et al. 2003; Ors and Sinayuc 2014)

COMBINED PROCESS

The thermal Stimulation method can also be combined with the depressurization method to dissociate the gas hydrates. Various studies were reported to observe the effects of this combined method. Wu et al. 2012 used two methods (thermal stimulation and depressurization) under the equilibrium pressure for the dissociation of pure methane hydrates formed below the ice point (Wu et al. 2012). Song et.al 2015 provided the experimental investigation of the dissociation of gas from porous media by the injection of two-cycle warm water, depressurization, and combined method with different hydrate saturations. They concluded that the combined method is an effective and improved method for the dissociation of gas from methane gas hydrate formed in sediments. The combined process increases the average production rate of gas with increasing saturation (Song et al. 2015). Yuan et.al. 2013 studied hydrate dissociation using hot water, ethylene glycol, and brine solution with the continuous injection mode. The NaCl solution shows the highest amount of gas production (Yuan et al. 2013). Wang et al. 2014 examined the production behavior of methane gas hydrate by applying the three-dimensional pressure vessel and thermal stimulation method in the porous media. Using, the experimental conditions are manipulated by the set of scaling criteria for the gas hydrate reservoir (Wang et al. 2014). Wang et al. 2017 experimented with investigating the effect of thermal heat flow on the dissociation of gas hydrate by applying the depressurization method with the thermal stimulation (Wang et al. 2017a). Pang et al. 2009 showed the kinetic behavior of dissociation conducted by hot water. They showed that the rate-determining factors for hydrate dissociation are the heat transfer rate and the thermodynamic driving force of dissociation. The production of gas from hydrate can be increased by combing heating and depressurization (Pang et al. 2009). Liu et al. 2009 developed a mathematical model of dissociation of gas hydrate in the porous media by applying the depressurization and the thermal stimulation method. The Huff and Puff method was applied within the Prudhoe Bay of the Alaskan North Slope (Liu et al. 2009). The dissociation was carried using N_2-CO_2-CH_4 exchange to produce methane from the hydrate reservoir (Max and Johnson 2018).

ECONOMICS

The production of gas from hydrates is seemed to be more expensive as compared to conventional natural gas. Because of the higher cost of current dissociation techniques, it becomes essential to find the economically favorable dissociation techniques of gas hydrates based upon higher production rate, lower transportation cost to balance the higher wellhead cost as compared to the conventional gas production from gas fields (Cranganu and Nitzov 2009). Many researchers have suggested depressurization as the most economical dissociation techniques for gas hydrates (Makogon 1997a; Pooladi-Darvish 2004). Table 4 shows the economic comparison of different dissociation techniques of gas hydrates.

Table 4. Comparison of different dissociation techniques of gas hydrates (Collett 2002; Cranganu and Nitzov 2009; Arora et al. 2015b)

Parameter	Depressurization	Thermal Stimulation	Conventional Gas Production
Investment (million Rs.)	166000	254200	157500
Annual Cost (million Rs.)	125500	160000	100000
Total Production (million m^3/year)	1557.43	1274.26	1557.43
Production Cost (Rs. million/m^3)	4025.9	6356.6	3213.6
Breakeven Wellhead Price (Rs. million/m^3)	2532.3	7945.8	3972.9

CONCLUSION

Natural gas hydrates are a significant source of energy all over the world. It can replace the various other sources of natural gas to fulfill the energy requirement of the world in future. Hence, natural gas recovery from gas hydrates became a significant research area in the present days. Various techniques are available for the dissociation of gas hydrates, but the economical production of gas from gas hydrates is a significant issue nowadays. Various safety issues are associated with the recovery of hydrate energy source for example seafloor subsidence and green house effects. The production testing of hydrate is in progress in Canada, The United States of America, Japan and India. This paper reviews the present dissociation techniques of gas hydrates and their advantages and disadvantages. Hence, to overcome these disadvantages, there is a need to explore the new dissociation techniques for the economical production of gas from gas hydrates. Hydrates can contribute significantly to the conventional future energy demand of the world.

REFERENCES

Amit Arora, S. S. (2015). Techniques for Exploitation of Gas Hydrate (Clathrates) an Untapped Resource of Methane Gas. *Journal of Microbial & Biochemical Technology*, *7*(2), 108–111. doi:10.4172/1948-5948.1000190

Arora, A., Cameotra Ss, S., & Balomajumder, C. (2015a). Natural Gas Hydrate as an Upcoming Resource of Energy. *Journal of Petroleum & Environmental Biotechnology*, *6*(1), 1–6. doi:10.4172/2157-7463.1000199

Arora, A., Chandrajit, B., & Cameotra, S. S. (2015b). Natural Gas Hydrate (Clathrates) as an Untapped Resource of Natural Gas. *Journal of Petroleum & Environmental Biotechnology*, •••, 6.

Castaldi, M. J., Zhou, Y., & Yegulalp, T. M. (2007). Down-hole combustion method for gas production from methane hydrates. *Journal of Petroleum Science Engineering*, *56*(1-3), 176–185. doi:10.1016/j.petrol.2006.03.031

Chatterji, J., & Griffith, J.E. (1998). *Methods of decomposing gas hydrates*. US Pat No 5,713,416.

Chong, Z. R., Yang, S. H. B., Babu, P., Linga, P., & Li, X. S. (2016). Review of natural gas hydrates as an energy resource: Prospects and challenges. *Applied Energy, 162*, 1633–1652. doi:10.1016/j.apenergy.2014.12.061

Collett, T. S. (2002). Energy Resource Potential of Natural Gas Hydrates. *The American Association of Petroleum Geologists Bulletin, 86*, 1971–1992.

Cranganu, C. (2009). In-situ thermal stimulation of gas hydrates. *Journal of Petroleum Science Engineering, 65*(1-2), 76–80. doi:10.1016/j.petrol.2008.12.028

Cranganu, C., & Nitzov, B. (2009) Natural Gas Production from Gas Hydrates–An Economic Perspective. *Analele Ştiinţifice Ale Univ "AL I CUZA" IAŞI Geol Tomul LV, 1*, 299–312.

Demirbas, A. (2010). *Methane gas hydrate: as a natural gas source*. Springer London. doi:10.1007/978-1-84882-872-8

Demirbas, A. (2010a). Methane from gas hydrates in the Black Sea. *Energy Sources. Part A, Recovery, Utilization, and Environmental Effects, 32*(2), 165–171. doi:10.1080/15567030802463885

Espinoza, D. N., & Santamarina, J. C. (2011). P-wave monitoring of hydrate-bearing sand during CH4-CO2replacement. *International Journal of Greenhouse Gas Control, 5*(4), 1031–1038. doi:10.1016/j.ijggc.2011.02.006

Fan, S., Zhang, Y., Tian, G., Liang, D., & Li, D. (2006). Natural gas hydrate dissociation by presence of ethylene glycol. *Energy & Fuels, 20*(1), 324–326. doi:10.1021/ef0502204

Fatykhov, M. A., & Bagautdinov, N. Y. (2005). Experimental investigations of decomposition of gas hydrate in a pipe under the impact of a microwave electromagnetic field. *High Temperature, 43*(4), 614–619. doi:10.100710740-005-0103-7

Fitzgerald, G. C., & Castaldi, M. J. (2013). Thermal stimulation based methane production from hydrate bearing quartz sediment. *Industrial & Engineering Chemistry Research, 52*(19), 6571–6581. doi:10.1021/ie400025f

Gabitto, J. F., & Tsouris, C. (2010). Physical Properties of Gas Hydrates: A Review. *Journal of Thermodynamics, 2010*, 1–12. doi:10.1155/2010/271291

Ginsburg, G. D., Guseynov, R. A., Dadashev, A. A., Ivanova, G. A., Kazantsev, S. A., Solov'yev, V. A., Telepnev, E. V., Askeri-Nasirov, R. Y., Yesikov, A. D., Mal'tseva, V. I., Mashirov, Y. G., & Shabayeva, Y. (1992). Gas Hydrates of the Southern Caspian. *International Geology Review, 34*(8), 765–782. doi:10.1080/00206819209465635

Ginsburg, G. D., & Soloviev, V. A. (1994). Mud Volcano Gas hydrates in the Caspian Sea. *Bulletin of the Geological Society of Denmark, 41*, 95–100. doi:10.37570/bgsd-1995-41-10

Goel, N., Wiggins, M., & Shah, S. (2001). Analytical modeling of gas recovery from in situ hydrates dissociation. *Journal of Petroleum Science and Engineering, 29*(2), 115 – 127.

Graue, A., Kvamme, B., Zornes, D.R., Stevens, J.C., Howard, J.J., & Baldwin, B.A. (2007). *Production of free gas by gas hydrate conversion*. US Pat No 7,222,673. doi:10.1126/science.Liquids

Gullapalli, I., Jones, E., & Moridis, G. (2009). *Method for gas production from gas hydrate reservoirs.* US Pat No 7,537,058.

Guo, B., Bretz, R.E., & Lee, R.L. (1995). *Method and apparatus for generating, transporting and dissociating gas hydrates.* US Pat No 5,473,904.

Haligva, C., Linga, P., Ripmeester, J. A., & Englezos, P. (2010). Recovery of methane from a variable-volume bed of silica sand/hydrate by depressurization. *Energy & Fuels*, *24*(5), 2947–2955. doi:10.1021/ef901220m

Hancock, S.H., Carle, D., Weatherill, B., Dallimore, S.R., Collett, T.S., Satoh, T., & Inoue, T. (2005) Overview of pressure-drawdown production-test results for the JAPEX/JNOC/GSC et al. Mallik 5L-38 gas hydrate production research well. *Bull Surv Canada 585*:134

Handa, Y. P. (1986). Calorimetric determinations of the compositions, enthalpies of dissociation, and heat capacities in the range 85 to 270 K for clathrate hydrates of xenon and krypton. *The Journal of Chemical Thermodynamics*, *18*(9), 891–902. doi:10.1016/0021-9614(86)90124-2

He, S., Liang, D., Li, D., & Ma, L. (2011). Experimental investigation on the dissociation behavior of methane gas hydrate in an unconsolidated sediment by microwave stimulation. *Energy & Fuels*, *25*(1), 33–41. doi:10.1021/ef1011116

Heinemann, R.F., Huang, D.D-T., Long, J., & Saeger, R. (2001). *Method for recovering gas from hydrates.* US Pat No 6,214,175. doi:10.1016/S0141

Huang, Y. J., Wu, C. Y., & Hsieh, B. Z. (2016). Effect of Fluid Saturation on Gas Recovery from Class-3 Hydrate Accumulations Using Depressurization: A Case Study of the Yuan-an Ridge Site in Southwestern Offshore Taiwan. *Energy Procedia*, *97*, 310–317. doi:10.1016/j.egypro.2016.10.006

Islam, M. R. (1994). A new recovery technique for gas production from Alaskan gas hydrates. *Journal of Petroleum Science Engineering*, *11*(4), 267–281. doi:10.1016/0920-4105(94)90046-9

Ji, C., Ahmadi, G., & Smith, D. H. (2001). Natural Gas Production from Hydrate Decomposition by Depressurization. *Chemical Engineering Science*, *56*(20), 5801–5814. doi:10.1016/S0009-2509(01)00265-2

Jung, J. W., Espinoza, D. N., & Santamarina, J. C. (2010). Properties and phenomena relevant to CH4-CO2replacement in hydrate-bearing sediments. *Journal of Geophysical Research. Solid Earth*, *115*(B10), 1–16. doi:10.1029/2009JB000812

Kawamura, T., Ohtake, M., Sakamoto, Y., Yamamoto, Y., Haneda, H., Komai, T., & Higuchi, S. (2007). Experimental study on steam injection method using methane hydrate core samples. *7th Int Offshore Polar Eng Conf (ISOPE), Ocean Min Symp*, 83–86.

Kawamura, T., Sakamoto, Y., Ohtake, M., Yamamoto, Y., Haneda, H., & Komai, T. (2006). Dissociation Behavior of Hydrate Core Sample Using Thermodynamic Inhibitor—Part 2 : Experimental Investigation Using Long Core Samples. *International Journal of Offshore and Polar Engineering*, *16*, 5–9.

Kawamura, T., Yamamoto, Y., Yoon, J. H., Haned, H., Ohga, K., & Higuchi, K. (2003) Dissociation behavior of methane gas hydrate in ethylene glycol and silicone oil. *Proc ISOPE Ocean Min Symp, 2*, 181–184

Khlebnikov, V. N., Antonov, S. V., Mishin, A. S., Bakulin, D. A., Khamidullina, I. V., Liang, M., Vinokurov, V. A., & Gushchin, P. A. (2016). A new method for the replacement of CH 4 with CO 2 in natural gas hydrate production. *Nat Gas Ind B*, *3*(5), 445–451. doi:10.1016/j.ngib.2017.02.005

Koh, C. A., Sum, A. K., & Sloan, E. D. (2009). Gas hydrates: Unlocking the energy from icy cages. *Journal of Applied Physics*, *106*(6), 061101. Advance online publication. doi:10.1063/1.3216463

Kono, H. O., Narasimhan, S., Song, F., & Smith, D. H. (2002). Synthesis of Methane Gas Hydrate in Porous Sediments and its Dissociation by Depressurizing. *Powder Technology*, *122*(2-3), 239–246. doi:10.1016/S0032-5910(01)00420-X

Kumari, A., Khan, S. H., Misra, A. K., Majumder, C. B., & Arora, A. (2020). Hydrates of Binary Guest Mixtures: Fugacity Model Development and Experimental Validation. *Journal of Non-Equilibrium Thermodynamics*, *45*(1), 39–58. doi:10.1515/jnet-2019-0062

Kvenvolden, K.A. (1998). A primer on the geological occurrence of gas hydrate. *Geol Soc London, 137*(9).

Kvenvolden, K. A., & Rogers, B. W. (2005). Gaia's breath - Global methane exhalations. *Marine and Petroleum Geology*, *22*(4), 579–590. doi:10.1016/j.marpetgeo.2004.08.004

Lee, B. R., Koh, C. A., & Sum, A. K. (2014). Quantitative measurement and mechanisms for CH4production from hydrates with the injection of liquid CO2. *Physical Chemistry Chemical Physics*, *16*(28), 14922–14927. doi:10.1039/C4CP01780C PMID:24931508

Lee, H., Seo, Y., Seo, Y. T., Moudrakovski, I. L., & Ripmeester, J. A. (2003). Recovering Methane from Solid Methane Hydrate with Carbon Dioxide. *Angewandte Chemie International Edition*, *42*(41), 5048–5051. doi:10.1002/anie.200351489 PMID:14595627

Lee, J. (2010). Experimental study on the dissociation behavior and productivity of gas hydrate by brine injection scheme in porous rock. *Energy & Fuels*, *24*(1), 456–463. doi:10.1021/ef900791r

Lee, J., Park, S., & Sung, W. (2010). An experimental study on the productivity of dissociated gas from gas hydrate by depressurization scheme. *Energy Conversion and Management*, *51*(12), 2510–2515. doi:10.1016/j.enconman.2010.05.015

Lee, Y., Kim, Y., Lee, J., Lee, H., & Seo, Y. (2015). CH4 recovery and CO2 sequestration using flue gas in natural gas hydrates as revealed by a micro-differential scanning calorimeter. *Applied Energy*, *150*, 120–127. doi:10.1016/j.apenergy.2015.04.012

Li, G., Li, X.-S., Tang, L.-G., & Zhang, Y. (2007). Experimental investigation of production behavior of methane hydrate under ethylene glycol injection in unconsolidated sediment. *Energy & Fuels*, *21*(6), 3388–3393. doi:10.1021/ef060644d

Li, X.S., Wan, L.H., Li, G., Li, Q.P., Chen, Z.Y., & Yan, K.F. (2008). *Experimental Investigation into the Production Behavior of Methane Hydrate in Porous Sediment with Hot Brine Stimulation*. doi:10.1021/acs.energyfuels.7b00464

Li, X. S., Wang, Y., Li, G., Zhang, Y., & Chen, Z. Y. (2011). Experimental investigation into methane hydrate decomposition during three-dimensional thermal huff and puff. *Energy & Fuels*, *25*(4), 1650–1658. doi:10.1021/ef101731g

Li, X. S., Yang, B., Duan, L. P., Li, G., Huang, N. S., & Zhang, Y. (2013). Experimental study on gas production from methane hydrate in porous media by SAGD method. *Applied Energy*, *112*, 1233–1240. doi:10.1016/j.apenergy.2013.02.007

Liang, D., He, S., & Li, D. (2009). Effect of microwave on formation/decomposition of natural gas hydrate. *Chinese Science Bulletin*, *54*(6), 965–971. doi:10.100711434-009-0116-4

Liu, J., Yan, Y., Xu, J., Li, S., Chen, G., & Zhang, J. (2016). Replacement micro-mechanism of CH 4 hydrate by N 2 /CO 2 mixture revealed by ab initio studies. *Computational Materials Science*, *123*, 106–110. doi:10.1016/j.commatsci.2016.06.025

Liu, Y., Strumendo, M., & Arastoopour, H. (2009). Simulation of methane production from hydrates by depressurization and thermal stimulation. *Industrial & Engineering Chemistry Research*, *48*(5), 2451–2464. doi:10.1021/ie8005275

Lu, H., Lorenson, T. D., Moudrakovski, I. L., Ripmeester, J. A., Collett, T. S., Hunter, R. B., & Ratcliffe, C. I. (2011). The characteristics of gas hydrates recovered from the Mount Elbert Gas Hydrate Stratigraphic Test Well, Alaska North Slope. *Marine and Petroleum Geology*, *28*(2), 411–418. doi:10.1016/j.marpetgeo.2010.01.002

Makogon, Y. (1997a). *Hydrates of Hydrocarbons*. Academic Press.

Makogon, Y. F. (1997b). *Hydrates of natural gas* (1st ed.). Penn Well Publ Co Tulsa.

Makogon, Y. F. (2010). Natural gas hydrates - A promising source of energy. *Journal of Natural Gas Science and Engineering*, *2*(1), 49–59. doi:10.1016/j.jngse.2009.12.004

Makogon, Y. F., Holditch, S. A., & Makogon, T. Y. (2007). Natural gas-hydrates - A potential energy source for the 21st Century. *Journal of Petroleum Science Engineering*, *56*(1-3), 14–31. doi:10.1016/j.petrol.2005.10.009

Max, M. D., & Cruickshank, M. J. (1999) Extraction of methane from oceanic hydrate system deposits. *Proc Annu Offshore Technol Conf*, *1*, 71–78. 10.4043/10727-MS

Max, M. D., & Johnson, A. H. (2018). *Exploration and production of oceanic natural gas hydrate: Critical factors for commercialization*. Springer.

Moridis, G. J., Kowalsky, M., & Pruess, K. (2005). Depressurization-Induced Gas Production From Class 1 Hydrate Deposits. *Lawrence Berkeley Natl Lab*, *10*, 458–481. doi:10.2118/97266-MS

Moridis, G. J., & Reagan, M. T. (2007). *Strategies for Gas Production From Oceanic Class 3 Hydrate Accumulations*. Offshore Technol Conf. doi:10.4043/18865-MS

Nasyrov, N. M., Nizaeva, I. G., & Sayakhov, F. L. (1997). Mathematical simulation of heat-and-mass transfer in gas hydrate deposits in a high-frequency electromagnetic field. *Journal of Applied Mechanics and Technical Physics*, *38*(6), 895–905. doi:10.1007/BF02467955

Ning, F., Wu, N., Jiang, G., Zhang, L., Guan, J. A., Yu, Y., & Tang, F. (2010). A method to use solar energy for the production of gas from marine hydrate-bearing sediments: A case study on the Shenhu Area. *Energies*, *3*(12), 1861–1879. doi:10.3390/en3121861

Ors, O., & Sinayuc, C. (2014). An experimental study on the CO2-CH4 swap process between gaseous CO2 and CH4 hydrate in porous media. *Journal of Petroleum Science Engineering*, *119*, 156–162. doi:10.1016/j.petrol.2014.05.003

Ota, M., Morohashi, K., Abe, Y., Watanabe, M., Smith, R. L. Jr, & Inomata, H. (2005). Replacement of CH4in the hydrate by use of liquid CO2. *Energy Conversion and Management*, *46*(11-12), 1680–1691. doi:10.1016/j.enconman.2004.10.002

Pang, W. X., Xu, W. Y., Sun, C. Y., Zhang, C. L., & Chen, G. J. (2009). Methane hydrate dissociation experiment in a middle-sized quiescent reactor using thermal method. *Fuel*, *88*(3), 497–503. doi:10.1016/j.fuel.2008.11.002

Phirani, J., & Mohanty, K. K. (2009). Warm water flooding of confined gas hydrate reservoirs. *Chemical Engineering Science*, *64*(10), 2361–2369. doi:10.1016/j.ces.2009.02.019

Pohlman, J. W., Kaneko, M., Heuer, V. B., Coffin, R. B., & Whiticar, M. (2009). Methane sources and production in the northern Cascadia margin gas hydrate system. *Earth and Planetary Science Letters*, *287*(3-4), 504–512. doi:10.1016/j.epsl.2009.08.037

Pooladi-Darvish, M. (2004). Gas Production From Hydrate Reservoirs and Its Modeling. *Soc Pet Eng*, 90–97. doi:10.2118/13340-PA

Rahim, I., Nomura, S., Mukasa, S., & Toyota, H. (2015). Decomposition of methane hydrate for hydrogen production using microwave and radio frequency in-liquid plasma methods. *Applied Thermal Engineering*, *90*, 120–126. doi:10.1016/j.applthermaleng.2015.06.074

Sakamoto, Y., Komai, T., Kawabe, Y., Tenma, N., & Yamaguchi, T. (2004) Gas Hydrate Extraction from Marine Sediments by Heat Stimulation Method. *Proc Fourteenth Int Offshore Polar Eng Conf*, 52.

Sasaki, K., Ono, S., & Sugai, Y. (2007) Gas Production System From Methane Hydrates Layers by Hot Water Injection Using Dual Horizontal Wells. *Can Int Petroelum Conf, 48*, 1–5. 10.2118/2007-195

Sawyer, W.K., Boyer, C.M., Frantz, J.H., & Yost, A.B. (2000). Comparative Assessment of Natural Gas Hydrate Production Models. *SPE/CERI Gas Technol Symp*, 1–9. doi:10.2118/62513-MS

Seo, Y., Lee, S., & Lee, J. (2013). Experimental Verification of Methane Replacement in Gas Hydrates by Carbon Dioxide. *Ital Assoc Chem Eng*, *47*, 131184–13190.

Seo, Y. J., Park, S., Kang, H., Ahn, Y. H., Lim, D., Kim, S. J., Lee, J., Lee, J. Y., Ahn, T., Seo, Y., & Lee, H. (2016). Isostructural and cage-specific replacement occurring in sII hydrate with external CO 2 / N 2 gas and its implications for natural gas production and CO 2 storage. *Applied Energy*, *178*, 579–586. doi:10.1016/j.apenergy.2016.06.072

Serrano, P. T., Ulrich, S., Szymcek, P., McCallum, S., & Phelps, T. (2009). Multiphase, Microdispersion Reactor for the Continuous Production of Methane Gas Hydrate. *Industrial & Engineering Chemistry Research*, *48*(13), 6448–6452. doi:10.1021/ie8019517

Shashkov, P., Khomutov, G., & Yerokhin, A. U. S. (2012). *United States Patent*. Advance online publication. doi:10.1126cience.Liquids

Shimada, W., Takeya, S., Kamata, Y., Uchida, T., Nagao, J., Ebinuma, T., & Narita, H. (2005). Texture change of ice on anomalously preserved methane clathrate hydrate. *The Journal of Physical Chemistry B, 109*(12), 5802–5807. doi:10.1021/jp044624t PMID:16851632

Singhe, U.W. (2013). *Production of Methane from Abundant Hydrate Deposits.* US 2013/0341179 A1. doi:10.1037/t24245-000

Sira, J. H., Pat, H. S. L., & Kamath, V. (1990) Study of Hydrate Dissociation by Methanol and Glycol Injection. *PE Annual Technical Conference and Exhibition*, 977–984. 10.2118/20770-MS

Song, Y., Cheng, C., Zhao, J., Zhu, Z., Liu, W., Yang, M., & Xue, K. (2015). Evaluation of gas production from methane hydrates using depressurization, thermal stimulation and combined methods. *Applied Energy, 145*, 265–277. doi:10.1016/j.apenergy.2015.02.040

Song, Y., Zhang, L., Lv, Q., Yang, M., Ling, Z., & Zhao, J. (2016). Assessment of gas production from natural gas hydrate using depressurization, thermal stimulation and combined methods. *RSC Advances, 6*(53), 47357–47367. doi:10.1039/C6RA05526E

Su, Z., He, Y., Wu, N., Zhang, K., & Moridis, G. J. (2012). Evaluation on gas production potential from laminar hydrate deposits in Shenhu Area of South China Sea through depressurization using vertical wells. *Journal of Petroleum Science Engineering, 86–87*, 87–98. doi:10.1016/j.petrol.2012.03.008

Sung, W., Lee, H., Lee, H., & Lee, C. (2002). Numerical Study for Production Performances of a Methane Hydrate Reservoir Stimulated by Inhibitor Injection. *Energy Sources, 8312*(6), 37–41. doi:10.1080/00908310290086527

Tang, L. G., Li, X. S., Feng, Z. P., Li, G., & Fan, S. S. (2007). Control mechanisms for gas hydrate production by depressurization in different scale hydrate reservoirs. *Energy & Fuels, 21*(1), 227–233. doi:10.1021/ef0601869

Tsypkin, G. G. (2005). Effect of decomposition of a gas hydrate on the gas recovery from a reservoir containing hydrate and gas in the free state. *Fluid Dynamics, 40*(1), 117–125. doi:10.100710697-005-0050-3

u, T., Guan, G., Abudula, A., Yoshida, A., Wang, D., & Song, Y. (2019). Gas recovery enhancement from methane hydrate reservoir in the Nankai Trough using vertical wells. *Energy, 166*, 834–844.

Wang, B., Dong, H., Fan, Z., Zhao, J., & Song, Y. (2017a). Gas Production from Methane Hydrate Deposits Induced by Depressurization in Conjunction with Thermal Stimulation. *Energy Procedia, 105*, 4713–4717. doi:10.1016/j.egypro.2017.03.1022

Wang, X. H., Sun, Y. F., Wang, Y. F., Li, N., Sun, C. Y., Chen, G. J., Liu, B., & Yang, L. Y. (2017b). Gas Production from Hydrates by CH4-CO2_H2 Replacement. *Applied Energy, 188*, 305–314. doi:10.1016/j.apenergy.2016.12.021

Wang, Y., Li, X. S., Li, G., Huang, N. S., & Feng, J. C. (2014). Experimental study on the hydrate dissociation in porous media by five-spot thermal huff and puff method. *Fuel, 117*, 688–696. doi:10.1016/j.fuel.2013.09.088

Williams, T. E., Millheim, K., & Liddell, B. (2005). *Methane hydrate production from Alaskan permafrost.* Maurer Technology Inc.

Wu, Q., Wang, Y., & Zhan, J. (2012). Effect of rapidly depressurizing and rising temperature on methane hydrate dissociation. *Journal of Natural Gas Chemistry*, *21*(1), 91–97. doi:10.1016/S1003-9953(11)60338-1

Xu, C. G., Cai, J., Lin, F. H., Chen, Z. Y., & Li, X. S. (2015). Raman analysis on methane production from natural gas hydrate by carbon dioxide–methane replacement. *Energy*, *79*, 111–116. doi:10.1016/j.energy.2014.10.068

Xu, C. G., Cai, J., Yu, Y. S., Yan, K. F., & Li, X. S. (2018). Effect of pressure on methane recovery from natural gas hydrates by methane-carbon dioxide replacement. *Applied Energy*, *217*, 527–536. doi:10.1016/j.apenergy.2018.02.109

Yang, J., Okwananke, A., Tohidi, B., Chuvilin, E., Maerle, K., Istomin, V., Bukhanov, B., & Cheremisin, A. (2017). Flue gas injection into gas hydrate reservoirs for methane recovery and carbon dioxide sequestration. *Energy Conversion and Management*, *136*, 431–438. doi:10.1016/j.enconman.2017.01.043

Yang, X., Sun, C. Y., Yuan, Q., Ma, P. C., & Chen, G. J. (2010). Experimental study on gas production from methane hydrate-bearing sand by hot-water cyclic injection. *Energy & Fuels*, *24*(11), 5912–5920. doi:10.1021/ef100367a

Yoon, J.-H., Kawamura, T., Yamamoto, Y., & Komai, T. (2004). Transformation of Methane Hydrate to Carbon Dioxide Hydrate: In Situ Raman Spectroscopic Observations. *The Journal of Physical Chemistry A*, *108*(23), 5057–5059. doi:10.1021/jp049683l

Yuan, Q., Sun, C. Y., Liu, B., Wang, X., Ma, Z. W., Ma, Q. L., Yang, L. Y., Chen, G. J., Li, Q. P., Li, S., & Zhang, K. (2013). Methane recovery from natural gas hydrate in porous sediment using pressurized liquid CO2. *Energy Conversion and Management*, *67*, 257–264. https://doi.org/10.1016/j.enconman.2012.11.018

Zhang, H.-Q., Brill, T., & Sarica, C. (2009). *Extracting gas hydrates from marine sediments*. US Pat No 7,546,880. doi:10.1038/incomms1464

Zhang, L., Yang, L., Wang, J., Zhao, J., Dong, H., Yang, M., Liu, Y., & Song, Y. (2017). Enhanced CH4 recovery and CO2 storage via thermal stimulation in the CH4/CO2 replacement of methane hydrate. *Chemical Engineering Journal*, *308*, 40–49. https://doi.org/10.1016/j.cej.2016.09.047

Zhao, J., Fan, Z., Wang, B., Dong, H., Liu, Y., & Song, Y. (2016). Simulation of microwave stimulation for the production of gas from methane hydrate sediment. *Applied Energy*, *168*, 25–37. https://doi.org/10.1016/j.apenergy.2016.01.091

Zhou, Y., Castaldi, M. J., & Yegulalp, T. M. (2009). Experimental Investigation of Methane Gas Production from Methane Hydrate. *Industrial & Engineering Chemistry Research*, *48*, 3142–3149. https://doi.org/10.1021/ie801004z

Chapter 11
Recent Advancements in Microalgae–Biofuel Generation Employing Nano–Additives

Mahesh Pattabhiramaiah
Bangalore University, India

Bhargavi Rajarathinam
Centre for Applied Genetics, Department of Zoology, Bangalore University, Bangalore, India

Shanthala Mallikarjunaiah
https://orcid.org/0000-0002-4699-7809
Bangalore University, India

ABSTRACT

Biofuels have gained significant interest as an alternative fuel in recent years owing to their environmental sustainability, cost-effectiveness, and the ability to blend with traditional fuels like gasoline without requiring engine modifications. The use of microalgae for biofuel production is universally preferred due to its energy efficiency and environmental sustainability; current research is geared towards enhancing the production of microalgae-biofuel from the initial stages to the final product as it is a cost-effective fuel option. The use of different types of nano-additives at different stages of microalgae cultivation and incorporation into biofuel produced noteworthy improvements in the final product. This chapter focuses on the prospective uses of nano-additives in microalgae cultivation, microalgal biomass conversion to biofuels, and biofuel combustion improvement for revolutionary advancements in biofuel technology.

INTRODUCTION

Photosynthetic microorganisms like green microalgae, diatoms, and cyanobacteria offer significant advantages over terrestrial plants providing an abundance of various biomolecules to be used for food, feed, and fuel applications. Microalgae or microphytes are microscopic algae. They are naturally found in freshwater bodies and oceans. Microalgae are unicellular species that can exist individually or in

DOI: 10.4018/978-1-6684-5269-1.ch011

chains or groups. Microalgae can vary in size depending on the species, from a few micrometers (μm) to a few hundred micrometers. Microalgae are efficient photosynthesizers and are crucial for life on earth producing around half of the atmospheric oxygen it also performs its role as a photoautotroph consuming greenhouse gases like carbon dioxide to grow. Forming the base of the food web Microalgae and bacteria provide energy for all the trophic levels above them (Peter *et al.*, 2021). There exists a rich biodiversity of microalgae which represents an abundant, almost untapped resource. There are about 200,000-800,000 species in different genera of microalgae of which about 50,000 species are characterized and described. More than 15,000 novel compounds originating from algal biomass have been identified including carotenoids, antioxidants, fatty acids, enzymes, polymers, peptides, toxins, and sterols.

Microalgae form an abundance of carbon-rich biomass, that can be tapped for the production of biofuels, cosmetics, health supplements, and pharmaceuticals. They also have numerous applications in wastewater treatment and sequestration of atmospheric CO2. A wide range of bioproducts, including polysaccharides, lipids, pigments, bioactive compounds, proteins, vitamins, and antioxidants are also derived from microalgae. There is a rapidly growing interest in the use of microalgae as a renewable and sustainable feedstock for biofuels production fueling new research in the field of biorefinery using microalgae. The cultivation of microalgae on an industrial scale to produce biofuels and bioproducts has increased radically in the last couple of decades (Low *et al.*, 2021).

Figure 1. Different biofuels from microalgae and conversion processes
(Hossain et al, 2009)

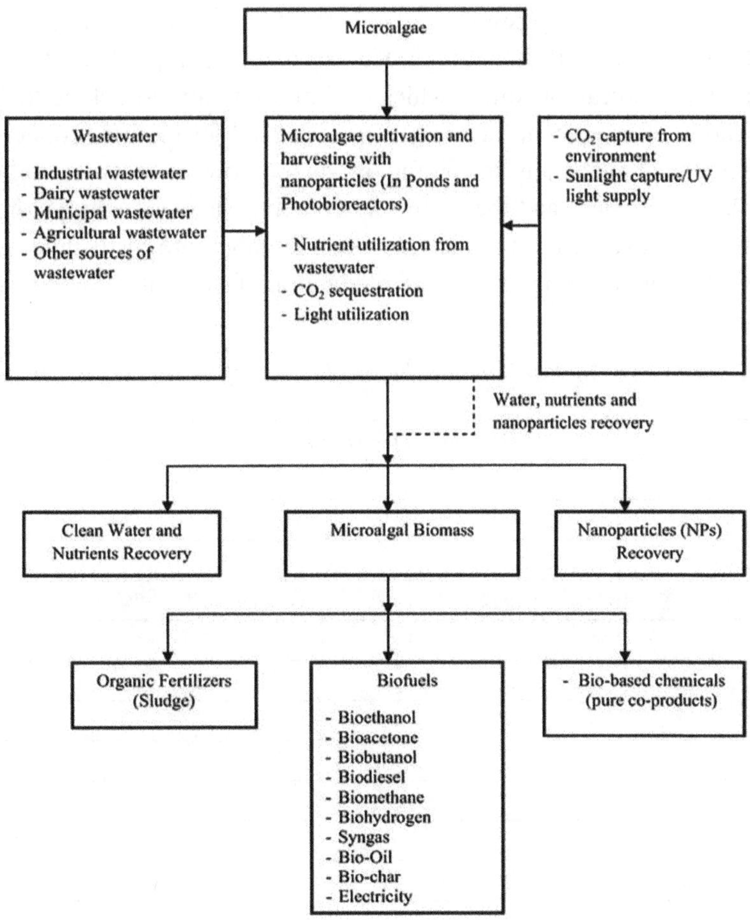

MICROALGAE AS A BIOFUEL

The increasing global demand for fuel energy, the intensive use of fossil fuels worldwide leads to its depletion and economic crises and global tension around the control of fossil fuels heralds the need for alternative, environment-friendly and sustainable fuel sources. Thus, biofuels are increasingly viewed as an alternative to fossil fuels. Some developed countries have already started the mass production of biofuels. Biofuels such as biodiesel and bioethanol are superb alternative fuels and can be produced from a number of sources of biomass, including garbage, food crops, crop wastes or fruits, woody parts of plants, and algae (Harun, Danquah and Forde, 2010; Ho, Chen and Chang, 2010). The advantages of the use of biofuels from biomass are renewability and a significantly smaller contribution to environmental pollution and global warming. The release of CO2 from the burning of fossil fuels is one of the main causes of global warming. 29 gigatons/year of CO2 is released with a total of 35.3 billion tons of CO2 released to date from the burning of fossil fuel (Paul *et al.,* 2014). Biofuels like algal fuels have oxygen levels of 10–45% and negligible levels of sulphur emission while petroleum-based fuels have no oxygen levels and considerable sulphur emission. Microalgae fuels are known for being environment-friendly, non-toxic, and capable of fixing global CO2. Gendy *et al.,* 2013 reported that 1 kg of algal biomass is capable of fixing 1.83 kg of CO2 additionally certain species are known to utilize and remove SOx and NOx along with CO2.

Various applications have been exhaustively studied for different species of microalgae. Table 1 summarizes various species of microalgae, ideal culture settings (cultivation environment and water quality), availability, and cellular requirements (Hossain, Zaini, Mahlia, and Azad, 2019). Carbohydrates, proteins and lipids are the primary constituents of complex biofuels (e.g., bio-oil, biodiesel, biobutanol, etc.,) (Harun and Danquah, 2011, Pathak, Singh, Kothari and Chopra, 2014). Microalgae contain large fractions of lipid, protein, and carbohydrates which are important for biofuel production. Table 2 details the cellular composition of prominent microalgae species used for biofuel production (Hossain *et al.,* 2019). Microalgae are ideal candidates for biofuel production owing to its feasibility to be grown and harvested on both land and water and the high lipid content. Nanotechnology comes into the picture because lipase immobilization and lipid extraction from microalgae are costly processes. To address this, nano-additives are increasingly being used in the production of microalgal-based biofuels.

Table 1. Prospective species of microalgae for the production of biofuel (Hossain et al., 2019).

Name of the Microalgae
Chlamydomonas reinhardtii
Symbiodinium sp.
Phaeodactylum tricornutum
Chlorella vulgaris
Saccharina japonica
Spirulina platensis
Ostreococcus tauri
Chlorella sp.
Nannochloropsis sp.
Botryococcus braunii
Chlorococum sp. and *Spirogyra* sp.

Table 2. Microalgae species with optimal physiochemical composition for biofuel generation (Hossain et al., 2019).

Species of Microalgae	Total Sugars Content (%)	Protein Content (%)	Lipid Content (%)	Ash and Other Content (%)
Bellerochea sp.	3	24	15	3
Chaetoceros sp.	2	18	18	3
Chlamydomonas reinhardtii	48	17	21	14
Chlorella sp.	56	22	19	3
Dunaliella salina	57	32	6	5
Porphyridium cruentum	35	50	11	4
Rhodomonas sp.	9	74	15	2
Scenedesmus sp.	18	56	12	–
Spirogyra sp.	20	55	16	9
Spirulina platensis	60	12	8	20

Numerous biofuels like biodiesel, bioethanol, biomethane, bio-oil, bio-hydrogen, etc., are produced from microalgae (Brennan and Owende, 2010; Hossain *et al.,* 2019). Biodiesel is increasingly used as a replacement for fossil-based diesel as a sustainable biofuel. Despite the many benefits of using biodiesel, properties such as lower fuel atomization and high viscosity, result in BTE reductions. Innovation in the field of nanotech have resulted in the development of nano-molecules with improved thermal properties, which help in combustion, and increase volume-surface ratio and results in improved fuel characteristics and better engine performance and lower emissions. Nano-particles can be added during various stages of microalgal growth (e.g., cell suspension, cell separation, and cell harvesting), biofuel conversion stage and fuel application stage for increased total yield in each stage.

Figure 2. Microalgal conversion of atmospheric CO2 to carbohydrates, lipids and other valuable bio-products by using light
(Khan et al, 2018)

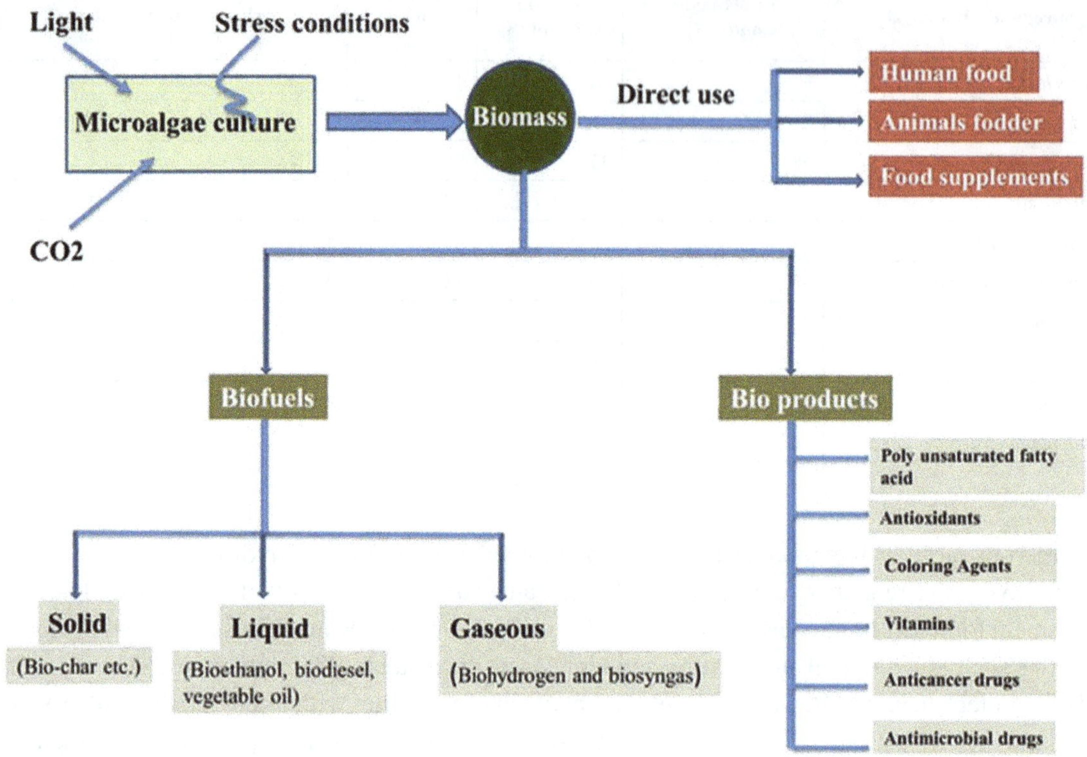

MICROALGAE-BIOFUEL WITH NANO-ADDITIVES

Nanoparticles and Nano-Additives

Nanoparticles (NPs) include a collection of materials that have at least one dimension less than 100 nm. Nanomaterials can be 0D (zero dimensional), 1D, 2D, or 3D depending on the general shape (Tiwari, Tiwari and Kim, 2012). NPs are usually made up of three layers i.e. (a) The surface layer- This can be made functional by using various small molecules, metal ions, surfactants, and polymers. (b) The shell layer- this has a different chemical composition from the core, and (c) The core- the central portion of the Nanoparticle (Shin, Cho, Kannan, Lee, and Kim, 2016). Due to these unique characteristics, nanomaterials garnered a great deal of intrigue and interest of researchers in various disciplines and fields. The NPs have applications in drug delivery (Lee, Lee, Kim, Kim, Hyeon, 2011), as biosenors and chemical sensing (Barrak, Saied, Chevallier, Laroche, M'nif, and Hamzaoui, 2019), gas sensing (Mansha, Qurashi, Ullah, Bakare, Khan, and Yamani, 2016; Rawal and Kaur, 2013), CO2 capture and sequestration (Ganesh, Hemalatha, Peng, and Jang, 2017; Ramacharyulu, Muhammad, Kumar, Prasad, and Mohanty, 2015) and various other applications (Shaalan, Saleh, El-Mahdy, and El-Matbouli, 2016). Nanomaterials are used as additives to improve the biofuel yield and facilitate biofuel utilization in petrol and diesel engines.

Nano-additives such as nano-magnets, nano-crystals, nano-droplets, nano-ðbres, etc., Figure 3 depicts the application areas of nano-additives in microalgae cultivation, and biofuel production.

Figure 3. Representation of biofuel types and their sources
(Arya et al, 2021)

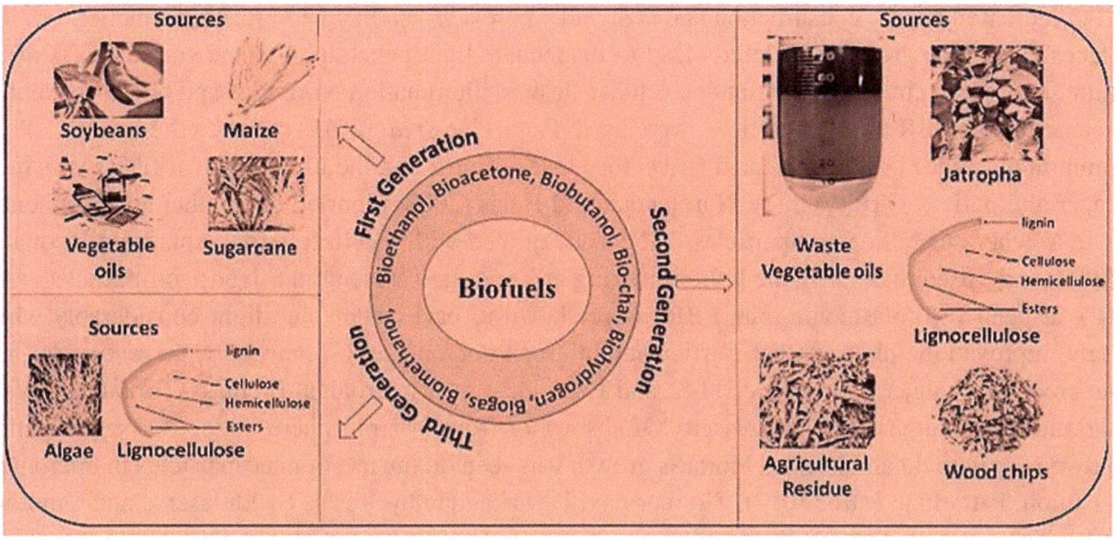

Nanoadditives can be added at various stages of biofuel production from microalgae cultivation to biofuel application in engines due to their durability, recyclability, crystalline nature, adsorption efficiency, stability, catalytic activity, high storage capacity, superb biofuel yield, economical advantage and eco-friendly characteristics. (Hossain *et al.,* 2019). Use of Nano-technology has been shown to improve microalgae cultivation, maximize the yield of many microalgal biofuels and facilitate the use of such microalgal-biofuels in petrol and diesel engines and enhance the performance of biofuel-diesel hybrid fuels. Various nano-materials, e.g., nano-fibers, nano-sheets, nano-particles, nano-tubes and other nanostructures, have been investigated as effective nano-catalysts in biofuel production (e.g., bioethanol, biodiesel, biomethane, and others) and yield enhancement (Hasannuddin *et al.,* 2018; Karthikeyan and Prathima, 2017; Kim, Jia and Wang, 2006; Safarik, Prochazkova, Pospiskova, and Branyik, 2016).

Nano-Additives for Microalgal Growth

Microalgal growth rate remains a limiting factor in the lipid accumulation rate by microalgae and ultimately the mass production of biofuel from microalgae. Growth rates can be significantly accelerated by the use of nanomaterials. Sarma *et al.* (2014) studied the effects of Mg nanoparticles (NPs) on the microalgal growth rate, biomass production, and accumulation of intracellular compounds when Mg NPs were integrated into microalgae culture at different concentrations. They observed an improvement in their growth and lipid production upon use of the Mg NPs as nutritional additives. Nanoparticles added to wastewater may supplement nutritional deficits. Nanoparticles in culture media can also compete for nutrient absorption, inducing microalgae to absorb nutrients and accumulate lipids at a higher rate. An added benefit of using nanoparticles is that they possess antibacterial and antifungal properties to the culture media, inhibiting the growth of and killing bacteria and fungi which compete with microalgae for nutrients.

Nano-particles added during microalgae culture in a Photobioreactor (PBR) can boost and improve light conversion efficiency. Recently there has been an increased use of light-emitting diodes (LEDs) equipped with nano-materials in PBRs especially in the PBRs which use artificial light sources that lead to increased energy consumption and cost and where self-shading and biofilm formation on PBR surfaces during the growth of biomass lead to inadequate light penetration to the various parts of the culture. In order to circumvent this and to achieve desired illumination properties and photo-conversion efficiency in the PBR the nanoparticles were used. Pattarkine *et al.* (2012) reported the use of Gallium aluminum arsenide (GAA)-fabricated LEDs for algae cultures and the use of optical fibers which are cheaper alternatives. Steele, Grady, Nordlander, and Halas (2007) reported on another advancement in this field where metallic nano-particles (MNPs) integrated with localized surface plasmon resonance (LSPRs) were used to amplify the light scattering at a certain wavelength. Silver nano-particles' (Ag-NPs') suspended in plasmonic mini-PBRs were shown to backscatter blue light considerably which greatly improved the photosynthetic efficiency in green microalgae, *Chlamydomonas reinhardtii*, and blue–green microalgae, Cyanothece 51142, and a 30% increase in microalgal-biomass (Torkamani, Wani, Tang, and Sureshkumar, 2010). Increased CO2 absorption from the atmosphere, better CO2 sequestration, improvement in yield and boosted biomass growth was seen on the use of nano-particles in microalgae cultivation. Pattarkine *et al.* (2012), Zimmerman, Hewakandamby, Tesař, Bandulasena, and Omotowa (2009); Zimmerman, Tesař, and Bandulasena (2011) reported the use of stable nano-bubbles that help algal biomass float into the culture thus facilitating high mass transfer efficiency, improved biomass density, improved CO2 absorption, O2 stripping, and slight buoyancy. Nano-bubbles were shown to suspend the biomass in an airlift-loop bioreactor (ALB) and use less energy than micro-bubbles. For this purpose a uniform nonporous membrane of ALB was used which was capable of producing 100 nm sized bubbles. Safarik *et al.* (2016) delineated that nano-additives are very important for the flocculation and separation process before biofuel production besides microalgae harvesting.

Figure 4. Nano-adhesives used for the enhancement of microalgae cultivation and biofuel production and implementation
(Hossain et al, 2019)

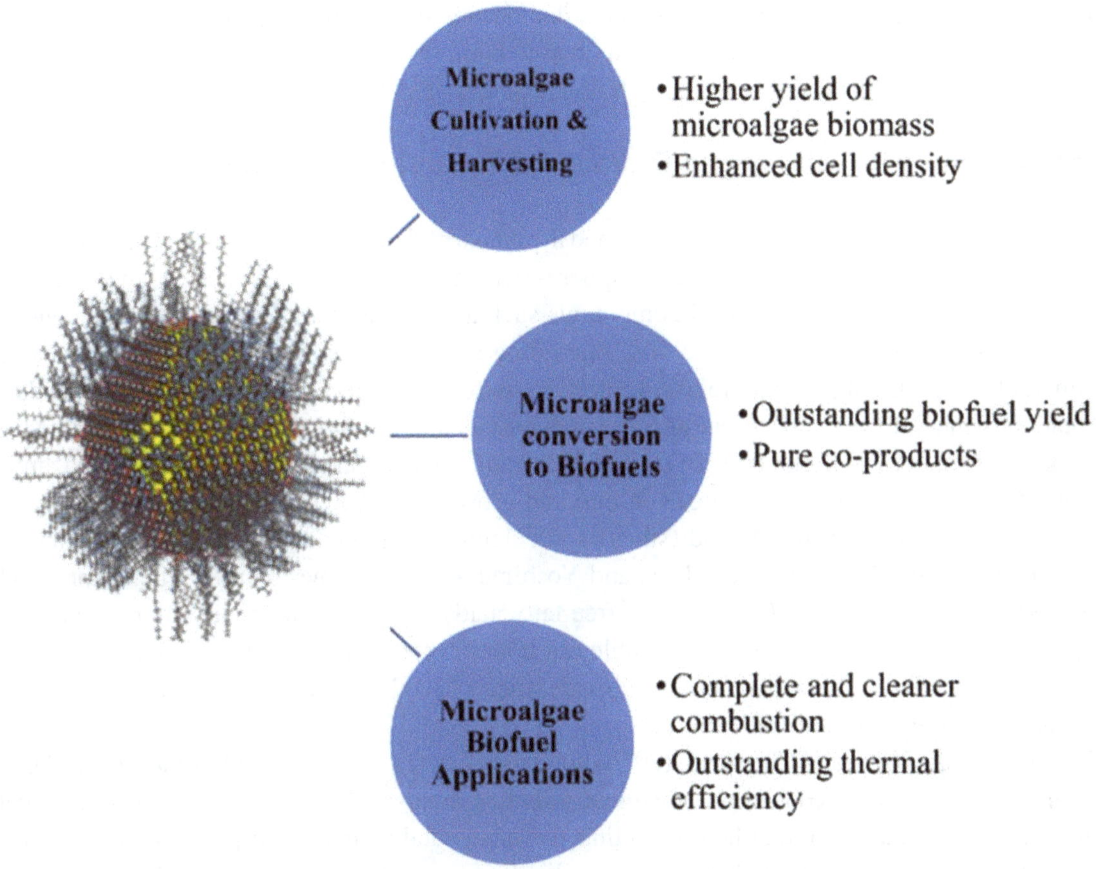

Nano-Additives for Microalgal Lipid Production

Microalgae typically accumulate large amounts of carbon sources when stressed in the form of pigments, starch, lipids, or moieties for self-protection from damage These results mainly from nutrient deprivation or shortage and can be exploited to abet the accumulation of vast amounts of carbon sources like sugars, starch, lipids and lipid moieties (Kadar, Rooks, Lakey, and White, 2012). These stress responses however result in decreased cellular growth. Nanoparticles added to microalgal culture media act as nutritional supplements to enhance growth and lipid accumulation. In the species *Trachydiscus minutes* a lipid increment from 26.20% to 35% was seen when 5.1 mg L−1 zero-valent iron nanoparticles (nZVI) were added (Pádrova *et al.* 2015). *Desmodesmus suspicious* showed a similar 38% increase in lipid accumulation when the identical concentration of nZVI was used. Ren *et al.* (2020) reported a similar occurrence when SiC nanoparticles were added to *Scenedesmus sp.* and the lipid content was enhanced. They also observed a biomass production of 3.18 g L−1 and lipid content of 40.26% with a 150 mg L−1 concentration of SiC nanoparticles under xenon light. Acetyl-coenzyme A carboxylase (ACCase) which is a key rate-limiting enzyme that catalyzes the first step in the biosynthesis of fatty acids was increased which likely increased the lipid content showing that SiC nanoparticles stimulate the activity of ACCase under xenon light, that lead to increased lipids synthesis. Several studies were conducted on the application of Mg, Se, Fe, Cu, C, Pb, Zn and Si nanomaterials in *C. vulgaris, S. obliquus, C. pyrenoidosa, T. minutus, S. rubescens, P. kessleri*, and *T. suecica* demonstrated NP application which led to accelerated microalgal growth, higher lipid contents and increased biomass production (as shown in Table 3) which can be exploited to achieve higher yields of biofuel (Sarkar *et al.* 2021).

Nano-Additives for Microalgae Biomass Conversion to Biofuels

Biodiesel, a form of microalgae biofuel, has gained in popularity as a commercial biofuel in the automotive fuel industry. Acidic and basic nano-catalyst spheres, which are recyclable and have a beneficial impact on the economy, can replace chemical compounds such as sodium methoxide in biodiesel generation by interacting with free fatty acids and oils. According to Trindade et al., 2011, using nano-catalysts minimises the amount of pollutant released into the environment generated by sodium methoxide, and the reactions may also be carried out at low temperatures and pressures. Silitonga, Atabani, Mahlia, Masjuki, Badruddin, and Mekhilef (2011); Trindade *et al.* (2011); Palaniappan (2017) and Nizami and Rehan (2018) described that Microalgae culture and growth may be considerably enhanced by using spherical nano-particles made of sand (silica) and calcium compounds without affecting harvesting. According to Chen, Mochizuki, Abe, Toba and Yoshimura (2014), a mesoporous silica nano-catalyst, Ti-loaded SBA-15, produced 10 times more free fatty acids (FFA) and tolerated water than any other catalyst for biodiesel generation from vegetable oil. It also outperformed other effective nano-catalysts like titanium silicalite-1 (TS-1) and titanium dioxide silicate (TiO2-S) and may thus be utilised to produce high-quality biodiesel from diverse sources.

Trindade et al. (2011) and Chen et al. (2014) described how Ti-loaded Santa Barbara Amorphous-15 (SBA-15) nano-catalysts reduced the chemical (alkaline catalyst, NaOH) cost of transesterification during biodiesel production, and how recycling the nano-catalyst made the process eco sustainable. Silva et al., 2017 demonstrated higher microalgae-biodiesel output when Niobia (N2O5) was combined with SBA-15 to produce biodiesel from biomass via esterification. Palaniappan et al. (2017) found that combining an enzyme extracted from the bacteria *Pseudomonas cepacia* with nano-particles such as polyacrylonitrile (PAN) nanofibre, Fe3O4, and nanoporous gold; lipase enzyme from *Rhizopus miehei* with silica nano-particles; ferric silica and the transesterification procedure used magnetic nanoparticles with Burkholderia sp. lipase and *Thermomyces lanuginosa* to create biodiesel from various bio-oils. Liu *et al.,* 2018 and Palaniappan et al. (2017) used a nano-magnetic biocatalyst of KF/CaO–Fe3O4, Li (lithium)-doped CaO, Fe2O3–CaO, sulphate (SO4) integrating Zi (zirconium), sodium titanate, and carbon-based nano-tubes and nano-particles from different kinds of biomass and biodegradable waste to produce 95 percent biodiesel. During the transesterification process, a zeolite NP (an alumina silicate mineral), has been employed as a commercial absorbent for increasing biodiesel output efficiency. Carrero et al. (2011) demonstrated the synthesis of biodiesel from *Nannochloropsis* microalga oil using a hierarchical h-Beta zeolite catalyst. In the process to produce FAMEs from Nannochloropsis oil, the catalyst had the maximum activity. The use of hierarchical catalysts demonstrates equal acid strength and number of acid sites for normal zeolites, but with a larger exterior surface area and the inclusion of a secondary micro-mesoporosity that minimizes lipid transport barriers to the zeolite acid active sites. Table 4 illustrates the use of nano-additives to enhance biodiesel yields during microalgae to biofuel conversion, as well as the relevant conversion procedures and efficiencies.

Table 3. Lipid accumulation following Nano Particle treatment in several algal biofuels (Sarkar et al., 2021)

Microalgae Species	NP (Size in nm)		References
		Lipids	
C. vulgaris	Cu (100)	-	Mykhaylenko and Zolotareva. (2017)
C. vulgaris	Se (100)	-	Mykhaylenko and Zolotareva. (2017)
S. obliquus	C (<2)	8.9	He *et al.* (2017)
P. kessleri	Fe (-)	14	Pádrova *et al.*, 2015
C. pyrenoidosa	Fe_2O_3 (<50)	15.29	Rana *et al.* (2020)
S. obliquus	MgO (<50)	18.5	He *et al.* (2017)
S. rubescens	ZnO (50–70)	27.27	Aravantinou *et al.* (2020)
T. minutus	Fe (-)	33.58	Pádrova *et al.*, 2015
Scenedesmus. sp.	SiC (25–100)	36.46	Ren *et al.* (2020)
S. obliquus	Fe_2O_3 (<30)	39.6	He *et al.* (2017)
T. suecica	Fe (50)	41.9	Kadar *et al.*, 2012
C. vulgaris	Cu (89)	86.67	Sibi *et al.* (2017)
C. vulgaris	$MgSO_4$ (100)	118.23	Sarma *et al.* (2014)
C. vulgaris	Pb (76)	206.67	Sibi *et al.* (2017)
C. vulgaris	Zn (92)	333.33	Sibi *et al.* (2017)
C. vulgaris	Mg (82)	393.33	Sibi *et al.* (2017)

Table 4. Employment of nano-additives to improve biodiesel yield

Nano-additives Employed	Processes Involved	Conversion Efficacy	References
Blends of calcium oxide nanoparticles (CaO-NPs)	Catalytic transesterification	91%	(Pattarkine *et al.*, 2012)
KF/CaO–Fe3O4, Li(lithium)-doped CaO, Fe2O3–CaO, sulphate (SO4) infused Zi (zirconium), sodium titanate, and carbon-based nano-tubes and nano-particles	Transesterification	95%	(Palaniappan *et al.*, 2017; Liu *et al.*, 2018)
Ti-loaded SBA-15 mesoporous silica nanocatalyst	Transesterification	More than ten times the efficacy of other catalysts. 3 times more efficient than TS-1 and TiO2–S nanocatalysts	(Trindade *et al.*, 2011; Chen *et al.*, 2014)
SBA-15 integrated with Niobia (N2O5)	Esterification	Biodiesel yields have increased significantly.	(Silva *et al.*, 2017)
PAN nanofibre, Fe3O4 and integration of nanoporous gold, silica nanoparticles, integration of ferric silica and magnetic nanoparticles, transesterification of polyacrylonitrile nano fibre	Transesterification	Biodiesel productivity has increased significantly.	(Palaniappan *et al.*, 2017)

Nano-Additives for Microalgal CO2 Absorption and Sequestration

Photosynthesis is one of the most common methods to utilize CO2 that has been sequestered. The largest contributors of biological CO2 fixation are microalgae. According to Huang et al. (2014), 1 tonne of microalgae can sequester 1.8 tonnes of CO2. Enhanced CO2 fixation resulted in improved nutrition supply to microalgae, resulting in raised biomass production during the log phase of development and product accumulation (Ling, Guo, Liu, Zhang, Wang, Lu, and Ng, 2015; Da Silva Vaz, da Silveira Mastrantonio, Costa and de Morais, 2019; Choi et al. 2012). Choi et al. (2012) revealed that when the quantity of dissolved gas in the liquid was enhanced well above natural balance of the algal culture, there was an elevation in CO2 fixation by microalgae. As adsorbent materials, nanoporous materials such as polymeric nanofibers have been increasingly popular in recent years (Vaz et al. 2018). Nanofibers are incredibly beneficial since they can be readily isolated from microalgal biomass, according to the researchers. In microalgae production, polymeric nanofibers can be employed as CO2 absorbents to enhance permeance in the medium, contact duration between the microalgae and the CO2, and CO2 mass transfer to the cells (Da Silva Vaz et al. (2019); da Silva Vaz, Costa and de Morais, 2020). Da Silva Vaz et al. (2019); Vaz, Costa and Morais (2018) revealed the synthesis of functional composite nanostructures by integrating metal nanoparticles or metal oxides into nanofibers, leading to improved adsorption efficiencies and molecule transport and interactions. The inclusion of nanoparticles to microalgal growth medium by Li et al. (2016) resulted in a microorganism-particle assembly for Co2 sequestration and in-situ conversion. The gas adsorbed on the nanofibers is progressively desorbed into the culture medium as a result of this. Vaz et al. (2018) investigated the impacts of different Fe2O3 nanoparticle concentrations and observed that 4 percent (w/v) Fe2O3 nanoparticles had a better CO2 adsorption capability. The smaller nanofibers and increased porosity are responsible for the enhanced CO2 adsorption/capture efficiency. When Chlorella fusca LEB 111 was grown in nanofibers with 4% Fe2O3 nanoparticles, a higher biomass concentration of 794 mg L−1 was observed, leading to the conclusion that nanofibers with Fe2O3 nanoparticles increased the specific surface area and CO2 adsorption capacity of the nanostructures, leading to greater gas availability in the microalgal medium, advancing production and enhance fixation.

Da Silva Vaz et al. (2019) investigated the addition of 4 percent Fe2O3 nanoparticles in their free form and encased in a stainless-steel screen support to *C. fusca* LEB 111 growth medium. Due to the desorption of the gas adsorbed on the nanostructures, the free form nanofibers supported a threefold buildup of CO2 in the medium, resulting in a 45 percent rise in fixation by the microalgae and an elevation in the (Dissolved inorganic carbon) DIC. Vaz et al, (2019) studied CO2 fixation in *C. fusca* LEB 111 grown in outdoor and indoor environments using polymeric nanofibers containing 4% Fe2O3 nanoparticles. Outdoor experiments showed increased CO2 fixation, which might be attributed to an increase in biomass content as the nanofibers-maintained CO2 for a longer period of time.

Nano-Additives for Microalgae-Biofuel Applications

Sinha et al. (2017) and Silitonga et al. (2018) reported enhanced fuel lubricity, catalytic efficiency, chemical processes, mass and heat transfer efficiency, cetane number, combustion rate, fire/flash point, water co-solvency, and reduced retention times, resulting in a more comprehensive and cleaner combustion of microalgae biofuel, while investigating solid nano-particles, nano-fluids, or nano-droplets with biofuel and fossil fuel. According to Trindade et al. (2011); Sinha et al. (2017); and Damanik, Ong, Tong,

Mahlia, and Silitonga, (2018), nano-additives can improve microalgal-biofuel combustion efficiency while lowering smoke, NOx, soot, HC, CO_2, and CO emissions into the atmosphere by 72 percent.

Karthikeyan and Prathima (2017), Sinha et al. (2017), Karthikeyan, Elango, Prathima and Raja (2014) and Karthikeyan and Dharma Prabhakaran (2018) reported an 11 percent rise in engine power, torque, and braking system thermal efficiency of biodiesel (derived from microalgae as well as other biomass) in CI and DI engines when solid nano-additives such as alumina (Al_2O_3) An research study with DI engines using nano-particles combined with biodiesel and also a diesel–biodiesel combination revealed outstanding performance and higher efficiencies than conventional catalysts (Prabu, 2018, Ong et al. 2018). Dhinesh and Annamalai (2018) found that nano-particles of CeO_2 integrated into a biofuel emulsion employing sol–gel combustion technology functioned admirably in mono-cylinder 4 stroke direct CI and DI engines with no hardware modifications. In general, nanoparticles in biofuels boosted the calorific value of the fuel, expedited evaporation, enhanced brake-specific fuel consumption and thermal efficiency, and minimized greenhouse gases (GHGs) including CO, NOx, and smoke, as well as unburned hydrocarbons.

Scientists have successfully developed unique nanofluids by combining several nanoparticles. Because of their capacity to enhance the chemical and thermophysical characteristics of single-phase nanofluids, hybrid nanofluids have piqued the interest of many researchers (Hussain et al. (2020); Karisathan Sundararajan and Ammal, 2018). Selvan et al. (2014) investigated engine efficiency and emission of diesel-biodiesel-ethanol mixtures including cerium oxide nanoparticles and carbon nanotubes. To oxidise CO and diminish nitrogen oxides, cerium oxide nanoparticles served as oxygen supply catalysts. The activation of cerium oxide, which reduced carbon deposits in the cylinder, resulted in a considerable decline in HC and smoke emissions. As a result, the employment of both nanoparticles together can result in cleaner combustion and lower emissions.

Arunprasad, Sener, Thirugnanasambantham, Elango and Bothichandar (2022) conducted study on the performance and emissions of diesel engines incorporating a spirulina microalgae biodiesel mixture (B20) containing magnesium oxide nanoparticles. Because of its low calorific value, Spirulina microalgae biodiesel has poor brake thermal efficiency (BTE) and brake-specific fuel consumption (BSFC) fuel characteristics in general. Conversely, in their study, MgO nanoparticles coupled with spirulina methyl ester significantly enhanced its catalytic action, raising BTE and decreasing BSFC. This paves the way for including spirulina biodiesel enriched with magnesium oxide nanoparticles in diesel engines to substitute diesel without incurring any additional costs or necessitating hardware alterations. Thirugnanasambantham and Elango, 2020 investigated B20, a combination of biodiesel (20%) and diesel (80%), as a way to augment diesel fuel in IC engines with less expensive algal biofuels while lowering emissions. In a four-stroke diesel engine with no alterations, TiO_2 nanoparticles added to the B20 fuel resulted in significant reductions in HC, CO, and smoke emissions.

Figure 5. Effect of nano-particle addition to microalgae culture
(Vargas-Estrada et al, 2020)

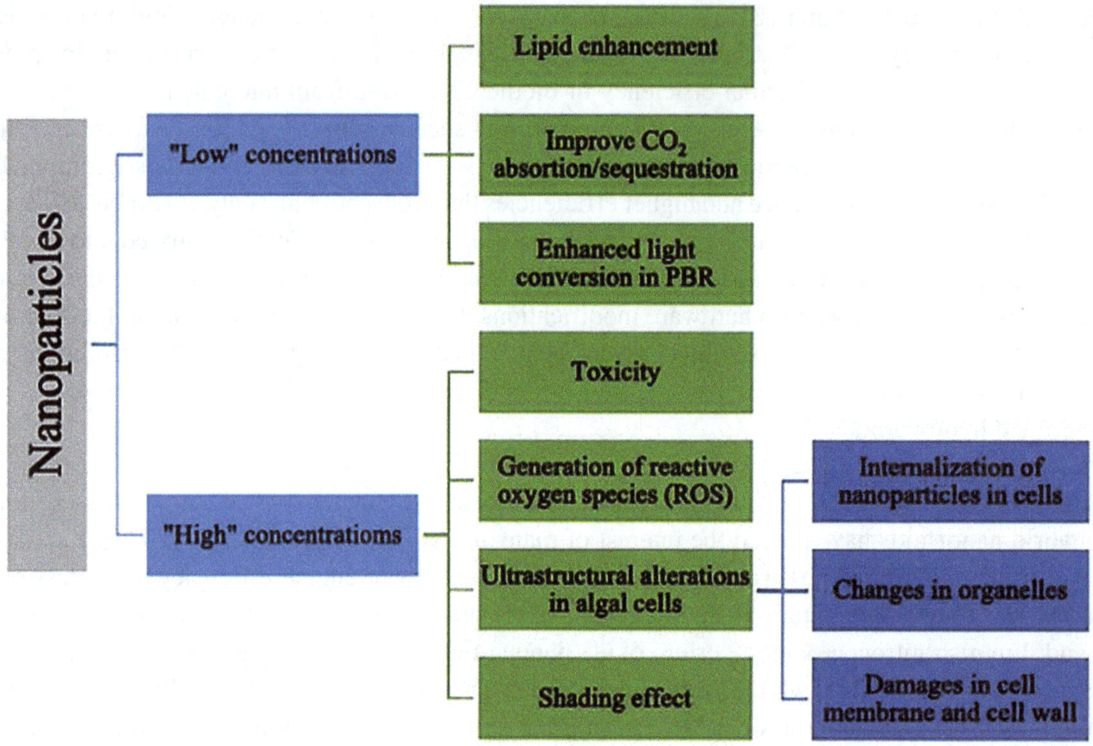

CONCLUSION

One of the major advantages of employing microalgae for biofuel production is the abundance of diversified, lipid-accumulating microalgal species in nature. Large-scale microalgal production for biofuel and several other significant products, as well as the selection of suitable microalgal strains, may be achieved in the future by integrating conventional and cutting-edge processes. Biodiesel may be generated using heterotrophic microalgae as a feedstock. Unfortunately, the exorbitant expense of lipid extraction makes this method difficult. Nanomaterials can be intended to safeguard microalgal cells from damage, and nanoparticles may also be used to harvest microalgae. Because nanoparticles alter the lipid metabolism of microalgal cells, adding them to the culture medium can enhance lipid buildup. Nanotechnology might help enhance biodiesel synthesis from heterotrophic microalgae, however this process is still in its nascent phases. Future research might consider the effect of lipid buildup produced by the inclusion of nano silica and iron oxide to heterotrophic microalgae's growth medium. Strategies to gene manipulation that are careful and meticulous in their usage of genetically - modified strains offer a promising path to long-term and economical microalgal based biodiesel synthesis. The fabrication of nanomaterials different functional groups essential for lipid disintegration and cell wall breaking/weakening should be given more attention. Immobilization of whole-cell lipase on nanoparticles should be optimised and evaluated. There is a need for technological advancement, technical knowledge, and the identification of relevant microalgae species with optimal growth and production characteristics. Further changes to the development of genetically enhanced species should be made so that the appropriate quantity of biodiesel may be generated in a shorter period and with less energy usage.

REFERENCES

Aravantinou, A. F., Andreou, F., & Manariotis, I. D. (2020). Long-term toxicity of ZnO nanoparticles on Scenedesmus rubescens cultivated in semi-batch mode. *Nanomaterials (Basel, Switzerland)*, *10*(11), 2262.

Arunprasad, J., Sener, A. S., Thirugnanasambantham, R., Elango, T., & Bothichandar, T. (2022). Experimental Investigation of Spirulina Microalgae Biodiesel with Metal Nanoadditive on Single-Cylinder Diesel Engine. *Journal of Nanomaterials*.

Arya, I., Poona, A., Dikshit, P. K., Pandit, S., Kumar, J., Singh, H. N., ... Kumar, S. (2021). Current Trends and Future Prospects of Nanotechnology in Biofuel Production. *Catalysts*, *11*(11), 1308.

Barrak, H., Saied, T., Chevallier, P., Laroche, G., M'nif, A., & Hamzaoui, A. H. (2019). Synthesis, characterization, and functionalization of ZnO nanoparticles by N-(trimethoxysilylpropyl) ethylenediamine triacetic acid (TMSEDTA): Investigation of the interactions between Phloroglucinol and ZnO@ TMSEDTA. *Arabian Journal of Chemistry*, *12*(8), 4340–4347.

Brennan, L., & Owende, P. (2010). Biofuels from microalgae—A review of technologies for production, processing, and extractions of biofuels and co-products. *Renewable & Sustainable Energy Reviews*, *14*(2), 557–577.

Carrero, A., Vicente, G., Rodríguez, R., Linares, M., & Del Peso, G. L. (2011). Hierarchical zeolites as catalysts for biodiesel production from Nannochloropsis microalga oil. *Catalysis Today*, *167*(1), 148–153.

Chen, S. Y., Mochizuki, T., Abe, Y., Toba, M., & Yoshimura, Y. (2014). Ti-incorporated SBA-15 mesoporous silica as an efficient and robust Lewis solid acid catalyst for the production of high-quality biodiesel fuels. *Applied Catalysis B: Environmental*, *148*, 344–356.

Choi, W., Kim, G., & Lee, K. (2012). Influence of the CO2 absorbent monoethanolamine on growth and carbon fixation by the green alga Scenedesmus sp. *Bioresource Technology*, *120*, 295–299.

da Silva Vaz, B., Costa, J. A. V., & de Morais, M. G. (2020). Physical and biological fixation of CO2 with polymeric nanofibers in outdoor cultivations of Chlorella fusca LEB 111. *International Journal of Biological Macromolecules*, *151*, 1332–1339.

da Silva Vaz, B., da Silveira Mastrantonio, D. J., Costa, J. A. V., & de Morais, M. G. (2019). Green alga cultivation with nanofibers as physical adsorbents of carbon dioxide: Evaluation of gas biofixation and macromolecule production. *Bioresource Technology*, *287*, 121406.

Damanik, N., Ong, H. C., Tong, C. W., Mahlia, T. M. I., & Silitonga, A. S. (2018). A review on the engine performance and exhaust emission characteristics of diesel engines fueled with biodiesel blends. *Environmental Science and Pollution Research International*, *25*(16), 15307–15325.

Dhinesh, B., & Annamalai, M. (2018). A study on performance, combustion and emission behaviour of diesel engine powered by novel nano nerium oleander biofuel. *Journal of Cleaner Production*, *196*, 74–83.

Ganesh, M., Hemalatha, P., Peng, M. M., & Jang, H. T. (2017). One pot synthesized Li, Zr doped porous silica nanoparticle for low temperature CO2 adsorption. *Arabian Journal of Chemistry*, *10*, S1501–S1505.

Gendy, T. S., & El-Temtamy, S. A. (2013). Commercialization potential aspects of microalgae for biofuel production: An overview. *Egyptian Journal of Petroleum*, *22*(1), 43–51.

Harun, R., & Danquah, M. K. (2011). Influence of acid pre-treatment on microalgal biomass for bioethanol production. *Process Biochemistry*, *46*(1), 304–309.

Harun, R., Danquah, M. K., & Forde, G. M. (2010). Microalgal biomass as a fermentation feedstock for bioethanol production. *Journal of Chemical Technology and Biotechnology (Oxford, Oxfordshire)*, *85*(2), 199–203.

Hasannuddin, A. K., Yahya, W. J., Sarah, S., Ithnin, A. M., Syahrullail, S., Sidik, N. A. C., ... Ramlan, N. A. (2018). Nano-additives incorporated water in diesel emulsion fuel: Fuel properties, performance and emission characteristics assessment. *Energy Conversion and Management*, *169*, 291–314.

Ho, S. H., Chen, W. M., & Chang, J. S. (2010). Scenedesmus obliquus CNW-N as a potential candidate for CO2 mitigation and biodiesel production. *Bioresource Technology*, *101*(22), 8725–8730.

Hossain, N., Mahlia, T. M. I., & Saidur, R. (2019). Latest development in microalgae-biofuel production with nano-additives. *Biotechnology for Biofuels*, *12*(1), 1–16.

Hossain, N., Zaini, J., Mahlia, T. M. I., & Azad, A. K. (2019). Elemental, morphological and thermal analysis of mixed microalgae species from drain water. *Renewable Energy*, *131*, 617–624.

Huang, C. H., & Tan, C. S. (2014). A review: CO2 utilization. *Aerosol and Air Quality Research*, *14*(2), 480–499.

Hussain, F., Soudagar, M. E. M., Afzal, A., Mujtaba, M. A., Fattah, I. M., Naik, B., ... Rahman, S. M. (2020). Enhancement in combustion, performance, and emission characteristics of a diesel engine fueled with Ce-ZnO nanoparticle additive added to soybean biodiesel blends. *Energies*, *13*(17), 4578.

Kadar, E., Rooks, P., Lakey, C., & White, D. A. (2012). The effect of engineered iron nanoparticles on growth and metabolic status of marine microalgae cultures. *The Science of the Total Environment*, *439*, 8–17.

Karisathan Sundararajan, N., & Ammal, A. R. B. (2018). Improvement studies on emission and combustion characteristics of DICI engine fuelled with colloidal emulsion of diesel distillate of plastic oil, TiO2 nanoparticles and water. *Environmental Science and Pollution Research International*, *25*(12), 11595–11613.

Karthikeyan, S., & Dharma Prabhakaran, T. (2018). Emission analysis of Botryococcus braunii algal biofuel using Ni-Doped ZnO nano additives for IC engines. *Energy Sources. Part A, Recovery, Utilization, and Environmental Effects*, *40*(9), 1060–1067.

Karthikeyan, S., Elango, A., Prathima, A., & Raja, K. (2014). Environmental effects of nano additive Co3O4 in grape seed oil biofuel fuelled in CI engine. *Research Journal of Chemistry and Environment*, *18*, 5.

Karthikeyan, S., & Prathima, A. (2017). Analysis of emissions from use of an algae biofuel with nano-ZrO_2. *Energy Sources. Part A, Recovery, Utilization, and Environmental Effects*, *39*(5), 473–479.

Karthikeyan, S., & Prathima, A. (2017). Microalgae biofuel with CeO_2 nano additives as an eco-friendly fuel for CI engine. *Energy Sources. Part A, Recovery, Utilization, and Environmental Effects*, *39*(13), 1332–1338.

Khan, M. I., Shin, J. H., & Kim, J. D. (2018). The promising future of microalgae: Current status, challenges, and optimization of a sustainable and renewable industry for biofuels, feed, and other products. *Microbial Cell Factories*, *17*(1), 1–21.

Kim, J., Jia, H., & Wang, P. (2006). Challenges in biocatalysis for enzyme-based biofuel cells. *Biotechnology Advances*, *24*(3), 296–308.

Lee, J. E., Lee, N., Kim, T., Kim, J., & Hyeon, T. (2011). Multifunctional mesoporous silica nanocomposite nanoparticles for theranostic applications. *Accounts of Chemical Research*, *44*(10), 893–902.

Ling, X., Guo, J., Liu, X., Zhang, X., Wang, N., Lu, Y., & Ng, I. S. (2015). Impact of carbon and nitrogen feeding strategy on high production of biomass and docosahexaenoic acid (DHA) by Schizochytrium sp. LU310. *Bioresource Technology*, *184*, 139–147.

Liu, G., Liao, Y., Wu, Y., & Ma, X. (2018). Synthesis gas production from microalgae gasification in the presence of Fe_2O_3 oxygen carrier and CaO additive. *Applied Energy*, *212*, 955–965.

Low, S. S., Bong, K. X., Mubashir, M., Cheng, C. K., Lam, M. K., Lim, J. W., ... Show, P. L. (2021). Microalgae cultivation in palm oil mill effluent (POME) treatment and biofuel production. *Sustainability*, *13*(6), 3247.

Mansha, M., Qurashi, A., Ullah, N., Bakare, F. O., Khan, I., & Yamani, Z. H. (2016). Synthesis of In_2O_3/graphene heterostructure and their hydrogen gas sensing properties. *Ceramics International*, *42*(9), 11490–11495.

Mykhaylenko, N. F., & Zolotareva, E. K. (2017). The effect of copper and selenium nanocarboxylates on biomass accumulation and photosynthetic energy transduction efficiency of the green algae *Chlorella vulgaris*. *Nanoscale Research Letters*, *12*(1), 1–8.

Nizami, A. S., & Rehan, M. (2018). Towards nanotechnology-based biofuel industry. *Biofuel Res. J.*, *5*, 798–799.

Ong, H. C., Milano, J., Silitonga, A. S., Hassan, M. H., Wang, C. T., Mahlia, T. M. I., ... Sutrisno, J. (2019). Biodiesel production from Calophyllum inophyllum-Ceiba pentandra oil mixture: Optimization and characterization. *Journal of Cleaner Production*, *219*, 183–198.

Pádrová, K., Lukavský, J., Nedbalová, L., Čejková, A., Cajthaml, T., Sigler, K., ... Řezanka, T. (2015). Trace concentrations of iron nanoparticles cause overproduction of biomass and lipids during cultivation of cyanobacteria and microalgae. *Journal of Applied Phycology*, *27*(4), 1443–1451.

Palaniappan, K. (2017). An overview of applications of nanotechnology in biofuel production. *World Appl Sci J*, *35*(8), 1305-1311.

Pathak, V. V., Singh, D. P., Kothari, R., & Chopra, A. K. (2014). Phycoremediation of textile wastewater by unicellular microalga Chlorella pyrenoidosa. *Cellular and Molecular Biology*, *60*(5), 35–40.

Pattarkine, M. V., & Pattarkine, V. M. (2012). Nanotechnology for algal biofuels. In *The science of algal fuels* (pp. 147–163). Springer.

Paul Abishek, M., Patel, J., & Prem Rajan, A. (2014). Algae oil: A sustainable renewable fuel of future. *Biotechnology Research International.*

Peter, A. P., Khoo, K. S., Chew, K. W., Ling, T. C., Ho, S. H., Chang, J. S., & Show, P. L. (2021). Microalgae for biofuels, wastewater treatment and environmental monitoring. *Environmental Chemistry Letters*, *19*(4), 2891–2904.

Prabu, A. (2018). Nanoparticles as additive in biodiesel on the working characteristics of a DI diesel engine. *Ain Shams Engineering Journal*, *9*(4), 2343–2349.

Ramacharyulu, P. V. R. K., Muhammad, R., Kumar, J. P., Prasad, G. K., & Mohanty, P. (2015). Iron phthalocyanine modified mesoporous titania nanoparticles for photocatalytic activity and CO 2 capture applications. *Physical Chemistry Chemical Physics*, *17*(39), 26456–26462.

Rana, M. S., Bhushan, S., Sudhakar, D. R., & Prajapati, S. K. (2020). Effect of iron oxide nanoparticles on growth and biofuel potential of Chlorella spp. *Algal Research*, *49*, 101942.

Rawal, I., & Kaur, A. (2013). Synthesis of mesoporous polypyrrole nanowires/nanoparticles for ammonia gas sensing application. *Sensors and Actuators. A, Physical*, *203*, 92–102.

Ren, H. Y., Dai, Y. Q., Kong, F., Xing, D., Zhao, L., Ren, N. Q., ... Liu, B. F. (2020). Enhanced microalgal growth and lipid accumulation by addition of different nanoparticles under xenon lamp illumination. *Bioresource Technology*, *297*, 122409.

Safarik, I., Prochazkova, G., Pospiskova, K., & Branyik, T. (2016). Magnetically modified microalgae and their applications. *Critical Reviews in Biotechnology*, *36*(5), 931–941.

Sarkar, R. D., Singh, H. B., & Kalita, M. C. (2021). Enhanced lipid accumulation in microalgae through nanoparticle-mediated approach, for biodiesel production: A mini-review. *Heliyon*, *7*(9), e08057.

Sarma, S. J., Das, R. K., Brar, S. K., Le Bihan, Y., Buelna, G., Verma, M., & Soccol, C. R. (2014). Application of magnesium sulfate and its nanoparticles for enhanced lipid production by mixotrophic cultivation of algae using biodiesel waste. *Energy*, *78*, 16–22.

Selvan, V. A. M., Anand, R. B., & Udayakumar, M. (2014). Effect of cerium oxide nanoparticles and carbon nanotubes as fuel-borne additives in diesterol blends on the performance, combustion and emission characteristics of a variable compression ratio engine. *Fuel*, *130*, 160–167.

Shaalan, M., Saleh, M., El-Mahdy, M., & El-Matbouli, M. (2016). Recent progress in applications of nanoparticles in fish medicine: A review. *Nanomedicine; Nanotechnology, Biology, and Medicine*, *12*(3), 701–710.

Shin, W. K., Cho, J., Kannan, A. G., Lee, Y. S., & Kim, D. W. (2016). Cross-linked composite gel polymer electrolyte using mesoporous methacrylate-functionalized SiO2 nanoparticles for lithium-ion polymer batteries. *Scientific Reports*, *6*(1), 1–10.

Sibi, G., Ananda Kumar, D., Gopal, T., Harinath, K., Banupriya, S., & Chaitra, S. (2017). Metal Nanoparticle Triggered Growth and Lipid Production in Chlorella Vulgaris. *Int J Sci Res Environ Sci Toxicol*, *2*(1), 1–8.

Silitonga, A. S., Atabani, A. E., Mahlia, T. M. I., Masjuki, H. H., Badruddin, I. A., & Mekhilef, S. (2011). A review on prospect of *Jatropha curcas* for biodiesel in Indonesia. *Renewable & Sustainable Energy Reviews*, *15*(8), 3733–3756.

Silitonga, A. S., Masjuki, H. H., Ong, H. C., Sebayang, A. H., Dharma, S., Kusumo, F., ... Sugiyanto, B. (2018). Evaluation of the engine performance and exhaust emissions of biodiesel-bioethanol-diesel blends using kernel-based extreme learning machine. *Energy*, *159*, 1075–1087.

Silva, Â., Wilson, K., Lee, A. F., dos Santos, V. C., Bacilla, A. C. C., Mantovani, K. M., & Nakagaki, S. (2017). Nb2O5/SBA-15 catalyzed propanoic acid esterification. *Applied Catalysis B: Environmental*, *205*, 498–504.

Sinha, V. K. (2017). Sharlin. A review approach on exhaust emission reduction by nanoparticle additives in diesel engine. *Int J Innov Res Sci Eng Technol*, *6*, 15046–15050.

Steele, J. M., Grady, N. K., Nordlander, P., & Halas, N. J. (2007). Plasmon hybridization in complex nanostructures. In *Surface plasmon nanophotonics* (pp. 183–196). Springer.

Thirugnanasambantham, R., & Elango, T. (2020). Emission Analysis of Chlorella sp. Microalgae Biodiesel with Oxide Nano Additives in Diesel Engine. *Journal of Scientific and Industrial Research*, *79*(11), 1031–1034.

Tiwari, J. N., Tiwari, R. N., & Kim, K. S. (2012). Zero-dimensional, one-dimensional, two-dimensional and three-dimensional nanostructured materials for advanced electrochemical energy devices. *Progress in Materials Science*, *57*(4), 724–803.

Torkamani, S., Wani, S. N., Tang, Y. J., & Sureshkumar, R. (2010). Plasmon-enhanced microalgal growth in miniphotobioreactors. *Applied Physics Letters*, *97*(4), 043703.

Trindade, S. C. (2011). *Nanotech biofuels and fuel additives. Biofuel's Engineering Process Technology*. InTech.

Vaz, B. D. S., Costa, J. A. V., & Morais, M. G. D. (2018). Innovative nanofiber technology to improve carbon dioxide biofixation in microalgae cultivation. *Bioresource Technology*, *273*, 592–598.

Zimmerman, W. B., Hewakandamby, B. N., Tesař, V., Bandulasena, H. H., & Omotowa, O. A. (2009). On the design and simulation of an airlift loop bioreactor with microbubble generation by fluidic oscillation. *Food and Bioproducts Processing*, *87*(3), 215–227.

Zimmerman, W. B., Tesař, V., & Bandulasena, H. H. (2011). Towards energy efficient nanobubble generation with fluidic oscillation. *Current Opinion in Colloid & Interface Science*, *16*(4), 350–356.

Chapter 12
Macroalgae-Based Bioethanol

Sakshi Kolpe
Ajeenkya D.Y. Patil University, India

Shankar Mukundrao Khade
https://orcid.org/0000-0002-1974-0709
Ajeenkya D.Y. Patil University, India

Shivraj Nile
Zhejiang Chinese Medical University, Hangzhou, China

ABSTRACT

The chapter discusses the methods effective for the extraction of useful energy from the macroalgae biomass including liquefaction, anaerobic digestion, fermentation to biobutanol, trans-esterification to biodiesel, pyrolysis, direct combustion, fermentation to bioethanol, and gasification. However, if the algae are suited for the production of biodiesel, they can be studied from the content of their triacylglycerols (TAGs). Due to having high fatty acid content, they have a high conversion rate to biodiesel, and the lack of sulphur, phosphorus, and nitrogen also aids in the conversion. This chapter highlights the limitations and suitability of macroalgae for the conversion process in reference to chemical composition, process optimization, and cost effectiveness. It is concluded that bio-oils and bioethanol produced from wet macroalgae are considered over biodiesel production because of high lipid content of microalgae biomass. Moreover, the chapter considers electricity production from the dry mass as it would turn profitable, and this can be achieved from fast-growing macroalgae like "Ulva."

INTRODUCTION

Fossil fuel hydrocarbons are being overly used which results in their scarcity and being expensive. The focus of research on converting biomass to biofuels has shifted to third-generation biomass, such as algae. The first-generation feedstock is divisive due to the food-fuel conflict, whereas the high-cost lignin removal technique limits the second-generation biomass (A Bayu, 2018). As algae can repair greenhouse gases (CO2) through photosynthesis, it is a promising renewable energy source. The typical photosynthetic productivity of marine biomass is 6–8%, which is significantly greater than the 1.8–2.2% of terrestrial

DOI: 10.4018/978-1-6684-5269-1.ch012

biomass (Adams, J.M., 2009 & Junying, 2015). Marine macroalgae are the source of aquatic biomass which shows significant opportunities to become renewable energy sources. Aquatic biomass has 6–8% of photosynthetic efficiency, which is significantly greater than terrestrial biomass (1.8–2.2%). For the conversion of solar energy and production of biofuels, either marine macroalgae such as seaweed or microalgae might be employed (Hirano, A, 1997).

Macroalgae are large, fast-growing marine and freshwater plants with a maximum length of 60 metres. Due to the lack of water limitations, the marine macroalgae growth rates outpace the growth rate of terrestrial biomass (K. Sudhakar, 2018). The principal marine macroalgae have higher yearly production rates (grams C m2 yr1) than most terrestrial biomass. Phytocolloids like agar, alginates (derived from Phaeophyta) and carrageenan (derived from Rhodophyta), are mostly generated from seaweeds. The by-products of such processing can likewise be used as a renewable energy source (Kawai, 2016 & Notoya, M., 2010). Macroalgae are primarily of three kinds green (Phaeophyta), Red (Rhodophyta) and brown (Phaeophyta) depending upon their photosynthetic pigmentation differences (Chlorophyta). Red has the most species (6000), green (4500), and brown (2000) (Inn Shi Tan, 2020; Mohsen Ghadiryanfar, 2016). Red algae grow in the intertropical zones while the brown algae thrive in temperate to cold or extremely cold seas. Green algae may be found in a range of aquatic environments. In Asia, macroalgae are now farmed for food, fertilisers, and hydrocolloid extraction, (72%) majority-owned by Japan, Korea, Philippines and China worldwide yearly output. Macroalgae productivity varies from 150 to 600 t per hectare per year in fresh weight, with a total global production of 12 million tonnes of dry matter per year (Ashok Pandey, 2011; Huihui Chen, 2015). The extraction from macroalgae biomass of biofuels, on the other hand, received not much attention. Hydrocolloid extraction is presently the only significant industrial product derived from macroalgae. However, macro-algae may be used to extract a variety of high-value speciality goods.

Figure 1. Stepwise representation of bioethanol production from algal biomass

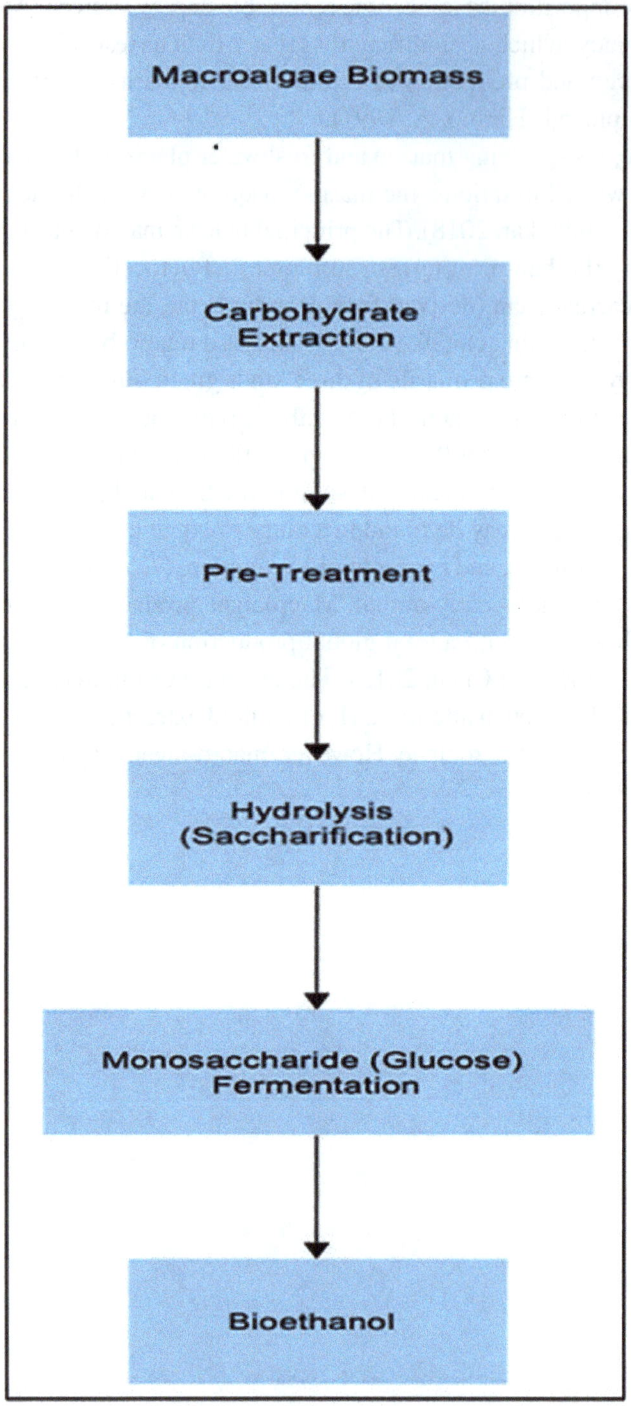

Chemical Composition of Macroalgae

Macroalgae have a dry matter content of only 10–15%. Large levels of carbohydrates may be found in dried macroalgae (approximately 60%). Carbohydrates are mostly obtained from cell wall components of macroalgae (Milledge, J.J., 2014). Red and green algae have an abundance of cellulose in them which is preferable for biofuel production. Seasonal changes in organic constitution and ash content differ correspondingly. For instance, the quantity of carbohydrates stored extends to the peak, in the fall. Mannan, ulvan, carrageenan, agar, laminarin, mannitol, alginate, fucoidin, fucose, and uronic acid distinguish macroalgae from microalgae and lignocellulosic biomass. Furthermore, lignin concentration as low as 0.03 g/kg of dry matter allows for downstream processing including fermentation or anaerobic digestion without the need for expensive pre-treatment.

Protein concentration in macroalgae can range from 7 to 31% dry weight, with lipid content of 2 to 13% dry weight. Macroalgae contain a significant quantity of carbohydrates (up to 32–60% of dry weight) (Gao, G., 2020). Macroalgae, being a photosynthetic organism is able to transfer into chemical energy from solar energy, usually as carbohydrates (mono- and polysaccharides). The intricacies of the aquatic environment, on the other hand, cause their metabolism to produce saccharides with distinct synthetic properties than those produced by terrestrial plants. For example, cellulose (35-50%) and hemicelluloses (20-35%), are made up of simple sugars such as ribose, glucose, mannose, xylose, galactose and arabinose, which are the most abundant carbohydrate components in all terrestrial plants. Furthermore, macroalgae contain a wide range of carbohydrates, including monomeric sugars of various kinds. Brown macroalgae have significant polysaccharides such as laminaran, fucoidan, and alginate, whereas red and green macroalgae have galactans (e.g. carrageenan and agar) and ulvan, respectively.

Carrageenan and agar are two polymers of modified galactose found in red macroalgae. Palmaria palmata has the greatest carrageenan content, with a concentration of 354 mg/g, and this high carrageenan content allows for a high bioethanol output. Brown algae, on contrary, is high in alginate and contains substantial amounts of laminarin, a significant amount of mannitol in the fall, and few commonly used sugars. The lipid content of brown and red algae is lower than 5%, which makes them unsuitable for biodiesel production. Green algae species vary in their cellulose content (Valonia types can have up to 70% cellulose) and starch content (20–30%). Furthermore, alkali metal and halogen levels in macroalgae are greater (0.5–11%) than in terrestrial biomass (1–1.5%). Nutrient content (N, P) is largely determined by macrophyte shape, followed by nutrient contamination in the environment.

Table 1. Algal sources suitable for bioethanol production with reference to their starch content

Algal Source	Starch Content of the Dry Matter	Reference
Ulva ohnoi	1.59% to 21.44% rising with nutrient starvation	(Meghanath Prabhu, 2019)
Saccharina latissima	50.0 (reserve food material)	(Adams, J.M., 2009)
Laminaria hyperborea	55.0 (reserve food material)	(Greetham, D., 2020 & Roesijadi, Guritno, 2010)
Chlamydomonas reinhardtii UTEX 90	53.0 (starch)	(Kim, M.S., 2006)
Chlorella vulgaris	12.0–17.0 (starch)	(Spolaore, P., 2006)
C. vulgaris	37.0 (starch)	(Horn, S.J., 2000)
Synechococcus sp.	15.0 (starch)	(Yusuke Sasaki, 2018)
Spirulina fusiformis	37.3–56.1 (starch)	(Rafiqul I.M., 2003)

Cultivation

The quantity of bioethanol that may be produced, both in absolute and relative terms, varies greatly across algae families and strains within identical ones. This is due to the fact that algae species and strains differ significantly relating to prolificity and multiplication, nutrition and light requirements, capacity to collect lipids or other desired chemicals, ability to adjust to unfavourable circumstances, and so forth. As a result, strains that have the ability to produce the most bioethanol, either directly or via biomass assembly, are frequently chosen.

Brennan and Owende (2010) compiled a list of desired features for strains of algae regarded as biofuel prospects, including:

1. Have tolerance for a broad range of temperatures due to seasonal and diurnal changes;
2. In open pond production settings, capable of dominating wild strains;
3. High ability to absorb CO2;
4. Nutritional needs are limited;
5. Possibility of providing useful co-products;
6. Quick production cycle;
7. Photosynthetic efficiency is high, and
8. Self- flocculation

As starch is used in ethanol fermentation as a substrate, a strain providing a large quantity of renewable energy due to high starch concentration is the characteristic feature when screening for algal strain used in the production of bioethanol. The total dry weight of biomass may be calculated, and biochemical tests can be used to assess the biomass starch/cellulose content.

Direct harvesting of seaweed from the maritime environment, wrack collecting from the beach, and culture of chosen seaweed species are the three principal sources of macroalgal biomass for downstream processing. Offshore, onshore, and integrated seaweed farming are the various ways of seaweed cultivation. Direct culture of macroalgae in saltwater is done using tethered ropes off the coast. It's also possible to integrate it with onshore aquaculture and offshore wind farm locations. Offshore farming provides for more controlled growing and harvesting circumstances, resulting in better efficiency and output than natural growth. To extract nutrients directly from the saltwater, in sheltered bays or lagoons macroalgae can be cultivated. Traditional seaweed growing methods include long lines, permanent off-bottom farming, and rock-based farming. For biomass production, large brown kelps are generally favoured. Smaller species, comparatively, are preferred for high-value components.

Enclosed photobioreactors and open ponds are the most prevalent production technologies for algae culture. Controlling downstream processing features, growth parameters, pollution, productivity, capital and operational expenses, water evaporation and so forth are all different aspects of production systems.

Harvesting of Algal Biomass

Harvesting the grown biomass can be done in two ways. It has long been a custom in many coastal areas. Techniques for harvesting include:

- Manual (Hand)

- Mechanical (Dredge, moving boat, mesh conveyor)

Three times a year, harvesting is usually carried out. Mechanical harvesting has a negative impact on marine ecosystems and stunts macroalgae development. In other cases, however, eliminating surplus nutrients from the environment by reducing prolific macroalgae growth is advantageous. Harvesting macroalgae blooms hasn't proven to be economically viable anywhere on the planet (Notoya, M., 2010).

Extraction of Carbohydrates

The extraction of energy from macroalgae may be classified in several ways, one of which being whether or not an initial drying phase is necessary. As a result, two separate groupings of processes emerge:

1. Energy extraction from dry macroalgae
   a. Pyrolysis;
   b. gasification (conventional);
   c. direct combustion;
   d. trans-esterification to biodiesel.
2. Energy extraction from wet macroalgae
   a. anaerobic digestion;
   b. fermentation to bioethanol or biobutanol;
   c. hydrothermal treatments.

Unfortunately, due to problems in processing technology and scaling up, bioethanol from second-generation feedstock has met with a lot of opposition. Cellulases degrade cellulose and hemicelluloses while lignin can prevent biodegradation, which is a major technical problem. Due to the lack of lignin in macroalgae, the conversion to ethanol from polysaccharides is more rapid in macroalgae. As a result, bioethanol determined from macroalgae are procuring popularity.

Brown, green, and red macroalgae are all forms of macroalgae that may be found in the ocean and have been used to make bioethanol. Because of their high carbohydrate content and ease of mass-cultivation, the predominant feedstock for bioethanol synthesis appears to be brown macroalgae. For example, a synthetic yeast platform can produce ethanol from brown macroalgal sugars at up to 83% of the theoretical maximum, but ethanol conversion from macroalgae at up to 50% of the theoretical maximum is regarded as ambitious (Elamathi Vimali, 2022).

Pre-treatment of the Seaweed Biomass and Hydrolysis of the Polysaccharides

Freshly harvested brown seaweed contains 15–20% carbs by weight, or roughly for every kg of wet weight it has 200 g carbohydrates, that is an acceptable concentration of substrate for conversion of microbial activities (Godvin Sharmila V, 2021). Moreover, seaweed does not require the pre-treatment that ligno-cellulosic biomass does to extract fermentable sugars since they don't contain lignin. Water is an easy way to extract laminaran and mannitol. Alginates (polymer blocks of Uronic and Guuronic acid) make up 30–40% of the macroalgae biomass. Calcium salt can be used to remove sodium alginate from the original extraction solution. This forms calcium alginate, which has a fibrous structure and is insoluble in water but may be isolated from it. To convert calcium alginate to alginic acid, the calcium alginate is separated and suspended in water along with acid. The fibrous alginic acid formed is easy to separate and mix with alcohol and the other sodium carbonate is slowly added to the paste to the endpoint when all alginic acid is transferred into sodium alginate. Sodium alginate paste is extruded occasionally into pellets, which are later dried and ground. Dilution with a substantial volume of water is necessary due to the high viscosity. Alginate concentrations of less than 2% are used in the procedure. Processes involving the use of alginate as a fermentation substrate cannot be done with such a dilution. Water should be avoided if at all possible, as it increases the downstream processing costs.

Enzymatic hydrolysis and fermentation can be used together to produce bioethanol from cellulosic sources. Hydrolysis is more difficult for wet seaweed as alginates do not extract easily, resulting in enzymatic destruction of untreated organic materials from macroalgae, and it will make hydrolysis a rate-limiting step in the fermentation method. As a result, hydrolysis should be included in the biomass pre-treatment process. This hydrolysis can be done physically using grinders and emulsifiers, chemically with acid or alkali, or enzymatically.

Alginates can be degraded by a number of alginate lyases discovered by marine microorganisms. There are several ways for partial or total decomposition of alginate in the biomass through mechanical pre-treatment. Chemical hydrolysis should be based on current technologies, such as modifying and adapting techniques used for wood biomass pre-treatment while combining with steam treatment using acid and alkali procedure (Adams, J.M., 2009).

Dihydroxy benzene-driven Fenton reaction and brown-rot fungus, are the other known techniques, which partly depolymerize oxidise phenols, carbohydrates and other fermentation inhibitors as they create highly reactive hydroxyl radicals and alike oxidants. Nonetheless, combining mechanical and enzymatic hydrolysis to pre-treat algal biomass is the simplest and most environmentally friendly method.

Algal Biomass to Bioethanol

There are three ways of producing bioethanol from algae. Algae may absorb a large quantity of starch/cellulose from the biomass, which is transformed into sugars used for fermentation, which can then be turned into bioethanol by ethanol production. During dark fermentation, certain algae can produce ethanol. The present techniques for the production of bioethanol from macroalgae speices: SHF (separate hydrolysis and fermentation) and SSF (simultaneous saccharification and fermentation) are two types of seaweed hydrolysis and fermentation. The fermentation procedure is broken down into many parts. Seaweed biomass is hydrolysed first and then fermented by bacteria or yeast in distinct units in the case of SHF. SSF, on the other hand, does both hydrolysis and fermentation in one step. One of the most common strategies for obtaining high sugar yields from lignocellulosic biomass is to use acid pretreatment. By acid hydrolysis (0.4 M H2SO4 at 125 1C for 25 minutes), red algae, Palmaria palmata, yielded glucose, galactose, and carbohydrates, which were subsequently fermented to ethanol.

Despite the fact that acidic hydrolysis of biomass has been claimed to be the most economical technology until today, glucose breakdown happened while in the process. The by-products of acid hydrolyses, such as 5-hydroxy-methyl-furfural and levulinic acid, have a significant influence on the production of ethanol capability. Enzyme hydrolysis, like acid hydrolysis of carbohydrates, has various drawbacks, including difficulties extracting the derived enzyme from the acquired products and a protracted hydrolysis period. Saccharification has recently been aided by the use of a two-step biomass treatment procedure that combines acid and enzymatic pretreatment. Pretreatment with 0.06% sulfuric acid at 170 1C for 15 minutes, followed by enzymatic pretreatment, resulted in a maximum glucan concentration of

29.10%, which was four times greater than Saccharina japonica (raw). The digestibility of total biomass was 83.96%. At the moment, suitable enzymes are required for increasing hydrolysis effectiveness, and enzyme redemption and utilisation in biorefining processes also offer a guarantee. Separation of the membrane has been shown to be a reliable method for recovering enzymes.

Most polysaccharides and disaccharides are converted to monosaccharides after the effective hydrolysis phase; for example, sucrose can be catalysed by invertase enzyme into fructose and glucose, and subsequently, they are converted to ethanol by Saccharomyces ceveresiae. In red algae hydrolysates, the coexistence of glucose and galactose restricts total ethanol generation, while microorganisms in brown algae can't utilise several sugars at the same time as the capacity to utilise alginate is absent, therefore ethanol production is limited. As a result, the critical concern for bioethanol production from macroalgae monosaccharides through ethanol fermentation is an economical hydrolysis process.

CONCLUSION

In the coming future, research into prospective molecular processing using a range of different macroalgae species has found potential biochemical processing processes for the production of renewable energy. Macroalgae biomass varies depending upon the geographical area, species, and season, as do yields and productivity, which are heavily dependent on processing technology. As a result, macroalgae have the potential to make a large contribution to a sustainable and renewable economy. Conclusively, marine macroalgae are habitual of high salinity, huge temperature fluctuations, enormous nutrient dynamics, and extreme UV sunshine exposure. Due to the severe environment, their metabolism produces primary and secondary metabolites with distinct chemical structures in order to survive. The conversion of chemical energy from solar energy by macroalgae, mostly in the form of carbohydrates, is a normal photosynthetic organism. Because of their high carbohydrate content, they may be used to make alcohol or gaseous fuels. Furthermore, the distinctiveness of their chemical structure allows for the production of extensive carbohydrate-based compounds, including sulphated oligosaccharides and uncommon sugars, that have a potential market in the pharmaceutical and nutraceutical sectors. It also has the potential to be used in the synthesis of new materials with novel physicochemical features. Polyphenols, essential amino acids, and pigments are all found in macroalgae, as well as other important compounds. Integrating bioenergy production with valuable chemicals should make macroalgal bioenergy development more economically appealing since it provides possible additional revenue, high efficiency, and better environmental outcomes.

REFERENCES

Adams, J. M., Gallagher, J. A., & Donnison, I. S. (2009). Fermentation study on Saccharina latissima for bioethanol production considering variable pre-treatments. *Journal of Applied Phycology, 21*(5), 569–574. doi:10.100710811-008-9384-7

Borines, de Leon, & Cuello. (2013). Bioethanol production from the macroalgae Sargassum spp. *Bioresource Technology, 138*, 22-29. doi:10.1016/j.biortech.2013.03.108

Chen. (2015). Prospects for Bioethanol Production from Macroalgae. *Trends in Renewable Energy, 1*(3), 185-197. Available at: http://www.futureenergysp.com/index.php/tre/article/view/16

Chen, H., Zhou, D., Luo, G., Zhang, S., & Chen, J. (2015). Macroalgae for biofuels production: Progress and perspectives. *Renewable and Sustainable Energy Reviews, 47*, 427-437. doi:10.1016/j.rser.2015.03.086

Gao, G., Burgess, J. G., Wu, M., Wang, S., & Gao, K. (2020). Using macroalgae as biofuel: Current opportunities and challenges. *Botanica Marina, 0*(0), 355–370. Advance online publication. doi:10.1515/bot-2019-0065

Ghadiryanfar, M., Rosentrater, K. A., Keyhani, A., & Omid, M. (2016). A review of macroalgae production, with potential applications in biofuels and bioenergy. *Renewable and Sustainable Energy Reviews, 54*, 473-481. doi:10.1016/j.rser.2015.10.022

Godvin Sharmila, Dinesh Kumar, Pugazhendi, Bajhaiya, Gugulothu, & Banu. (2021). Biofuel production from Macroalgae: Present scenario and future scope. *Bioengineered, 12*(2), 9216-9238. doi:10.1080/21655979.2021.1996019

Greetham, D., Adams, J. M., & Du, C. (2020). The utilization of seawater for the hydrolysis of macroalgae and subsequent bioethanol fermentation. *Scientific Reports, 10*(1), 9728. doi:10.103841598-020-66610-9 PMID:32546695

Hirano, A., Ueda, R., Hirayama, S., & Ogushi, Y. (1997). CO2 fixation and ethanol production with microalgal photosynthesis and intracellular anaerobic fermentation. *Energy, 22*(2-3), 137–142. doi:10.1016/S0360-5442(96)00123-5

Horn, S. J., Aasen, I. M., & Ostgaard, K. (2000). Production of ethanol from mannitol by Zymobacter palmae. *Journal of Industrial Microbiology & Biotechnology, 24*, 51–57. doi:10.1038j.jim.2900771

Hossain, A. B. M. S., Saleh, A. A., Aishah, S., Boyce, A. N., Chowdhury, P. P., & Naqiuddin, M. (2008). Bioethanol production from agricultural waste biomass as a renewable bioenergy resource in biomaterials. In N. A. A. Osman, F. Ibrahim, W. A. B. W. Abas, H. S. A. Rahman, & H. N. Ting (Eds.), *Biomed 2008* (Vol. 21, pp. 300–305). Proceedings. doi:10.1007/978-3-540-69139-6_77

Kawai, S., & Murata, K. (2016). Biofuel Production Based on Carbohydrates from Both Brown and Red Macroalgae: Recent Developments in Key Biotechnologies. *International Journal of Molecular Sciences, 17*(2), 145. doi:10.3390/ijms17020145 PMID:26861307

Kim, M.-S., Baek, J.-S., Yun, Y.-S., Sim, S. J., Park, S., & Kim, S.-C. (2006). Hydrogen production from Chlamydomonas reinhardtii biomass using a two-step conversion process: Anaerobic conversion and photosynthetic fermentation. *International Journal of Hydrogen Energy, 31*(6), 812–816. doi:10.1016/j.ijhydene.2005.06.009

Kraan, S. (2013). Mass-cultivation of carbohydrate rich macroalgae, a possible solution for sustainable biofuel production. *Mitigation and Adaptation Strategies for Global Change, 18*(1), 27–46. doi:10.100711027-010-9275-5

Kumar, K., Ghosh, S., Angelidaki, I., Holdt, S. L., Karakashev, D. B., Morales, M. A., & Das, D. (2016). Recent developments on biofuels production from microalgae and macroalgae. *Renewable and Sustainable Energy Reviews, 65*, 235-249. doi:10.1016/j.rser.2016.06.055

Milledge, J. J., Smith, B., Dyer, P. W., & Harvey, P. (2014). Macroalgae-Derived Biofuel: A Review of Methods of Energy Extraction from Seaweed Biomass. *Energies, 7*(11), 7194–7222. doi:10.3390/en7117194

Notoya, M. (2010). Production of Biofuel by Macroalgae with Preservation of Marine Resources and Environment. In J. Seckbach, R. Einav, & A. Israel (Eds.), *Seaweeds and their Role in Globally Changing Environments. Cellular Origin, Life in Extreme Habitats and Astrobiology* (Vol. 15). Springer., doi:10.1007/978-90-481-8569-6_13

Øverland, M., Mydland, L. T., & Skrede, A. (2019). Marine macroalgae as sources of protein and bioactive compounds in feed for monogastric animals. *Journal of the Science of Food and Agriculture, 99*(1), 13–24. doi:10.1002/jsfa.9143 PMID:29797494

Prabhu, M., Chemodanov, A., Gottlieb, R., Kazir, M., Nahor, O., Gozin, M., Israel, A., Livney, Y. D., & Golberg, A. (2019). Starch from the sea: The green macroalga Ulva ohnoi as a potential source for sustainable starch production in the marine biorefinery. *Algal Research, 37*, 215-227. doi:10.1016/j.algal.2018.11.007

Rafiqul, I. M., Hassan, A., Sulebele, G., Orosco, C. A., Roustaian, P., & Jalal, K. C. A. (2003). Salt stress culture of blue-green algae Spirulina fusiformis. *Pakistan Journal of Biological Sciences, 6*(7), 648–650. doi:10.3923/pjbs.2003.648.650

Roesijadi, G., Jones, S. B., Snowden-Swan, L. J., & Zhu, Y. (2010). *Macroalgae as a Biomass Feedstock: A Preliminary Analysis*. doi:10.2172/1006310

Rojan, Anisha, Nampoothiri, & Pandey. (2011). Micro and macroalgal biomass: A renewable source for bioethanol. *Bioresource Technology, 102*(1), 186-193. doi:10.1016/j.biortech.2010.06.139

Sasaki, Y., Takagi, T., Motone, K., Shibata, T., Kuroda, K., & Ueda, M. (2018, August 3). Direct bioethanol production from brown macroalgae by co-culture of two engineered *Saccharomyces cerevisiae* strains. *Bioscience, Biotechnology, and Biochemistry, 82*(8), 1459–1462. doi:10.1080/09168451.2018.1467262 PMID:29708475

Spolaore, P., Joannis-Cassan, C., Duran, E., & Isambert, A. (2006). Review: Commercial application of microalgae. *Journal of Bioscience and Bioengineering, 101*(2), 87–96. doi:10.1263/jbb.101.87 PMID:16569602

Sudhakar, K., Mamat, R., Samykano, M., Azmi, W. H., & Ishak, W. F. W. (2018). An overview of marine macroalgae as bioresource. *Renewable and Sustainable Energy Reviews, 91*, 165-179. doi:10.1016/j.rser.2018.03.100

Suganya, T., Varman, M., Masjuki, H. H., & Renganathan, S. (2016). Macroalgae and microalgae as a potential source for commercial applications along with biofuels production: A biorefinery approach. *Renewable and Sustainable Energy Reviews, 55*, 909-941. doi:10.1016/j.rser.2015.11.026

Tan, I. S., Lam, M. K., Henry, C. Y. F., Lim, S., & Lee, K. T. (2020). Advances of macroalgae biomass for the third generation of bioethanol production. *Chinese Journal of Chemical Engineering, 28*(2), 502-517. doi:10.1016/j.cjche.2019.05.012

Tan, I. S., Lam, M. K., & Lee, K. T. (2013). Hydrolysis of macroalgae using heterogeneous catalyst for bioethanol production. *Carbohydrate Polymers, 94*(1), 561-566. doi:10.1016/j.carbpol.2013.01.042

Vimali, E., Gunaseelan, S., Devi, V. C., Mothil, S., Arumugam, M., Ashokkumar, B., Innasi, M. G. M., Pugazhendhi, A., & Varalakshmi, P. (2022). Comparative study of different catalysts mediated FAME conversion from macroalga Padina tetrastromatica biomass and hydrothermal liquefaction facilitated bio-oil production. *Chemosphere, 292*, 133485. doi:10.1016/j.chemosphere.2021.133485 PMID:34979211

Wei, N., Quarterman, J., & Jin, Y.-S. (2013). Marine macroalgae: an untapped resource for producing fuels and chemicals. *Trends in Biotechnology, 31*(2), 70-77. doi:10.1016/j.tibtech.2012.10.009

Chapter 13
Microalgae Biofuels: Challenges and Potential

Naresh Tanwer
Maharshi Dayanand University, India

Vaishali Arora
Maharshi Dayanad University, India

Priyanka Bumbra
Maharshi Dayanand University, India

Kiran Grewal
Maharshi Dayanand University, India

Jitender Singh Laura
Maharshi Dayanand University, India

Babita Khosla
Maharshi Dayanand University, India

ABSTRACT

Development of sustainable energy resources is a major concern in front of the scientific community and stakeholders in this field. Invariably, fossil fuels are utilized to meet the major part of energy demand all over the world, but their stocks are limited. Burning of fossil fuels also causes significant environmental pollution that results in drastic climate change. There is an emergent need of an alternative to fulfill the current energy demand in a sustainable way along with sorting out the pollution issues. Microbial biofuels could be the most appropriate choice in this regard because it bears enormous potential that needs to be exploited. But it has some challenging issues to employ it to meet out the current energy demands. This chapter illustrates the microbial biofuels, harvesting techniques, different process, and potential to utilize them as sustainable energy sources along with considerable challenges that need to be addressed to make it an eco-friendly and economic fuel choice to pave the path of sustainable development.

DOI: 10.4018/978-1-6684-5269-1.ch013

INTRODUCTION

The scarcity of fossil fuel resources and the catastrophic environmental consequences caused by their use are major global concerns today. Fossil fuels have resulted in massive and long-term damage to our environment, including acid rain, global warming, and climate change. Greenhouse gas emissions continued to increase, and in 2010, they reached up to 30.6 gigatonnes (GT). The OECD (Organization for Economic Co-operation and Development) suggested that without implementing more effective policies than in force today, it will increase by another 50% by the end of 2050. It is anticipated that without taking appropriate measures, the atmospheric concentration of greenhouse gases would be reaching 685 ppm CO_2-equivalents by 2050. This is well above the 450ppm level of CO_2 required to have at least a 50% chance of increasing the global average temperature to 2°C. As per OECD predictions, the global average temperature is likely to exceed an increment of 2°C by 2050 and rises to 3 to 6°C greater than the pre-industrial levels by the end of this century. The United Nations Environment Programme (UNEP) "Emission Gap Report 2019", unless the greenhouse gas emissions must fall by 7.6% each year from 2020 to 2030, the world will fail to meet the 1.5 °C temperature reduction goal of the Paris agreement. (UNEP, 2019). Such a high rate of temperature rise would continue to change rain patterns and the melting of glaciers, leading to rising sea levels and worsening weather events. It could also exceed some critical "tipping points", resulting in dramatic natural changes resulting in irreversible or catastrophic events for natural systems and society (OECD, 2012). Also, fossil fuel reserves are depleting very fast. The growing economy needs fossil fuels to function, from the production of plastics and agricultural fertilizers to the energy requirements for heating, lighting and transportation services. Rapidly growing industrialization and the exploding world population are the driving force to become increasingly dependent on fossil hydrocarbons for economic growth (Ahmed et al., 2021). The fossil fuel demand is predicted to grow by 40% from 2010 to 2040 (Bhore et al., 2019). Fossil fuel reserves are insufficient to meet rising demand at the current consumption rate and will be exhausted soon. An overreliance on fossil fuel supplies challenges the world's energy security resilience. It will lead to a dwindling of fossil fuel resources and is expected to exhaust in less than 80 years (Rafa et al., 2021). The majority of the world's energy demand is fulfilled by fossil fuels (approximately 80%), with biofuels (11.3%), nuclear energy (5.5%), and other renewable energy sources, such as wind, filling in the gaps (2.2%) (Kopplaar, 2012). According to the International Energy Agency (IEA), biofuels will be contributing more to total energy sources by 2030. It could be expanded significantly if new petroleum fields are not accessed or substantial new alternatives are not identified (Hannon et al., 2010). Biofuels are among the most sustainable options. Biofuels and other renewable energy sources, such as wind and solar power, have attracted researchers, industrialists, and administrations to address energy shortages, rising crude oil costs, global climate change, and land and water degradation (Alam et al., 2015). This chapter entails a detailed description of microbial biofuels and their potential and contemporary challenges to make their most appropriate economically viable and sustainable choice of energy demands.

Biofuels

Biofuels have a zero-carbon footprint and are renewable fuels that are essential for environmental and economic sustainability. Any solid, liquid, or gaseous fuel derived from biomass (plants, animal waste, manure, sludge, etc.) stores solar energy through physical and chemical processes. This energy source can replace fossil fuels (Mobin & Alam, 2018). The usages of biofuels have several over conventional fossil fuels, as shown in **Fig. 1**. Biofuel can be converted to another variety of customized fuels as per the requirement. As reported by IEA in 2017, the annual global production of terrestrial and marine biofuels is estimated to be around 1,324 million tons of oil equivalents (International Energy Agency, 2017).

Figure 1. Benefits of biofuels over fossil fuels

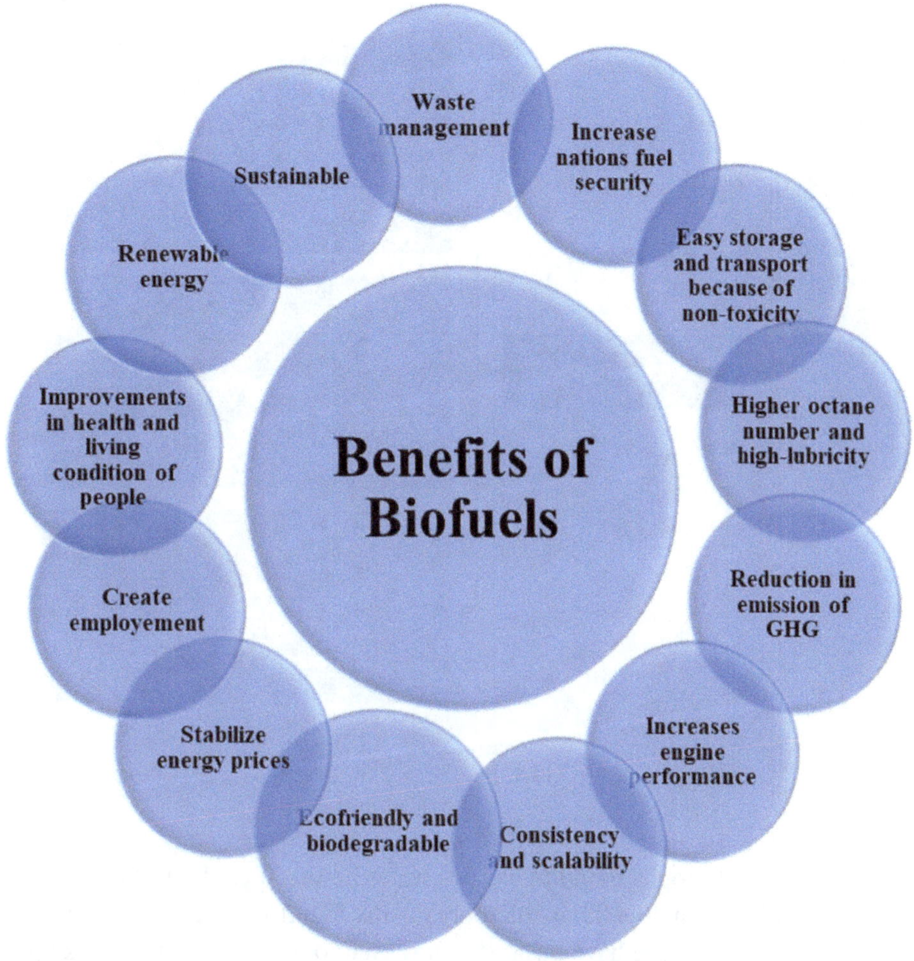

Classification of Biofuels

The classification of biofuels is commonly categorized into two groups, namely primary (1þ) biofuel and secondary (2þ) biofuel. There is a further classification of secondary biofuel into three groups (as shown in **Fig. 2**) (a) 1stgeneration biofuel, (b) 2ndgeneration biofuel and (c) 3rdgeneration biofuels. Biofuels produced from oil crops have great potential as renewable and carbon-neutral alternatives to fossil fuels. But biomass, in the case of oil crops, used to produce biofuels, is part of human food stock that competes with the human food chain and causes a shortage of food, especially in developing countries (Alam et al., 2012)

Figure 2. Types and sources of Biofuels
(Modified from Dragone et al., 2010)

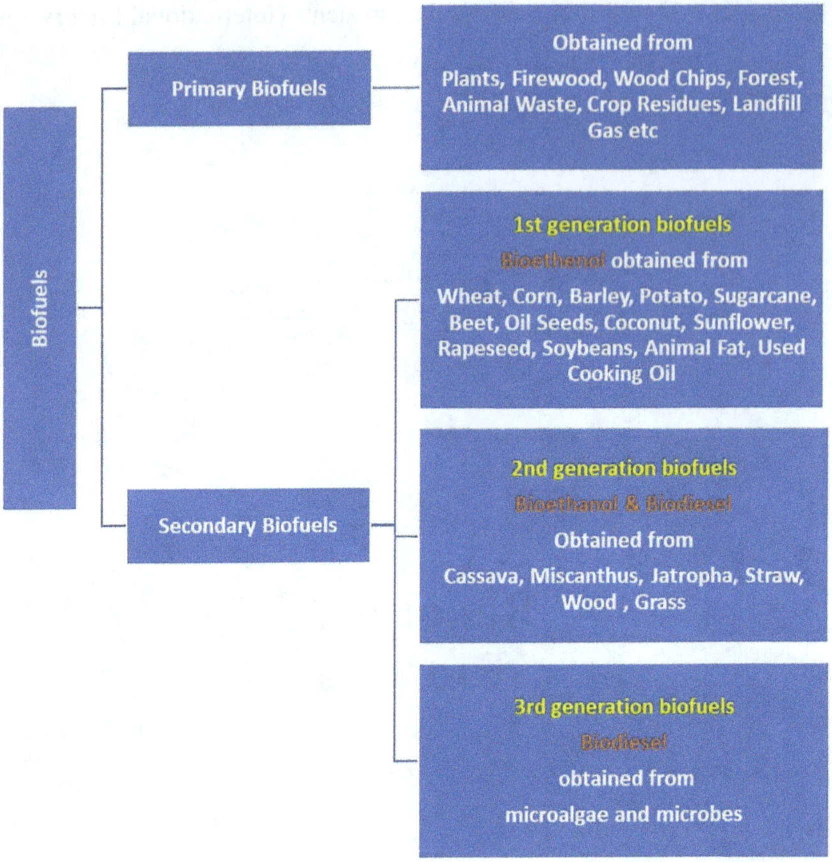

Microalgae

Algae are photosynthetic autotrophs found in various water basins, including rivers, lakes, and oceans. Microalgae are unicellular or multicellular microorganisms found in marine and freshwater habitats. In comparison to plants, microalgae have basic structures and organs. It can undertake photosynthesis using sunlight, carbon dioxide, and water due to the existence of chlorophyll pigments involved in photosynthesis. Microalgae have a substantially greater average photosynthetic efficiency (6–8%) than terrestrial biomass (1.8–2.2%) (Aresta et al., 2005). Growing microalgae have various advantages, as shown in **Fig. 3**.

Figure 3. Advantages of growing microalgae as biofuel resources

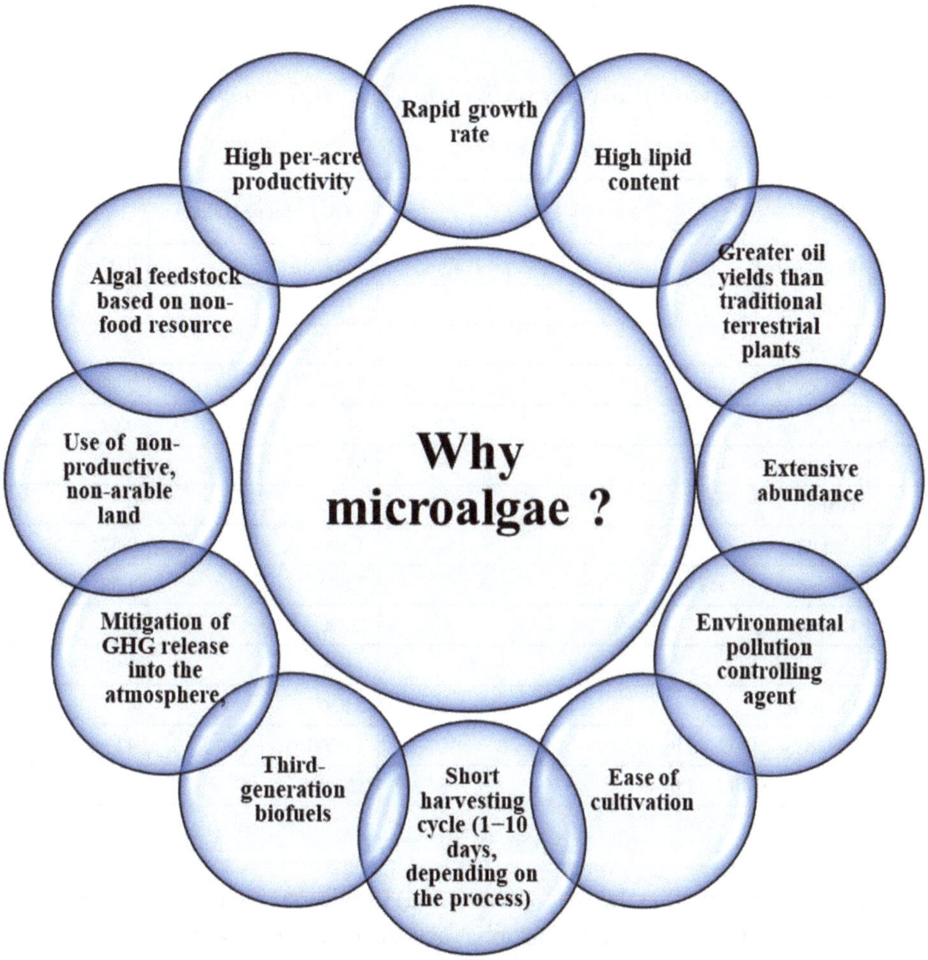

It is microscopic and found in aquatic and terrestrial environments in a large community that can survive varied environmental conditions. Microalgae can grow faster under various metabolic conditions, including autotrophic, heterotrophic, mixotrophic, and photoheterotrophic, and demand three essential components: sunlight, water, and carbon dioxide. Geographically, microalgae strains can be exploited to produce biomass grow in tropical and subtropical climates with abundant year-round solar radiation and water (either from the sea or as groundwater) (Mobin & Alam, 2018). So, the tropical countries have a relatively more significant advantage in growing microalgae as a sustainable source for biofuels production. Because of their high lipid content, simplicity of cultivation, rapid growth rate, and volume at which they can be cultivated, microalgae have become appealing oil sources for the production of biofuels (Chisti, 2007). Producing microalgae has several benefits, including the feature that it does not compete with human or animal crop production and may be produced in non-agricultural fields as well as in saline water. Microalgae are an excellent replacement for industrial raw materials for biofuels production because of their high photosynthetic capability and ability to assemble many bio-products in their cells (Li & Yu, 2011). Microalgae have been employed as a biodiesel feedstock since the 1970s due to their widespread abundance and higher oil content than regular terrestrial plants (Demirbas, 2009). A comparative account of microalgae with various other biofuels sources is given in **Table 1**.

Table 1. The table shows a comparative account of different sources of biofuels and their oil content(modified from Najafi et al., 2011).

S. No.	Crops	Oil Content (L/hac)	Oil Yield Relative to Microalgae (%)
1	Rubber seed	100	0.07
2	Corn	172	0.13
3	Soybean	446	0.33
4	Safflower	779	0.57
6	Camelina	915	0.67
7	Sunflower	952	0.70
8	Peanut	1,059	0.77
9	Canola	1,190	0.87
12	Jatropha	1,892	1.38
13	Karanj	2,590	1.89
14	Coconut	2,689	1.96
15	Oil palm	5,950	4.35
16	Microalgae (30% oil by wt)	58,700	42.88
17	Microalgae (70% oil by wt)	136,900	100.00

It is well clear from the table that the oil content (L/ha) in microalgae is much higher than other feedstocks available such as oil palm, coconut, jatropha, avocado and rapeseed/canola (Maity et al., 2014).

Types and Composition of Microalgae

Algae are divided into three main classes, i.e., Chlorophyceae, Phaeophyceae and Rhodophyceae. Chlorophyceae are also known as green algae due to chlorophyll a and b, e.g. *Chlamydomonas* and *Volvox*. Phaeophyceae are also known as brown algae. They possess chlorophyll a, Chlorophyll c, carotenoids and xanthophylls. Rhodophyceae is also known as red algae because of red pigment, i.e., r-phycoerythrin. Marine algae are commonly known as seaweed. Algae are featured as microalgae due to their microscopic size. Microalgae are single cellular photosynthetic cells. Microalgae are either prokaryotic such as cyanobacteria *(Chloroxybacteria),* or eukaryotic such as green algae *(Chlorophyta)*(Mathushika & Gomes, 2022). At least 30,000 species of microalgae have been identified. (Mathushika & Gomes, 2022). Different types of associations of microalgae form large structures called colonies. Colonies may be heterogeneous groups formed by eukaryotic species like diatoms and bacteria and prokaryotic organisms like cyanobacteria.

Microalgae species can produce up to 58,700 L of oil per hectare (Mobin & Alam, 2018). Subsequent studies have identified the micro-algal species, such as *Botryococcusbraunii, Scenedesmus obliquus, Chlorella vulgaris, Chlamydomonas reinhardtii, and Nannochloropsisoculata,* that are recognized as promising species for the production of biofuel (Veeramuthu & Chawalit Ngamcharussrivichai, 2020). All three biochemical fractions of microalgae (lipids, carbohydrates, and proteins) can be converted into fuels (Ghasemi et al., 2012). The fatty acid profile of microalgae species greatly favours biodiesel production (Qari et al., 2017). The major constituents of microalgae are lipids varying from 2–23%, proteins (6–52%), carbohydrates(5–23%) and some fat (Brown et al., 1997). Green micro-algae have a significantly higher lipid synthesis rate (mg/L/day) than red or blood-red micro-algae (Mobin& Alam, 2018). Microalgae lipid fraction consists of both neutral lipids and polar lipids. It accumulates triacylglycerols under stress conditions. Microalgae are rich in polar lipids in the exponential stage of development. Some microalgae have high protein concentrations in terms of quantity, up to 58% for *Chlorella vulgaris* on a dry weight basis. Carbohydrates (complex polysaccharides) are the main structural characteristics of the cell wall. The energy storage products of microalgae are glucose, starch and fat. Microalgae contain both saturated and unsaturated fatty acids. The lipid content of microalgae species varies from 20% to 50% of dry biomass (w/w). The most productive and cultivable microalgae and their respective lipid content are in **Fig. 4.**

Figure 4. Figure depicting different algal species and their respective lipid content (in percent) (modified from Maity et al., 2014)

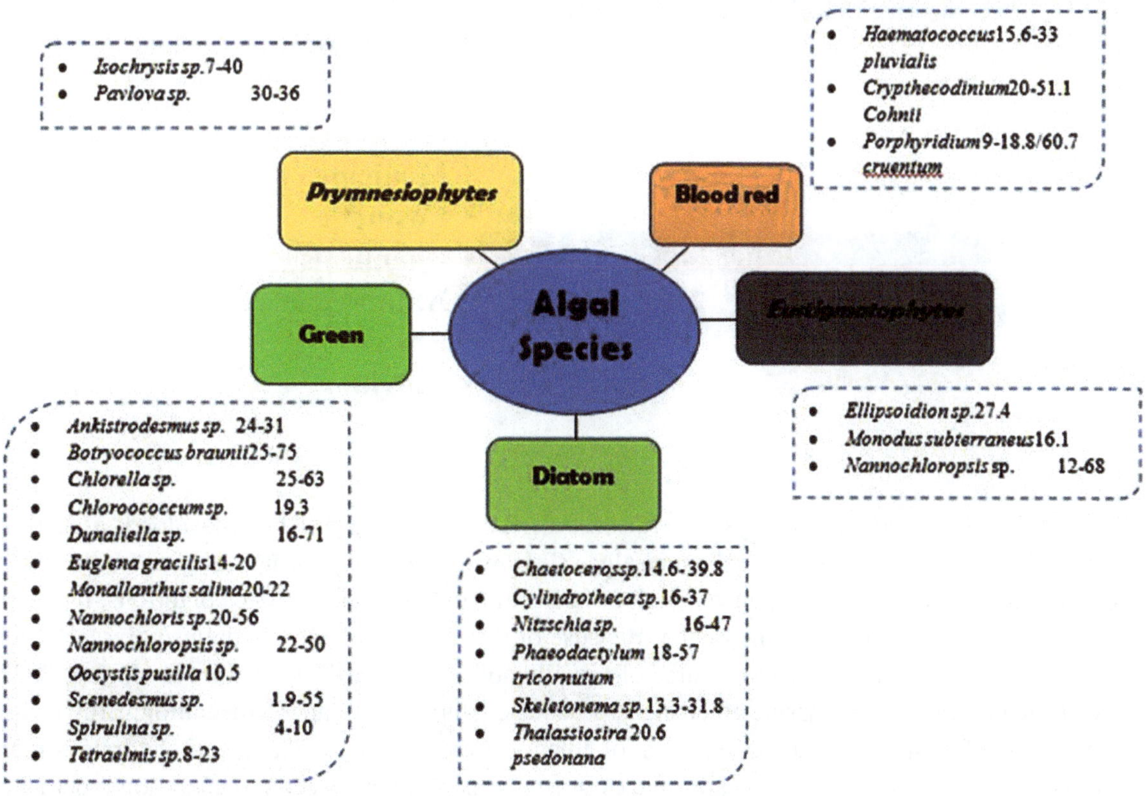

Complementary Algal By-Products

Along with the production of biofuels, microalgae have a variety of by-products with significant economic value. Following lipid extraction, the residual microalgae, i.e., dry biomass/cake, can be used as livestock feed, organic fertilizer due to its high nitrogen and phosphorus content, or merely burned for fuel co-generation (Jain et al., 2004). In addition, the resulting dry algal cake can be used to reduce the quantity of coal needed in thermal power plants. Furthermore, algal biomass has a better calorific value and lower SO_x concentration in flue gas when compared to coal. The calorific value of algal biomass is around 28,000kJ/kg, whereas the calorific value of typical high-quality coal is 27,000kJ/kg (Sudhakar & Premalatha, 2012). The algal by-products are used in different sectors, as shown in **Fig. 5.**

Figure 5. Different kinds of biofuels and by-products are produced from microalgae

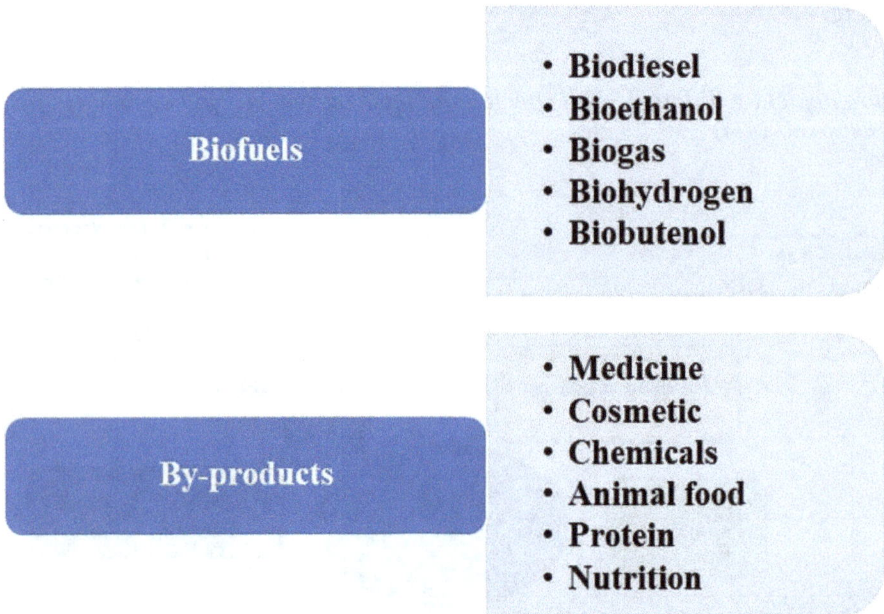

Direct photolysis, indirect photolysis, and an ATP-driven pathway have all been developed as methods for producing hydrogen. The single-cell green algae *Chlamydomonas reinhardtii* has been studied extensively for photo-biological hydrogen generation. In sulfur-limited and anaerobic conditions, the starch in algal cell walls is converted to hydrogen. Because of the quantity of carbohydrates/polysaccharides and thin cellulose, algae are a better source of bio-ethanol production (Mathushika & Gomes, 2022), e.g. *Chlorococcumlittorale*, marine green algae produce ethanol under dark fermentation. Brown algae are remarkable feedstock for the development of ethanol because they have high carbohydrate content. Algal bioethanol prevails over the disadvantages of first-generation and second-generation bioethanol.

Current Tools and Techniques to Grow and Harvest Microalgae

Currently, industrial production of microalgae is small, estimated to be around 6000 tons (dry weight) per year worldwide (Maity et al., 2014).

The equation below shows that microalgae require essential environmental components to grow and synthesize raw material for biofuels.

Light $+CO2_{+}\ H2O_{+}$nutrients lipids+proteins+carbohydrates

Essential requirements of nutrients from the environment are carbon, phosphorus, nitrogen, and iron, and some minor are calcium, potassium, magnesium, manganese, sulphur, copper, zinc, and cobalt (Chisti, 2007). The temperature requirement must be within the range of 20°C- 30°C. The microalgae cultivation methods are mainly categorized into ways 1. Suspended cultivation. It includes 1a) open ponds and 1b) closed photobioreactors 2. Immobile cultivation. It includes 2a) immobile matrix cultivation and 2b) bioðlms cultivation. Different types of techniques used to cultivate microalgae are illustrated in Fig. **6. The** industrial-scale of microalgae production utilizes a continuous suspended culture technique during the daytime. This method needs to feed a new culture and continuously remove the same quantity of culture from the system. The biomass is generated during the day; out of that, approximately 25%biomass lost during nighttime due to respiration failure and sufficient unavailability of growth nutrients, temp. and sunlight (Chisti, 2007). Open ponds or Raceways are also the most widely used methods of microalgae production on a large scale because this system is a shallow and open, equipped circulating wheel which provides adequate mixing and circulation of nutrients. These setups are economically suitable to build and run but face relatively lower production due to uncontrolled conditions(Mata et al., 2010). These systems have been utilizing microalgae biomass production for the last sixty years (Tredici & Materassi, 1992). The open ponds are shallow, not deeper than 30 cm, to allow sunlight for microalgae growth. Therefore they become easy to operate and economically viable but confront some disadvantages (Chisti, 2007). A paddlewheel creates mixing and circulation in raceways, and it works continuously to prevent sedimentation. However, open systems are generally simple to operate and use solar energy as an energy source; they have several drawbacks, including the need for a vast area to cultivate, unregulated environmental conditions which change seasonally and diurnally, water evaporation, comparatively unequal carbon dioxide and light distribution in the system, and high exposure to contamination or infection that is difficult to control. To grow in such fluctuating environmental conditions, the extremophilic strain of microalgae is required to cultivate. Normal species having a high production rate face low productivity in the case of an open pond system.

Figure 6. Techniques used to grow microalgae

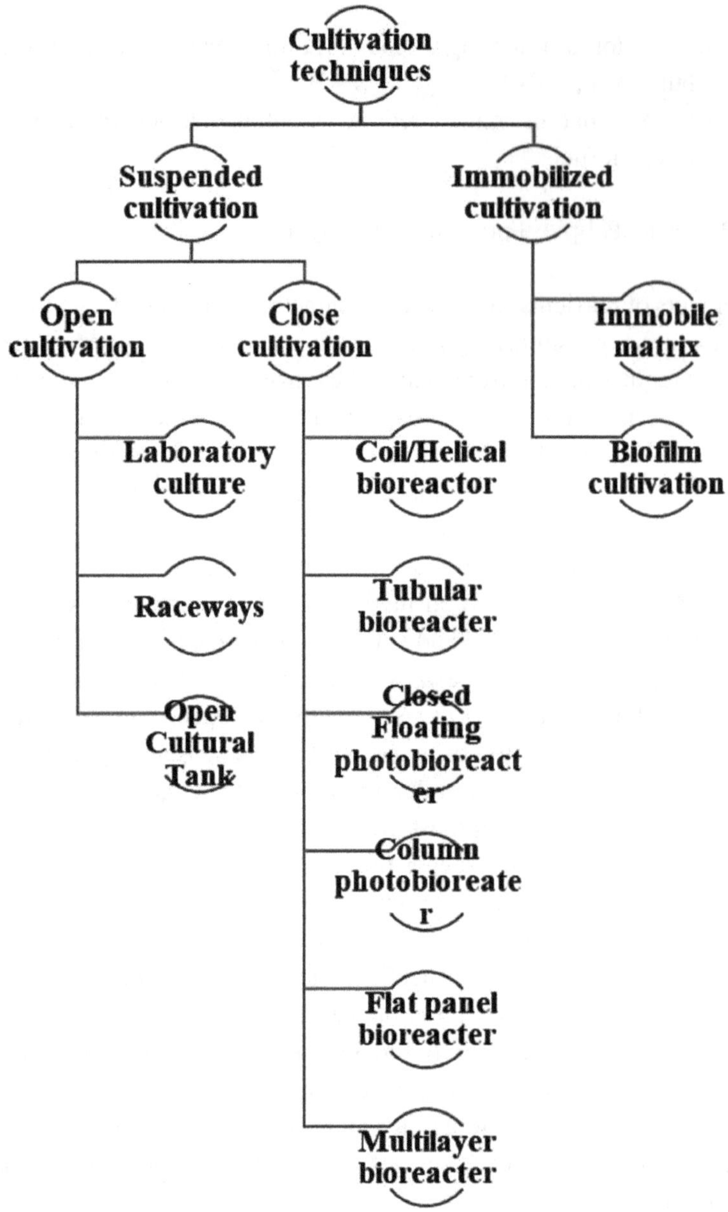

Photobioreactors are considered to be most productive for microalgae cultivation as compared to open pond or raceways systems because of better controlled medium parameters like temperature, pH, nutrient addition, efficient mixing, better control over contamination or infection, low water evaporation and a large number of cells per unit area (Lundquist et al., 2010, Carvalho et al., 2006, Kunjapur & Eldridge, 2010, Darzins et al., 2010, Lam et al., 2019). Photo-bioreactors can also be customised per the specific strain of microalgae to enhance productivity in every possible way and better control environmental fluctuation. But these reactors are energy-intensive and expensive to build, making biofuels economically infeasible (Mata et al., 2010). Photobioreactors are made up of transparent glass or plastic tubes with £ 0.1 m in diameter. Based on different closed structures, photobioreactors are of the following type 1) vertical, in that tubes are parallelly arranged vatical to ground. 2) Horizontal, in that tubes or plates are parallel to one another but horizontal to the ground 3) helical, in that tubes are making coiled structure around a supporting frame. They are considered easy to increase the production, up to 13 times more than raceways (Chisti, 2007). But every system has its advantages and disadvantages. To increase the production of microalgae, the surface area needs to be improved, increasing sunlight exposure. This could be achieved by making tubes thinner and plates broader, increasing the surface area to volume ratio. Zemke et al., 2016 reported that a doubling of volume had doubled the microalgal biomass production, but built-up cost increases only 59% of the initial. In the case of phototropic open culture doubling surface area increases the capital cost by 100%, but it has the advantage of using free energy; otherwise, it requires a significant investment cost to provide it (Zemke et al., 2016). The significant disadvantages of photobioreactor are the recovery of biomass, but this issue was resolved up to a great extent by biofilm or immobilized matrix cultivation system (Alam et al., 2015, Hoffmann, 1998). Open ponds or raceways are also preferable for microalgae cultivation due to their low capital cost and relative ease of operation. The current technologies are not economically suitable and capable enough to compete in the market. To cut down its production cost and improve harvesting and processing techniques, optimising the utilization of wastewater as algal culturing feed is required. Various studies have considered the most appropriate environment-friendly and economically viable method to produce biofuel and other complementary products with significant economic value.

But the use of wastewater as microalgae feed faces significant challenges, which include high turbidity causes improper penetration of sunlight in the medium, overdose of nutrients which also hinder the algal growth, the suitable strain of microalgae required to cope with up wastewater environment, and essential requirement of freshwater to dilute the waste etc. Different techniques, methods, and strategies are described (Mobin & Alam, 2014, Dragone et al., 2010, Chisti, 2007, Scott et al., 2010, Kumar & Bharadvaja, 2020). Harvesting has also been a challenging task after cultivating microalgae because it costs up to 30% of the total production cost of microalgae biofuel(Grima et al., 2003). Various methods presently employed to harvest microalgae include mechanical, physical, chemical, biological and electrical processes such as centrifugation, ultra-filtration, floatation, auto flotation and flocculation, etc. (Dragone et al., 2010). The flocculation is utilized when algae size is comparatively small. Flocculants like ferric chloride, aluminium sulphate and ferric sulphate are used—moreover, different combinations are employed for efficient harvesting. Centrifugation is used to separate suspended microalgae, and the medium is decanted. This is considered the most reliable method for separating microalgae from the medium(Christenson & Sims, 2011). Suspended culture is passed through the filter in filtration until a thick layer of microalgae biomass is formed. An electrical method is based on separating algal cells under an electric field, but the low-cost method is required to do this task efficiently.

Different Processes to Produce Microalgae Biofuel

Several processes are available to convert microalgae biomass into different types of energy. It has the following method (1) direct combustion, (2) chemical reaction, (3) thermochemical conversion (4) biochemical conversion. The different processes of microalgae biomass conversion and resulting products are shown in **Fig.7**_(Wang et al., 2008). Along with the production of biofuel, some other complementary products also can be produced, and their detailed description can be obtained from these sources(Dragone et al., 2010, Spolaore et al., 2006, Benemann et al., 1987, Eriksen et al., 1998, Harun et al., 2010, Mendes-Pinto et al., 2001).

Figure 7. The different processes of microalgae biomass conversion and resulting products
(Modified from Dragone et al., 2010)

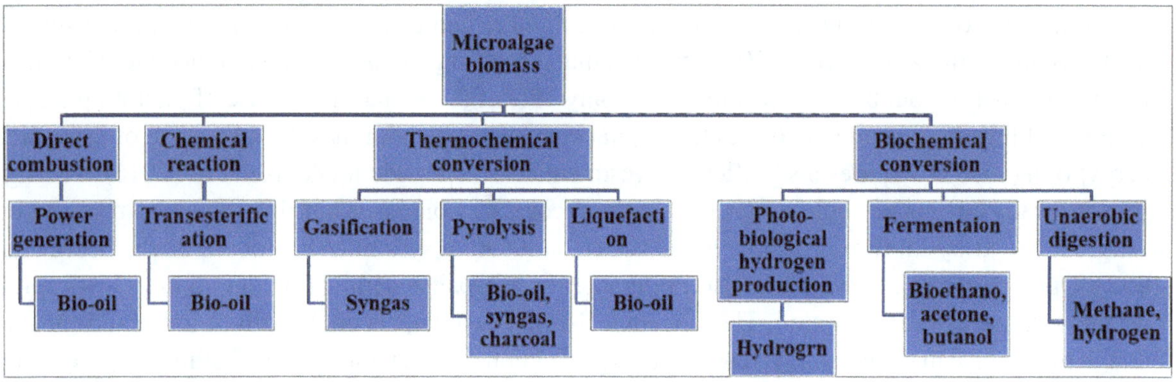

Potential and Challenges

Microalgae can utilise CO_2 emitted from thermal power plants with higher than 10% enhanced photosynthetic efficiency compared to 0.5% of terrestrial plants (Khan and Rashmi, 2008).

It is estimated that microalgal biomass of one kilogram can fix 4.8 kg of CO_2 (Baral et al., 2015). So, microalgal cultivation with sequestration of CO_2 emitted from the thermal power plant and production of value-added products, including biodiesel, make them an efficient and carbon-neutral alternative for energy production. Several pilot studies of microalgae cultivation using emitted CO2is going on worldwide, including Stanwell Power Plant in Queensland, Loy Yang Power Plant in Victoria, and Eraring Power Plant in New South Wales. Due to technical challenges and higher capital costs of solar and wind energy utilization, third-generation algal biofuel can play an essential role in filling the partial gap left by reducing fossil fuel utilization. However, some fundamental limitations must be solved to obtain the maximum potential of third-generation biofuels derived primarily from microalgal biomass.

The following are real challenges:

1. Technological constraints in microalgae selection, culture, harvesting, processing, and distribution techniques. The development of microalgae strains that are high yielding, broad temperature tolerant, and salinity tolerant is a challenge. Commercial-scale microalgal production lacks a cost-effective, consistent culture and harvesting process and system. The available mechanical and chemical technologies for extracting biofuels are not economically sustainable. Compared to conventional methods, the expense of infrastructure for biofuel distribution and retailing is also a challenge.
2. Microalgae cultivation also faces significant obstacles in terms of land availability and agricultural inputs. Commercial-scale microalgae cultivation cannot wholly utilise arable land and freshwater resources, which would pose a danger to human and animal food demand. Excessive use of agricultural additives (inorganic and organic fertiliser) in algal mass production can lead to eutrophication when phosphate and nitrate levels rise.

3. Gene modification remains a barrier, even though detailed information can assist. Comprehensive perception of microalgal biology, genetic manipulation, and gene cloning are all challenges that must be addressed to improve photosynthetic efficiency, metabolic activity, temperature tolerance, and algae biomass output and oil content.
4. Wastewater utilization for microalgae cultivation remains a concern due to the lack of a well-developed and cost-effective procedure for screening out toxic and undesired substances that can interfere with algal growth.
5. A significant amount of investigation and innovation is necessary to use value-added by-products such as bio-fertilizer, animal feed, cosmetics, colouring agents, chemical and pharmaceutical commodities, biogas masses, and biofuel generation.
6. Currently, biofuels' production cost is significantly higher than fossil fuel prices. So, it is highly challenging for biofuel to be competitive in the market.

CONCLUSION

Sustainable energy sources for the growth and development of the world are a mindboggling task in front of the scientific community. Fossil fuels are not environment friendly and have limited stocks, which will be exhausted soon. Biofuels are emerging areas of sustainable, potential energy sources that need to be exploited to meet growing energy demands in this chaos. Microalgae are a flexible biofuels source with significant potential to produce different biofuels. Microalgae are considered a carbon-neutral fuel due to the capability to utilize wastewater as a feed to grow, co2 from a thermal power plant with enhanced photosynthetic capacity, and the production of various complementary by-products with significant economic benefits values. Still, multiple gaps need to be addressed to commercialise fuel production using microalgae, including efficient technologies that harvest, process, and convert it into economically viable biofuels. Systematic and comprehensive research is required in this regard. Presently, microalgae biofuels face significant challenges, including high performance, broad range tolerating and high yielding microalgae strains, improved culturing techniques, reduction in production cost, utilisation of waste to compete in markets, and improved harvest and processing techniques to make economically viable source.

REFERENCES

Ahmed, S. F., Liu, G., Mofijur, M., Azad, A. K., Hazrat, M. A., & Chu, Y. M. (2021). Physical and hybrid modelling techniques for earth-air heat exchangers in reducing building energy consumption: Performance, applications, progress, and challenges. *Solar Energy, 216*(November), 274–294. doi:10.1016/j.solener.2021.01.022

Alam, F., Date, A., Rasjidin, R., Mobin, S., Moria, H., & Baqui, A. (2012). Biofuel from algae-Is it a viable alternative? *Procedia Engineering*, *49*(December), 221–227. doi:10.1016/j.proeng.2012.10.131

Alam, F., Mobin, S., & Chowdhury, H. (2015). Third generation biofuel from Algae. *Procedia Engineering, 105*, 763–768. doi:10.1016/j.proeng.2015.05.068

Aresta, M., Dibenedetto, A., & Barberio, G. (2005). Utilization of macro-algae for enhanced CO2 fixation and biofuels production: Development of a computing software for an LCA study. *Fuel Processing Technology*, *86*(14–15), 1679–1693. doi:10.1016/j.fuproc.2005.01.016

Baral, S. S., Singh, K., & Sharma, P. (2015). The potential of sustainable algal biofuel production using CO2 from thermal power plant in India. *Renewable & Sustainable Energy Reviews*, *49*, 1061–1074. doi:10.1016/j.rser.2015.04.181

Benemann, J. R., Tillett, D. M., & Weissman, J. C. (1987). Microalgae biotechnology. *Trends in Biotechnology*, *5*(2), 47–53. doi:10.1016/0167-7799(87)90037-0

Brown, M. R., Jeffrey, S. W., Volkman, J. K., & Dunstan, G. A. (1997). Nutritional properties of microalgae for mariculture. *Aquaculture (Amsterdam, Netherlands)*, *151*(1–4), 315–331. doi:10.1016/S0044-8486(96)01501-3

Carvalho, A. P., Meireles, L. A., & Malcata, F. X. (2006). Microalgal reactors: A review of enclosed system designs and performances. *Biotechnology Progress*, *22*(6), 1490–1506. doi:10.1002/bp060065r PMID:17137294

Chisti, Y. (2007). Biodiesel from microalgae. *Biotechnology Advances*, *25*(3), 294–306. doi:10.1016/j.biotechadv.2007.02.001 PMID:17350212

Christenson, L., & Sims, R. (2011). Production and harvesting of microalgae for wastewater treatment, biofuels, and bioproducts. *Biotechnology Advances*, *29*(6), 686–702. doi:10.1016/j.biotechadv.2011.05.015 PMID:21664266

Darzins, A., Pienkos, P., & Edye, L. (2010). Current Status and Potential for Algal Biofuels Production Al Darzins (NREL) Philip Pienkos (NREL) Les Edye (BioIndustry Partners). *NREL*. http://task39.org/files/2013/05/IEA-Task-39-Current-Status-and-Potential-of-Algal-biofuels0.pdf

Demirbas, A. (2009). Political, economic and environmental impacts of biofuels: A review. *Applied Energy*, *86*(Suppl. 1), S108–S117. doi:10.1016/j.apenergy.2009.04.036

Dragone, G., Fernandes, B., Vicente, A., & Teixeira, J. (2010). Third generation biofuels from microalgae. *Current Research, Technology and Education Topics in Applied Microbiology and Microbial Biotechnology,* 1355–1366. http://repositorium.sdum.uminho.pt/handle/1822/16807

Eriksen, N. T., Poulsen, B. R., & Iversen, J. J. L. (1998). Dual sparging laboratory-scale photobioreactor for continuous production of microalgae. *Journal of Applied Phycology*, *10*(4), 377–382. doi:10.1023/A:1008029521111

Ghasemi, Y., Rasoul-Amini, S., Naseri, A. T., Montazeri-Najafabady, N., Mobasher, M. A., & Dabbagh, F. (2012). Microalgae biofuel potentials [Review]. *Applied Biochemistry and Microbiology*, *48*(2), 126–144. doi:10.1134/S0003683812020068 PMID:22586908

Hannon, M., Gimpel, J., Tran, M., Rasala, B., & Mayfield, S. (2010). Biofuels from algae: Challenges and potential Importance & challenges of algal biofuels. *Biofuels*, *1*(5), 763–784. doi:10.4155/bfs.10.44 PMID:21833344

Harun, R., Singh, M., Forde, G. M., & Danquah, M. K. (2010). Bioprocess engineering of microalgae to produce a variety of consumer products. *Renewable & Sustainable Energy Reviews*, *14*(3), 1037–1047. doi:10.1016/j.rser.2009.11.004

Hoffmann, J. P. (1998). Wastewater treatment with suspended and nonsuspended algae. *Journal of Phycology*, *34*(5), 757–763. doi:10.1046/j.1529-8817.1998.340757.x

IAE. (2017). *International Energy Efficiency*. https://www.iea.org/reports/world-energy-outlook-2017

Jain, S. K., Prakash, V., & Kapoor, S. (2004). Flue gas treatment alternatives or enhancing ESP collection efficiency: NTPC experience. *Workshop on ESP performance: Role of Fly Ash Resistivity. IIT-DandNTPC*, 1–5.

Kumar, L., & Bharadvaja, N. (2020). A review on microalgae biofuel and biorefinery: Challenges and way forward. *Energy Sources. Part A, Recovery, Utilization, and Environmental Effects*, *00*(00), 1–24. doi:10.1080/15567036.2020.1836084

Kunjapur, A. M., & Eldridge, R. B. (2010). Photobioreactor design for commercial biofuel production from microalgae. *Industrial & Engineering Chemistry Research*, *49*(8), 3516–3526. doi:10.1021/ie901459u

Lam, M. K., Khoo, C. G., & Lee, K. T. (2019). Scale-up and commercialization of algal cultivation and biofuels production. In Biofuels from Algae (2nd ed.). Elsevier B.V. doi:10.1016/B978-0-444-64192-2.00019-6

Li, W. W., & Yu, H. Q. (2011). From wastewater to bioenergy and biochemicals via two-stage bioconversion processes: A future paradigm. *Biotechnology Advances*, *29*(6), 972–982. doi:10.1016/j.biotechadv.2011.08.012 PMID:21884782

Lundquist, T. J., Woertz, I. C., Quinn, N. W. T., & Benemann, J. R. (2010). *A Realistic Technology and Engineering Assessment of Algae Biofuel*. Academic Press.

Maity, J. P., Bundschuh, J., Chen, C. Y., & Bhattacharya, P. (2014). Microalgae for third generation biofuel production, mitigation of greenhouse gas emissions and wastewater treatment: Present and future perspectives - A mini-review. *Energy*, *78*, 104–113. doi:10.1016/j.energy.2014.04.003

Mata, T. M., Martins, A. A., & Caetano, N. S. (2010). Microalgae for biodiesel production and other applications: A review. *Renewable & Sustainable Energy Reviews*, *14*(1), 217–232. doi:10.1016/j.rser.2009.07.020

Mathushika, J., & Gomes, C. (2022). Development of microalgae-based biofuels as a viable green energy source: Challenges and future perspectives. *Biointerface Research in Applied Chemistry*, *12*(3), 3849–3882. doi:10.33263/BRIAC123.38493882

Mendes-Pinto, M. M., Raposo, M. F. J., Bowen, J., Young, A. J., & Morais, R. (2001). Evaluation of different cell disruption processes on encysted cells of Haematococcus pluvialis. *Journal of Applied Phycology*, *13*(1), 19–24. doi:10.1023/A:1008183429747

Mobin, S., & Alam, F. (2014). Biofuel production from algae utilizing wastewater. *Proceedings of the 19th Australasian Fluid Mechanics Conference, AFMC 2014*.

Mobin, S. M. A., & Alam, F. (2018). A review of microalgal biofuels, challenges and future directions. *Green Energy and Technology*, 83–108. doi:10.1007/978-981-10-0697-5_4

Molina Grima, E., Belarbi, E. H., Acién Fernández, F. G., Robles Medina, A., & Chisti, Y. (2003). Recovery of microalgal biomass and metabolites: Process options and economics. *Biotechnology Advances*, *20*(7–8), 491–515. doi:10.1016/S0734-9750(02)00050-2 PMID:14550018

Najafi, G., Ghobadian, B., & Yusaf, T. F. (2011). Algae as a sustainable energy source for biofuel production in Iran: A case study. *Renewable & Sustainable Energy Reviews*, *15*(8), 3870–3876. doi:10.1016/j.rser.2011.07.010

OECD. (2012). *Organization for Economic Co-operation and Development*. https://www.oecd.org/env/cc/Outlook%20to%202050_Climate%20Change%20Chapter_HIGLIGHTS-FINA-8pager-UPDATED%20NOV2012.pdf

Qari, H., Rehan, M., & Nizami, A. S. (2017). Key Issues in Microalgae Biofuels: A Short Review. *Energy Procedia*, *142*, 898–903. doi:10.1016/j.egypro.2017.12.144

Rafa, N., Ahmed, S. F., Badruddin, I. A., Mofijur, M., & Kamangar, S. (2021). Strategies to Produce Cost-Effective Third-Generation Biofuel From Microalgae. *Frontiers in Energy Research*, *9*(September), 1–11. doi:10.3389/fenrg.2021.749968

Scott, S. A., Davey, M. P., Dennis, J. S., Horst, I., Howe, C. J., Lea-Smith, D. J., & Smith, A. G. (2010). Biodiesel from algae: Challenges and prospects. *Current Opinion in Biotechnology*, *21*(3), 277–286. doi:10.1016/j.copbio.2010.03.005 PMID:20399634

Spolaore, P., Joannis-Cassan, C., Duran, E., & Isambert, A. (2006). Commercial applications of microalgae. *Journal of Bioscience and Bioengineering*, *101*(2), 87–96. doi:10.1263/jbb.101.87 PMID:16569602

Sudhakar, K., & Premalatha, M. (2012). Energy Balance and Exergy analysis of large scale algal biomass production 1. Goal Definition 2. Life Cycle Inventory 4. *Interpretation (Tulsa)*, *1454*(June), 13–15.

Tredici, M. R., & Materassi, R. (1992). From open ponds to vertical alveolar panels: The Italian experience in the development of reactors for the mass cultivation of phototrophic microorganisms. *Journal of Applied Phycology*, *4*(3), 221–231. doi:10.1007/BF02161208

UNEP. (2019). *United Nation Environment Programme, Emmision Gap Report*. UNEP.

Veeramuthu, A., & Ngamcharussrivichai, C. (2020). Potential of Microalgal Biodiesel: Challenges and Applications. *Potential of Microalgal Biodiesel: Challenges and Applications*, 1–14.

Wang, B., Li, Y., Wu, N., & Lan, C. Q. (2008). CO2 bio-mitigation using microalgae. *Applied Microbiology and Biotechnology*, *79*(5), 707–718. doi:10.100700253-008-1518-y PMID:18483734

Chapter 14
Microbial Cellulase in the Production of Second Generation Biofuels:
State-of-the-Art and Beyond

Jovana Trbojević-Ivić
https://orcid.org/0000-0003-0984-9193
Innovative Centre, Faculty of Chemistry, Belgrade, Ltd., Serbia

ABSTRACT

Bioethanol from inedible cellulose-rich biomass is the most promising candidate to replace fast depleting, environmentally polluting fossil fuels. Hydrolysis of cellulose to glucose is the crucial step in its biotransformation to bioethanol. Enzymatic hydrolysis is favored over acid hydrolysis, as enzymes are eco-friendly biocatalysts with high substrate specificity and superior catalytic efficiency in mild reaction conditions. Complete hydrolysis of cellulose is achieved by cellulase. Higher cellulase production yield, stability, and catalytic efficiency are the main attentive points for the successful implementation in industrial production of bioethanol. This chapter will highlight general characteristics of microbial cellulases and their role in the bioconversion of cellulose to biofuels, economic sustainability of cellulose-based biofuels, and the latest innovations in cellulase immobilization as the most comprehensive strategy for improvement of enzyme stability, activity, and reusability for cost-effective large-scale application.

INTRODUCTION

In 2016 the use of bioethanol-gasoline blends in transportation reduced the emission of carbon dioxide by 43.5 million tons, the equivalent of removing 9.3 million cars from the road for the entire year (Robak & Balcerek, 2018). Lignocellulose biomass is the most abundant and renewable source of carbohydrates in the biosphere. It is obtained from inedible energy crops, agricultural and industrial waste (Kricka et al., 2015; Verma & Kumar, 2021). Bioethanol from lignocellulose biomass is a second-generation biofuel, whose production does not compete against food supplies, in contrast to the first-generation bioethanol,

DOI: 10.4018/978-1-6684-5269-1.ch014

produced from edible seed-derived starch. Accordingly, second-generation bioethanol is the most promising candidate to replace conventional gasoline in the near future (Hoffman et al., 2021; Horn et al., 2012). Enzymes are green biocatalysts with remarkable substrate specificity and catalytic efficiency in mild reaction conditions. They can be genetically or chemically modified to meet the specific requests of the biotechnological process in which they are employed (Ejaz et al., 2021; Trbojević et al., 2013). Enzymatic hydrolysis of cellulose to glucose is the critical step in production of the second-generation bioethanol. Complete hydrolysis of cellulose is achieved by concerted activity of endoglucanase, exoglucanase (cellobiohydrolase) and β-glucosidase, which form a multi-enzyme cellulase complex. Mesophilic and thermophilic bacteria and fungi are the major industrial cellulase producers (Acharya & Chaudhary, 2012; Bhardwaj et al., 2021; Contreras et al., 2020; Horn et al., 2012; Menendez et al., 2015; Srivastava et al., 2018; Verma & Kumar, 2021). Commercial cellulase cocktails have contributed to the significant cost-reduction of cellulose bioconversion to bioethanol (Bhardwaj et al., 2021; Contreras et al., 2020; Lamichhane et al., 2021). However, further decrease in cellulase price is necessary by improving cellulase production yield, stability, specific activity and catalytic efficiency for economically viable bioethanol production.

Immobilization is the strategy of choice for improving catalytic performance of industrial enzymes, since it encompasses significant biocatalyst stabilization and reusability. Enzyme reusability significantly reduces the cost of the biocatalyst application at industrial scale and can be achieved exclusively by immobilization (Milosavic et al., 2016; Trbojević et al., 2013). With myriad of possibilities regarding immobilization methods and the choice of immobilization support, cellulases can be tailored for cost-effective and optimum catalytic performance. This chapter is a comprehensive overview of current state-of-the-art in the industrial application of cellulases in biofuel production. It will provide detailed fundamental understanding of microbial cellulases and their critical role in bioconversion of cellulose to bioethanol. Furthermore, the techno-economical aspect of the cellulose-derived biofuels will be discussed, with additional analysis of COVID-19 impact. The separate section of the chapter will be dedicated to cellulase immobilization and its effect on enzyme stability, activity and reusability. It will specifically deal with nanomaterials as innovative immobilization carriers, having in mind their increasing popularity in this area of research and application.

BACKGROUND

Lignocellulose biomass is a complex mixture of carbohydrate and non-carbohydrate compounds. Cellulose is the dominant component (40-60% of the total biomass weight), followed by hemicellulose (20-40%) and lignin (20–30%). It also contains proteins, lipids, pectin, minerals and soluble sugars (Adegboye et al., 2021; Horn et al., 2012; Kricka et al., 2015). Structure of major constituents of lignocellulose biomass is represented in Figure 1.

Figure 1. Constituents of lignocellulose biomass
Source: (Adegboye et al., 2021; Horn et al., 2012; Kricka et al., 2015; Mohanram et al., 2013; Zoghlami & Paës, 2019)

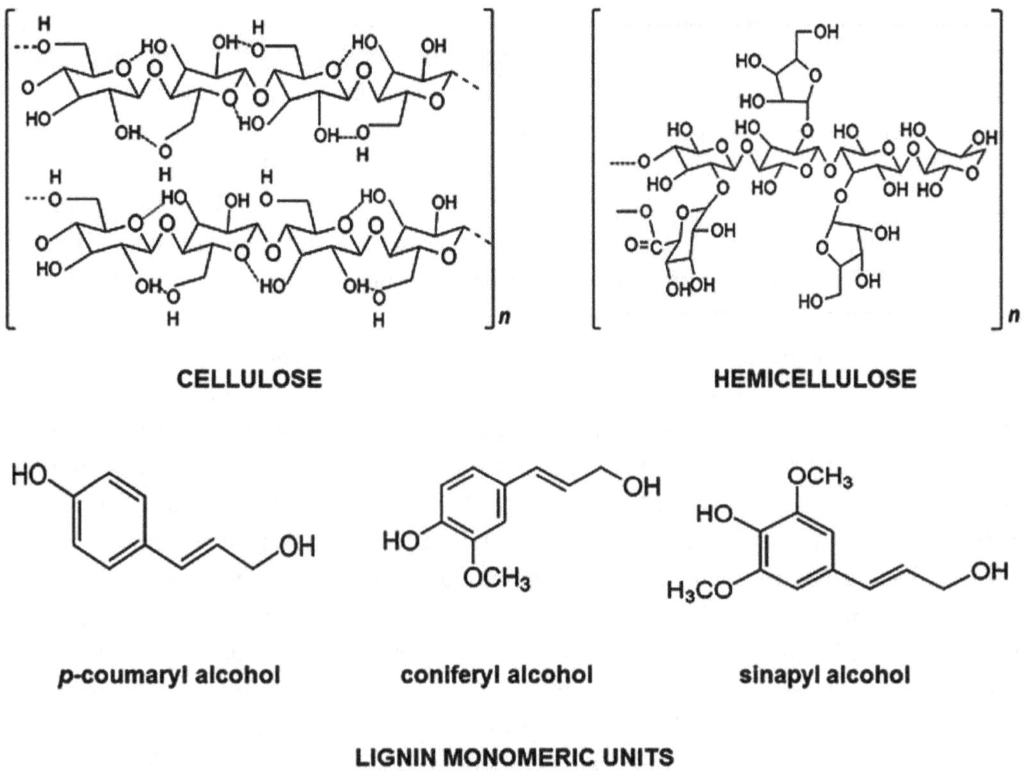

Cellulose is a linear homopolysaccharide of β-D-glucopyranose units linked by β-(1,4)-glycosidic bond. Its structure is stabilized into microfibrils by numerous hydrogen bonds and van der Waals forces. Each microfibril contains up to 36 glucan chains with thousands of glucose units (Mohanram et al., 2013). Consecutive sugars along chains in crystalline cellulose are rotated by 180°, meaning that cellobiose is the fundamental building block (Kricka et al., 2015). Cellulose can be crystalline or amorphous. Individual glucan fibrils in crystalline cellulose are tightly packed and hardly accessible to enzymes or solvents (Adegboye et al., 2021). Depolymerization of cellulose induces changes in its crystallinity and porosity (Zoghlami & Paës, 2019).

Hemicellulose is heteropolysaccharide, composed of xylose, arabinose, rhamnose, galactose and mannose. It can also contain glucuronic, galacturonic and methylgalacturonic acid (Adegboye et al., 2021). Common hemicelluloses are xylan, mannan and xyloglucan. Hardwood xylans contain β-(1,4)-linked xylose backbone with a high amount of methylgalacturonic acid, while grass xylan is more complex with high amounts of arabinose and ester modifications with hydroxycinnamic acid. Hydroxycinnamic acids can also cross-link adjacent xylan chains, building a more complex and rigid structure (Horn et al., 2012). Degree of hemicellulose polymerization is lower than of cellulose, therefore hemicellulose is amorphous with little physical strength (Zoghlami & Paës, 2019). Lignin is hydrophobic heteropolymer of coniferyl, sinapyl and p-coumaryl alcohol, linked *via* C-C and C-O-C links. In lignocellulose biomass lignin is cross-linked with carbohydrates by ether or ester linkages which provides structural integrity of plant cell walls and hinders the enzymatic hydrolysis of biomass, a phenomenon known as biomass recalcitrance (Adegboye et al., 2021; Horn et al., 2012; Verma & Kumar, 2021).

Endoglucanase (EC 3.2.1.4, systematic name: 1,4-β-D-glucan-4-glucanohydrolase), exoglucanase (EC 3.2.1.91, systematic name: 4-β-D-glucan cellobiohydrolase; EC 3.2.1.74, systematic name: 4-β-D-glucan glucohydrolase) and β-glucosidase (EC 3.2.1.21; systematic name: β-D-glucoside glucohydrolase) form multi-enzyme machinery for complete hydrolysis of cellulose. Aerobic bacteria and fungi produce individual cellulases, which are further combined in biocatalytic cocktails with a precise share of each individual enzyme for efficient hydrolysis (Acharya & Chaudhary, 2012). Microbial cellulases have a two-domain structure, consisting of cellulose-binding domain and catalytic domain, connected by flexible hinge (Ejaz et al., 2021). Hinge region is rich in proline, threonine and serine and enables the enzyme to move along cellulose fibril in a caterpillar-like mode. Detailed data of microbial cellulase structure were obtained from model systems: endoglucanase from *Thermoascus aurantiacus* (Cel5A), cellobiohydrolase from *Phanerochaete chrysosporium* (Cel7D) and β-glucosidase A from *Bacillus polymyxa*. Polypeptide chain of Cel5A is organized into a very compact eight-fold β/α-barrel with short loops and it contains only catalytic domain. Its active site contains acid/base glutamate 133 and nucleophilic glutamate 240. Substrate-binding site of Cel5A is shallow and long, delineated with tryptophans 170, 174, 273, 278 and 279 (Lo Leggio & Larsen, 2002). Catalytic domain of Cel7D is organized in a β-sandwich with antiparallel β-sheets, stabilized by hydrophobic interactions and disulfide bridges. In comparison to *T. reesei*, active site is more open, however not as open as in endoglucanase. Nucleophilic glutamate 207, acid/base glutamate 212 and proton donor aspartate 209 form catalytic triad of Cel7D, located in the tunnel-shaped active site, delineated with tryptophan and additional aromatic residue from tyrosine 47 at the tunnel entrance (Muñoz et al., 2001). β-Glucosidase from *B. polymyxa* is octamer, organized into α/β barrel, stabilized by fourteen hydrogen bonds. Loops A, B and C connect individual parts of the α/β barrel. Length and chemical nature of these loops governs substrate specificity of β-glucosidase. Most significant differences were observed in the loop C, a part of the substrate-binding cavity whose end is in a close proximity to the non-reducing end of cellobiose. Substrate-binding cavity of β-glucosidase A is wider and more extended, so that it can accommodate larger substrates than cellobiose. Interesting mechanistic attribute of glucosidases, including β-glucosidase, is the ability to catalyze synthesis of disaccharides and oligosaccharides from glucose in transglucosylation. Hydrolysis *vs* transglucosylation is regulated by the water content in the reaction mixture (Pavlović et al., 2013; Veličković et al., 2014). Concerted action of individual cellulases in cellulose hydrolysis is depicted in Figure 2.

Figure 2. Enzymatic hydrolysis of cellulose
Source: (Adegboye et al., 2021; Barcelos et al., 2015; Ejaz et al., 2021; Horn et al., 2012; Mohanram et al., 2013; Tiwari et al., 2018)

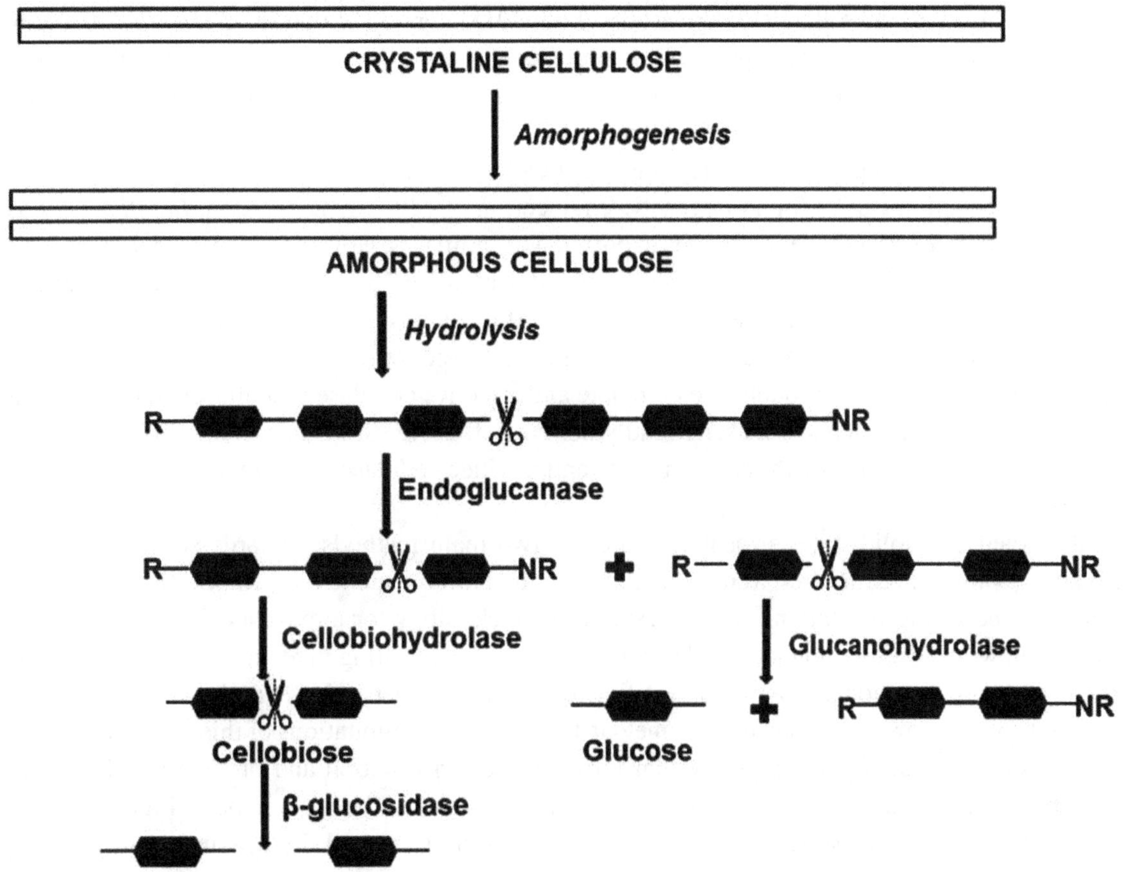

Cellulose hydrolysis is initiated by random attacks of internal sites in amorphous regions in cellulose by endoglucanase, resulting in rapid depolymerization (Barcelos et al., 2015; Ejaz et al., 2021). Exoglucanases then attack non-reducing or reducing ends of the glucan chain, releasing cellobiose or glucose, respectively (Adegboye et al., 2021). Exoglucanases acting on non-reducing ends are called cellobiohydrolases and they are the dominant component of cellulase cocktails (Adegboye et al., 2021; Horn et al., 2012). Exoglucanases, acting on reducing ends are called glucanohydrolases (Mohanram et al., 2013). In the final step, β-glucosidase attacks the non-reducing end of cellobiose, yielding glucose. Hydrolysis of cellobiose is the rate-limiting step of the overall process, making β-glucosidase the key regulatory enzyme in cellulose hydrolysis (Tiwari et al., 2018). β-Glucosidase is strongly inhibited by glucose, with the exception of β-glucosidase from *Aspergillus oryzae* (Barcelos et al., 2015). Synergism of cellulases is influenced by physical and chemical heterogeneity of the substrate, access of the substrate to the active site of the cellulose-processing enzymes and ratio of individual cellulases in the enzyme cocktail. Low content of endoglucanase and high content of cellobiohydrolase provide the best synergy in the catalytic process and the highest glucose yield (Contreras et al., 2020).

Natural and genetically modified fungi and bacteria have been heavily exploited for production of industrial cellulases. Fungal cellulases are more common, due to high productivity of engineered strains (>100g protein/liter of cell culture), which secrete the enzyme to the medium for easy extraction and purification (Mohanram et al., 2013; Srivastava et al., 2015). On top of that, *Escherichia coli*, the most common host for heterologous expression of proteins, suffers from several significant limitations in cellulase production. Thick outer membrane of *E. coli* or impaired transport across the membrane due to degradation of linker sequence result in limited secretion of the target enzyme. Formation of inclusion bodies further complicates isolation and purification of the target enzyme and often leads to its inactivation. Solution can be found in genetic engineering of cellulolytic bacteria, *Bacillus subtilis* or *Clostridium thermocellum* for homologous expression of cellulases. However, high production costs and need for enriched media limit this approach (Menendez et al., 2015). On the other hand, a significant drawback of fungal systems is the deficiency of a specific cellulase. *Hypocrea jecorina* (*T. reesei*) cellulase mixture is rich in cellobiohydrolase, but low in β-glucosidase. The opposite is true for *Aspergillus niger*. Therefore, in order to produce the complete and effective cellulase system, several fungal strains need to be employed, which is not user-friendly nor cost-effective (Srivastava et al., 2015). Apart from that, bacteria have higher growth rate than fungi and produce cellulases of higher stability (Mohanram et al., 2013).

Submerged and solid-state fermentation are the two main methods for large-scale production of cellulases, using lignocellulose biomass as a growth medium for inducible cellulolytic microbial producers. Submerged fermentation is a well-established technology for large-scale production of fungal cellulases in which water is abundant in fermentation mixture, resulting in reduced oxygen, temperature and nutrient concentration gradients and facilitated enzyme recovery. However, low enzyme yield and additional costs of dewatering during enzyme purification are key limitations of this strategy. Solid state fermentation is essentially a one-pot system for cellulase fermentation and production of bioethanol from lignocellulose biomass. It is performed without free water in the system, but with ample amount of moisture. Accordingly, solid state fermentation is cost-effective and time-saving method for high-yield production of bioethanol in lower reactor volume, lower enzyme loading and lower requirement for sterile conditions. The major bottleneck of this method is reaction temperature: in saccharification step temperature is maintained at 50°C, while during bioethanol production temperature is maintained between 28 and 37°C. Therefore, thermotolerant enzymes are necessary to endure this process (Bhardwaj et al., 2021; Dey et al., 2020). Table 1 summarizes the leading market suppliers and characteristics of commercial cellulase preparations.

Table 1. Commercial cellulase preparations

Market Name	Supplier	Source	Characteristics
Accellerase® 1500	Genencor-Danisco, USA DuPont Industrial Biosciences	*T. reesei*	82mg/mL cellulase 51mg/mL xylanase Stable at 50-65°C and pH 4.0-5.0
Bio-feed beta L	Novozymes, Denmark	*Trichoderma longibrachiatum* *T. reesei*	<5 filter paper units/mL 12U/mL β-glucosidase
Biocellulase A	Quest Intl., USA	*A. niger*	32U/mL β-glucosidase High stability at pH 5 and 55°C
Cellubrix (Celluclast)	Novozymes, Denmark	*T. longibrachiatum* *A. niger*	56 filter paper units/mL 136U/mL β-glucosidase
Cellic (CTec2)	Sigma Aldrich, USA	*T. reesei* *A. niger*	Variable efficiency
Cellulase 2000L	Rhodia-Danisco, France	*T. longibrachiatum* *T. reesei*	10 filter paper units/mL
Cellulyve 50L	Lyven, France	*T. longibrachiatum* *T. reesei*	24 filter paper units/mL
Cellulase AP 30 K	Amano Enzyme Inc., Japan	*A. niger*	60U/mL β-glucosidase High stability at pH 4.5 and 60°C
Cellulase NLT	Zhongbei Bio-Chem Industry Co., Ltd, China	*Trichoderma spp.*	10 000U/g cellulase Optimum activity at pH 4.5 and 40–50°C
Energex L	Novozymes, Denmark	*T. longibrachiatum* *T. reesei*	<5 filter paper units/mL 19U/mL β-glucosidase
GC220	Genencor-Danisco, USA	*T. longibrachiatum* *T. reesei*	116 filter paper units/mL 215U/mL β-glucosidase
GC440			<5 filter paper units/mL 70U/mL β-glucosidase
GC880			<5 filter paper units/mL 86U/mL β-glucosidase
Novozyme 188	Novozymes, Denmark	*A. niger*	1116 filter paper units/mL
Rohament CL	Rhom-AB Enzymes, Finland	*T. longibrachiatum* *T. reesei*	51 filter paper units/mL 28U/mL β-glucosidase
Shearzyme Plus	Novozymes, Denmark	*A. oryzae* *T. reesei*	250U/g xylanase 350U/g cellulase β-glucosidase side activity
Spezyme CP	Genencor-Danisco, USA	*T. longibrachiatum* *T. reesei*	49 filter paper units/mL
Ultraflo L	Novozymes, Denmark	*T. longibrachiatum* *T. reesei*	<5 filter paper units/mL 20U/mL β-glucosidase
Ultraflo XL		*Humicola insolens* *Bacillus amyloliquefaciens*	1500U/g xylanase 500U/g β-glucanase α-amylase side activity
Viscoferm		Not defined	222U/g β-glucanase Other activities: cellulase, xylanase Active at 30-80°C and pH 5.5-6.5
Viscozyme L		*T. longibrachiatum* *T. reesei*	<5 filter paper units/mL 23U/mL β-glucosidase
Viscostar 150L	Dyadic, USA	*T. longibrachiatum* *T. reesei*	33 filter paper units/mL 111U/mL β-glucosidase

Source: (Acharya & Chaudhary, 2012; Contreras et al., 2020; Srivastava et al., 2015)

MAIN FOCUS OF THE CHAPTER

In this section following topics will be discussed in detail:

- Bioconversion of lignocellulose biomass to bioethanol
    - Fundamental aspects and the main phases of bioethanol production
    - Process parameters influencing lignocellulose bioethanol yield
- Techno-economic analysis of bioethanol as an alternative fuel
    - Economic segments of global bioethanol market
    - Impact of COVID-19 on economic viability of bioethanol
    - Co-relation of global cellulase market and global bioethanol market.

Accordingly, this section will address actual issues, which need to be taken into an account from both scientific community and industry when considering bioethanol as transportation fuel of the future.

Bioconversion of Lignocellulose Biomass to Bioethanol

Bioconversion is a complete transformation of complex lignocellulose biomass to bioethanol as a final product. It is a multiphase process, as presented at Figure 3.

Figure 3. Bioconversion of lignocellulose biomass to bioethanol
Source:(Devi et al., 2022; Dey et al., 2020; Mohanram et al., 2013; Robak & Balcerek, 2020)

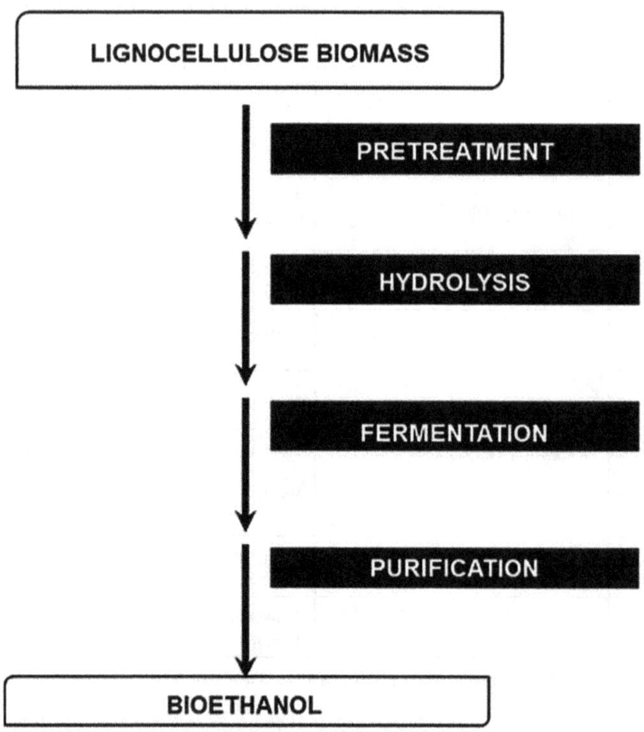

Pretreatment of Lignocellulose Biomass

The main purpose of pretreatment is to reduce the biomass recalcitrance by removal of lignin and hemicellulose and to promote amorphogenesis of cellulose to make it more accessible for hydrolysis. Methods of lignocellulose biomass pretreatment can be physical, chemical, physicochemical and biological, as indicated in Table 2.

Table 2. Pretreatment of lignocellulose biomass

Type of Pretreatment	Method	Reaction Conditions	Remarks
Physical	Mechanical comminution	Chipping, grinding, milling	Reduction of particle size to 0.2-2mm
	Pyrolysis	Intense heating above 300°C, followed by cooling and condensation	Rapid decomposition of cellulose to H_2, CO and char Leaching of char to obtain fermentable sugar
	Microwave pretreatment	High heating in a microwave oven	Inefficient separation of hemicellulose and lignin from cellulose Energy consuming and expensive for industrial applications
	Electron-beam irradiation	Continuous change of magnetic field	
Chemical	Alkaline hydrolysis	Dilute NaOH or $Ca(OH)_2$, aided with H_2O_2 60°C for NaOH, 120°C for $Ca(OH)_2$	Sequestration of lignin and hemicellulose into alkali-soluble fraction Reduction of cellulose crystallinity Lower reactor cost than in acid hydrolysis
	Acid hydrolysis with diluted inorganic acids	0.75–5% H_2SO_4, HCl, or HNO_3 120–200°C 1 MPa 2–10min	Lignin is not hydrolyzed pH neutralization is necessary High production of toxic inhibitors
	Acid hydrolysis with concentrated inorganic acids	10-30% H_2SO_4 170–190°C	Equipment corrosion Acid recovery is necessary Environmentally challenging High production of toxic inhibitors
	Organosolv	Organic solvents mixed with water, inorganic salts and inorganic acids >120°C, 20-60min	Removal of lignin and hemicellulose Environmentally challenging
	Ionic liquids	/	Efficient solubilization and removal of lignin Low hemicellulose and cellulose loss High price of ionic liquids
Physicochemical methods	Liquid hot water	170–230°C >5MPa 1-45min	Cellulose depolymerization Partial lignin solubilization High xylose recovery No toxic inhibitors
	Steam explosion (autohydrolysis)	160–300°C 0.69–4.85MPa Several seconds to minutes	High solid loads Formation of inhibitors Lignin is not solubilized High degree of hemicellulose hydrolysis
	Ammonia fiber explosion	1–2kg ammonia/kg dry biomass 90°C 1.12–1.36MPa 30min	Ammonia recovery is required No inhibitors Partial lignin solubilization
Biological methods	Enzymatic delignification	Enzymatic cocktail of laccase, lignin peroxidase and manganese peroxidases from white rot fungi	No inhibitors formed Mild reaction conditions Environmental friendly Long pretreatment time

Source: (Dey et al., 2020; Robak & Balcerek, 2020)

Mechanical comminution, alkaline hydrolysis, steam explosion and liquid hot water method are the most common physical, chemical, or physicochemical methods of lignocellulose biomass pretreatment, respectively. The most significant limitation of all pretreatment methods, except biological, is formation of toxic inhibitory byproducts: phenolic compounds, furfural, 5-hydroxymethylfurfural, formic, acetic and levulinic acid. These compounds interfere with microbial growth, inhibit yeast alcohol dehydrogenase, induce cell membrane damage and chromatin destruction, compromising fermentation productivity and bioethanol yield. Hence, it is of utmost importance to detoxify the substrate mixture prior to hydrolysis by membrane filtration, absorption on activated charcoal, solvent extraction or ion-exchange (Dey et al., 2020; Robak & Balcerek, 2020).

Hydrolysis of Polysaccharides

Hydrolysis of cellulose and hemicellulose is otherwise known as saccharification, indicating depolymerization and hydrolysis of polysaccharides into a mixture of corresponding monosaccharides (fermentable sugar). Hydrolysis can be chemical and enzymatic. In chemical hydrolysis polysaccharide mixture is treated with inorganic acids. Diluted acid hydrolysis facilitates step-by-step separation of hemicellulose and cellulose monomers, whereas hydrolysis with concentrated acids mainly yields hemicellulose monomers. Acid hydrolysis is characterized by high saccharification yield (70-100%, depending on the acid concentration). However, environmental concerns, equipment corrosion and glucose decomposition are significant limitations of this method (Dey et al., 2020).

Enzymatic hydrolysis provides selective conversion of lignocellulose polysaccharides in mild reaction conditions with a saccharification yield of 75–85%. Depending on the pretreatment technique, cellulases or mixtures of cellulases and hemicellulase are necessary for complete hydrolysis. Hemicellulases are complex enzyme cocktails that hydrolyze hemicellulose to a mixture of pentoses (xylose, arabinose) and hexoses (galactose, glucose, mannose). Commercial development of hemicellulases is not as advanced as for cellulases, since the predominant substrate for enzymatic hydrolysis is the biomass pretreated with diluted inorganic acids, where hemicellulose is removed before saccharification (Mohanram et al., 2013). Large-scale enzymatic saccharification of lignocellulose polysaccharides is most commonly performed in fed-batch bioreactor, which relies on sequential time-dependent addition of substrates for complete hydrolysis in each step. Most common disadvantages of this method are batch-to-batch oscillations, frequent startup and shutdown problems, high cost of labor and problem of enzyme recovery (Dey et al., 2020).

Thermostable cellulases are ideal biocatalysts for cellulose bioprocessing, because higher temperatures increase the reaction rate, diffusion coefficient, bioavailability and solubility of the substrate, while decreasing viscosity and minimizing the risk of microbial contamination (Srivastava et al., 2015; Trbojević Ivić et al., 2016). Thus far, *Thermoascus aurantiacus*, *Humicola grisea* var *thermoidea*, *Chaetomium thermophilum*, *Geobacillus* sp. HTA426, *Sporotrichum thermophile*, *Thermomonospora curvata*, *Scytalidium thermophilum*, *Clostridium staminisolvens* and *Pyrococcus furiosus* were reported as producers of thermostable cellulases (Ejaz et al., 2021).

Fermentation of Monosaccharides to Bioethanol

Industrial bioethanol fermentation is usually performed with *S. cerevisiae* and bacteria *Zymomonas mobilis* (Robak & Balcerek, 2020). Significant disadvantage of *S. cerevisiae* is its inability to ferment pentoses, therefore, for efficient fermentation of mixed pentose and hexose substrate, a microbial co-culture would be preferred (Lamichhane et al., 2021; Verma & Kumar, 2021). Also, with the upgrade of fermentation techniques implementation of thermostable microorganisms can result in better bioethanol yield.

The typical method of bioethanol production in batch fermentation is the Melle-Boinot process. It consists in broth preparation (pH adjustment, 14–22% by mass) and sterilization, followed by yeast fermentation. Fermented broth is centrifuged and bioethanol is separated from supernatant, while the yeast biomass is recycled for the next fermentation. The process can be repeated until the activity and viability of yeast cells is lost after which fresh inoculum is added (Bušić et al., 2018). From this basic set-up, several upgraded technologies emerged, summarized in Table 3.

Table 3. Advancement in bioethanol fermentation techniques

Fermentation Method	Characteristic	Remark
Separate hydrolysis and fermentation	Independent enzymatic hydrolysis and fermentation	Each stage is performed under its optimal conditions Enzyme hydrolysis is unaffected by bioethanol inhibition Surfactants must be added to reduce β-glucosidase inhibition by glucose High cost of separate bioreactor system
Simultaneous saccharification and fermentation	Simultaneous enzymatic hydrolysis and fermentation	Lower production cost Reduced enzyme inhibition by simultaneous end product removal Lower requirement for sterile conditions (bioethanol is produced immediately with glucose conversion) Thermotolerant microbes must be used
Simultaneous saccharification and fermentation with short pre-saccharification	Pre-hydrolysis at 50°C, followed by fermentation	Reduced viscosity of fermentation substrate Higher productivity of fermentation
Simultaneous saccharification and co-fermentation	Simultaneous saccharification and fermentation of pentose/hexose mixture	High bioethanol yield Pentoses and hexoses can be fermented Recombinant thermotolerant microbes, capable of fermenting both monosaccharides are necessary
Consolidated bioprocessing	One-pot production of hydrolytic enzymes, hydrolysis of pretreated biomass and fermentation of all types of soluble sugars	Cost-effective More efficient hydrolysis Reduced risk of microbial contamination Specialized microbes must be used (recombinant *E. coli* or *Fusarium oxysporum*)

Source: (Dadwal et al., 2020; Dey et al., 2020; Hoffman et al., 2021; Lamichhane et al., 2021; Liu et al., 2016).

After fermentation bioethanol is extracted from fermentation broth and purified. Distillation is the conventional method for separation and partial purification of bioethanol from aqueous fermentation broth, based on the difference of the boiling points of bioethanol and water (78.3°C and 100°C, respectively). Continuous distillation column system is conventionally used in large-scale processes (Robak & Balcerek, 2020). Lately, membrane technology has been employed in bioethanol recovery. Pervaporation was used to purify bioethanol from bioethanol-water mixture, but also from more complex lignocellulose broth (Abels et al., 2013; Gaykawad et al., 2013). Membrane extraction of bioethanol from fermentation broth was performed with around 80% of the product recovery (Chang et al., 1992). High purity bioethanol (99.7% v/v) is obtained only by dehydration (Robak & Balcerek, 2018).

Factors Influencing Bioethanol Yield

Composition of lignocellulose biomass, enzyme and substrate loading and mixing are the major factors to consider in the hydrolysis step. Lignin and hemicellulose are the main contributors to biomass resistance to hydrolysis, therefore better results could be achieved by the careful selection of raw material with lower percentage of lignin and hemicellulose, such as rice husk, saw dust or pinewood (Preethi et al., 2021). Higher enzyme and solid loadings favor higher saccharification yield. However, high solid loading can also increase viscosity, heat and mass transfer limitations and inhibition by toxic components during fermentation (Dey et al., 2020; Robak & Balcerek, 2020). Inappropriate mixing will negatively affect saccharification yield, either due to insufficient aeration when the mixing is poor, either due to enzyme denaturation by the shearing force. Non-ionic surfactant Tween 80, non-catalytic protein bovine serum albumin or polyethylene glycol prevent non-productive adsorption of cellulase to lignin, with no effect on enzyme catalytic activity and are common additives during hydrolysis and fermentation (Robak & Balcerek, 2020).

Fermentation requires more complex regulation, since it must provide optimum conditions for growth of microbial cell mass and utilization of fermentable sugars. Temperature and pH range of industrial fermentation are adapted to *S. cerevisiae* and have to be optimized for any other microorganism. Higher glucan loading exhibits positive dose-dependence to the bioethanol yield. However, high sugar concentrations induce osmotic pressure on yeast and plasmolysis. Steering speed and reaction time are inversely proportional variables: at higher agitation speed fermentation will be completed in shorter reaction time. Too intense agitation can cause microbial cell disruption. At a low biomass density (optical density at 600 nm=1) the yeast cell viability is compromised by the substrate's inhibitory effect, while at the excessively high biomass density (optical density at 600 nm=8) yeasts consume more glucose, which is converted into the microbial biomass instead of bioethanol (Robak & Balcerek, 2020). Bacterial contamination during fermentation is a common cause of low bioethanol yield. Antibiotics are traditionally used to control contamination; however, they can be detected in the final product. Alternatively, microbial contamination can be prevented by using antimicrobial natural products, lysozyme and bacteriophages and by the proper selection of fermentation method (*i. e.* simultaneous saccharification and fermentation, Table 3) (Lamichhane et al., 2021; Verma & Kumar, 2021).

Techno-economic Analysis

Bioethanol market is segmented by feedstock type, application, and geography. In 2020 approximately 100 billion liters of bioethanol were produced globally, with projected compound annual growth rate (CAGR) >4.5% between 2021 and 2026 (Mordor Intelligence, 2016).

Bioethanol Market by Feedstock Type

Majority of global bioethanol production arises from food crops: maize (60% share, main producer USA), sugarcane (25% share, main producer Brazil), molasses (7% share, main producer India), cereals and sugar beet (4%, main producer the European Union) (Flach et al., 2019; Ibeto et al., 2011). Global production of lignocellulose-derived second-generation bioethanol is 400 million liters per year, but is expected to increase in the future (Mizik et al., 2020). Significant step forward is the opening of processing plants for production of bioethanol from bagasse, rice straw, wood chips, corn stover and straw in Finland, Italy, Australia and Serbia, respectively (Dey et al., 2020; Martinov et al., 2015).

The price of feedstock represents 40-75% of the total bioethanol production costs. According to Devi, only 3% of global annual production of lignocellulose biomass accounts for all of its derivatives (bioethanol, bio-based chemicals, non-food products) (Devi et al., 2022). Paradoxically, food crops are more cost-effective feedstock than lignocellulose biomass (US$0.2 - US$0.3/L for bioethanol from sugar cane, compared to US$0.8 – US$1.2/L for bioethanol from lignocellulose material) (Bušić et al., 2018). The rationale behind this is the non-uniform composition of the lignocellulose biomass components, depending on the source. Biomass with higher lignin content requires additional costs of pretreatment for lignin removal, than biomass in which cellulose and hemicellulose are more abundant (Preethi et al., 2021; Verma & Kumar, 2021).

Bioethanol Market by Application

Approximately 80% of globally produced bioethanol is used as transportation fuel, followed by the food, pharmaceutical, and cosmetics sector (Mizik et al., 2020). As a transportation fuel, bioethanol is mixed with gasoline at volume fractions 5, 10 or 85% of bioethanol (fuel codes E5-85). Only E5 and E10 can be directly used in the existing motors without modifications, hence they represent the largest share in the market (Bušić et al., 2018). Recently, efforts are also being made to implement bioethanol as bio-jet fuel in aviation transport (Han et al., 2019).

Bioethanol Market by Geography

The leading global producers of bioethanol are USA, Brazil, the European Union, China, India and Canada (Bušić et al., 2018; Mordor Intelligence, 2016). In Europe, Sweden, Germany, the United Kingdom, France and Italy have primacy in bioethanol production. Collectively, European countries produce annually more than 2 billion liters of bioethanol, while in the USA production capacity is 66 billion liters bioethanol/year (Robak & Balcerek, 2018). The USA and Brazil are the leading global consumers of bioethanol, whereas the top three European bioethanol consumers are Germany, the United Kingdom and France (Flach et al., 2019). The Asia-Pacific region is considered to be the fastest-growing market, with China and Japan as regional leaders.

Impact of COVID-19 on Global Bioethanol Market

The COVID-19 pandemic has affected the bioethanol economy dramatically. From April-June 2020 global consumption and bioethanol price decreased by 40%, or US$3.4 billion (Renewable Fuels Association, 2021). The USA was most seriously affected, with US$2.9 billion losses in the bioethanol sector. Pandemics had no measurable impact on the European bioethanol market (Mizik et al., 2020). This interesting observation can be explained by the regional differences in feedstock for bioethanol production: USA relies predominantly on corn, while in Europe cereals and sugar beet are the dominant raw material. Reduction in biofuels demand has also sharply increased market prices of bioethanol value-added by-products as a result of a cutback in supplies (Renewable Fuels Association, 2021). The negative effect of the pandemics is even more pronounced in the middle-income countries where agriculture is a more significant contributor to the GDP, however the long-term effect of the pandemic will be comprehended only after the COVID-19 crisis passes (Shruthi & Ramani, 2021).

Co-relation of Global Cellulase Market and Global Bioethanol Market

It is estimated that the global biofuel enzymes market will reach US$2.0 billion by 2026. Cellulase market follows the similar pattern as bioethanol market. Around 40–49% of the total cost of bioethanol production is attributed to cellulase. Novozymes and Genecor, market leaders in the production of industrial enzymes, have contributed to the significant decrease in price of industrial cellulases (Srivastava et al., 2015). Nevertheless, to improve bioethanol production economy, further decrease of the price is necessary to the projected value of US$0.026/L (Acharya & Chaudhary, 2012). Cost-reduction can also be achieved by the careful choice of the fermentation method. By switching from separate to simultaneous saccharification and fermentation (Table 3) costs could be reduced from US$1.2/L to US$0.6/L (Dey et al., 2020).

SOLUTIONS AND RECOMMENDATIONS

Solutions and perspectives for the issues presented in the previous section will be considered in this section from the following aspects:

- Increasing the economic value of lignocellulose biomass
- Tailoring of cellulases for efficient production of bioethanol.

Increasing the Economic Value of Lignocellulose Biomass

As discussed previously, the price of lignocellulose raw material contributes most significantly to the overall price of bioethanol. These expenses can be compensated by production of other valuable by-products. Accordingly, subject of great attention in recent decades is design and development of biorefineries for integrated production of bioethanol and other valuable products from cellulose, hemicellulose and lignin, as a base of circular bioeconomy. Other than bioethanol, cellulose is a source of sorbitol, levulinic and succinic acid, additives in food and pharmaceutical industry. Lactic acid, xylitol, xylo-oligosaccharides, furfural, hydroxymethylfurfural and polyhydroxyalkanoates, obtained from hemicellulose, are used in pharmaceutical, chemical and food industry. Furfural and hydroxymethylfurfural are precursors of pesticides and fertilizers. As mentioned previously in this chapter, these two compounds are inhibitors of cellulose hydrolysis. Therefore, their sequestration represents a double gain. Lignin can be used directly as fuel to boilers for pulp production, but it is even more important as a source of lignol and vanillyl alcohol (Devi et al., 2022). Vanillyl alcohol exhibits antioxidant and antimicrobial activity, but also can be used as a flavoring agent in food and cosmetic industry. In a recent study it was used for synthesis of capsinoids, small molecules with wide array of physiological activity (Trbojević-Ivić et al., 2017).

Tailoring of Cellulases for Efficient Production of Bioethanol

Biorefinery process requires biocatalysts with high catalytic efficiency on insoluble substrate, stable at higher temperature and pH, and resistant to end-product inhibition (Mohanram et al., 2013). This section will highlight how these requirements can be met by enzyme immobilization.

Fundamental Aspects of Enzyme Immobilization

Enzyme immobilization is sequestration of enzymes in a limited, well-defined space with preservation of its catalytic activity. This is unique technique of biocatalyst tailoring that combines simultaneous stabilization, reusability and easy separation of the enzyme from the reaction mixture. Additionally, multiple enzymes with different catalytic activities can be simultaneously immobilized into unique biocatalyst machinery for one-pot biotransformation. An overview of enzyme immobilization methods is represented at Figure 4.

Figure 4. Techniques of enzyme immobilization
Source: (Datta et al., 2013; Sheldon, 2007)

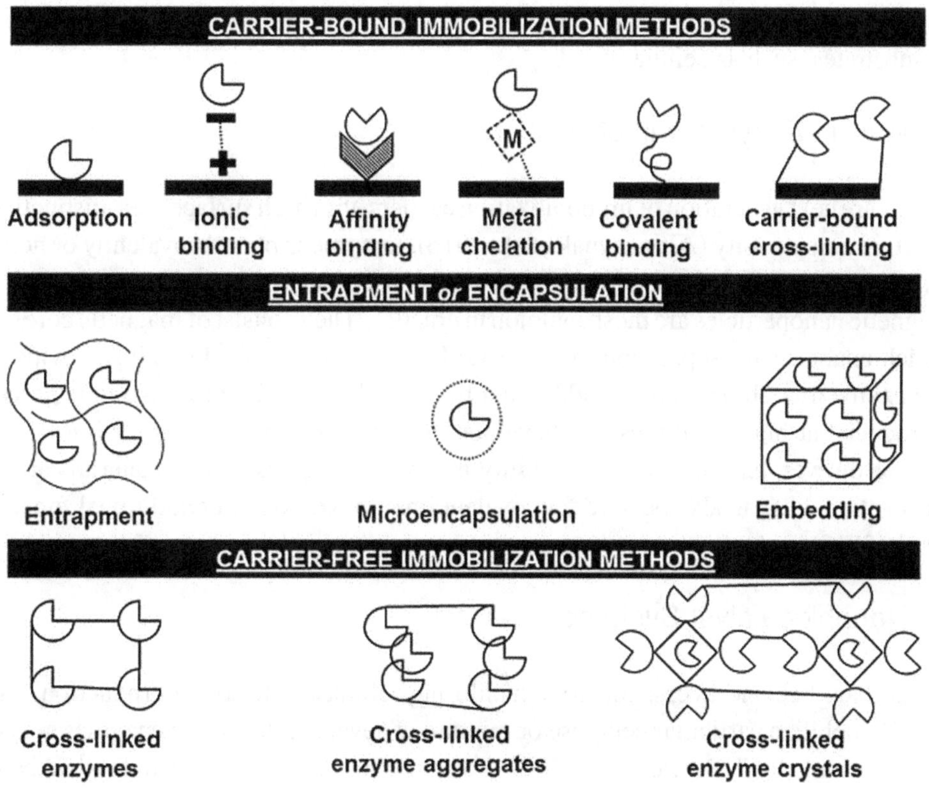

Carrier-bound techniques and entrapment have the most practical significance. In physical adsorption, ionic and affinity binding and metal chelation enzymes are reversibly bound to the carrier *via* weak non-covalent bonds. Enzyme immobilization carriers can be of different origin, chemical structure, particle morphology and size. Natural immobilization carriers, polysaccharides and proteins, are biocompatible and non-toxic, which is relevant for medical applications, pharmaceutical or food industry (Trbojević-Ivić et al., 2018). Eupergit-C, Sepabeads-EC, Amberlit and Purolite A109 are frequently used synthetic polymers for enzyme immobilization (Mihailović et al., 2016; Milosavić et al., 2012; Trbojević-Ivić et al., 2018; Žuža et al., 2009). Among inorganic materials, silica-gel, metal oxides and ceramics are most common. Recently, immobilization of lipase from *Candida rugosa* on hydroxyapatite *via* metal chelation resulted in an improved biocatalyst with interesting novel applications (Trbojević-Ivić et al., 2017; Trbojevic Ivic et al., 2016). Entrapment is an irreversible immobilization method, in which an enzyme is confined inside the carrier matrix but the two entities do not interact chemically. Alginate gels are the most popular matrixes for entrapment (Dimitrijević et al., 2011). In microencapsulation concentrated enzyme solution is confined in micelles or reverse micelles, which enables better catalytic efficiency (Trbojević-Ivić et al., 2018).

Immobilization efficiency is governed by the immobilization method (Figure 4) and immobilization carrier characteristics. In covalent immobilization, enzymes are irreversibly chemically attached to the carrier so enzyme leakage is minimized. However, problems arise if enzymes are bound *via* an active site. Physical adsorption is a popular technique, because high immobilization yield is achieved in a short time. However, a significant limitation of this method is enzyme leakage. In encapsulation and entrapment enzyme leakage is minimum, however these methods are unsuitable for enzymes processing polymeric substrates, such as cellulases.

Nanomaterials in Enzyme Immobilization

Nanomaterials are next generation of immobilization carriers with high surface area, distinctive morphology, architecture and porosity (Altinkaynak et al., 2016). Enzyme is bound covalently or non-covalently to silica-based, polymer-based, metal-based or carbon-based nanomaterials, forming nanobiocatalyst system. Magnetic nanoparticles are most common in practice. They consist of magnetic core (iron, nickel, cobalt), which enables easy separation in magnetic field, while chemical coating provides functional groups for enzyme immobilization and additional functionalization. Polymer-based nanomaterials can be easily produced at high quantities and have plethora of functional groups for immobilization and modification. However, their mechanical stability is low. Carbon-based nanomaterials are suitable for physical adsorption, but can also be used for covalent immobilization after additional functionalization (Misson et al., 2015).

Cellulase Nanobiocatalyst Systems

Nanomaterials have become an essential tool in tailoring cellulase activity for production of bioethanol. Strategies to immobilize particular cellulase or mixture of several cellulolytic enzymes are described in the literature (Husain, 2017; Perwez et al., 2017; Zanuso et al., 2022). Among individual cellulases, the most common are those from *T. reesei*, *A. niger*, *A. fumigates*, *A. nidulans* or *T. viride* and commercial cellulases Cellusoft CR, Cellic CTec2 (Husain, 2017; Jabasingh, 2011; Lupoi & Smith, 2011; Simon et al., 2018). In multi-enzyme preparations, endoglucanase is co-immobilized with β-glucosidase, cellobiohydrolase and xylanase, while β-glucosidase can also be co-immobilized with glucose oxidase or cellobiohydrolase. Silica-based, magnetic and metal-oxide nanoparticles are the most common carriers for immobilization of cellulases. Functionalization of these materials with polymers, polyalcohols, glutaraldehyde, carboxyl and amino groups enables efficient covalent binding or physical adsorption. Nanomaterials with small pores and encapsulation are unsuitable for cellulase immobilization, due to significant diffusional limitations of bulk substrate, such as pretreated biomass (Husain, 2017).

Efficiency of the applied immobilization protocol is determined through immobilization yield, temperature and pH stability and optima and reusability. In most cases, regardless of the method and support choice, immobilization yield was between 60–95% (Husain, 2017; Jabasingh, 2011; Manasa et al., 2017; Sillu & Agnihotri, 2020; Simon et al., 2018). Immobilization commonly resulted in the shift of the optimum pH and temperature to higher values and improved temperature stability (Lupoi & Smith, 2011; Manasa et al., 2017; Sillu & Agnihotri, 2020). The most prominent feature of immobilized cellulase is reusability: immobilizate could be used in more than ten consecutive cycles of carboxymethyl cellulose hydrolysis without significant loss of activity (Husain, 2017; Jabasingh, 2011; Lupoi & Smith, 2011; Manasa et al., 2017; Zanuso et al., 2022). Storage stability of cellulase was also significantly improved at 4°C upon immobilization (Simon et al., 2018; Zanuso et al., 2022).

Industrial potential of immobilized cellulase was assessed by hydrolysis of lignocellulose biomass after different pretreatments. *T. reesei* cellulase covalently immobilized on zinc ferrite nanoparticles effectively hydrolyzed sonified and alkali-pretreated sun hemp biomass, with hydrolysis yield of 53% (Manasa et al., 2017). Cellic CTec2 cocktail, immobilized on chitosan-coated magnetic nanoparticles, hydrolyzed hydrothermally pretreated corn cob with hydrolysis yield of 65% (Zanuso et al., 2022). Cellulase-halloysite-metal-oxide nanobiocomposites hydrolyzed bagasse extracted with ionic liquids with 50.2% hydrolysis yield (Sillu & Agnihotri, 2020), while cellulase from *T. viride*, physically adsorbed on silica nanoparticles, was successfully implemented in simulation of simultaneous saccharification and fermentation of bioethanol from microcrystalline cellulose (Lupoi & Smith, 2011).

FUTURE RESEARCH DIRECTIONS

Bioethanol is technologically, economically and environmentally viable alternative to fossil fuels. However, major issue is the production of bioethanol at a competitive price. Efficient selection and pretreatment of raw material, together with circular bioeconomy concept will reduce the total cost of lignocellulose bioethanol production by co-generating other valuable products. However, a good biorefinery is the one that leaves minimum or no waste during a process. Therefore, zero-waste biorefinery needs the skills and proficiency from different interdisciplinary fields such as chemistry, biochemistry, chemical engineering, biology, biomolecular engineering and materials sciences. Genetic modification of microbial producers of cellulase and biocatalyst tailoring are aimed at reducing biocatalyst price by increasing its catalytic efficiency and stability for industrial reaction conditions. Immobilization provides an endless field of possibilities to improve cost-effectiveness of cellulases in bioconversion of lignocellulose biomass to second-generation bioethanol. Plethora of well-established immobilization protocols, innovative immobilization carriers and development of nanotechnology will lead to design of reusable nanobiocatalysts of high stability and activity for large-scale processes. It is believed that these solutions combined will significantly contribute to price competitiveness of bioethanol on the global fuel market.

CONCLUSION

Modern society must find a long-lasting solution to depletion and environmental concerns caused by extensive use of fossil fuels. Lignocellulose biomass is the most abundant raw material in the biosphere, giving rise to second-generation bioethanol, the most promising alternative to fossil fuels. Bioconversion of lignocellulose to bioethanol is a complex process of selection and pretreatment of biomass, hydrolysis of cellulose, fermentation and purification of bioethanol. Hydrolysis of cellulose is the key phase of the process and is governed by synergistic activity of cellulase multi-enzyme complex. Mesophilic and thermophilic fungi and bacteria are the main industrial producers of cellulases. Price of raw material and cellulase affect the overall cost of bioethanol production from lignocellulose biomass and need to be reduced to make bioethanol more concurrent on the global fuel market. Reducing cellulase price is considered crucial to increase bioethanol market competitiveness. Important feature, which contributes to cost-effectiveness is biocatalyst recycling or reusability. This can be achieved only by enzyme immobilization. Immobilization of enzymes is a well-known and well-established strategy with a myriad of elaborate protocols and immobilization methods available to choose from. By proper selection of the method and carrier a superior catalyst can be designed with high stability and activity, which can be reused many times without significant activity loss and can be easily separated from the product. Recently, nanomaterials are considered as innovative immobilization carriers with unique characteristics and a whole new field of research and application of nanobiocatalysts has emerged. Silica, magnetic and metal-oxide nanoparticles are the most common carriers for cellulase covalent and non-covalent immobilization. Upon immobilization, temperature stability and reusability of cellulases were significantly improved, which resulted in successful pilot-scale implementation of immobilized cellulases in hydrolysis of biomass with glucose yield above 50%. Issues addressed in this chapter can provide a good starting point to research and applied sectors in devising a production strategy for more efficient and cost-effective production of cellulosic biofuels.

ACKNOWLEDGMENT

This research was supported by the Ministry of Education, Science and Technological Development of the Republic of Serbia [Contract number: 451-03-68/2022-14/200288].

REFERENCES

Abels, C., Carstensen, F., & Wessling, M. (2013). Membrane processes in biorefinery applications. *Journal of Membrane Science*, *444*, 285–317. doi:10.1016/j.memsci.2013.05.030

Acharya, S., & Chaudhary, A. (2012). Bioprospecting thermophiles for cellulase production: A review. *Brazilian Journal of Microbiology*, *43*(3), 844–856. doi:10.1590/S1517-83822012000300001 PMID:24031898

Adegboye, M. F., Ojuederie, O. B., Talia, P. M., & Babalola, O. O. (2021). Bioprospecting of microbial strains for biofuel production: Metabolic engineering, applications, and challenges. *Biotechnology for Biofuels*, *14*(1), 5. Advance online publication. doi:10.118613068-020-01853-2 PMID:33407786

Altinkaynak, C., Tavlasoglu, S., Özdemir, N., & Ocsoy, I. (2016). A new generation approach in enzyme immobilization: Organic-inorganic hybrid nanoflowers with enhanced catalytic activity and stability. *Enzyme and Microbial Technology*, *93–94*, 105–112. doi:10.1016/j.enzmictec.2016.06.011 PMID:27702469

Barcelos, C. A., Rocha, V. A., Groposo, C., de Castro, A. M., & Pereira, N. (2015). Enzymes and accessory proteins involved in the hydrolysis of lignocellulosic biomass for bioethanol production. In R. N. Silva (Ed.), *Mycology: Current and future developments* (pp. 23–56). Bentham Books. doi:10.2174/97816810807411150100005

Bhardwaj, N., Kumar, B., Agrawal, K., & Verma, P. (2021). Current perspective on production and applications of microbial cellulases: A review. *Bioresources and Bioprocessing*, *8*(1), 95. Advance online publication. doi:10.118640643-021-00447-6

Bušić, A., Mardetko, N., Kundas, S., Morzak, G., Belskaya, H., Šantek, M. I., Komes, D., Novak, S., & Šantek, B. (2018). Bioethanol production from renewable raw materials and its separation and purification: A review. *Food Technology and Biotechnology*, *56*(3), 289–311. doi:10.17113/ftb.56.03.18.5546 PMID:30510474

Chang, H. N., Yang, J. W., Park, Y. S., Kim, D. J., & Han, K. C. (1992). Extractive ethanol production in a membrane cell recycle bioreactor. *Journal of Biotechnology*, *24*(3), 329–343. doi:10.1016/0168-1656(92)90041-7 PMID:1367988

Contreras, F., Pramanik, S., Rozhkova, A. M., Zorov, I. N., Korotkova, O., Sinitsyn, A. P., Schwaneberg, U., & Davari, M. D. (2020). Engineering robust cellulases for tailored lignocellulosic degradation cocktails. *International Journal of Molecular Sciences*, *21*(5), 1589. doi:10.3390/ijms21051589 PMID:32111065

Dadwal, A., Sharma, S., & Satyanarayana, T. (2020). Progress in ameliorating beneficial characteristics of microbial cellulases by genetic engineering approaches for cellulose saccharification. *Frontiers in Microbiology*, *11*, 1387. doi:10.3389/fmicb.2020.01387 PMID:32670240

Datta, S., Christena, L. R., & Rajaram, Y. R. S. (2013). Enzyme immobilization: An overview on techniques and support materials. *3 Biotech, 3*(1), 1–9. doi:10.1007/s13205-012-0071-7

Devi, A., Bajar, S., Kour, H., Kothari, R., Pant, D., & Singh, A. (2022). Lignocellulosic biomass valorization for bioethanol production: A circular bioeconomy approach. *BioEnergy Research*. Advance online publication. doi:10.100712155-022-10401-9 PMID:35154558

Dey, P., Pal, P., Kevin, J. D., & Das, D. B. (2020). Lignocellulosic bioethanol production: Prospects of emerging membrane technologies to improve the process - A critical review. *Reviews in Chemical Engineering*, *36*(3), 333–367. doi:10.1515/revce-2018-0014

Ejaz, U., Sohail, M., & Ghanemi, A. (2021). Cellulases: From bioactivity to a variety of industrial applications. *Biomimetics*, *6*(3), 44. doi:10.3390/biomimetics6030044 PMID:34287227

Flach, B., Lieberz, S., & Bolla, S. (2019). *EU biofuels annual 2019*. Global Agriculturan Information Network. https://apps.fas.usda.gov/newgainapi/api/report/downloadreportbyfilename?filename=Biofu els Annual_The Hague_EU-28_7-15-2019.pdf

Gaykawad, S. S., Zha, Y., Punt, P. J., van Groenestijn, J. W., van der Wielen, L. A. M., & Straathof, A. J. J. (2013). Pervaporation of ethanol from lignocellulosic fermentation broth. *Bioresource Technology*, *129*, 469–476. doi:10.1016/j.biortech.2012.11.104 PMID:23266848

Han, G. B., Jang, J. H., Ahn, M. H., & Jung, B. H. (2019). Recent application of bio-alcohol: Bio-jet fuel. In Y. Yun (Ed.), *Alcohol fuels - Current technologies and future prospect*. IntechOpen.

Hoffman, S. M., Alvarez, M., Alfassi, G., Rein, D. M., Garcia-Echauri, S., Cohen, Y., & Avalos, J. L. (2021). Cellulosic biofuel production using emulsified simultaneous saccharification and fermentation (eSSF) with conventional and thermotolerant yeasts. *Biotechnology for Biofuels*, *14*(1), 157. doi:10.118613068-021-02008-7 PMID:34274018

Horn, S. J., Vaaje-Kolstad, G., Westereng, B., & Eijsink, V. G. (2012). Novel enzymes for the degradation of cellulose. *Biotechnology for Biofuels*, *5*(1), 45. doi:10.1186/1754-6834-5-45 PMID:22747961

Husain, Q. (2017). Nanomaterials immobilized cellulolytic enzymes and their industrial applications : A literature review. *JSM Biochemistry and Molecular Biology*, *4*(3), 1029.

Ibeto, C. N., Ofoefule, A. U., & Agbo, K. E. (2011). A global overview of biomass potentials for bioethanol production: A renewable alternative fuel. *Trends in Applied Sciences Research*, *6*(5), 410–425. doi:10.3923/tasr.2011.410.425

Intelligence, M. (2016). *Global bio-ethanol market 2022-2027*. https://www.mordorintelligence.com/industry-reports/bio-ethanol-market

Jabasingh, S. A. (2011). Optimization and kinetics of cellulase immobilization on modified chitin using response surface methodology. *Adsorption Science and Technology*, *29*(9), 897–916. doi:10.1260/0263-6174.29.9.897

Kricka, W., Fitzpatrick, J., & Bond, U. (2015). Challenges for the production of bioethanol from biomass using recombinant yeasts. *Advances in Applied Microbiology*, *92*, 89–125. doi:10.1016/bs.aambs.2015.02.003 PMID:26003934

Lamichhane, G., Acharya, A., Poudel, D. K., Aryal, B., Gyawali, N., Niraula, P., Phuyal, S. R., Budhathoki, P., Bk, G., & Parajuli, N. (2021). Recent advances in bioethanol production from lignocellulosic biomass. *International Journal of Green Energy*, *18*(7), 731–744. doi:10.1080/15435075.2021.1880910

Liu, Z., Ho, S. H., Sasaki, K., Den Haan, R., Inokuma, K., Ogino, C., Van Zyl, W. H., Hasunuma, T., & Kondo, A. (2016). Engineering of a novel cellulose-adherent cellulolytic Saccharomyces cerevisiae for cellulosic biofuel production. *Scientific Reports*, *6*(1), 24550. doi:10.1038rep24550 PMID:27079382

Lo Leggio, L., & Larsen, S. (2002). The 1.62 Å structure of Thermoascus aurantiacus endoglucanase: Completing the structural picture of subfamilies in glycoside hydrolase family 5. *FEBS Letters*, *523*(1–3), 103–108. doi:10.1016/S0014-5793(02)02954-X PMID:12123813

Lupoi, J. S., & Smith, E. A. (2011). Evaluation of nanoparticle-immobilized cellulase for improved ethanol yield in simultaneous saccharification and fermentation reactions. *Biotechnology and Bioengineering*, *108*(12), 2835–2843. doi:10.1002/bit.23246 PMID:21702028

Manasa, P., Saroj, P., & Korrapati, N. (2017). Immobilization of cellulase enzyme on zinc ferrite nanoparticles in increasing enzymatic hydrolysis on ultrasound-assisted alkaline pretreated Crotalaria juncea biomass. *Indian Journal of Science and Technology*, *10*(24), 1–7. doi:10.17485/ijst/2017/v10i24/112798

Martinov, M., Djatkov, D., Golub, M., Viskovic, M., Bojic, S., & Krstic, J. (2015). *Plant for lignocellulosic bioethanol. Case study*. Department of Environmental Engineering, Faculty of Technical Sciences.

Menendez, E., Garcia-Fraile, P., & Rivas, R. (2015). Biotechnological applications of bacterial cellulases. *AIMS Bioengineering*, *2*(3), 163–182. doi:10.3934/bioeng.2015.3.163

Mihailović, M., Trbojević-Ivić, J., Banjanac, K., Milosavić, N., Velicković, D., Carević, M., & Bezbradica, D. (2016). Immobilization of maltase from Saccharomyces cerevisiae on thiosulfonate supports. *Journal of the Serbian Chemical Society*, *81*(12), 1371–1382. doi:10.2298/JSC160730099M

Milosavić, N., Bogdanović Pristov, J., Veličković, D. V., Dimitrijević, A. S., Kalauzi, A., & Radotić, K. (2012). Study of the covalently immobilized amyloglucosidase on macroporous polymer by mathematical modeling of the pH optima. *Journal of Chemical Technology and Biotechnology (Oxford, Oxfordshire)*, *87*(10), 1450–1457. doi:10.1002/jctb.3768

Milosavic, N., Trbojevic Ivic, J., Veličković, D., Dimitrijević, A., & Bezbradica, D. (2016). Immobilized lipases: An old-fashioned twist for a new generation of industrial biocatalysts. *Journal of Microbial & Biochemical Technology*, *8*(6), 8–9. doi:10.4172/1948-5948.1000e127

Misson, M., Zhang, H., & Jin, B. (2015). Nanobiocatalyst advancements and bioprocessing applications. *Journal of the Royal Society, Interface*, *12*(102), 20140891. Advance online publication. doi:10.1098/rsif.2014.0891 PMID:25392397

Mizik, T., Nagy, L., Gabnai, Z., & Bai, A. (2020). The major driving forces of the EU and US ethanol. *Energies*, *13*(21), 5614. doi:10.3390/en13215614

Mohanram, S., Amat, D., Choudhary, J., Arora, A., & Nain, L. (2013). Novel perspectives for evolving enzyme cocktails for lignocellulose hydrolysis in biorefineries. *Sustainable Chemical Processes*, *1*(1), 15. doi:10.1186/2043-7129-1-15

Muñoz, I. G., Ubhayasekera, W., Henriksson, H., Szabó, I., Pettersson, G., Johansson, G., Mowbray, S. L., & Ståhlberg, J. (2001). Family 7 cellobiohydrolases from Phanerochaete chrysosporium: Crystal structure of the catalytic module of Cel7D (CBH58) at 1.32 Å resolution and homology models of the isozymes. *Journal of Molecular Biology*, *314*(5), 1097–1111. doi:10.1006/jmbi.2000.5180 PMID:11743726

Pavlović, M., Dimitrijević, A., Trbojević, J., Milosavić, N., Gavrović-Jankulović, M., Bezbradica, D., & Veličković, D. (2013). A study of transglucosylation kinetic in an enzymatic synthesis of benzyl alcohol glucoside by α-glucosidase from S. cerevisiae. *Russian Journal of Physical Chemistry A*, *87*(13), 2285–2288. doi:10.1134/S0036024413130207

Perwez, M., Ahmad, R., & Sardar, M. (2017). A reusable multipurpose magnetic nanobiocatalyst for industrial applications. *International Journal of Biological Macromolecules*, *103*, 16–24. doi:10.1016/j.ijbiomac.2017.05.029 PMID:28495621

Preethi, G., M, G., Kumar, G., Karthikeyan, O. P., Varjani, S., & J, R. B. (2021). Lignocellulosic biomass as an optimistic feedstock for the production of biofuels as valuable energy source: Techno-economic analysis, environmental impact analysis, breakthrough and perspectives. *Environmental Technology and Innovation*, *24*, 102080. doi:10.1016/j.eti.2021.102080

Renewable Fuels Association. (2021). *The economic impact of Covid-19 on the ethanol industry*. https://ethanolrfa.org/file/1753/COVID-19-Ethanol-Industry-Economic-Impact-July-2020-Update.pdf

Robak, K., & Balcerek, M. (2018). Review of second generation bioethanol production from residual biomass. *Food Technology and Biotechnology*, *56*(2), 174–187. doi:10.17113/ftb.56.02.18.5428 PMID:30228792

Robak, K., & Balcerek, M. (2020). Current state-of-the-art in ethanol production from lignocellulosic feedstocks. *Microbiological Research*, *240*, 126534. doi:10.1016/j.micres.2020.126534 PMID:32683278

Sheldon, R. A. (2007). Enzyme immobilization: The quest for optimum performance. *Advanced Synthesis & Catalysis*, *349*(8–9), 1289–1307. doi:10.1002/adsc.200700082

Shruthi, M. S., & Ramani, D. (2021). Statistical analysis of impact of COVID 19 on India commodity markets. *Materials Today: Proceedings*, *37*(Part 2), 2306–2311. doi:10.1016/j.matpr.2020.07.729 PMID:32837924

Sillu, D., & Agnihotri, S. (2020). Cellulase immobilization onto magnetic halloysite nanotubes: Enhanced enzyme activity and stability with high cellulose saccharification. *ACS Sustainable Chemistry & Engineering*, *8*(2), 900–913. doi:10.1021/acssuschemeng.9b05400

Simon, P., Lima, J. S., Valério, A., De Oliveira, D., Araújo, P. H. H., Sayer, C., De Souza, A. A. U., Guelli, U., & De Souza, S. M. A. (2018). Cellulase immobilization on poly(methyl methacrylate) nanoparticles by miniemulsion polymerization. *Brazilian Journal of Chemical Engineering*, *35*(2), 649–657. doi:10.1590/0104-6632.20180352s20160094

Srivastava, N., Srivastava, M., Mishra, P. K., Gupta, V. K., Molina, G., Rodriguez-Couto, S., Manikanta, A., & Ramteke, P. W. (2018). Applications of fungal cellulases in biofuel production: Advances and limitations. *Renewable & Sustainable Energy Reviews*, *82*, 2379–2386. doi:10.1016/j.rser.2017.08.074

Srivastava, N., Srivastava, M., Mishra, P. K., Singh, P., & Ramteke, P. W. (2015). Application of cellulases in biofuels industries: An overview. *Journal of Biofuels and Bioenergy*, *1*(1), 55–63. doi:10.5958/2454-8618.2015.00007.3

Tiwari, R., Nain, L., Labrou, N. E., & Shukla, P. (2018). Bioprospecting of functional cellulases from metagenome for second generation biofuel production: A review. *Critical Reviews in Microbiology*, *44*(2), 244–257. doi:10.1080/1040841X.2017.1337713 PMID:28609211

Trbojević, J. N., Dimitrijević, A. S., Veličković, D. V., Gavrović-Jankulović, M., & Milosavić, N. B. (2013). Isolation of Candida rugosa lipase isoforms. *Chemistry & Industry*, *67*(5), 703–706. doi:10.2298/HEMIND120828113T

Trbojevic Ivic, J., Dimitrijevic, A., Milosavic, N., Bezbradica, D., & Drakulic, B. J. (2016). Assessment of the interacting mechanism between Candida rugosa lipases and hydroxyapatite and identification of the hydroxyapatite-binding sequence through proteomics and molecular modelling. *RSC Advances*, *41*(41), 34818–34824. doi:10.1039/C6RA07521E

Trbojević-Ivić, J., Milosavić, N., Dimitrijević, A., Gavrović-Jankulović, M., Bezbradica, D., Kolarski, D., & Veličković, D. (2017). Synthesis of medium-chain length capsinoids from coconut oil catalyzed by Candida rugosa lipases. *Food Chemistry*, *218*, 505–508. doi:10.1016/j.foodchem.2016.09.049 PMID:27719942

Trbojević Ivić, J., Veličković, D., Dimitrijević, A., Bezbradica, D., Dragačević, V., Gavrović Jankulović, M., & Milosavić, N. (2016). Design of biocompatible immobilized Candida rugosa lipase with potential application in food industry. *Journal of the Science of Food and Agriculture*, *96*(12), 4281–4287. doi:10.1002/jsfa.7641 PMID:26801832

Trbojević-Ivić, J., Veličković, D., Dimitrijević, A., & Milosavić, N. (2018). Immobilization and stabilization of microbial lipases. In H. Broeken (Ed.), *Lipases* (pp. 55–107). Nova Science Publishers, Inc.

Veličković, D., Milosavić, N., Bezbradica, D., Bihelović, F., Segal, A. M., Šegan, D., Trbojević, J., & Dimitijević, A. (2014). The specificity of α-glucosidase from Saccharomyces cerevisiae differs depending on the type of reaction: Hydrolysis versus transglucosylation. *Applied Microbiology and Biotechnology*, *98*(14), 6317–6328. doi:10.100700253-014-5587-9 PMID:24682477

Verma, N., & Kumar, V. (2021). Microbial conversion of waste biomass into bioethanol: Current challenges and future prospects. *Biomass Conversion and Biorefinery*. Advance online publication. doi:10.100713399-021-01824-z

Zanuso, E., Ruiz, H. A., Domingues, L., & Teixeira, J. A. (2022). Magnetic nanoparticles as support for cellulase immobilization strategy for enzymatic hydrolysis using hydrothermally pretreated corn cob biomass. *BioEnergy Research*. Advance online publication. doi:10.100712155-021-10384-z

Zoghlami, A., & Paës, G. (2019). Lignocellulosic biomass: Understanding recalcitrance and predicting H hydrolysis. *Frontiers in Chemistry*, *7*, 874. Advance online publication. doi:10.3389/fchem.2019.00874 PMID:31921787

Žuža, M., Milosavić, N., & Knežević-Jugović, Z. (2009). Immobilization of modified penicillin G acylase on Sepabeads carriers. *Chemical Papers, 63*(2), 117–124. doi:10.2478/s11696-009-0012-z

ADDITIONAL READING

Golan, A. E. (Ed.). (2011). *Cellulase: Types, and action, mechanism, and uses*. Nova Science Publishers.

Guisan, J. M., Lopez-Gallego, F., Bolivar, J. M., Rocha-Martin, J., & Fernandez-Lorente, G. (2020). The science of enzyme immobilization. In *Immobilization of enzymes and cells: Methods and protocols* (pp. 1–26). Humana Press. doi:10.1007/978-1-0716-0215-7_1

Klein-Marcuschamer, D., Oleskowicz-Popiel, P., Simmons, B. S., & Blanch, H. W. (2012). The challenge of enzyme cost in the production of lignocellulose biofuels. *Biotechnology and Bioengineering*, *109*(4), 1083–1087. doi:10.1002/bit.24370 PMID:22095526

Ohmiya, K., Sakka, K., Karita, S., & Kimura, T. (1997). Structure of cellulases and their applications. *Biotechnology & Genetic Engineering Reviews*, *14*(1), 365–414. doi:10.1080/02648725.1997.10647949 PMID:9188160

Olkiewicz, M., Tylkowski, B., Montornés, J. M., Garcia-Valls, R., & Gulaczyk, I. (2020). *Modelling of enzyme kinetics: Cellulose enzymatic hydrolysis case*. Physical Sciences Reviews. doi:10.1515/psr-2020-0039

Payne, C. M., Knott, B. C., Mayes, H. B., Hansson, H., Himmel, M. E., Sandgren, M., Stahlberg, J., & Beckham, G. T. (2015). Fungal cellulases. *Chemical Reviews*, *115*(3), 1308–1448. doi:10.1021/cr500351c PMID:25629559

Perez, S., & Mazeau, K. (2005). Conformations, Structures, and Morphologies of Celluloses. In S. Dimitriu (Ed.), *Polysaccharides* (pp. 41–68). Marcel Decker, Inc.

Rajnish, K. N., Samuel, M. S., Ashwini, J. J., Datta, S., Chandrasekar, N., Balaji, R., Jose, S., & Selvarajan, E. (2021). Immobilization of cellulase enzymes on nano and micro-materials for breakdown of cellulose for biofuel production - A narrative review. *International Journal of Biological Macromolecules*, *182*, 1793–1802. doi:10.1016/j.ijbiomac.2021.05.176 PMID:34058212

Singh, N., Dhanya, B. S., & Verma, M. L. (2020). Nano-immobilized biocatalysts and their potential biotechnological applications in bioenergy production. *Materials Science for Energy Technologies*, *3*, 808–824. doi:10.1016/j.mset.2020.09.006

Wyman, C. E. (Ed.). (1996). *Handbook on bioethanol: Production and utilization*. Taylor & Francis.

KEY TERMS AND DEFINITIONS

Cellulase: A multi-enzyme complex, consisting of endoglucanase, exoglucanase, and β-glucosidase, which act synergistically for efficient hydrolysis of cellulose to glucose.

Enzyme: A protein of unique structure, which accelerates chemical reactions in biological systems with high substrate specificity. Also referred to as biocatalyst.

Enzyme Immobilization: Confinement of enzyme in a limited, well-defined space, with preserved catalytic activity, resulting in enzyme stabilization, activation, and easy continuous use.

Enzyme Reusability: Continuous use of enzyme in more than one reaction cycle without significant loss in activity. The most significant parameter of efficiency of the enzyme immobilization method.

Fermentable Sugar: Mixture of monosaccharides, obtained after hydrolysis of carbohydrate polymers in lignocellulose biomass. They are converted to bioethanol by metabolic machinery of specialized microorganisms in well-defined, tightly controlled conditions.

Lignocellulose Biomass: A complex material, composed of lignin, cellulose, and hemicellulose, which strengthens and protects plant cells from pathogens and mechanical damage. The most abundant and renewable raw material for production of bioethanol.

Nanobiocatalyst: An assembly of enzyme immobilized on nanomaterial with high catalytic efficiency, operational and storage stability, which can be easily removed from the reaction mixture and reused in the next reaction cycle.

Chapter 15
Comparative Studies on Neem and Jatropha Oil–Derived Biodiesels

Sunil Kulkarni
https://orcid.org/0000-0002-5988-3448
Gharda Institute of Technology, Lavel, India

Ajaygiri Goswami
University Institute of Chemical Technology, Jalgaon, India

Ghayas Usmani
University Institute of Chemical Technology, Jalgaon, India

ABSTRACT

Sustainable development is a key aspect of modern-day development. Reducing fuel cost and emissions is considered as the most important research area in sustainable development. Biodiesel has the potential to reduce the requirement of petroleum diesel if blended in suitable proportion. Various feedstocks are being explored for biodiesel production. Oils derived from neem and jatropha can be utilized for biodiesel production. In the current investigation, optimum parameters for the production, namely catalyst proportion, temperature, and oil-to-alcohol ratio were optimized. Experiments were carried out at different blend proportions to study fuel properties, namely kinematic viscosity, aniline point, diesel index, flash and fire points, specific and API gravity, cetane number, and ASTM distillation characteristics.

INTRODUCTION

Sustainable development is key aspect of modern-day development. Reducing fuel cost and emissions is considered as the most important research area in sustainable development. Most commonly fuels used across the world for vehicles, railways, boilers and majority of industrial applications are derived from fossil fuel. These fossil fuels are non-renewable. Alternative fuel sources are being explored such as wind and solar energy, hydrogen fuel, tidal energy and biodiesel from renewable feedstocks like

DOI: 10.4018/978-1-6684-5269-1.ch015

vegetable oil, cellulose, fats and chitosan. Viability of biodiesel synthesis depends on raw material and biodiesel properties. Waste food, vegetables, nonedible crops are suitable raw materials for biodiesel synthesis. Advantages of biodiesel include reduced toxicity, improved emission, rapid degradation and high flash point (Romano and Sorichetti, 2011). Predojevic et al. utilized Linoleic and oleic sunflower oils for biodiesel by (Predojevic et al.,2012). They used calcium oxide as solid base catalyst (Predojevic et al.,2012). Low-cost feed stocks are important for economic viability of biodiesel synthesis. Waste minimization through energy synthesis can serve the sustainable development goal (Ray et al., 2013; Stephane Bungener, 2016). Selection of appropriate catalyst is also very crucial aspect. Use of strong basic catalyst such as zeolite is being explored as promising method for transesterification (Ejikeme et al., 2010). Sodium methoxide and sodium hydroxide found to have good catalytic properties(Foon et al., 2004). Use of waste cooking oil for biodiesel can serve twin purpose, one of disposal and other of the raw material (Raqeeb and Bhargavi, 2015).

The fatty acid alkyl esters have characteristics very similar to diesel derived from fossil fuel (European Tech. and Innovation Platform, 2022). Acid or base catalysts can be used for the transesterification (Patel and Shah, 2015). The functional group of the catalyst attacks the carbonyl carbon. Catalytic and non-catalytic pathways are available for transesterification (Patel and Shah, 2015). Transesterification temperature depends on the raw material (Mekala et al., 2014). High temperature favors reaction but need to be controlled to avoid denaturation.

Neem is a tree in a mahogany family found in abundance in India (Elkadi et al., 2013). It has pesticidal and insecticidal application. Neem seeds find application in remedial therapy. Neem oil is extracted from the fruits and seeds of the neem. This neem oil is being explored for its use as a feedstock for synthesis as the alkyl ester by transesterification process. The composition of the oil depends on processing method. The neem oil yield that can be obtained from neem seed kernels varies widely from 25% to 45% (Anya et al.,2013; Sathya and Manivannan, 2013). Jatropha grows widespread in world in the tropical and subtropical areas. Jatropha oil finds applications in soapmaking. *Jatropha curcas* oil plays important role in cosmetic industry (Pandey et al., 2012). For transesterification, homogeneous catalysts such as NaOH, KOH can be employed (De Oliveira et al., 2009; Wang et al., 2006, Berchaman et al., 2010). The two-step method with an acid and basic-catalysed reactions can be used to produce biodiesel from oil containing high amount of free fatty acid (FFA) such as Jatropha oil (Choudhury, 2013; Ngo et al., 2008). Before the investigations, parameters for the production, namely catalyst proportion, temperature, oil to alcohol ratio were optimized. Experiments were carried out at these optimized parameters with different blend proportions to study fuel properties namely kinematic viscosity, aniline point, diesel index, flash and fire points, specific and API gravity, cetane number and ASTM distillation characteristics.

METHODOLOGY

Oil Extraction

Neem Oil

Transesterification by alkali catalyst is one of the most effective methods (Ibrahim et al., 2021.; Ramasubramanian *et al.,* 2017). Hexane was used as solvent and Soxhlet apparatus was used for extraction. 200 ml of hexane and neem powder (70 gm thimble) was taken in Whitman paper. Round bottom flask and condenser were installed in Soxhlet apparatus. The solvent was heated to maintain the temperature of 65-70 ^{0}C. As the liquid boils, the vapors are condensed. The condensate of n-hexane falls down and wets the thimble. n-hexane extracts the neem oil. The mixture n-hexane and oil is collected in the same round bottom flask. Four cycles were repeated for extracting more neem oil. The oil was separated from the solvent by using distillation. This oil was used for the transesterification process. Oil was heated in a three neck round bottom flask to 110þC for removal of water and then cooled. 0.5 gm of KOH and 300 gm of methanol were added to the oil. The mixture was stirred to blend with KOH. The oil was warmed to 60þC & mixed with potassium methoxide. The reaction was carried out for 5 hours at temperature 67þC. The resultant mixture changed to light green shade. The fishy scent indicated formation of ester and glycerol layers. Traces of glycerol, unreacted methanol and impurities were removed by adding 200 ml of warm water. Finally, a clear light biodiesel was obtained. Neem oil biodiesel blends with diesel were prepared. The blends were analyzed for specific gravity, density, kinematic viscosity, aniline point, pour point, flash point and cetane number.

Jatropha Oil

Pellets of potassium hydroxide were taken in a measuring cylinder and methanol was added in required proportion. Proper blending of oil with the methanol was ensured by shaking, warming and stirring. Blending and warming was continued for 2 hrs at temperature of 64 ^{0}C. The mixture changes to a turbid orange tan color in 5-10 minutes. Then it changed to a reasonably transparent dark shade. Sweet smell indicated the finishing stage. Blending and warming was continued for another 60 minutes. Two distinct phases are formed once blending and warming is halted. In the ester phase, likely impurities are glycerol, unreacted methanol and traces of soap. These impurities are removed by adding 100 ml of warm water to the ester. Then it was passed through the calcium hydroxide pellet for removing moisture content properly. The other phase was glycerol as a byproduct. The jatropha ester was blended with petroleum diesel fuel in several proportions and the analysis was carried out. The blends were analyzed for acid value, specific gravity, density, kinematic viscosity, aniline point, pour point, flash point and cetane number.

As indicated in table 1, optimum parameters obtained from optimization experiments were used to carry out experiments for studying properties of oil derived biodiesels and their different blends.

Table 1. Optimum parameters

Parameter/Factor	Parameter Value
Neem oil	
process temperature	67°C
Catalyst concentration	0.5 gm
Oil: methanol ratio	1:6.5
Jatropha oil	
process temperature	64°C
Catalyst concentration	0.35 gm
Oil: methanol ratio	1:5.5

RESULTS AND DISCUSSION

Calorific values for neem oil biodiesel and jatropha oil biodiesels were 7439.91 kcal/kg and 9121.8 kcal/kg respectively. Blends of these biodiesels with petroleum diesel can be used with improved calorific values. The diesel fuel has calorific value 10800 kcal/kg. The BIS specified range is 9900-10,500 kcal/kg. Other fuel properties are discussed in following sections or biodiesels and their blends.

Kinematic Viscosity

The kinematic viscosity at 40 ^{0}C for high-speed diesel must 2.5 to 7.5 cSt (centistokes) as per the standard specification of BIS. For pure neem oil, kinematic viscosity was found to be 5.34 cSt, well within the permissible range. Pure diesel has kinematic viscosity 2.75 cSt while neem oil biodiesel, 6.32 cSt. The viscosity of 10%, 20%, 30%, 40%, 60%, 80% blends of methyl ester of pure neem with diesel are 3.28, 3.36, 3.48, 4.53, 5.15, and 6.12 cSt respectively(Figure1). Viscosity increased with increase in percentage of biodiesel in the blend. For pure biodiesel, viscosity was within the range, but high. To avoid improper ignition, lower blends (20%, 30%) having the kinematic viscosities bellow 4 cSt can be used. The kinematic viscosity of the jatropha oil methyl ester was 5.5 cSt, higher compared to diesel(2.7 cSt). The viscosities were observed to be 3.63, 3.76, 3.91, 4.81, 5.11 and 5.22cSt for 10%, 20%, 30%, 40%, 60% and 80% blends respectively (figure 1). First three blends had suitable viscosity for fuel application. It can be observed from the figure that pure biodiesel from neem oils had higher viscosity. Need oil biodiesel blends had lesser viscosity than corresponding jatropha oil biodiesel for 10, 20 and 30% biodiesel blends. For higher blends, jatropha oil biodiesel had lesser viscosity than neem oil biodiesel.

Figure 1. Kinematic viscosity for different blend compositions

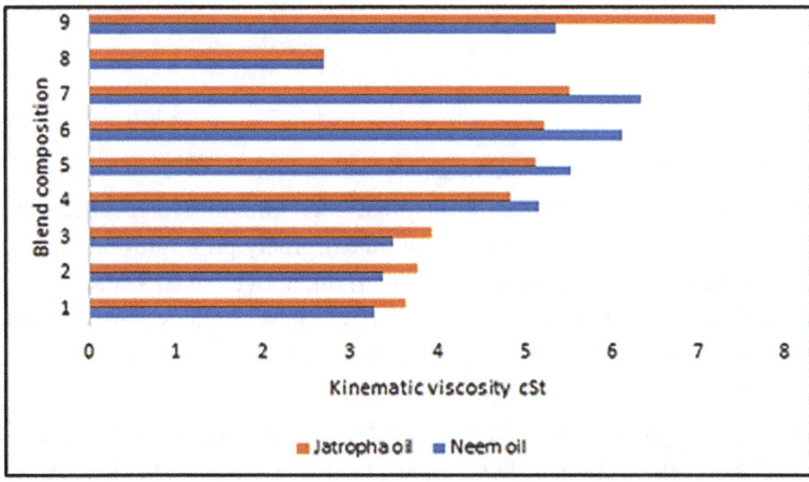

1	10% Biodiesel
2	20% Biodiesel
3	30% Biodiesel
4	40% Biodiesel
5	60% Biodiesel
6	80% Biodiesel
7	100% Biodiesel
8	100 % Diesel
9	Oil

Aniline Point

High aniline point indicates that the fuel is highly paraffinic and has high diesel index indicating good ignition quality. High speed diesel has an aniline point of 68 ^{0}C whereas 100% biodiesel has an aniline point of 77 ^{0}C. Blends of neem oil methyl ester and high speed diesel have a higher aniline point when compared to high speed diesel. It can be seen from Figure 2 that the aniline point increases with biodiesel percentage in the blend. Thus the blending is advantagious as per aniline point characterization.

In case of jatropha oil derived biodiesel, 100% biodiesel had high aniline point (134 ^{0}C). The blends of jatropha methyl ester and its blends had higher aniline point compared to high speed diesel. Aniline points for various blends have been shown in figure 2. Larger the aniline point, larger the diesel index, which leads to superior quality of ignition in diesel engine. When the fuel is rich in aromatic content, it has low values of aniline point and the ignition is poor. Jatropha oil derived biodiesel is better alternative from ignition point of view.

Figure 2. Aniline point for different blend compositions

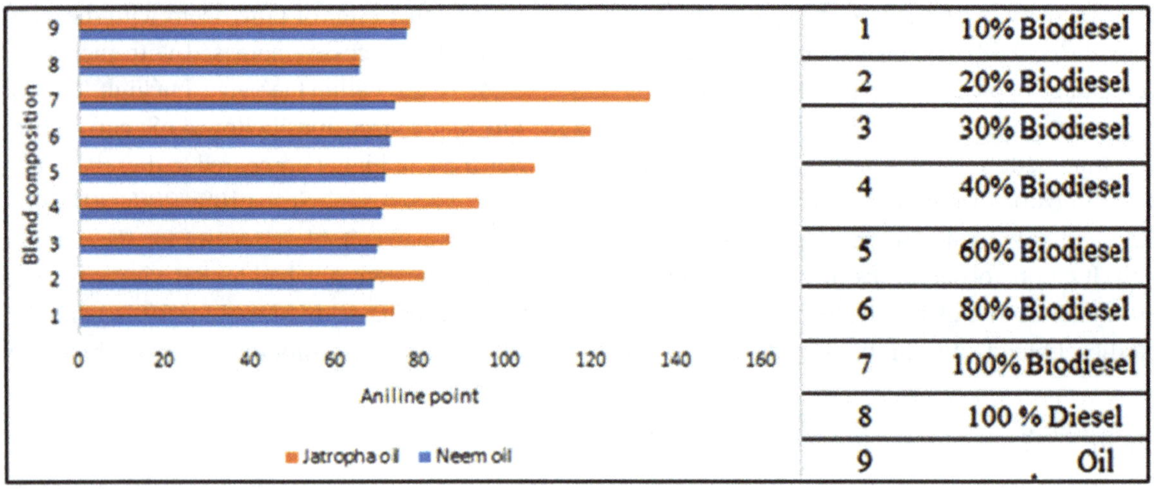

Diesel Index

The BIS specified value of diesel index is 45. The high speed diesel has a diesel index of 57.14. The diesel index of the neem oil is approximately 37.75 which is far below the prescribed limit of BIS specification. Consequently utilization of pure neem oil as a fuel will bring about an extremely poor ignition quality. A diesel index of 51.72 is obtained by using 100% methyl ester of neem oil, above the BIS specified of 45. By blending different amount of diesel and methyl ester from neem oil, diesel indices were found to decrease slightly. The diesel index for blends containing 10%, 20%, 30%, 40%, 60% and 80% biodiesel were 53.65, 57.40, 56.80, 56.50, 55.86 and 55.53 respectively, well above the specified value,45.

Pure jatropha biodiesel diesel index is approximately 81.03. For the different blends of 10%, 20%, 30%, 40%, 60% and 80% of methyl ester, the obtained diesel index values are above specified value and hence had better ignition property(figure3). Jatropha oil biodiesel was better choice than neem oil biodiesel considering diesel index.

Figure 3. Diesel index for different blend compositions

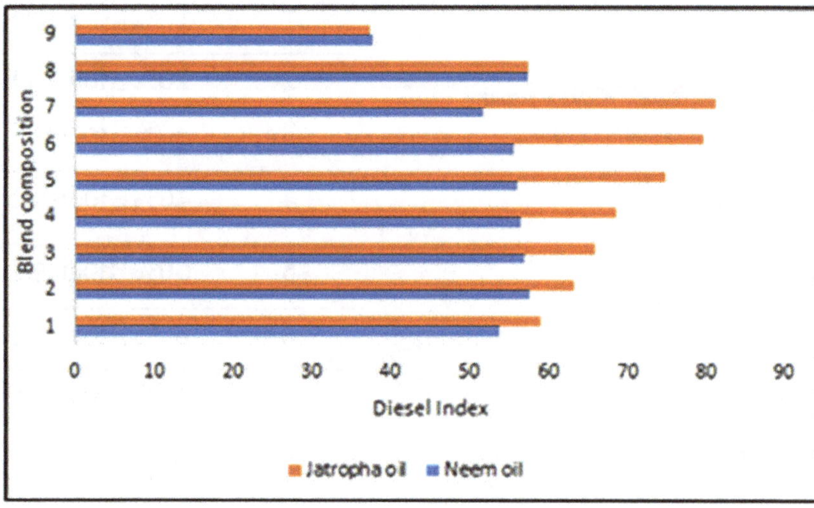

1	10% Biodiesel
2	20% Biodiesel
3	30% Biodiesel
4	40% Biodiesel
5	60% Biodiesel
6	80% Biodiesel
7	100% Biodiesel
8	100 % Diesel
9	Oil

Flash and Fire Point

The flash and fire points indicate possible fire hazards during transportation, handling and storage of diesel fuel. For the diesel fuel, the flash and fire points was 68 ^{0}C and 71^{0}C respectively. For 100% neem oil methyl ester, these value were 181^{0}C & 184^{0}C.As the blend percentage increases, the flash and fire point also increases. For 10, 20, 30, 40, 60 & 80%, the values of flash and fire point was obtained as 79 & 82, 89 & 92, 105 & 108, 126 & 128, 142 & 146, 162 & 165^{0}C respectively. The blending of biodiesel in petroleum diesel improves the flash point properties which is needed for better performance of fuel in terms of storage, handling and transportation.

Flash and fire points of jatropha biodiesel was 153 & 156 ^{0}C respectively. Increase in flash and fire point temperature is observed as jatropha biodiesel concentration in diesel blend increases. Flash and fire points for 10%, 20%, 30%, 40%, 60% and 80% blends were 81 ^{0}C & 84 ^{0}C, 87 ^{0}C & 89 ^{0}C, 92 ^{0}C & 95 ^{0}C, 98 ^{0}C & 101 ^{0}C, 118 ^{0}C & 120 ^{0}C, 130 ^{0}C & 133 ^{0}C respectively. The flash point study suggests that use of jatropha oil methyl ester in diesel will improve fuel efficiency. Jatropha oil biodiesel had lower flash and fire points compared to corresponding neem oil biodiesel blends. Figure 4 and 5 depicts flash and fire points for biodiesel derived from two oils.

Figure 4. Flash point for different blend compositions

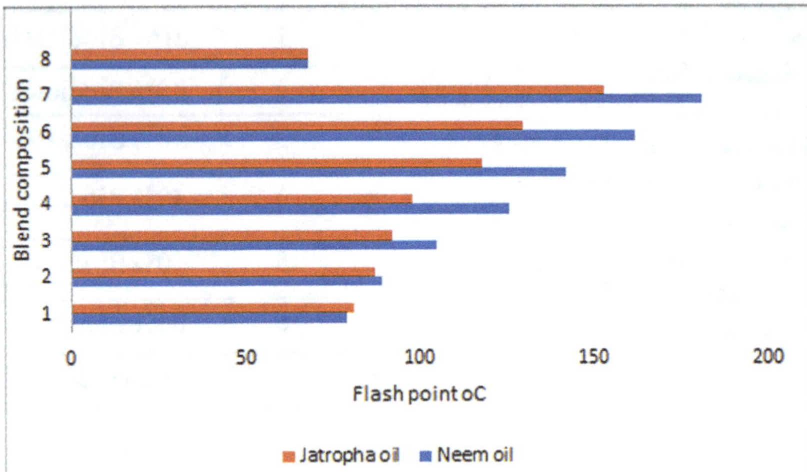

1	10% Biodiesel
2	20% Biodiesel
3	30% Biodiesel
4	40% Biodiesel
5	60% Biodiesel
6	80% Biodiesel
7	100% Biodiesel
8	100 % Diesel

Figure 5. Fire point for different blend compositions

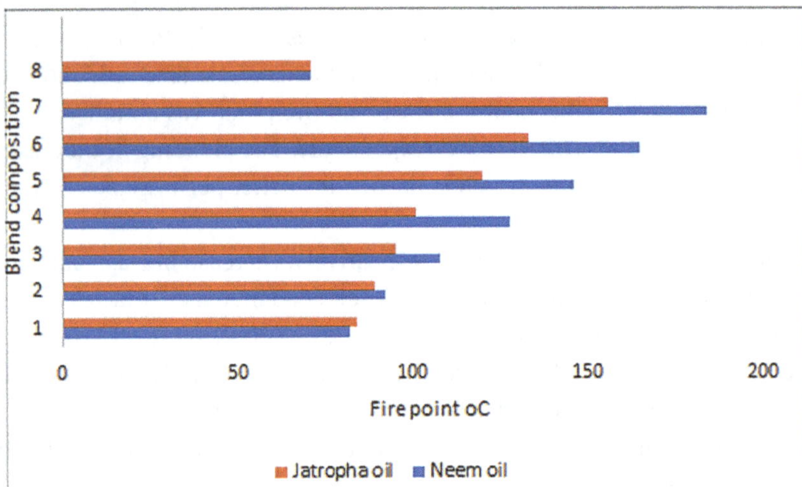

1	10% Biodiesel
2	20% Biodiesel
3	30% Biodiesel
4	40% Biodiesel
5	60% Biodiesel
6	80% Biodiesel
7	100% Biodiesel
8	100 % Diesel

Specific and API Gravity

The specific gravity of the diesel was noted 0.835. Its value for 100% jatropha oil methyl ester was 0.89 and that for neem oil biodiesel was 0.87. The specific gravity values increased marginally with the blends proportion. Values for 10, 20, 30, 40, 60 and 80% biodiesel blend in diesel were found to be 0.838, 0.842, 0.846, 0.855, 0.862, 0.872 and 0.839,0.841,0.845,0.848,0.852,0.855 respectively for jatropha and neem oil biodiesel(figure 6 and 7). The specific gravity values are used for calculating API gravity. High specific gravity indicates high diesel index.The API gravity and aniline point values are useful for calculating diesel index by well-known relation

Diesel Index=(0.018×Aniline point °C+0.32)×API gravity

For neem oil, The API gravity pure biodiesel was found 37.96 and 31.14.Here the reduction of API gravity with blend proportion rise were observed. Jatropha oil biodiesel was better option considering this property.

Figure 6. Specific gravity for different blend compositions

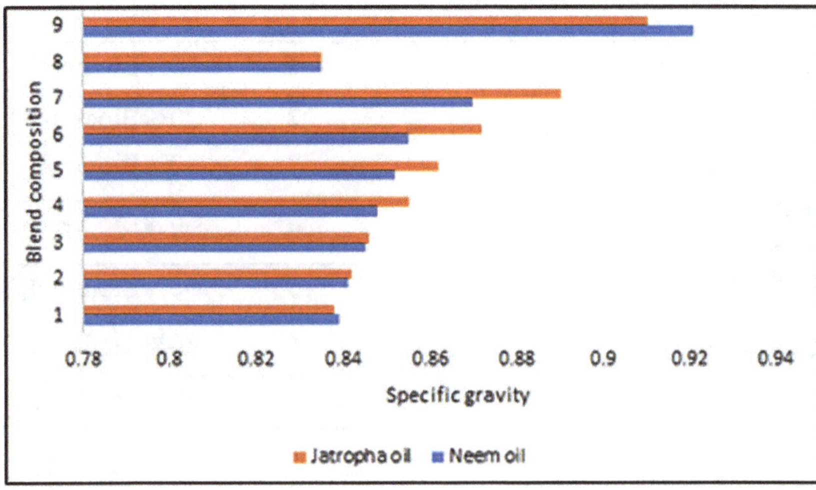

1	10% Biodiesel
2	20% Biodiesel
3	30% Biodiesel
4	40% Biodiesel
5	60% Biodiesel
6	80% Biodiesel
7	100% Biodiesel
8	100 % Diesel
9	Oil

Figure 7. API gravity for different blend compositions

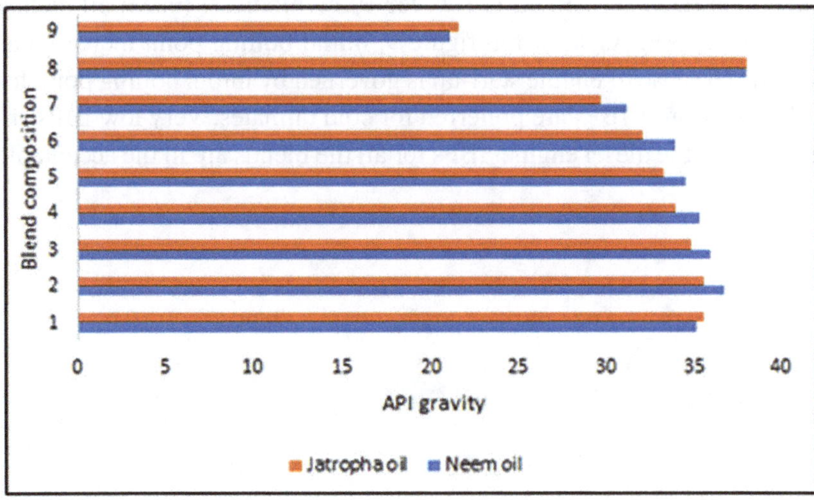

1	10% Biodiesel
2	20% Biodiesel
3	30% Biodiesel
4	40% Biodiesel
5	60% Biodiesel
6	80% Biodiesel
7	100% Biodiesel
8	100 % Diesel
9	Oil

Cetane Number

The Cetane number is directly proportional to diesel index. The cetane numbers for pure diesel and neem oil biodiesel were 50 and 47 respectively(figure 8). For jatropha oil biodiesel it was 67. As the blend percent increased from 10 to 80%, Cetane values decreased. Cetane numbers for neem oil biodiesel almost remained same for all blends. Cetane number for all blends was satisfactory compared to pure diesel.

Figure 8. Cetane number for different blend compositions

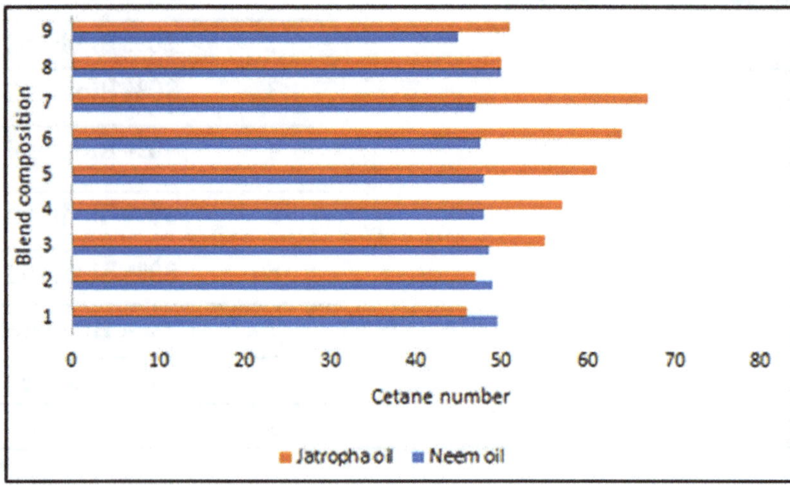

1	10% Biodiesel
2	20% Biodiesel
3	30% Biodiesel
4	40% Biodiesel
5	60% Biodiesel
6	80% Biodiesel
7	100% Biodiesel
8	100 % Diesel
9	Oil

Initial Boiling Point

Initial boiling point(IBP) was 92 ^{0}C for 100% diesel and 125 ^{0}C for 100% biodiesel (neem oil methyl ester). It was 132^{0}C for jatropha oil biodiesel. As shown in figure 9, initial boiling point increased as the proportion of biodiesel increased. The ease of engine start-up is governed by initial boiling point for the 10% recovery in ASTM distillation. Lower IBPs are preferred for cold climates. Very low IBPs are responsible for vapor locking tendency in the diesel engine. IBPs for all the blends are in the acceptable limits for diesel fuel.

Figure 9. Initial boiling point for different blend compositions

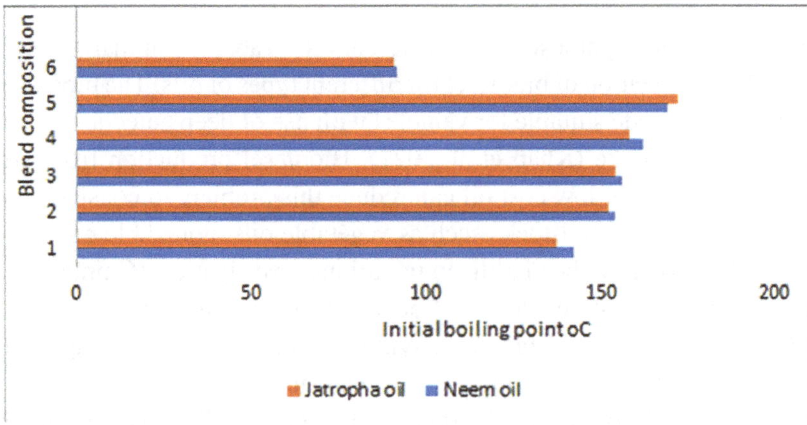

1	10% Biodiesel
2	20% Biodiesel
3	30% Biodiesel
4	40% Biodiesel
5	100% Biodiesel
6	100 % Diesel

Final Boiling Point

Need oil biodiesel blends for both 10% biodiesel and 20% biodiesel yielded 92% recovery and the final boiling points(FBP) 368 and 369^{0}C respectively indicating good distillation characteristics(Figure10). While for higher blends and pure neem and jatropha oil methyl esters, the FBP was not obtained due to cracking.

From the experimental study it was observed that for 10% jatropha biodiesel blend, the FBP was 363^{0}C with distillate recovery 92%, for 20% biodiesel blend, FBP was 365 ^{0}C with distillate recovery 92%(figure 10). For higher blends cracking was observed. As per ASTM distillation technique only 10% and 20% jatropha oil methyl ester blends in diesel is feasible.

Figure 10. Final boiling point for different blend compositions

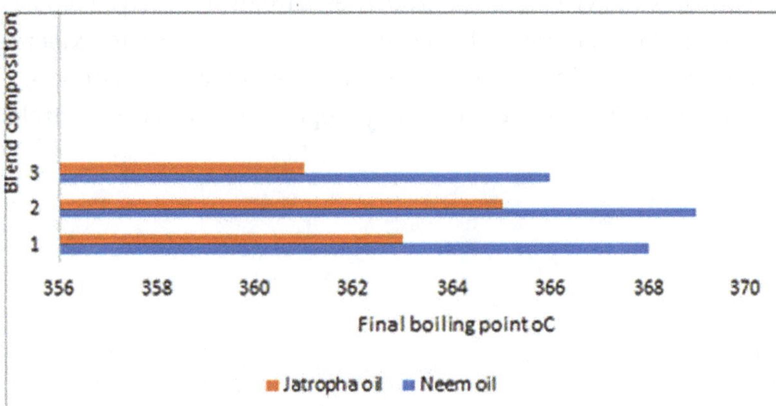

1	10% Biodiesel
2	20% Biodiesel
3	100 % Diesel

Comparison With Past Work

Many investigators have investigated biodiesel synthesis using different oils. Table 2 summarizes significant past work for synthesis and characterization of biodiesel for different types of oils. The biodiesel synthesized using waste cooking oil can be made suitable for vehicles with use of deep eutectic solvent (DES) in the local moderate temperature climate (Khan et al.,2021). Biodiesel production from edible oil faces challenges due to cost of oil. Use of low cost oil can reduce this problem to considerable extent (Abdelrahman, 2020; Shaah et al., 2021).Feedstocks such as vegetable oils, nonedible oils, oleaginous microalgae, fungi, yeast, and bacteria have been utilized in various investigations(Chintagunta et al., 2021). Management of waste frying oil can be additional benefit of its use as starting material for biodiesel synthesis(Topi, 2020).Various aspects of biodiesel synthesis such as process design and simulation, sustainability evaluation, optimization and supply chain management needs to be studied and investigated further to render viability to the oil derived biodiesel(Pasha et al., 2021).There is higher emphasis for sustainability in fuel synthesis in the production of the third generation of biofuels(Vignesh et al, 2021). Advanced oil extraction technologies for biodiesel along with supercritical and microwave-assisted transesterification and novel catalysts are being explored(Zulqarnain et al., 2021; Jamil et al., 2021; Pikula et al., 2020).

CONCLUSION

Utilizing 100% methyl ester of neem oil or jatropha oil in diesel motor is unrealistic. Studies in recent past indicate that provide up to 40% oil yield per seed based on weight(Riayatsyah et al., 2022).Blends of 10% and 20% methyl esters of neem oil and jatropha with diesel meet the kinematic viscosity, pour point, diesel index and ASTM refining temperatures according to BIS values. So utilizing 10% and 20% blends of methyl ester of neem oil or jatropha oil with diesel may reduce 20% diesel requirement of the nation. Blends of 30% methyl ester of with diesel meet the kinematic viscosity(neem and jatropha oil biodiesel) and pour point(neem oil biodiesel, ASTM refining properties(jatropha oil derived biodiesel) according to BIS. However, it doesn't meet the diesel index(neem oil derived biodiesels), kinematic viscosity(jatropha oil biodiesel) and ASTM refining properties(neem and jatropha oil biodiesel). Use of 20% biodiesel blends of both types of biodiesels(neem oil and jatropha oil) can reduce petroleum diesel requirement by 20%.

Table 2. The past work for synthesis and characterization of biodiesel for different types of oils

Oil	Parameters			Biodiesel Characterization				Authors
	Oil: Alcohol Ratio	Catalyst Loading	Temperature and Time	Specific Gravity	Kinematic Viscosity	Flash point /fire point	Cetane Number/ Calorific Value	
Waste cooking oil (WCO)	1:8	1% by wt.	reaction time 90 minutes	----------	----------	----------	----------	Elkady et al., 2015
Second-used cooking oil	1:6	1%	temperature 66.5 °C	0.90	4.9 cSt	flash point 180.5 °C	----------	Kawentara and Budiman 2013
Waste cooking oil (WCO)	1:8	1%	reaction time90 minutes	----------	----------	--------	----------	Degfie et al., 2019
Canola cooking oil	1:1	0.5%	temperature 55 °C, time 160 minutes	----------	----------	--------	----------	Hossain and Mekhled, 2010
Soyabean oil	1:6	0.5%	reaction temperature 60 °C, 100 minutes reaction time	----------	----------	--------	----------	Churasia et al. 2015
Soyabean oil	1:6	1%	reaction temperature 40 °C, 180 minutes reaction time	----------	----------	--------	----------	Hossain and Al-saif 2010
Sunflower oil		1 gm/l	Temperature 49 °C	0.68	6 cSt	----------	47	Thirumarimurugan et al. 2012
Rapeseed	1:9	0.7 w/w	60 to 80 °C temperature, 90 to 100 minutes time	0.86 to 0.87	4.7 to 4.9 cSt	flash point 175 to 178 °C	43-44	Encinaret al. 2018
Palm oil	1:6	1%	60 to 80 °C temperature	0.85 to 0.89,	3.05 to 6.2 cSt	flash point from 56 to 164 °C	Calorific value 42800 to 38050 kJ/kg	Mamilla et al.2012
Jatropha oil	1:6	1%,	reaction temperature 20 to 25°C (room temperature)	0.87	4 to 5.3 cSt	flash point 93 to 95°C	48-50	Kywe and Oo 2009
Cotton seed oil		------------	-----------	0.89	3.6 cSt	flash point 155 °C	50.4, Calorific value 40000 kJ/kg	Singh et al.2015

REFERENCES

Abdelrahman, M., Yassin, A. A., Hussein, I. H., Mirghani, M. E. S., & Karama, A. B. (2020). Biodiesel Production from Vegetable Oil: A Review. *J Marine Biol Environ Sci*, *2*(1), 1–11.

Anya, A. U., Chioma, N. N., & Obinna, O. (2012). Optimized reduction of free fatty acid content on neem seed oil for biodiesel production. *J. Basic. Appl. Chem.*, *2*, 21–28.

Berchamans, H. J., Morishita, K., & Takarada, T. (2010). Kinetic study of hydroxide-catalyzed methanolysis of Jatropha curcas-waste food oil mixture for biodiesel production. *Fuel*, *104*, 46–52. doi:10.1016/j.fuel.2010.01.017

Chintagunta, A.D., Zuccaro, G., Kumar, M., Kumar, S.P.J., Garlapati, V.K., Postemsky, P.D., Kumar, N.S.S., Chandel, A.K., & Simal-Gandara, J. (2021). Biodiesel production from lignocellulosic biomass using oleaginous microbes: prospects for integrated biofuel production. *Front. Microbiol., 12*, 658284. doi:.2021.658284 doi:10.3389/fmicb

Choudhury, H. A., Malani, R. S., & Moholkar, V. S. (2013). Acid catalysed biodiesel synthesis from Jatropha oil: Mechanistic aspects of ultrasonic intensification. *Chemical Engineering Journal, 231*, 262–272. doi:10.1016/j.cej.2013.06.107

Churasia, A., Singh, J., & Kumar, A. (2016). Production of biodiesel from soybean oil biomass as renewable energy source. *Journal of Environmental Biology, 37*, 1303–1307. PMID:29261252

De Oliveira, J. S., Leite, P. M., de Souza, L. B., Mello, V. M., Silva, E. C., Rubim, J. C., Meneghetti, S. M. P., & Suarez, P. A. Z. (2009). Characteristics and composition of Jatropha gossypiifolia and Jatropha curcas L. oils and application for biodiesel production. *Biomass and Bioenergy, 33*(3), 449–453. doi:10.1016/j.biombioe.2008.08.006

Degfie, T. A., Mamo, T. T., & Mekonnen, Y. S. (2019). Optimized biodiesel production from waste cooking oil (WCO) using calcium oxide (CaO) nano-catalyst. *Scientific Reports, 9*(1), 18982. doi:10.103841598-019-55403-4 PMID:31831823

Ejikeme, P. M., Anyaogu, I. D., Ejikeme, C. L., Nwafor, N. P., Egbuonu, C. A. C., Ukogu, K., & Ibemesi, J. A. (2010). Catalysis in biodiesel production by transesterification processes-an insight. *E-Journal of Chemistry, 7*(4), 1120–1132. doi:10.1155/2010/689051

Elkadi, J. M., Pillay, A. E., Manuel, J., Stephen, S., & Khan, M. Z. (2013). Kinetic study of neem biodiesel producton. *British Biotechnology Journal, 3*(4), 500–508. doi:10.9734/BBJ/2013/4434

Elkady, M. F., Zaatout, A., & Balbaa, O. (2015). Production of biodiesel from waste vegetable oil via KM micromixer. *Journal of Chemistry, 2015*, 1–9. doi:10.1155/2015/630168

Encinar, J. M., Pardal, A., Sanchez, N., & Nogales, S. (2018). Biodiesel by transesterification of rapeseed oil using ultrasound: A kinetic study of base-catalysed reactions. *Energies, 11*(2229), 1–13. doi:10.3390/en11092229

Europian Tech. and Innovation Platform. (2022). *Trans esterification to biodiesel*. http://www. etip-bioenergy. eu/value-chains/conversion-technologies/convention al-technologies/ transesterifi cation-to-biodiesel 22

Foon, C. S., May, C. Y., Ngan, M. A., & Hock, C. C. (2004). Kinetics study on transesterification of palm oil. *Journal of Oil Palm Research, 16*(2), 19–29.

Hossain, A. B. M. S., & Al-saif, A. M. (2010). Biodiesel fuel production from soybean oil waste as agricultural bio-resource. *Australian Journal of Crop Science, 4*(7), 538–542.

Hossain, A. B. M. S., & Mekhled, M. A. (2010). Biodiesel fuel production from waste canola cooking oil as sustainable energy and environmental recycling process. *Australian Journal of Crop Science, 4*(7), 543–549.

Ibrahim, I. S., Abdullahi, I. T., & Muhammad, F. Y. (2021). Comparative analyses of biodiesel produced from neem and jatropha seed oil. *Bayero Journal of Pure and Applied Sciences*, *12*(2), 141–143. doi:10.4314/bajopas.v12i2.20

Jamil, F., Kumar, P.S.M., Al-Haj, L., Myint, M.T.Z., & Al-Muhtaseb, A.H. (2021). Heterogeneous carbon-based catalyst modified by alkaline earth metal oxides for biodiesel production: Parametric and kinetic study. *Energy Convers. Manag. X 2020, 10*, 10047. doi:10.1016/j.ecmx.2020.100047

Kawentara, W. A., & Budiman, A. (2013). Synthesis of biodiesel from second-used cooking oil. *Energy Procedia*, *32*, 190–199. doi:10.1016/j.egypro.2013.05.025

Khan, N., Park, S. H., Kadima, L., Bourdeau, C., Calina, E., Edmunds, C. W., & Pursell, D. P. (2021). Locally sustainable biodiesel production from waste cooking oil and grease using a deep eutectic solvent: Characterization, thermal properties, and blend performance. *ACS Omega*, *6*(13), 9204–9212. doi:10.1021/acsomega.1c00556 PMID:33842789

Kywe, T. T., & Oo, M. M. (2009). Production of biodiesel from jatropha oil (jatropha curcas) in pilot plant. *Proceedings of World Academy of Science, Engineering and Technology*, *38*, 481–489.

Mamilla, V. R., Mallikarjun, M. V., & Rao, G. L. (2012). Biodiesel Production from Palm Oil by Transesterification Method. *International Journal of Current Research*, *4*(8), 83–88.

Mekala, N. K., Potumarthi, R., Rajubaadhe, R., & Gupta, V. K. (2014). Current Bioenergy Researches: Strengths And Future Challenges. *Bioenergy Research: Advances And Applications*, *1*, 1–21. doi:10.1016/B978-0-444-59561-4.00001-2

Ngo, H. L., Zafiropoulos, N. A., Foglia, T. A., Samulski, E. T., & Lin, W. (2008). Efficient two-step synthesis of biodiesel from grease. *Energy & Fuels*, *22*(1), 626–634. doi:10.1021/ef700343b

Pandey, V. C., Singh, K., Singh, J. S., Kumar, A., Singh, B., & Singh, R. P. (2012). Jatropha curcas: A potential biofuel plant for sustainable environmental development. *Renewable & Sustainable Energy Reviews*, *16*(5), 2870–2883. doi:10.1016/j.rser.2012.02.004

Pasha, M. K., Dai, L., Liu, D., Guo, M., & Du, W. (2021). An overview to process design, simulation and sustainability evaluation of biodiesel production. *Biotechnology for Biofuels*, *14*(1), 129. Advance online publication. doi:10.118613068-021-01977-z PMID:34074344

Patel, K., & Shah, S. N. (2015). Biodiesel from plant oils. In *Food, Energy, and Water.* Elsevier. https: / /Doi. Org/10.1016/B978-0-12-800211-7.00011-9. 23.

Pikula, K., Zakharenko, A., Stratidakis, A., Razgonova, M., Nosyrev, A., Mezhuev, Y., Tsatsakis, A., & Golokhvast, K. (2020). The advances and limitations in biodiesel production: Feedstocks, oil extraction methods, production, and environmental life cycle assessment. *Green Chemistry Letters and Reviews*, *13*(4), 275–294. doi:10.1080/17518253.2020.1829099

Predojevic, Z., Skrbic, B., & Durisic-Mladenovicn, N. (2012). Transesterification of linoleic and oleic sunflower oils to biodiesel using Cao as a solid base catalyst. *Journal of the Serbian Chemical Society*, *77*(6), 815–832. doi:10.2298/JSC110824206P

Ramasubramanian, S., Manavalan, S., Gnanavel, C., & Balakrishnan, G. (2017). Processing of neem and jatropha methyl esters –alternative fuels from vegetable oil. *IOP Conf. Series: Materials Science and Engineering*, *183*, 012036. 10.1088/1757-899X/183/1/012036

Raqeeb, M. A., & Bhargavi, R. (2015). Biodiesel production from waste cooking oil. *Journal of Chemical and Pharmaceutical Research*, *7*(12), 670–681.

Ray, N. H. S., Mohanty, M. K., & Mohanty, R. C. (2013). Anaerobic digestion of kitchen wastes: Biogas production and pre-treatment of wastes, a review. *International Journal of Scientific and Research Publications*, *3*(11), 1–6.

Riayatsyah, T. M. I., Sebayang, A. H., Silitonga, A. S., Padli, Y., Fattah, I. M. R., Kusumo, F., Ong, H. C., & Mahlia, T. M. I. (2022). Current progress of jatropha curcascommoditisation as biodiesel feedstock: A comprehensive review. *Frontiers in Energy Research*, *9*, 815416. doi:10.3389/fenrg.2021.815416

Romano, S.D., & Sorichetti, P.A. (2011). Dielectric spectroscopy in biodiesel production and characterization. *Green Energy and Technology*, 7-27. doi:10.1007/978-1-84996-519-4_2

Sathya, T., & Manivannan, A. (2013). Biodiesel production from neem oil using two step transesterification. *International Journal of Engineering Research and Applications*, *3*, 488–492.

Shaah, M. A. H., Hossain, M. S., Allafi, F. A. S., Alsaedi, A., Ismail, N., Kadir, M. O. A., & Ahmad, M. I. (2021). A review on non-edible oil as a potential feedstock for biodiesel: Physicochemical properties and production technologies. *RSC Advances*, *11*(40), 25018–25037. doi:10.1039/D1RA04311K PMID:35481051

Singh, S., Sharma, S., & Mohapatra, S. K. (2015). A Production Of Biodiesel from Waste Cotton Seed Oil and Testing on Small Capacity Diesel Engine. *International Journal of Advance Research In Science And Engineering*, *4*(2), 172–176.

Stephane Bungener, L. G. (2016). *Energy efficiency and integration in the refining and petrochemical industries* [PhD thesis].

Thirumarimurugan, M., Sivakumar, V. M., Xavier, A. M., Prabhakaran, D., & Kannadasan, T. (2012). Preparation of biodiesel from sunflower oil by transesterification. *International Journal of Bioscience, Biochemistry, Bioinformatics*, *2*(6), 441–445. doi:10.7763/IJBBB.2012.V2.151

Topi, D. (2020). Transforming waste vegetable oils to biodiesel, establishing of a waste oil management system in Albania. *SN Applied Sciences*, *2*(513), 1–7. doi:10.100742452-020-2268-4

Vignesh, P., Kumar, A. R. P., Ganesh, N. S., Jayaseelan, V., & Sudhakar, K. (2021). Biodiesel and green diesel generation: An overview. *Oil & Gas Science and Technology – Rev. IFP Energies Nouvelles*, *76*, 1–15.

Wang, Y., Ou, S., Liu, P., Xue, F., & Tang, S. (2006). Comparison of two different processes to synthesize biodiesel by waste cooking oil. *Journal of Molecular Catalysis A Chemical*, *252*(1-2), 107–112. doi:10.1016/j.molcata.2006.02.047

Zulqarnain, A., Ayoub, M., Yusoff, M. H. M., Nazir, M. H., Zahid, I., Ameen, M., Sher, F., Floresyona, D., & Budi Nursanto, E. (2021). A comprehensive review on oil extraction and biodiesel production technologies. *Sustainability*, *13*(788), 1–28. doi:10.3390u13020788

Chapter 16
Characterization and Comparison of Various Blends of Honge Oil Methyl Ester (Biodiesel) With Diesel Fuel

Sunil Kulkarni
https://orcid.org/0000-0002-5988-3448
Gharda Institute of Technology, Lavel, India

Ajaygiri K. Goswami
University Institute of Chemical Technology, Jalgaon, India

Ghayas A. Usmani
University Institute of Chemical Technology, North Maharashtra University, Jalgaon, India

ABSTRACT

The Honge oil, known as pongamia oil, is one of the most widely used lamp oils around the world. It is also used in leather tanning, in soap making, and as a lubricant. The Honge oil methyl ester can be used as an alternative fuel for diesel. The transesterification of methanol and Honge oil in the presence of base catalyst results in Honge oil methyl ester. In the current work, the Honge oil methyl-ester known as biodiesel was produced. It was mixed with petroleum diesel fuel in different proportions. Various physical and petroleum properties were investigated for Honge oil biofuel and also for various diesel and ester mixtures. The optimum parameters were temperature was 65°C, catalyst quantity, 0.4 grams and oil: alcohol ratio, 1:6. The blends of 10%, 20%, and 30% compositions were most suitable as potential alternative fuel. Different proportions of biodiesel ester in diesel fuel have been also compared with petroleum diesel fuel to identify a suitable blend having good engine fuel potential.

DOI: 10.4018/978-1-6684-5269-1.ch016

INTRODUCTION

Worldwide economy runs on energy and with increase in population, the fuel demand is increasing rapidly (Natarajan and Venkatesh, 2012; Ahmad et al. 2009; Kumar et al., 2013; Patil et al., 2008; Hayyan et al., 2010). These needs, coupled with environmental changes, increasing costs and exhaustion of fossil fuel have motivated investigators to discover new energy assets (Georgianna and Mayfield, 2012; Ramadhas et al., 2004; A. Kumar and P. Kumar, 1984; Khan et al., 2007). Biodiesel has been considered to be one the best option forfuture fuel needs.Itcan be renewed because of its biodegradability (Bobade and Khyade, 2012; Santana et al., 2012; Ma and Hanna, 1999). Biodiesel creation from a renewable source can fulfill goals of a balanced environment and manageable monetary quality (Bobade and Khyade, 2012; Hossain and Salleh, 2008). Advantages of biodiesel are, its biodegradability and no requirement of alteration of engine to accommodate the fuel (Ahmad et al. 2009). Biodiesel can be smoothly blended (Anya et al., 2012; Saribiyik et al., 2010; Hribernik and Kegl, 2007). Any alternative fuel should be characteristically achievable, agreeable and financially feasible.

The source of Honge oil is the seeds of the Millettia pinnata tree. Karanja oil (in Hindi), Honge oil (in Kannada), Kanuga oil (in Telugu) and Pungai oil (in Tamil) are different terms used for this oil, Pongamia being the generic name. The Pongamia tree is around 40 feet tall after complete growth. Pink and white flower are grown on their branches. These mature and develop into brown seed pods (Allen O and Allen E., 1981). They grow in temperature range 0 to 40°C.Also they can survive in 5–25 cm annual rainfall. This tree grows wild on soil that is rough and hard (Meher, et al., 2004). The plant is nonedible, has harmful properties and can cause harmful effects on body if injected orally. However, it finds applications in the traditional medicines via its fruits, flowers and seeds (Bobade and Khyade, 2012). Honge plant derived oil and juices are effective on skin disease and have pest control abilities. Complete growth of this plant takes 10 years and it takes 5 years only to develop pods. A single tree can yield oil in the range of 10 to 50 kilograms depending on its size (Meher et al., 2004). Different methods like extraction and pressing (expeller and cold) can be used for Pongamia oil for its extraction from seeds. Pongamia oil is antiseptic and resistant to pests. The bitter flavonoid constituents results in disagreeable taste and odor and also has a high content of triglycerides.

In the recent years Pongamia oil industry has caught great attention in the global market. India, being one of the largest Pongamia producers, has a very important role in it. Pongamia pinnata (the Derris Indica trees) has been identified a definite source of next generation fuel. This plant can grow on barren and unused land. The plant has no food applications and hence its use as a source of biodiesel doesn't affect food requirement. High oil recovery is another important advantage of this source. All these advantages makes Honge oil, even more attractive alternative. This plant can minimize nitrogen requirement of soil as it can fix nitrogen from atmosphere.

Meng et al. (Meng et al., 2008) optimized the reaction temperature at 65^0C whereas Freedman et al. (Freedman et al.., 1986) suggested that optimal catalyst concentration is generally between 0.5 wt. % to 1.0 wt. %. The effect of ratio of oil to alcohol was also optimized experimentally at 1:6 by Alia and Taya (Alia and Taya, 2013). The major objective of the current investigation is to prepare the biodiesel from Honge oil and to study its petroleum properties, compare with pure petroleum diesel. Initially parametric studies were performed and parameters were optimized for experiments on different blend compositions. Prepared biodiesel and its blends were analyzed by comparing their properties with pure diesel. The words Honge oil and Pongamia oil are used in the investigation interchangeably.

MATERIAL AND METHOD

Material

Honge oil alkyl ester was synthesized with base catalyzed transesterification reaction. The oil was obtained from market and was provided by Jinendra Scientifics, Jalgaon. Potassium hydroxide used was manufactured by Hi-Pure Rankem Chemicals and Methanol provided was for laboratory use only of AR grade.

Method

The setup used for the synthesis of biodiesel is shown in figure 1. The synthesis of ester was carried out in the equipment using methanol and KOH. Optimum values of the parameters are indicated in table 1. By using waste vegetable oil(WVO), WVO oil to methanol ratio 1:12, 1% wt/wt of NaOH, reaction time 90 minutes were found to be optimum parameters by Elkady et al. (Elkady et al.,2015). Degfie et al. obtained 1:8 WCO(waste cooking oil) oil to methyl alcohol, 90 minutes reaction time, 0.01 weight fractionpotassium hydroxide loading rate as optimum parameters(Degfie et al., 2019). In another investigation, maximum yield was obtained at alcohol to oil molar ratio 6.18:1, 1% wt/wt KOH as catalyst and 66.5^{0}C(Kawentara and Budiman, 2013). The catalyst was properly shaken and mixed with the alcohol using stirrer. Proportion of hydroxide pellets and methanol was obtained after analysis of effect of various proportions on the yield. The Honge oil was exposed to high temperature to remove moisture content and mixed with methanol and potassium hydroxide mixture. The mixture was heated and agitatedfor 180 minutes and at 63 to 67 ^{o}C (Average 65 ^{o}C). The reaction mixture became turbid and changed to orange colour after 15 to 18 minutes. The colour became transparent with time. Completion of the reaction was identified from a sweet fragrance of the mixture. The mixture was then agitated and subjected to high temperaturefor 60 minutes more. The mixture was taken into a separating funnel to obtain layers ester and glycerol. Warm water was used to separate glycerol, soap and other impurities from the mixture. It was then treated with calcium hydroxide medium to minimize moisture level. The ester portion is the Honge oil biodiesel. The petroleum properties were studied of pure Honge oil ester and blends of this ester and diesel sample. The performance of blended samples was analyzed and compared with the pure diesel fuel properties.

Figure 1. Setup for biodiesel preparation

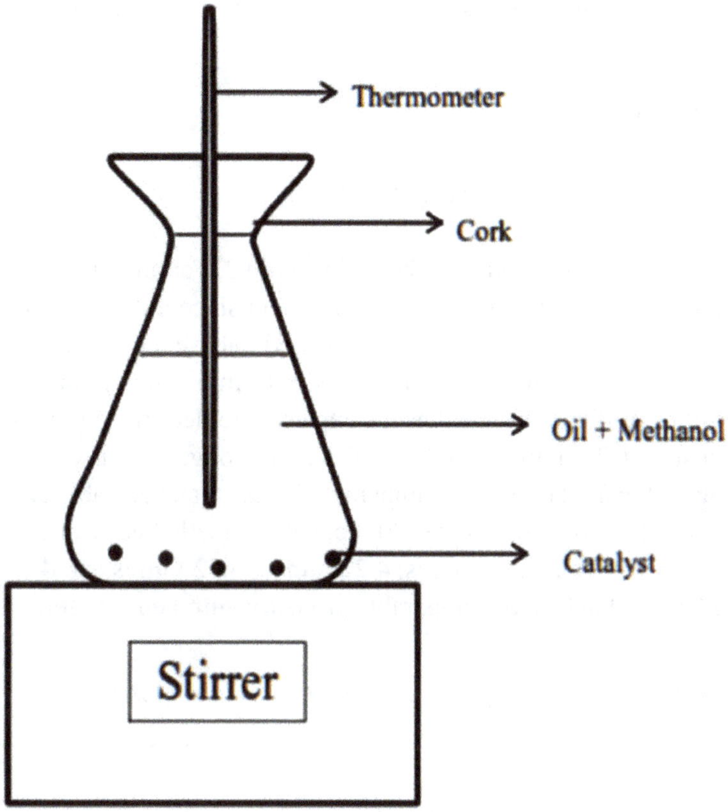

Table 1. Parametric optimization

Parameter/Factor	Optimum Value
catalyst	KOH, maximum yield 81.3%
time	3 hrs.
catalyst concentration	0.4 gm
process temperature	65 °C
oil-methanol ratio	1:6

Analysis of Biodiesel

The analysis of biodiesel and its various blends with diesel thus obtained is important in order to understand the properties and to compare them to diesel fuel in order to know its feasibility and viability. Various properties were studied from the analysis of biodiesel obtained from Honge oil. These include kinematic viscosities, diesel index, aniline point, flash and fire point, ASTM distillation properties, specific gravity, API gravity. The determinations of these properties and their values along with the methodologies have been described extensively in the result and discussion section.

RESULT AND DISCUSSION

Evaluating Kinematic Viscosity Requirement

Kinematic viscosities for different blends are shown in figure 2. For any fuel used in diesel engine, the desired range of viscosity is necessity and prime requirement since it directly affects the ignition characteristics. According to ASTM D6571, the liquid fuel need to have the viscosity in the specified limit. The limit for diesel for its efficient and safe use is viscosity 1.9mm^2/s minimum to 6.0 mm^2/s maximum. The kinematic viscosities of diesel and biodiesel samples were determined by viscometer. The diesel has a kinematic viscosity of 2.5 mm^2/s at 40^{0}C while pure Honge oil has a viscosity of 7.25 mm^2/s. Viscosity of the Honge oil methyl ester was comparatively less than vegetable Honge oil (5.03 mm^2/s). The viscosities of 10%, 20%, 30%, 40%, 60%, 80% blends of methyl ester of Honge with diesel were obtained as 2.91mm^2/s, 3.26 mm^2/s, 3.81mm^2/s, 4.22 mm^2/s, 4.52 mm^2/s and 4.87 mm^2/s respectively. Up to 30%, Honge oil ester blends have comparable viscosity with Petro diesel.

Figure 2. Change in kinematic viscosity for change in biodiesel composition

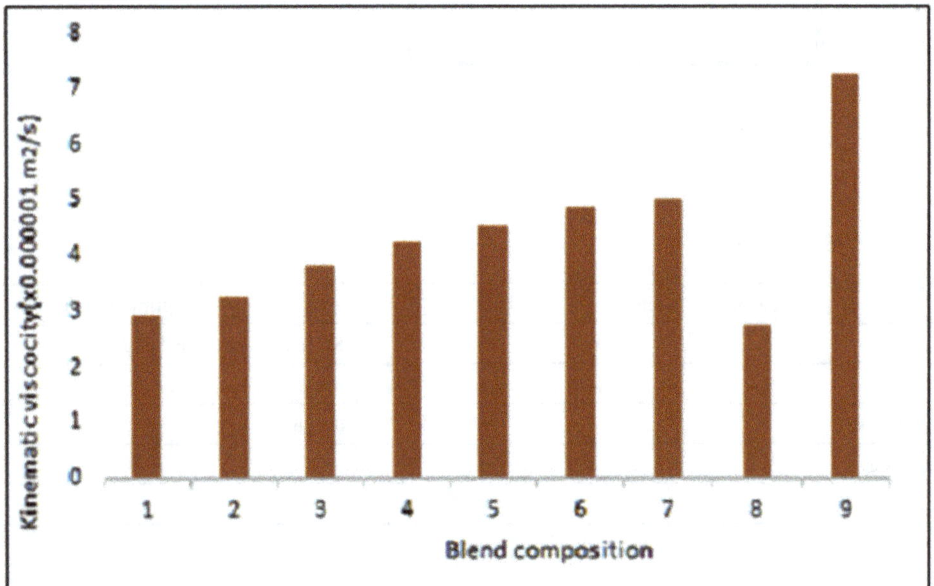

No.	1	2	3	4	5	6	7	8	9
Blend Composition	10% Biodiesel	20% Biodiesel	30% Biodiesel	40% Biodiesel	60% Biodiesel	80% Biodiesel	100% Biodiesel	100 % Diesel	Pongamia Oil

Evaluation of Ignition Quality With the Help of Diesel Index

Ignition property of the liquid fuel is well represented by the diesel index. The optimum value should be closed to 45 as per mentioned specification of BIS. By knowing aniline point and specific as well as API gravity, diesel index can be calculated as:

Diesel Index value ≐ Aniline point^{0}F x degree API Gravity / 100

Market purchased diesel fuel has a diesel index between 58 to 60 which is well above the specified value of 45. On the other hand, the diesel index of the Honge oil is approximately 16.2 which is far below the specification. Diesel index values for different blends are plotted in figure 3. A diesel index of 23.1 was observed for pure ester synthesized from Honge oil. Thus, use of pure Honge oil or its methyl ester in place of diesel will result in very poor ignition characteristics. Hence, Honge oil and its methyl ester alone doesn't have appropriate fuel characteristic from ignition perspective. It can be blended with diesel fuel. An addition of 10% methyl ester of Honge oil results in diesel index to 51.27 and 20% methyl ester of Honge oil has a diesel index of 48.1 which is above the specified value, indicating superior ignition property. For the different blends of 60% and 80% of methyl ester, the obtained diesel index is below the specified value and hence will result in very poor ignition property.

Figure 3. Change in Diesel Index for change in biodiesel composition

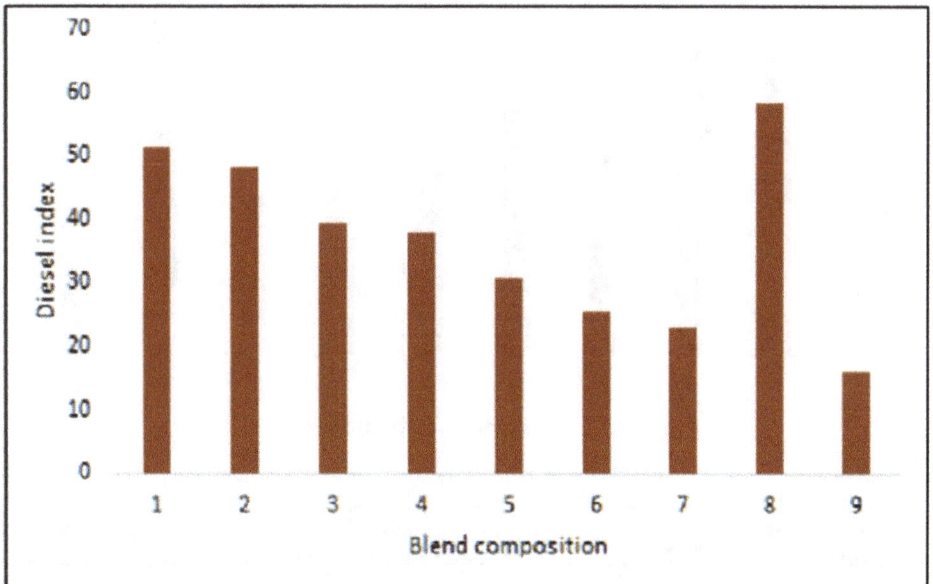

No.	1	2	3	4	5	6	7	8	9
Blend Composition	10% Biodiesel	20% Biodiesel	30% Biodiesel	40% Biodiesel	60% Biodiesel	80% Biodiesel	100% Biodiesel	100 % Diesel	Pongamia Oil

Evaluation of Aniline Point Property

Aniline point was determined to study fuel properties. A test tube with thermometer and stirrer was filled with dry sample and aniline. Sample and aniline mix completely. After cooling, the mixture become opaque. For determination of the aniline point, 5 ml of a carefully dried sample and aniline were filled in a test tube fitted with a thermometer and stirrer. On stirring, the sample and aniline are completely miscible. The apparatus is then cooled until the mixture is opaque. Then the temperature in increased slowly until the thermometer bulb is visible. Aniline point is the lowest temperature at which equal volume proportion mixtures of aniline and oil are completely miscible. Aniline point is indicator of aromatics present. Lower the aniline point, higher are aromatic compounds in the oil.

High values of aniline point temperature indicates that the fuel has high paraffinic hydrocarbon and hence has high diesel index, which is good ignition quality (figure 4). Determining aniline point is important in determining diesel index properties. Aniline point of diesel is 68 °C whereas that of 100% biodiesel is 29 °C. Aniline points of the blend of Honge oil methyl ester and high-speed diesel are lower compared to high-speed diesel. Compared to pure biodiesel obtained from Honge oil, these values were higher.

Figure 4. Change in Aniline point for change in biodiesel composition

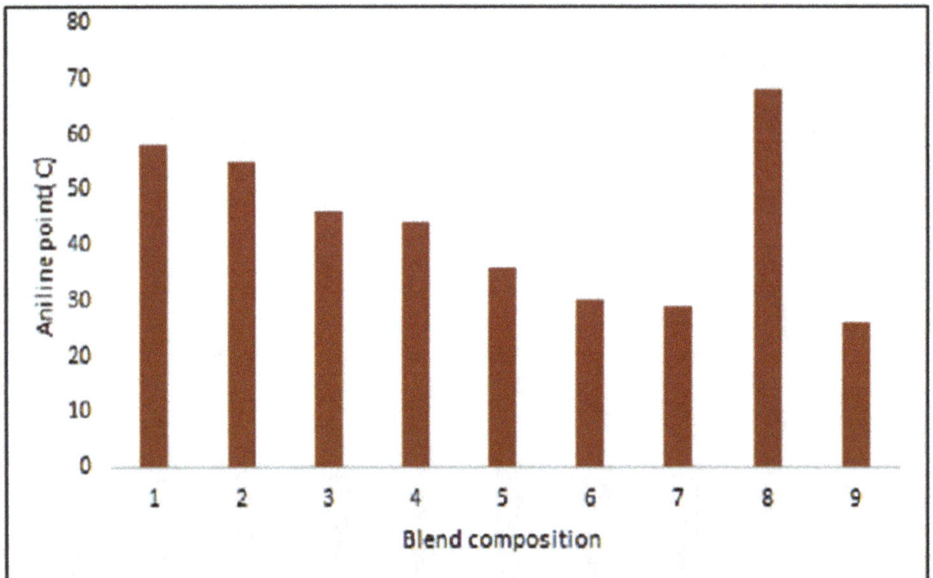

No.	1	2	3	4	5	6	7	8	9
Blend Composition	10% Biodiesel	20% Biodiesel	30% Biodiesel	40% Biodiesel	60% Biodiesel	80% Biodiesel	100% Biodiesel	100 % Diesel	Pongamia Oil

Evaluating Pour Point Properties

The pour point temperature evaluation is necessary for fuels to analyze their behavior in cold weather climatic conditions and their storage handling. The flow characteristics of liquid vanishes at pour point and liquid ceases to flow. A tube with flat bottom with dimensions 3 cm diameter and 12 cm height is used for measuring pour point characteristics. It is kept in air jacket. The jacked in placed in a jar that contains the freezing mixture (Ice + $CaCl_2$). The tube is filled up to 50 percent height. A thermometer is inserted in the tube through cork. The temperature falls with cooling due to freezing mixture. For every 1 ^{0}C drop, the mixture is taken out and observed for cloudiness. Cloudiness is indicator of decrease in flowability. After every 3^{0}C drop in temperature, the sample was titled to check oil flow characteristic. Pour point is identified when the oil doesn't flow, even after complete horizontal tilting of the tube for 5 seconds.The Honge oil biodiesel at the minimum temperature of ambient condition will be free flowing. Pour point of Honge oil was -5°C which is far below the requirements. Pour point of methyl ester of Honge oil (100%) and 10%, 20%, 30%, 40%, 60% and 80% blends of methyl ester of Honge oil with diesel were experimentally obtained as 2°C, 5°C, 4°C, 4°C, 4°C, 3°C and 3°C respectively. Thus, increasing amount of Honge oil biodiesel in diesel, decrease in the pour point temperature was estimated. So, values of any blend of pour point temperature can be used since all values are close to diesel pour point value which is 5°C.

Evaluating ASTM Distillation Properties

Using standard ASTM Distillation apparatus, the initial and final boiling points and the temperatures for various percent recoveries of distillates were determined. Initial boiling temperature for diesel was 88 ^{0}C while for Honge oil biofuel it was 128 ^{0}C. There was rise in initial boiling point for 10 to 40% Honge oil biofuel blends in diesel (figure 5). The initial boiling temperatures were 130^{0}C,140^{0}C,142^{0}C and 146^{0}C for 10 to 40% Honge oil biofuel blends respectively. The initial boiling temperature represents ease of starting of an engine. The lower value of initial boiling temperature is desirable especially in chilled weather climates. However, it causes vapor locking problems. Greater values of Honge oil biofuel blends protect the diesel engine from vapour locking. For the 10% blend of Honge oil biofuel, the final boiling temperature was 362^{0}C (recovery 92%), for 20% blend, the temperature was reduced to 348 ^{0}C at same distillate recovery and for 30% biofuel blend, maximum boiling temperature was found slightly higher 352 ^{0}C (recovery 90%). For the next proportion of 40 percent Honge oil biodiesel blend, 8% recovery of distillate was recorded at temperature 278^{0}C. For higher honge oil methyl ester blends, cracking phenomenon was observed. Only the lower blends of honge oil ester would have proper ignition in diesel engine. Figure 5 and 6 depict initial and final boiling points for different blends.

Figure 5. Change in Initial Boiling Point for change in biodiesel composition

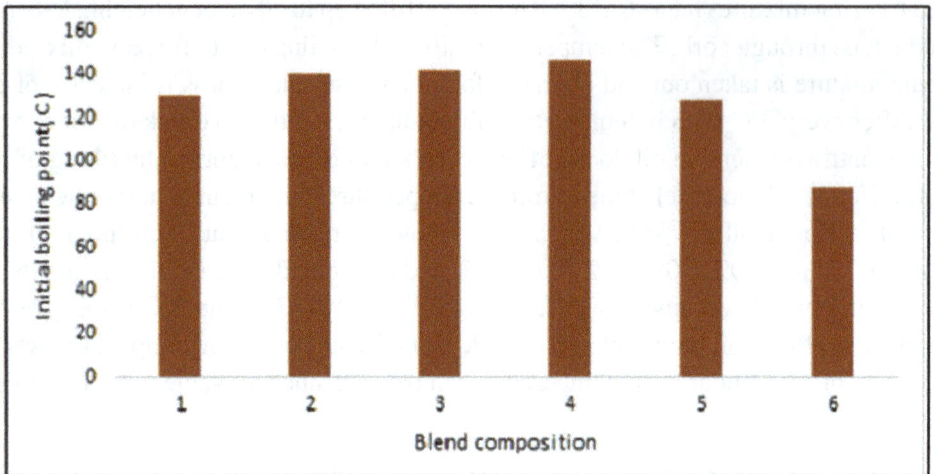

No.	1	2	3	4	5	6
Blend Composition	10% Biodiesel	20% Biodiesel	30% Biodiesel	40% Biodiesel	100 % Biodiesel	100% diesel

Figure 6. Change in Final Boiling Point for change in biodiesel composition

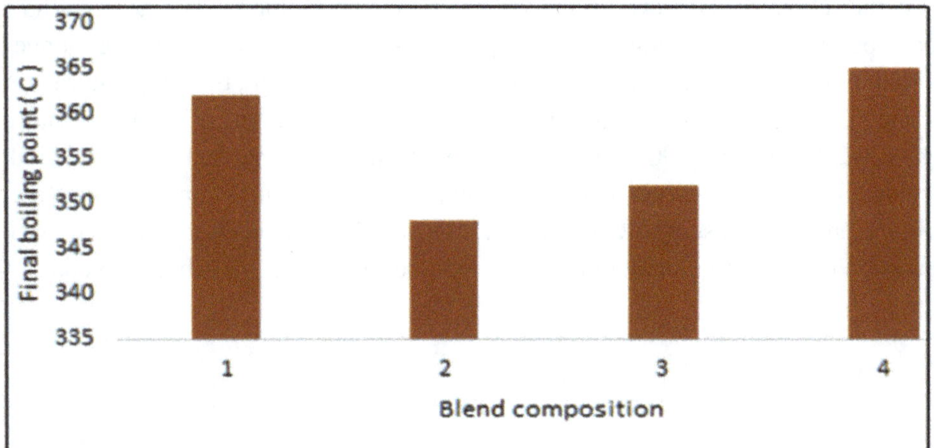

No.	1	2	3	4
Blend composition	10% Biodiesel	20% Biodiesel	30% biodiesel	100 % Diesel

Flash Point and Fire Point

Flash and fire points for different blends are shown in figure 7 and figure 8 respectively. The oil sample under test is filled up to the mark in the oil cup. It was heated in the air bath by a burner. Flame was introduced for a moment, for every degree rise in temperature. The flash point is identified by appearance of distinct sound and light. The test flame is applied and heating is continued for some more time. The fire point of the oil is obtained by observing the test flame which burns continuously for minimum five seconds. Flash point of oil is about 5 to 40 ^{0}C lower than its fire point. The diesel fuel flash and fire points by open cup method were 68^{0}Cand 70^{0}C respectively.

The pure Hong oil biodiesel had higher values of 130^{0}C and 134^{0}C respectively. The flash and fire point values increased as the Honge oil methyl ester blend percent in diesel increased by every ten percent like 78^{0}C and 81^{0}C, 84^{0}C and 87^{0}C, 91^{0}C and 95^{0}C, 98^{0}C and 101^{0}C, 112^{0}Cand115^{0}C, 125^{0}C and135^{0}C respectively. Flash and fire points are indicators of possible fire hazard during the transportation of the fuel material and also in the storage and handling condition.

Figure 7. Flash points for different blend compositions

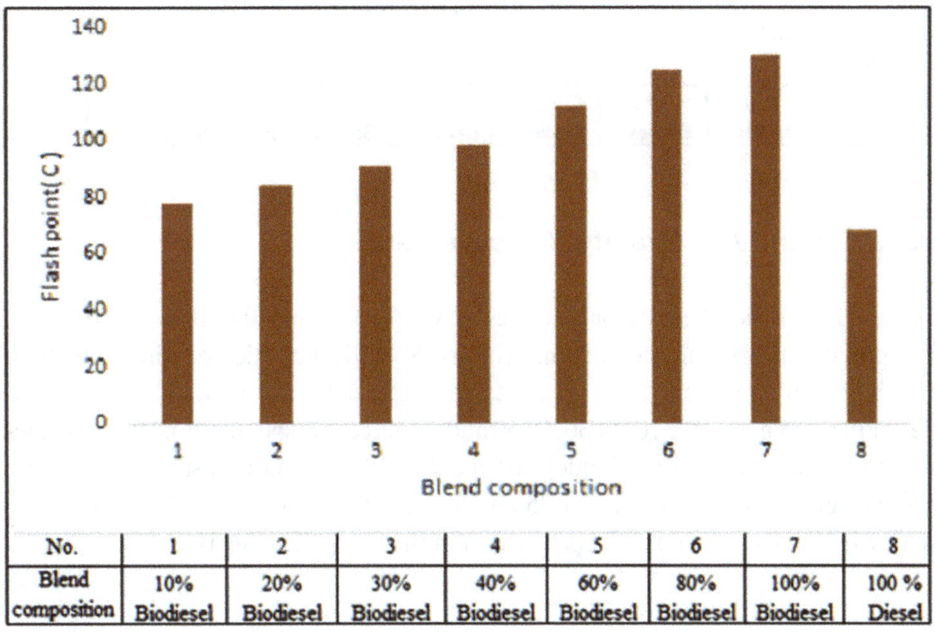

No.	1	2	3	4	5	6	7	8
Blend composition	10% Biodiesel	20% Biodiesel	30% Biodiesel	40% Biodiesel	60% Biodiesel	80% Biodiesel	100% Biodiesel	100 % Diesel

Figure 8. Fire points for different blend compositions

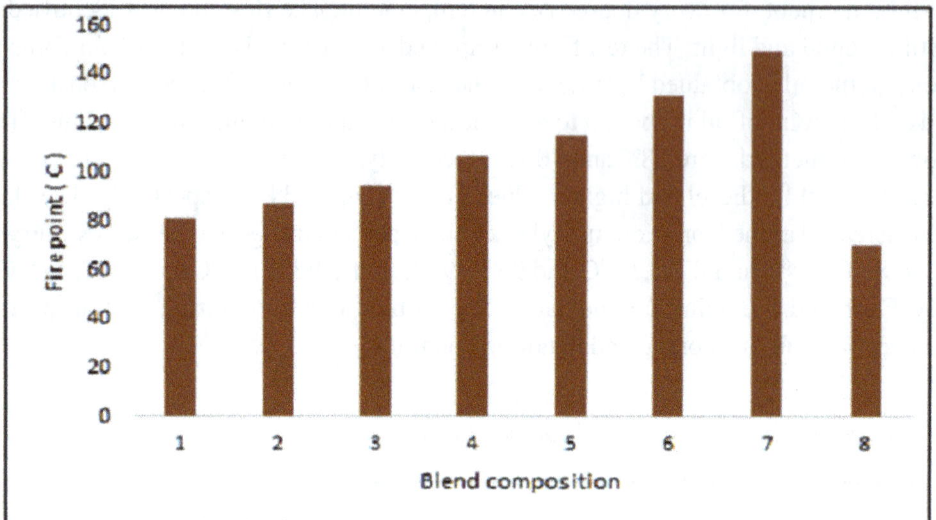

No.	1	2	3	4	5	6	7	8
Blend composition	10% Biodiesel	20% Biodiesel	30% Biodiesel	40% Biodiesel	60% Biodiesel	80% Biodiesel	100% Biodiesel	100 % Diesel

Specific Gravity and API Gravity Measurement

The specific gravity of diesel and Honge oil ester was 0.836 and 0.890 respectively. Figure 9 depicts specific gravity at different blend compositions. Specific gravities obtained for the blends were 0.837,0.841,0.853,0.867,0.878 and 0.890 in increasing order of their compositions(for every ten percent addition of ester in petrodiesel like 10,20,30,40,60 and 80 percent respectively). Higher values of specific gravity leads to higher values of diesel index. 10 and 20% blends were most suitable from diesel index analysis. The API gravity for diesel was 37.75 and for pure biodiesel it was 27.42. There was decrease in API gravity with increase in biodiesel proportion in the blend (Figure 10).

Figure 9. Specific gravity for different blend compositions

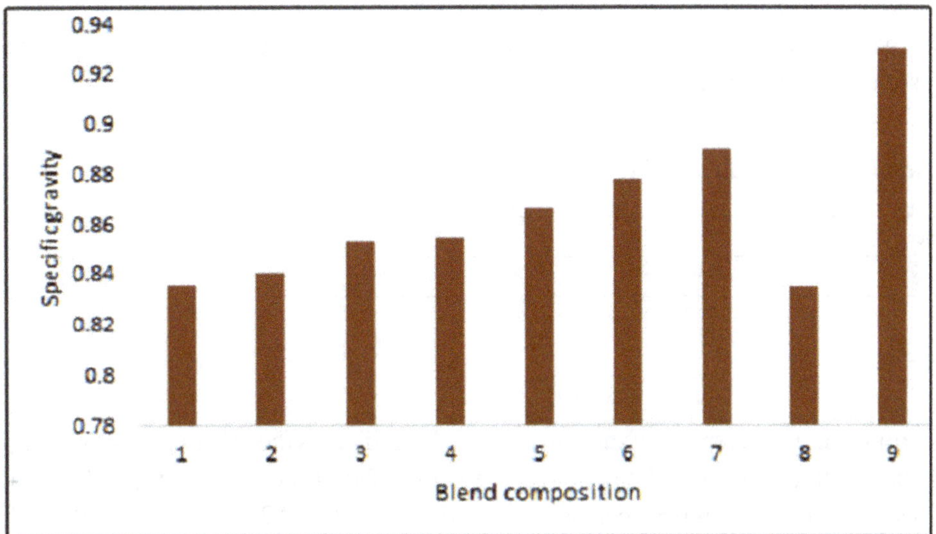

1	2	3	4	5	6	7	8	9
10% Biodiesel	20% Biodiesel	30% Biodiesel	40% Biodiesel	60% Biodiesel	80% Biodiesel	100% Biodiesel	100 % Diesel	Pongamia Oil

Figure 10. API gravity for different blend compositions

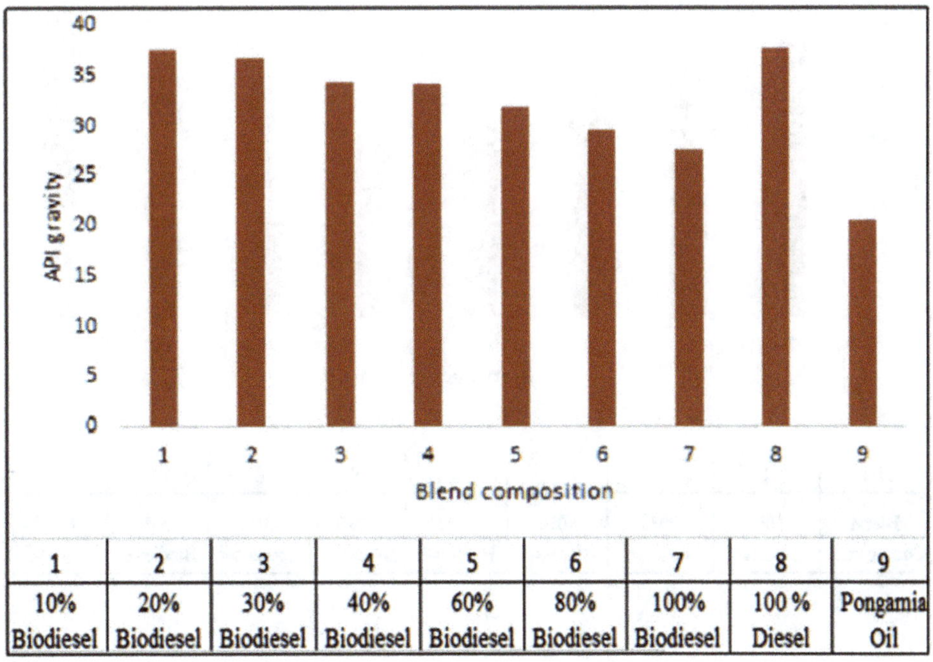

1	2	3	4	5	6	7	8	9
10% Biodiesel	20% Biodiesel	30% Biodiesel	40% Biodiesel	60% Biodiesel	80% Biodiesel	100% Biodiesel	100 % Diesel	Pongamia Oil

Cetane Number and Calorific Value Analysis

The samples were analysed for cetane number and calorific value(Table 2). These properties are very important as identification of good fuel.

Table 2. Cetane number and Calorific value

Cetane Number	52.1
Calorific Value	3780 kcal/kg

Cetane number, indicates speed of fuel burning. It also indicates the amount of compression required. For satisfactory operation of diesel engine, cetane values should be in the range, 40 to 55. Vegetable oil sources yield esters which have a cetane number range of 46 to 52. Size of the carbon chain is indicator of the calorific value. There is direct relation between the Cetane number and diesel index.Pure diesel has Cetane number of 50. Cetane values decreased with increase in biodiesel proportion in the blend(figure 11). For first four samples, up to 40% biodiesel content, the cetane number was found to be within satisfactory range(near 40).

Figure 11. Cetane numbers at various blend compositions

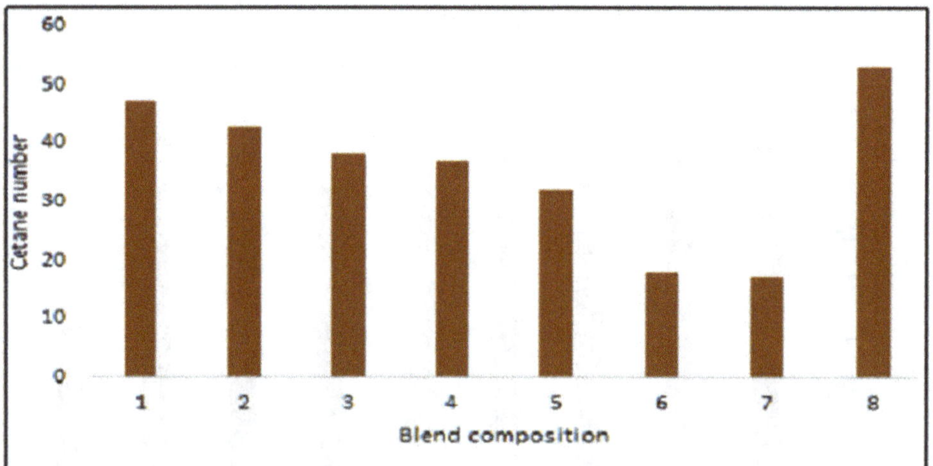

No.	1	2	3	4	5	6	7	8
Blend Composition	10% Biodiesel	20% Biodiesel	30% Biodiesel	40% Biodiesel	60% Biodiesel	80% Biodiesel	100% Biodiesel	100 % Diesel

Table 3 summarizes the past work for synthesis and characterization of biodiesel for different types of oils. It can be observed that the optimum parameters and biodiesel characteristics are in the range of the values obtained in these investigations. This indicates that the biodiesel synthesis and characterization is satisfactory and Honge oil is very good source of biodiesel.

Table 3. The past work for synthesis and characterization of biodiesel for different types of oils

Oil	Parameters	Biodiesel Characterization	Authors
Waste cooking oil (WCO)	WCO oil to methanol ratio 1:8, Catalyst loading 1% by wt., reaction time 90 minutes	-----------------	Elkady et al., 2015
Waste cooking oil (WCO)	1:8 WCO oil to methanol, 90 minutes reaction time, 1% by weight of catalyst loading rate	-----------------	Degfie et al., 2019
Second-used cooking oil	1:6.18 oil to methanol ratio, 1% KOH, temperature 66.5 °C,	Specific gravity 0.90 kinematic viscosity 4.9 mm²/sec, flash point 180.5 °C	Kawentara and Budiman, 2013
Soyabean oil	1:6 molar ratio oil to methanol, catalyst concentration 0.5%, 100 minutes reaction time, reaction temperature 60 °C	-----------------	Churasia et al., 2015
Soyabean oil	1:6 molar ratio oil to methanol, catalyst concentration 1%, 180 minutes reaction time, reaction temperature 40 °C	-----------------	Hossain and Al-saif, 2010
Sunflower oil	Temperature 49 °C, 1 gm/l catalyst loading	Kinematic viscosity 6 mm²/sec, Cetane number 47, specific gravity 0.68	Thirumarimurugan et al., 2012
Rapeseed	Methanol to oil ratio 1:9, 0.7 w/w catalyst concentration, 60 to 80 °C temperature, 90 to 100 minutes time	Flash point 175 to 178 °C, Cetane index 43 to 44, specific gravity 0.86 to 0.87, viscosity 4.7 to 4.9 mm²/sec.	Encinaret al., 2018
Canola cooking oil	catalyst concentration of 0.5%, temperature 55 °C, time 160 minutes, 1:1 molar ratio of methanol to oil	-----------------	Hossain and Mekhled, 2010
Jatropha oil	Methanol: oil ratio 1:6, potassium hydroxide catalyst concentration 1%, reaction temperature 20 to 25°C (room temperature)	Cetane number 48 to 50, flash point 93 to 95°C, specific gravity 0.87, kinematic viscosity 4 to 5.3 mm²/s	Kywe and Oo, 2009
Palm oil	1:6 ratio of oil to alcohol, 60 to 80 °C temp., 1% catalyst concentration	For 20 to 100% composition specific gravity varied from 0.85 to 0.89, Kinematic viscosity varied from 3.05 to 6.2 mm²/sec, Flash point from 56 to 164 °C, Calorific value 42800 to 38050 kJ/kg	Mamilla et al., 2012
Cotton seed oil	-----------------	Specific gravity0.890, Viscosity 3.6 mm²/sec, flash point 155 °C, Cetane number 50.4, Calorific value 40000 kJ/kg	Singh et al., 2015

CONCLUSION

Use of pure Honge oil methyl ester as an alternative fuel for diesel will not be advisable as 100% Honge oil biodiesel because of viscosity. It can be used as 10, 20 and 30 percent blends and can be very useful alternate fuel. The higher blends like percent 40 and above have high values of viscosity and can create ignition difficulties. The results obtained for kinematic viscosity, ASTM properties, specific gravity, aniline point, cetane number, flash point, fire point were in agreement with earlier investigations.(Natarajan and Venkatesh 2012; Ahmad et al. 2009; Kumar et al. 2013;Mamilla et al.2012). As per diesel index analysis and aniline point determination, the 10 and 20% blends were most useful. The specific gravity, flash and fire point and pour points of all the Honge oil ester blends were within the limits. The distillation property also recommends the use of lower blends up to 20 percent. The 20% Honge oil biodiesel has potential to reduce diesel usage by an equivalent amount and minimize environmental footprints.

ACKNOWLEDGMENT

Authors express their gratitude towards their affiliated institutes for their support and cooperation. Both the institutes (KBC NMU University Institute of Chemical Technology and Gharda Institute of Technology) have played motivational role in experimentation and preparation of this book chapter.

REFERENCES

Ahmad, M., Zafar, M., Khan, M. A., & Sultana, S. (2009). Biodiesel from pongamia pinnata l. oil: A promising alternative bioenergy source. *Energy Sources*, *31*(16), 1436–1442. doi:10.1080/15567030802092882

Ahmed, M., Khan, M. A., Zafar, M., Sultana, S., & Gulzar, S. (2007). Indigenous plant-based biodiesel resources in Pakistan. *J. Ethnobotany Leaflets*, *11*, 224–230.

Alia, E. N., & Taya, C. I. (2013). Characterization of biodiesel produced from palm oil via base catalyzed transesterification. *Procedia Engineering*, *53*, 7–20. doi:10.1016/j.proeng.2013.02.002

Allen, O. N., & Allen, E. K. (1981). *The Leguminosae*. University of Wisconsin Press. doi:10.1007/978-1-349-06142-6

Anya, A. U., Chioma, N. N., & Obinna, O. (2012). Optimized reduction of free fatty acid content on neem seed oil for biodiesel production. *J. Basic. Appl. Chem.*, *2*, 21–28.

Bobade, S. N., & Khyade, V. B. (2012a). Detail study on the properties of pongamia pinnata (karanja) for the production of biofuel. *Research Journal of Chemical Sciences*, *2*(7), 16–20.

Bobade, S. N., & Khyade, V. B. (2012b). Preparation of methyl ester (biodiesel) from Karanja (pongamia pinnata) oil. *Research Journal of Chemical Sciences*, *2*(8), 43–50.

Churasia, A., Singh, J., & Kumar, A. (2016). Production of biodiesel from soybean oil biomass as renewable energy source. *Journal of Environmental Biology*, *37*, 1303–1307. PMID:29261252

Degfie, T. A., Mamo, T. T., & Mekonnen, Y. S. (2019). Optimized biodiesel production from waste cooking oil (WCO) using calcium oxide (CaO) nano-catalyst. *Scientific Reports*, *9*(1), 18982. doi:10.103841598-019-55403-4 PMID:31831823

Elkady, M. F., Zaatout, A., & Balbaa, O. (2015). Production of biodiesel from waste vegetable oil via KM micromixer. *Journal of Chemistry*, *2015*, 1–9. doi:10.1155/2015/630168

Encinar, J.M., Pardal, A., Sanchez, N., & Nogales, S. (2018). Biodiesel by transesterification of rapeseed oil using ultrasound: a kinetic study of base-catalysed reactions. *Energies, 11*, 1–13. doi:10.3390/en11092229

Freedman, B., Butterfield, R. O., & Pryde, E. H. (1986). Transesterification kinetics of soybean oil. *Journal of the American Oil Chemists' Society*, *63*(10), 1375–1380. doi:10.1007/BF02679606

Georgianna, D. R. G., & Mayfield, S. P. (2012). Exploring diversity and synthetic biology for the production of algal biofuels. *Nature*, *448*(7411), 329–335. doi:10.1038/nature11479 PMID:22895338

Hayyan, A., Alam, M. Z., Mirghani, M. E. S., Kabbashi, A. A., Hakimi, N. I. N. M., Siran, Y. M., & Tahiruddin, S. (2010). Production of biodiesel from sludge palm oil by esterification process. *Journal for Energy and Power Engineering*, *4*(1), 11–17.

Hossain, A. B. M. S., & Aishah Salleh, A. (2008). Biodiesel fuel production from algae as renewable energy. *American Journal of Biochemistry and Biotechnology*, *4*(3), 250–254. doi:10.3844/ajbbsp.2008.250.254

Hossain, A. B. M. S., & Al-saif, A. M. (2010). Biodiesel fuel production from soybean oil waste as agricultural bio-resource. *Australian Journal of Crop Science*, *4*(7), 538–542.

Hossain, A. B. M. S., & Mekhled, M. A. (2010). Biodiesel fuel production from waste canola cooking oil as sustainable energy and environmental recycling process. *Australian Journal of Crop Science*, *4*(7), 543–549.

Hribernik, A., & Kegl, B. (2007). The influence of biodiesel on combustion and emission characteristicsof a diesel engine. *Jixie Gongcheng Xuebao*, *53*, 683–695.

Kawentara, W. A., & Budiman, A. (2013). Synthesis of biodiesel from second-used cooking oil. *Energy Procedia*, *32*, 190–199. doi:10.1016/j.egypro.2013.05.025

Kumar, A., & Kumar, P. (1984). Improving the productivity of the petrocrops in Rajasthan. *Proc. Bioenergy Soc., India, First convention and symposium,* 125.

Kumar, R. S., Manimaran, R., & Gopalakrishnan, V. (2013). Performance and emission analysis using pongamia oil biodiesel fuel with an artificial neural network. *Advanced Engineering and Applied Sciences*, *3*(1), 17–20.

Kywe, T. T., & Oo, M. M. (2009). Production of biodiesel from Jatropha Oil (Jatropha curcas) in pilot plant. *Proceedings of World Academy of Science, Engineering, and Technology*, *38*, 481–489.

Ma, F., & Hanna, M. A. (1999). Biodiesel production: A review. *Bioresource Technology*, *70*, 1–15. doi:10.1016/S0960-8524(99)00025-5

Mamilla, V. R., Mallikarjun, M. V., & Rao, G. L. (2012). Biodiesel production from palm oil by transesterification method. *International Journal of Current Research*, *4*(8), 83–88.

Meher, L. C., Naik, S. N., & Das, L. M. (2004). Metanalysis of pongamia pinnata (karanja) oil for production of biodiesel. *Journal of Scientific and Industrial Research*, *63*, 913–918.

Meng, X., Chen, G., & Wang, Y. (2008). Biodiesel production from waste cooking oil via alkali catalyst and its engine test. *Fuel Processing Technology*, *89*(9), 851–857. doi:10.1016/j.fuproc.2008.02.006

Natarajan, E., & Venkatesh, K.P. (2012). Production and characterization of biodiesel using pongamia oil by immobilized rhizopusoryzae. *Advanced Biotech, 11*(7), 32–34.

Patil, V., Tran, K. Q., & Giselrod, H. R. (2008). Towards sustainable production of biofuels. *International Journal of Molecular Sciences*, *9*(7), 1188–1195. doi:10.3390/ijms9071188 PMID:19325798

Ramadhas, A. S., Jayraj, S., & Muralidharan, C. (2004). Use of vegetable oil as engine fuels: A review. *Renewable Energy*, *29*(5), 727–742. doi:10.1016/j.renene.2003.09.008

Santana, A., Jesus, S., Larrayoz, M. A., & Filho, R. M. (2012). Supercritical carbon dioxide extraction of algae lipids for the biodiesel production. *Procedia Engineering*, *42*, 1755–1761. doi:10.1016/j.proeng.2012.07.569

Saribiyik, O. Y., Ozcanli, M., Serin, H., Serin, S., & Aydin, K. (2010). Biodiesel production from ricinus communis oil and its blends with soybean biodiesel. *Jixie Gongcheng Xuebao*, *56*, 811–816.

Singh, S., Sharma, S., & Mohapatra, S. K. (2015). A Production of biodiesel from waste cotton seed oil and testing on small capacity diesel engine. *International Journal of Advance Research in Science and Engineering*, *4*(2), 172–176.

Thirumarimurugan, M., Sivakumar, V. M., Xavier, A. M., Prabhakaran, D., & Kannadasan, T. (2012). Preparation of biodiesel from sunflower oil by transesterification. *International Journal of Bioscience, Biochemistry, Bioinformatics*, *2*(6), 441–445. doi:10.7763/IJBBB.2012.V2.151

Chapter 17
Explicit Conceptual Design Approach to Adapt a Biomass-Fed Anaerobic Digester and Status Indicators in Semi-Arid Areas

Yusto Mugisha Yustas
Sokoine University of Agriculture, Tanzania

ABSTRACT

The mitigation of adverse climate change requires holistic strategies. The strategies include the proper handling of biomass wastes such as cow dung. The handling of the wastes has to be efficient using appropriately designed anaerobic digesters. In addition, the operating status of these digesters has to be monitored for detection of any fault of the digesters. Unfortunately, existing approaches for designing and monitoring the working condition of digesters have limitations. This chapter presents the innovative and explicit conceptual design approach of adapting the biomass-fed anaerobic digester and the digester operating-status-indicators in semi-arid areas. The approach is recommended to the renewable energy system designers. Finally, the appropriate indicators are recommended to be applied by the biogas plant operators or users.

INTRODUCTION

The performance improvement of biomass-fed anaerobic digesters for sustainable bioenergy generation is essential. Several ways of employing engineering design approaches are available for digester design improvement. In addition, the approaches are useful in adapting the anaerobic digester (AD). Adaptation refers to modifying the existing AD design to get its improved design with high performance (Divya *et al.*, 2015). The existing designs of the biogas plants have some pitfalls, which render poor plant performance. The pitfalls oblige the adaptation of the biomass-fed AD design. The adaption process involves

DOI: 10.4018/978-1-6684-5269-1.ch017

comprehensive engineering designing procedures, including conceptual design explained by Andreasen *et al.* (2015). The shortcomings noted in the methodologies presented by researchers, including Wu (2013) and Al *et al.* (2019), who applied the approaches for AD designing, impose the investigation of the innovative but explicit conceptual design approach. The faults of biogas plants harm the financial base of users, e.g. failing to attain the predetermined net present value, and polluting the environment. Additionally, the pollution of the environment due to the biogas plant faults, like biogas leakage, exacerbate the adverse climate change. Wu *et al.* (2019) and others have proposed several methodologies of monitoring the biogas plant working conditions intending instantly to track the faults. However, the methodologies have some limitations, including late detection of the faults resulting in the search for novel biogas plant working-status-monitoring methodology with appropriate indicators. Therefore, this chapter aims to demonstrate i) the development and application of the innovative but explicit conceptual design approach for adapting biomass-fed digesters and ii) the establishment of the novel biogas plant working-status-monitoring appropriate indicators.

BACKGROUND

Biomass is a renewable energy resource that harvests solar energy and stores that energy in the form of chemical energy. It includes animal, agricultural, and municipal wastes. Various ways are available to convert biomass into bioenergy, e.g., biogas. The conversion of biomass may be done through the anaerobic digestion process. The process is run without air while employing microbial degradation to produce biogas, among other by-products. The process is a proven technology. Several factors, including temperature, influence the technology of sustainable biogas generation and the state of feedstock applied (Theuerl *et al.*, 2015). Other factors affecting the efficient production of biogas from anaerobic digesters (ADs) include substrate pH, Carbon-Nitrogen ratio of feedstock, feedstock loading rate and particle size, hydraulic retention time and additives. The essential and critical factor, especially in batch-fed ADs, is temperature. It significantly affects biogas generation and other factors (Adekunle and Okolie, 2015). Here below is the information on the temperature relative to the biogas generation. The mesophilic state requires AD temperature to be within 20°C to 45°C. However, keeping the AD at a constant temperature, around 35°C, increases biogas generation. The biogas microbes survive optimally at a temperature variation of 0.5°C/day; henceforth, keeping ADs at temperature variations approaching this value is important (Rowse, 2011). Usually, biogas microbes can sustain temperature variation of ± 3°C/day. Therefore, devising a simple temperature control system to reduce temperature variation of biogas digester toward 0.5°C/day is of paramount importance for sustainable biogas generation.

The feedstock employed in biogas generation may consist of dry matter of up to 15% wet basis for the liquid state digestion, i.e. liquid anaerobic digestion (LAD) or above 15% wet basis for the solid-state digestion, i.e. high solid anaerobic digestion (HSAD) (Li *et al.*, 2011). The HSAD feedstock has a much longer retention time and lacks control over the biological process (Li *et al.*, 2011). Therefore, it produces more biogas and needs less or no addition of water compared to the LAD feedstock. However, the HSAD feedstock needs a large amount of inoculum for exciting biogas generation (Yang *et al.*, 2015).

Biogas generation technology is essential for the sustainable energy supply to communities and conserving the environment (Surendra *et al.*, 2014). The technology is relevant in rural areas of economically growing countries of most sub-Saharan African regions. In these countries, Tanzania inclusive, the energy insufficiency rate, especially for cooking, is increasing (Kusekwa, 2011). The energy insufficiency is caused by dwindling forest acreages resulting from the following: an expanded search for fuelwood, a traditional source of energy; enormous pressure exerted by population growth for habitation, and the surging expansion of agricultural and pastoral land (Ordway *et al.*, 2017). The effects of energy insufficiency are exemplified in rural semi-arid areas (RSAAs). The areas are found in central regions, for the case of Tanzania.

Strategies to abate the effects of the energy insufficiency in Tanzania involved introducing biogas technology in RSAAs under a subsidy scheme facilitated by non-governmental organisations (Wawa, 2012). The introduced technology was the continuous fed underground fixed dome digester ranging from 4 to 6 m^3 in size. It was a modified Chinese-type digester design, mainly known in Tanzania as the Centre for Agricultural Mechanisation and Rural Technology (CAMARTEC) type digester (Mrema *et al.*, 2011). Figure 1 illustrates this design. In addition, CAMARTEC has two biogas plant designs, one for humid regions and another for semi-arid areas. Troubleshooting faults, like leakages, in this biogas plant is challenging as they are underground installed. Besides, they are fed with feedstock of biomass and water at a ratio of one to one (Oloche *et al.,* 2017).

Figure 1. CAMARTEC digester design found around the study site
(Melinda et al., 2013)

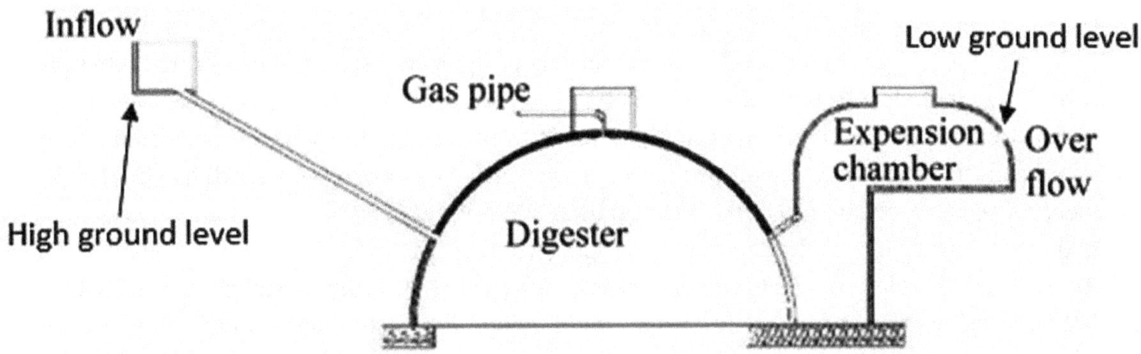

Apart from the large-sized digesters, portable small-sized digesters are available in Tanzania. For example, Kusekwa (2011) reported the portable small-sized plastic digester design of about 200 litres capacity developed by the Dar-es-Salaam Institute of Technology in Tanzania. However, its performance and dissemination of information are not available. Furthermore, in 2012 a Netherland-based company, SimGas BV with Silafrica Company, introduced in Tanzania the small-sized biogas system, alias GesiShamba (Figure 2), with a minimum capacity of 2 m^3 and fed continuously with biomass and water at a 1:1 ratio (SimGas, 2018). The GesiShamba is made of cheap plastic material, making it portable, unlike the conventional biogas plant.

Figure 2. GesiShamba system design
(WB, 2015)

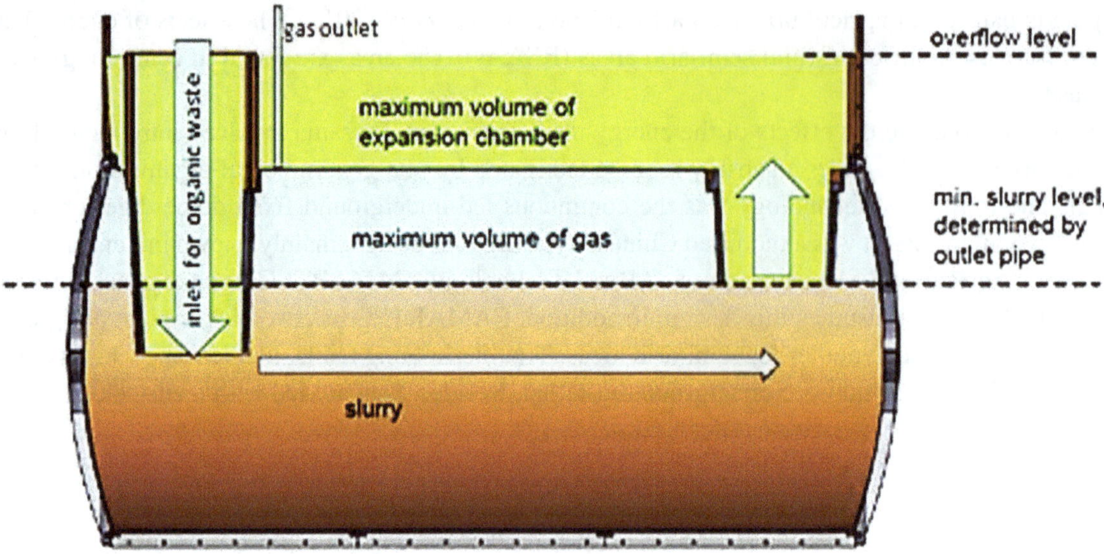

Aside from the designs mentioned above, other AD designs are used in Tanzania and other parts of the world. They include a floating drum (India type), separate storage (Sri-Lanka type), and plug flow digesters (Mrema *et al.*, 2011). Furthermore, by mode of operation, digesters are classified into batch-fed, semi-batch-fed, and continuous-type digesters. The batch-type digester for handling HSAD is simple and cheap in design compared to the others, although it has clogging problems (Kranert *et al.*, 2012).

The biogas, among others, contain methane (CH_4) content, which indicates biogas quality (Poulsen and Adelard, 2016). The high CH_4 content means high quality of biogas (Siefers *et al.*, 2010). Su *et al.* (2015) described techniques for improving the quality. The techniques include reducing impurities (Carbon dioxide, hydrogen sulphide, water vapour and ammonia) from biogas. The biogas plant auxiliary components for reducing the impurities include water traps and scrubbers filled with water, steel wools or other organic and inorganic materials like steel wools and activated carbon (Kulkarni and Ghanegaonkar, 2019; Awe *et al.*, 2017). Affordable auxiliary components facilitate the uptake of biogas technology in low-income communities in RSAAs. Enormous-sized biogas plants, for industrial application, of 100-m^3 capacity and 10 000-kPa pressure, support the biogas' purification up to 95% CH_4 content (Achinas *et al.*, 2017). The small domestic biogas plants of fixed dome or floating drum type mostly operate at low pressure, which may hardly reach 16 kPa because of being open to the atmospheric pressure at the substrate outlet (Zalm, 2017). However, for a batch-fed AD design, which is well sealed and the biogas outlet valve is closed, its pressure may rise above 16 kPa. Furthermore, the biogas flow rate in the low-pressure systems may decrease to 0.025 m^3/h. On the other hand, most applied standard biogas cooking facilities operate at a moderate biogas flow rate of 0.45 m^3/h (Zalm, 2017). Therefore, applying the low biogas flow rate directly while cooking is challenging. Therefore, cheap biogas storage is vital when operating small domestic biogas plants at low pressure and flow rate due to the prevailing weather conditions in semi-arid areas.

Several researchers have adapted the biogas plant designs using strategies explained by Divya *et al.* (2015). The adaptation involves connecting the mixing and recycling of leachate in the digester, adding fixed-film on the surface inside the digester periphery walls, and/or adding pieces of inert materials (plastics, nylon, and sponge) in the feedstock to increase the contact surface area of biogas microbes' growth. In addition, it includes minimising the abrupt changes in temperature inside the digesters. Furthermore, the digester designs alteration for biogas quality increment and handling may be done by adding improved auxiliary components like a water trap for carbon dioxide gas and water vapour screening, a scrubber for hydrogen sulphide trapping, and separate robust biogas storage for releasing the biogas efficiently (Muñoz *et al.*, 2015). However, some of these adaptations are expensive, energy-intensive, and result in biogas plant operating challenges (Karellas *et al.,* 2010). Moreover, on-ground installed small-sized (200 - 1000 litres) ADs operated in individual households are unavailable in RSAAs to address the above-stated developmental and research gaps. Besides, limited literature is available on the small-batch type system to easily release biogas from the interior of substrates of high solid anaerobic digestion. Therefore, adapting batch-type AD for domestic purposes, which addresses the fore-explained challenges, is considered.

The system adaption process involves two main design procedures: detailed and conceptual design approaches (Al *et al.,* 2019). The systems designers that deal with various types of ADs apply both approaches. However, the approaches presented by Sánchez and Cardona (2012) are obscure, involve sophisticated and complicated tools (Wu, 2013), and models which do not adequately reflect the existing physical condition of the system being developed (Al *et al.,* 2019). In some cases, results based on these approaches are prone to biases by the designers. Moreover, understanding the monitoring of biogas plant working conditions reduces the risk of plant upset to generate the biogas sustainably. Several researchers have proposed methodologies for monitoring the status of ADs (Wu *et al.,* 2019). They proposed methods to monitor the digester conditions, including checking the substrates properties, percentage of biogas constituents, physical condition of the plant, and pressure variation. Other methods include checking the physical condition of the plant for damages or biogas leakage. However, the methods are time-consuming before identification of the faults, while some are difficult to implement depending on the design and installation of the biogas plant (Iswanto *et al.,* 2021). In addition, there is a lack of standard digester pressure curves indicating the leakage and the proper indicators of biogas microbes' activities inside the small-sized digesters. The challenges related to the adaptation process necessitate developing an adaptation approach for the digester design that is relatively clear, simple but effective and efficient biogas to facilitate sustainable biogas generation. In addition, the approach, which indicates the biogas plant's working condition at the appropriate time, is essential.

Therefore, this chapter presents and demonstrates the explicit conceptual design approach of adapting biomass-fed AD and the novel working condition indicators of the adapted anaerobic digestion system. The adapted AD may fit into any other renewable energy system to produce useful energy and bio-slurry.

EXPLICIT CONCEPTUAL DESIGN APPROACH OF ADAPTING THE BIOMASS-FED ANAEROBIC DIGESTER AND THE DIGESTER OPERATING-STATUS INDICATORS

There are several ways of adapting biomass-fed AD to produce biogas and bio-slurry. However, the approaches for adapting the digester based on the engineering design methodology presented by many researchers have several shortcomings, as noted in Sánchez and Cardona (2012) and Al *et al.* (2019). Additionally, a quick and reliable approach to identifying the faults of biomass-fed ADs facilitates a rapid remedying of malfunctioning ADs. However, systematic methods proposed for monitoring the digester operating status have limitations, as Wu *et al.* (2019) and Iswanto *et al.* (2021) explain. Furthermore, there is a lack of standard digester pressure curves indicating the leakage and appropriate indicators of biogas microbes' activities inside the digesters. These indicators are vital, as they closely reflect the working condition of the digester. The solutions to these shortcomings and limitations reinforce the theme of Biomass and Bioenergy Solutions for Climate Change Mitigation and Sustainability.

With that briefing, this chapter provides the explicit conceptual design approach of adapting the biomass-fed AD and demonstrates its application. The approach's demonstration involves adapting the batch-fed AD for effective cow dung degradation for sustainable biogas generation. Besides, the appropriate digester operating-status indicators are established and presented.

The research area of this study included the Department of Engineering Sciences and Technology at Sokoine University of Agriculture (SUA), located in Morogoro, Tanzania. In addition, it involved Chamwino District, Dodoma region Tanzania. The region represents semi-arid areas.

Formulation and Demonstration of Conceptual Design Approach of Adapting the Biomass-Fed Anaerobic Digester

After a thorough review of literature related to engineering design approaches and frameworks, intuitively, the AD adaption approach was formulated. Figure 3 illustrates the approach. The approach employs the adapted engineering procedural and research techniques. These guide the adaptation process of the biomass-fed AD design. Authors executed several tasks during the adaptation process in the field than in laboratories and workshops of research centres. In addition, the approach involves input, processing and output phases. The inputs phase includes reviewing the literature on the functionality of various existing designs of digesters, developing design criteria to be achieved by the intended components or system to be adapted/re-designed, synthesising the working principles (WPs) or concepts as per design criteria, and selection of the best working principle (WP).

The processing phase comprises the conceptual design of the selected best WP, fabrication, fixing of the fabricated components, and assembly of the digester system using materials and tools available in the workshop. Finally, the assembled system enacted the output phase as the intended biomass-fed digester; the digester was adapted batch AD for this study.

Figure 3. The framework of adapting the biomass-fed AD

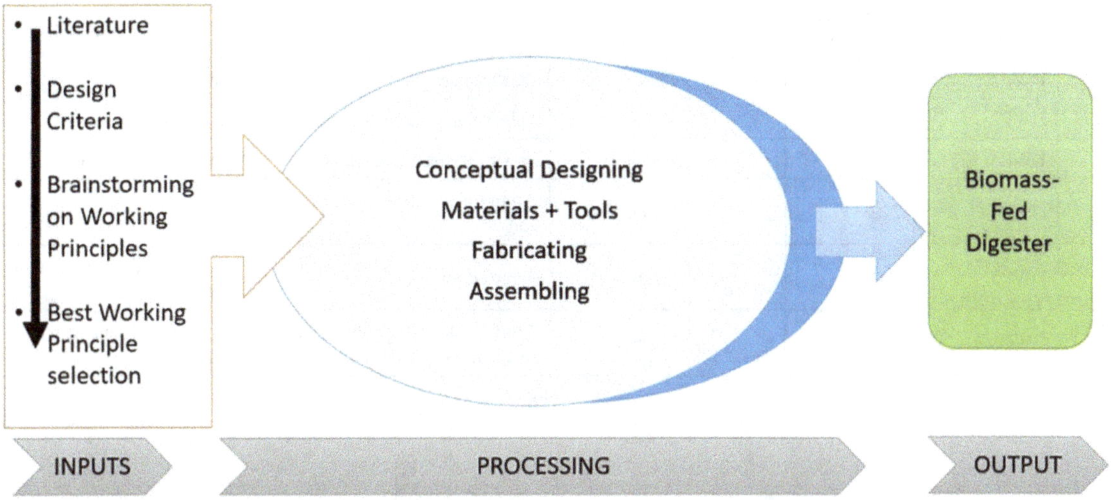

The Inputs Phase of the Adaption Framework

After a thorough literature review of various digesters' designs and operations (Garfi *et al.,* 2016), the authors selected the batch-fed anaerobic digestion system for use. The design simplicity, low cost and ease of operation guided the selection of the system. However, this system needed an adaptation to offer the intended service of increased generation of quantity and biogas quality. The basis of achieving this result was the design criteria explained in the subsequent sub-sections and the approach shown in Figure 3.

The Batch Anaerobic Digestion System Design Criteria

The criteria formulated as a basis for devising the intended system are the ability to withstand the pressure of about 16 kPa; biogas tight seal capacity/airtight sealing by 100%; self-mixing during operation; possible self-leachate re-circulation during operation, and increased contact surface area of biogas microbes. Other criteria include easy passage of biogas from the interior of substrates; easy to operate; ensuring no biogas outlet clogging; availability of raw materials locally; less cost involved; more durable and easy manufacturability.

Synthesis and Selection Approach of Best Working Principle

1. Synthesis of the Working Principles

Based on the criteria, the authors generated WPs of all batch anaerobic digestion system components. The components include the digester, the water trap, the scrubber, biogas storage and the piping system. Table 1 shows details of the generated digester WPs. The unscaled drawing in Figure 4 illustrates the generated digester WP with various types of top cover.

Table 1. Generated digester working principles

S/N	Working Principle	Volume (litres)		Drum Material	Seal Used	Other Components
1	DWP1	240		Plastic	Plastic sheet and lid	None
2	DWP2	240		Plastic	Tube rubber thread and lid	Sealed central core
3	DWP3	240		Plastic	Flexible plastic tube and lid	Perforated central core
4	DWP4	240		Plastic	2 mm thick gasket rubber and lid	The perforated four-winged central core
5	DWP5	500		Plastic	Screwed top lid	The perforated central core and four wings
6	DWP6	240		Plastic	2 mm thick gasket rubber, lid and 2 mm thick flat bar frame	The perforated four-winged central core

Figure 4. Unscaled drawing of the digester working principle

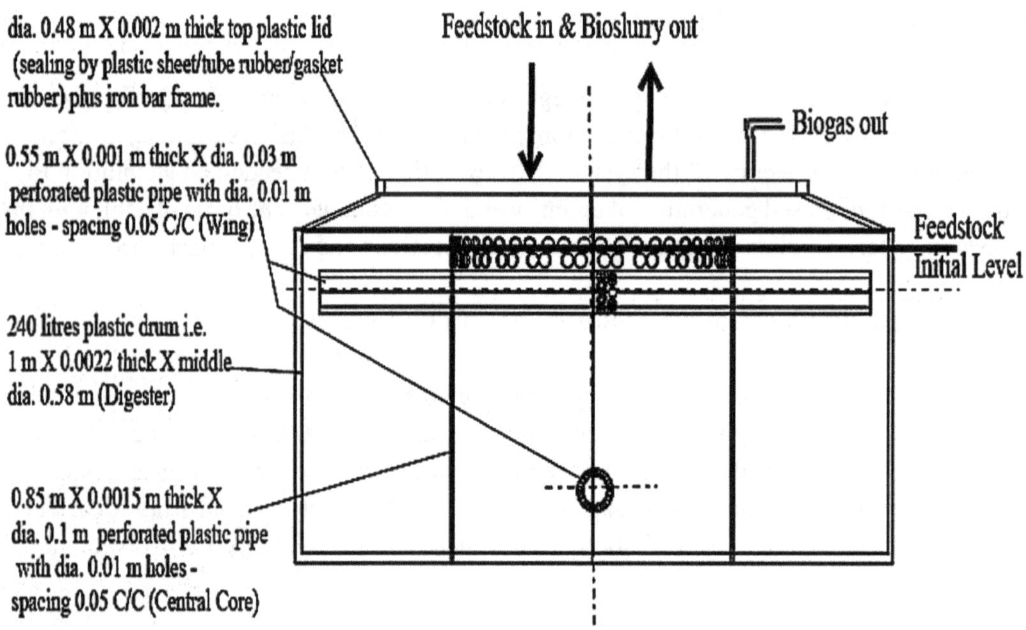

Table 2. Generated water trap working principles

S/N	Working Principle	Volume (litres)	Main Part	Important Dimension	Other Components
1	WWP1	5	Plastic jerry-can	1 mm-thick-wall	None
2	WWP2	3	Plastic-bucket	1.5 mm-thick-wall	None
3	WWP3	0.6	IPS Plastic-pipe	Length 500 mm and diameter 40 mm	None
4	WWP4	5	Plastic-can	0.7 mm-thick-wall	None
5	WWP5	12	Plastic-can	0.7 mm-thick-wall	None
6	WWP6	10	Plastic-bucket	1 mm-thick-wall	None
7	WWP7	20	Plastic-bucket	1 mm-thick-wall	None
8	WWP8	10	Plastic-jerry-can	1 mm-thick-wall	None
9	WWP9	20	Plastic-jerry-can	1 mm-thick-wall	None
10	WWP10	3	Plastic-bucket	1 mm-thick-wall	Fastening 2 mm-thick iron bar frame

Table 3. Generated scrubber working principles

S/N	Working Principle	Volume (litres)	Main Part	Important Specification	Other Components
1	SWP1	0.5	IPS-plastic-pipe	Length 400 mm and diameter 40 mm (monolithic)	None
2	SWP2	0.5	IPS-plastic-pipe	Length 400 mm and diameter 40 mm (non-monolithic)	50 mm-diameter IPS-Connector
3	SWP3	1.5	Plastic-bottle	0.5 mm-thick-wall	None
4	SWP4	1	Plastic-bottle	0.5 mm-thick-wall	None
5	SWP5	2	Plastic-jug	1 mm-thick-wall	None
6	SWP6	1	Plastic-jug	1 mm-thick-wall	None
7	SWP7	0.8	Iron-pipe-size IPS-plastic pipe	Length 400 mm and diameter 50 mm (monolithic)	None
8	SWP8	0.8	IPS-pipe	Length 400 mm and diameter 50 mm (non-monolithic)	40 mm-diameter IPS-Connector

The WPs of biogas storage thought of are the following: inflatable plastic bag (BSWP1), inner tyre tube - of various sizes (BSWP2) and inflatable plastic bag covered with clothes (BSWP3). The consideration of the WPs of the piping system was to comprise one, two or three types of pipe. Authors of this chapter assumed a total length of 1.7 m of the piping system without including fittings for computing the percentage of the length of the type of pipe in the whole system. The percentages and combinations of types of pipes differentiate the WPs generated. Table 4 shows the intuitively generated WPs of the piping system.

Table 4. Generated piping system working principles

S/N	Working Principle	Pipe Type Length Percentage to the Total Length of the Piping System			Fittings Material Types
		Horse Pipe (%)	Another Flexible Pipe (%)	IPS Plastic Pipe (%)	
1	PWP1	100	-	-	IPS-plastic-and-galvanised-steel
2	PWP2	70	30	-	IPS-plastic-and-galvanised-steel
3	PWP3	-	-	100	IPS-plastic-and-galvanised-steel
4	PWP4	70	30	-	IPS-plastic-and-galvanised-steel
5	PWP5	70	25	5	IPS-plastic-and-galvanised-steel

2. Selection Methods of the Best Working Principle

Best concept selection techniques are available. They involve Decision Matrix Analysis (DMA)/grid analysis, ABC analysis, and Pareto analysis; Analytic Hierarchy Process (AHP) based approach, anonymous voting technique, cost-benefit (CB) analysis, and Failure Modes and Effects Analysis (FMEA). Other concept selection techniques include Plus-Minus and Interesting (PMI) analysis, Kano model, nominal group technique, Sticking dots, and "Strength, Weakness, Opportunity and Threat" (SWOT) analysis (Mardani *et al.*, 2015). The following criteria are applied in choosing an appropriate selection technique of the best WP: its simplicity in application, ability to single out one best option, quick sorting of more than three options and diverse application in different disciplines, which include designing renewable energy systems.

With the procedure mentioned above, the authors found that CB analysis, SWOT analysis and Kano model are discipline-specific for financial, business, and marketing analysis, respectively. Additionally, the ABC analysis utilisation intends to prioritise one among the three options only and for controlling inventories. Furthermore, Pareto analysis uses the 80/20 rule, i.e. only 20% of options influence 80% of the same options. Therefore, it renders the analysis unable to single out one best option. The Analytic Hierarchy Process (AHP) based approach is complex to apply. It requires special software tools to perform the calculations, which involve the concept of Eigen-vectors, to get the best option.

The anonymous voting technique is very subjective and based on individuals' opinions. Additionally, FMEA was found to fit in identifying potential failure modes, determining their effect on the operation of the product, and identifying actions to mitigate the failures (Rebernik *et al.*, 2008). However, it is complex to apply, as it is impossible to anticipate every failure mode. The PMI analysis employs finding positive, negative and interesting points of an idea leading to its decision being entirely subjective regarding the most appropriate idea screening technique. The Nominal Group Technique is useful for prioritising ideas and projects rather than choosing one best option. At the same time, the Sticking dots analysis found is the quick method for determining priorities by voting. The Sticking dots analysis is useful for collecting opinions early in the idea selection process.

The review of DMA showed that the technique evaluates an idea as per criteria (Mardani *et al.*, 2015). It is simple and allows quick sorting of options/concepts by identifying their relative strengths and weaknesses. The DMA is applied in different areas of study, including designing renewable energy systems (Abu-Taha and Daim, 2011). However, the careful setting of scoring criteria is necessary for the expected useful results (Abu-Taha and Daim, 2011). Therefore, this study opted for the application of the DMA method.

3. Application of the DMA Method

The best WPs were selected using the DMA method as applied by Olabanji and Mpofu (2014). In this procedure, the weight values assigned to design criteria are based on their importance to ensure the component or the system needs to meet the requirement set. Experts in renewable energy, system design, and mechanical engineering assigned a weight to the design criteria. The selection of experts to participate in the weight values assignment exercise was made intuitively by considering the expert's experience to design and operate renewable energy systems and machines involving fluid systems. The authors contacted twenty experts to participate in this exercise, but only thirteen were effectively ready. Therefore, eight renewable energy system developers and five mechanical engineers were involved based on their readiness to participate in this task.

The fore-mentioned weight assignment process produced the weight values in brackets. The presentation of assignment results is as follows: the ability to withstand the pressure of about 16 kPa (0.21), biogas tight/airtight at 100% sealing (0.4), and self-mixing during operation (0.01). Others were the following: possible self-leachate re-circulation during operation (0.01), the increased surface area of contact for biogas microbes (0.03), easy passage of biogas from the interior of substrates (0.08), easy to operate (0.1), ensuring no biogas outlet clogging (0.05), availability of raw materials locally (0.04), less cost involved (0.03), more durable (0.02), and easy manufacturability (0.02).

Moreover, factors with their corresponding rating on Likert factor scale values in the bracket needed to determine the possibility of the WPs meeting each design criteria were intuitively set as follows: not at all able (1), lowly able (2), moderately able (3), highly able (4), and very highly able (5). Furthermore, the weighted score of the WPs relative to each design criterion was determined as the product of design criteria weight and the rating on Likert factor scale values. Finally, the total weighted score for each WP was determined by summing up the weighted scores.

Above mention analysis produced plots of the total weighted score against the WPs. The authors selected the WP with the highest total weighted score as the best. However, for those WPs that did not produce conclusive weighted scores, testing results interpretation of the prototypes was a basis for their selection. For instance, this happened when WPs assignment of the same rating of at least three for the highly weighted design criterion that weighs more than 0.21. The authors fabricated the prototypes as no cheap and reliable alternative way to evaluate them was feasible at that particular time and place.

4. Selection Results of the Best Working Principle

In this subsection, the authors present the results of selecting the best WP of the ADS components. The tested fabricated prototypes are the water trap, scrubber, and biogas storage, regarding the purpose of the best option selection. This exercise considered the leakage problems where the prototype was filled with air to a safe pressure of 8 kPa. Any sign of leakage was detected by immersing the prototype in water or pouring water on it. The selected best options resisted 100% of the leakage at the stated pressure. If more than one option passed the test, the authors applied the prototypes' total weighted score for the selection. Additionally, the authors considered another new WP for conceptual designing if all options failed the test.

1. Selection of the working principle of the digester

The results obtained and shown in Figure 5 reveal that a working principle DWP6 secured the highest total weighted score of 3.97 and was then selected, meaning it was the best of all WPs. Note that a frame of flat iron bars encases the plastic drum.

Figure 5. Decision matrix analysis results for digester working principles

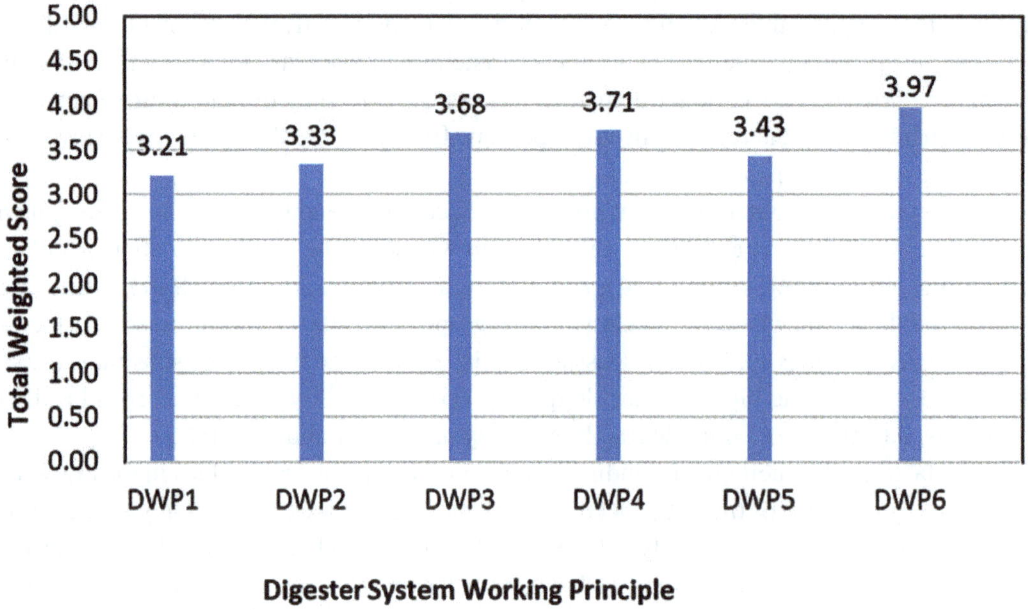

The score of 3.97 by DWP6 was based on the high likelihood of the following: the gasket rubber to offer an airtight sealing and withstanding 16 kPa pressure, and the central core and its four wings to allow easy passage of biogas from the interior of the substrates. The ease of passage of biogas from the interior reduces the possibility of building up a biogas mass at the bottom of the digester, pushing the substrate mass up, and leading to system clogging.

The central core offers a space for releasing the leachate into the bottom centre of the digester, as it is perforated and not filled with cow dung during digester feeding. Additionally, the leachate may rise to wet the above substrates through the perforated wall of that central core and its wings as the digestion process continues. Authors assumed that as the leachate mass finds space in the central core, the solid substrates drop down to occupy its space. In doing so, kneading solid substrates by the central core wings is possible.

Moreover, the provision of the central core and its wings results in increased contact surface area for the growth of the biogas-generating microbes in the central mass of the substrate. There was a likelihood of increased biogas microbes as they would be away from the possible rapid changes of temperature near the wall of the digester. The rapid temperature changes affect the performance of the microbes adversely.

The iron bar frame supports the gasket for airtight sealing as it holds together the top lid and the plastic drum of the digester, even above 16 kPa pressure. According to Tubecon (2018), a flat bar may withstand more than 16 kPa under tension, a factor of safety exclusive.

2. Selection of the water trap working principle

The DMA results were inconclusive regarding the high-weighted design criterion for the water trap WP. Therefore, the authors fabricated the three water-trap prototypes for leakage testing. The concepts in question are WWP1, WWP3, and WWP10. All three concepts had equal-weighted scores for the high-weighted design criterion of withstanding airtight sealing. Therefore, WWP3 was best option selected for sealing the leakages by 100% and having the highest-scoring of the total weighted score after the leakage test, as shown in Figure 6.

Figure 6. Decision-matrix analysis results for water trap working principles

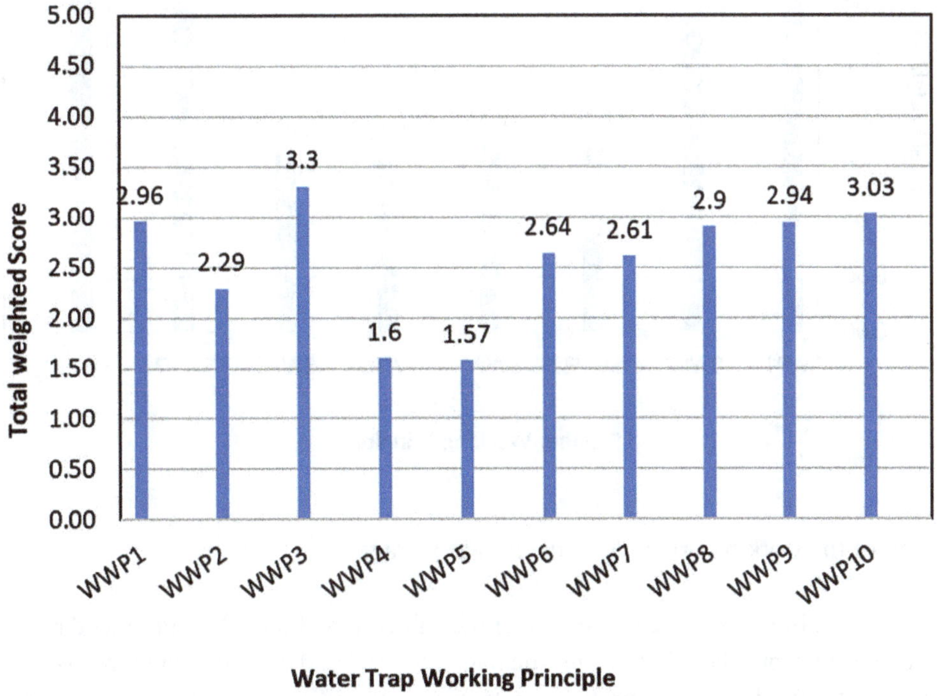

The authors filled the water trap with a 140 ml volume of freshwater. Then, the authors applied trial and error methods to determine the volume while ensuring no water and air inflow to the digester and outflow beyond the water trap on the accidental opening of the system to air. As a result, water enhances moisture and Carbon dioxide (CO_2) trapping. Note that, by assuming the annual mean atmospheric pressure of 101.9 kPa at the study site (TDA, 2018) and pressure and temperature of both 6 kPa and 28°C, respectively, in the water trap, the authors considered the solubility of CO_2 in water as 15 ml of CO_2 per one gram of water (Diamond and Akinfiev, 2003). Therefore, a process removes about 2.1 litres of CO_2 from the 7 litres of biogas (assuming biogas contains 30% of CO_2) passing through the water trap.

3. Selection of the working principle of the scrubber

The authors subjected the SWP1 and SWP2 to the leakage test while parked with iron wools. Both concepts were not leaking at all under testing conditions. Therefore, the selection is based on the total weighted score shown in Figure 7. Following these results, the authors selected the SWP2, with a score of 3.59, as the best scrubber design to be incorporated in the intended anaerobic digestion system.

Figure 7. Decision matrix analysis results for scrubber working principles

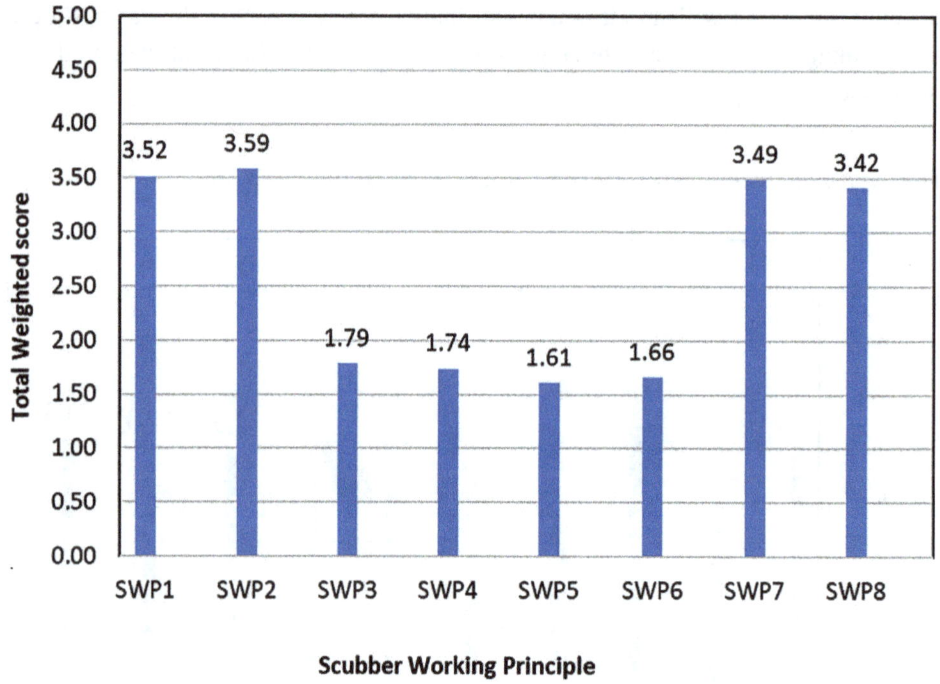

4. Selection of the working principle of the biogas storage

Three concepts of biogas storage facility under the subjection of the DMA produced results shown in Figure 8. The two concepts, the inflatable plastic bag (BSWP1) and the inner tyre tube (BSWP2), further passed leakage tests after fabrication prototypes. However, the BSWP1 performed poorly; therefore, the authors selected the BSWP2 as biogas storage's best option and included it in the list of components of the ADS. The used tyre tube sizes were 7.5-15 (TR174A), 8.25-16 (TR177A), 9-20 (TR177A) and 12-24 (TR177A.).

Figure 8. Decision matrix analysis results for biogas storage concepts

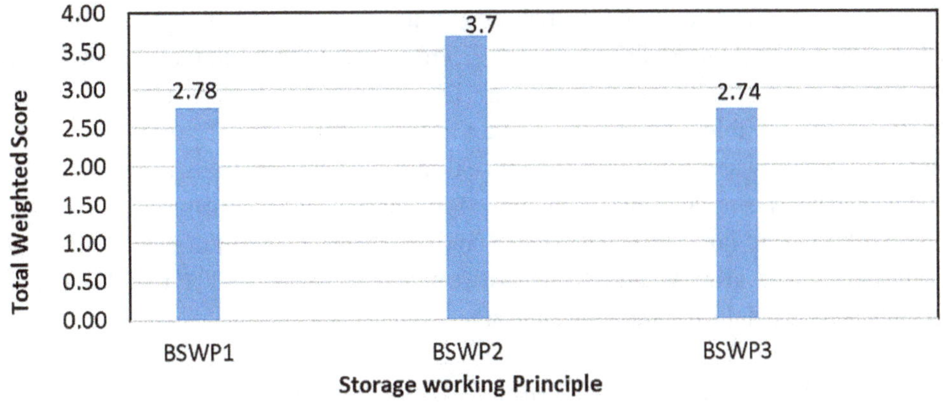

5. Selection of the working principle of the piping system design

The results from the DMA, as shown in Figure 9, guided the selection of the best piping system option. The option PWP5 (IPS plastic pipe (70%) + other flexibles (25%) + horse pipe (5%) + IPS fittings) secured the total weighted score of 3.76, which was the highest, leading to its selection as the best option.

Figure 9. Decision matrix analysis results for piping system concepts

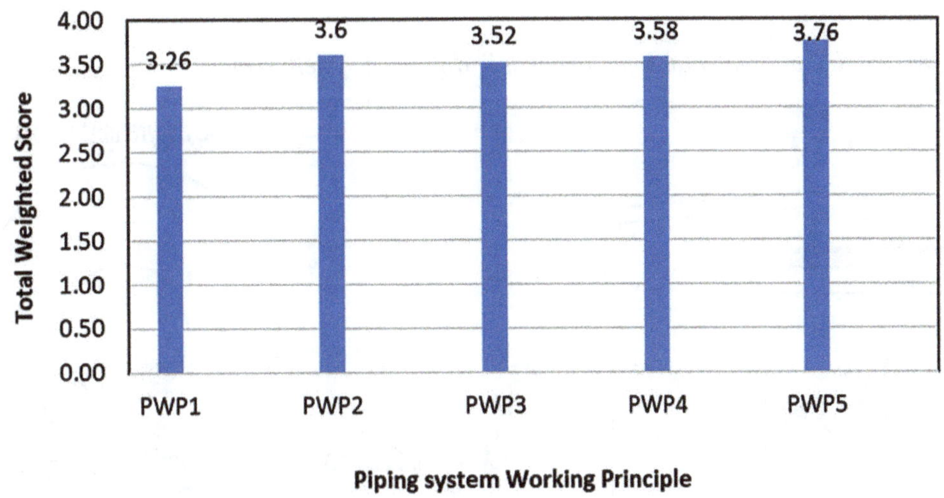

The Processing and Output Phases of the Adaption Framework

The authors did the conceptual designing of the selected WPs in the processing phase. After that, using predetermined materials and tools, they fabricated and assembled all the conceptual designed components to produce the wholly anaerobic digestion system. Finally, the fully assembled system activated the output phase.

1. Materials and Tools Applied for Fabrication of the System

The materials used were fabrication consumables and items, including IPS plastic pipes (50 mm, 38 mm, and 13 mm diameter); standard plastic pipes (25 mm and 100 mm diameter. Other materials were the 240-litre plastic drum (2.2 mm wall-thickness), IPS fittings (13-38 mm diameter), bolt and nuts of N10 (8 mm diameter), flexible tube clamps, and a Grade S235JR - 25 x 2 mm flat bar (EU standard); 13 mm diameter flexible water pipes and thread seal tapes. Other materials used were putty sealants, gasket rubbers, plastic sheets, and pieces of inner tyre tubes. Additional materials were the 2.5 mm square wires and butyl tyre inner tubes (Korean Standard, made in India) of sizes: 8.25- 16, 9-20, and 12-24, all with valves type TR177A and 7.5-15 with valve type TR175A. The metalwork and plumbing workbench tools were used during the fabrication.

2. Fabrication and Assembly of the ADS and Its Components

Authors fabricated all selected components of the system except those selected during the second stage of selecting the WP, as their prototypes were utilised directly. Then the final finishing of all components was done. Afterwards, they were assembled by connecting the batch-fed digester, water trap, scrubber, biogas storage and biogas cooking stove to produce the whole ADS. Next, the steel wool and 140 ml of soft water were filled into the scrubber and water trap to end the processing phase. Finally, the complete ADS with the digester, the water trap, the scrubber, the storage, and the piping system was ready for use in the output phase of the approach as the adapted batch-fed anaerobic digestion (ABFAD) system. Figure 10 illustrates the complete connected ABFAD system ready for use or incorporated into the integrated renewable energy system.

Figure 10. Assembled adapted batch-fed anaerobic digestion system sketch

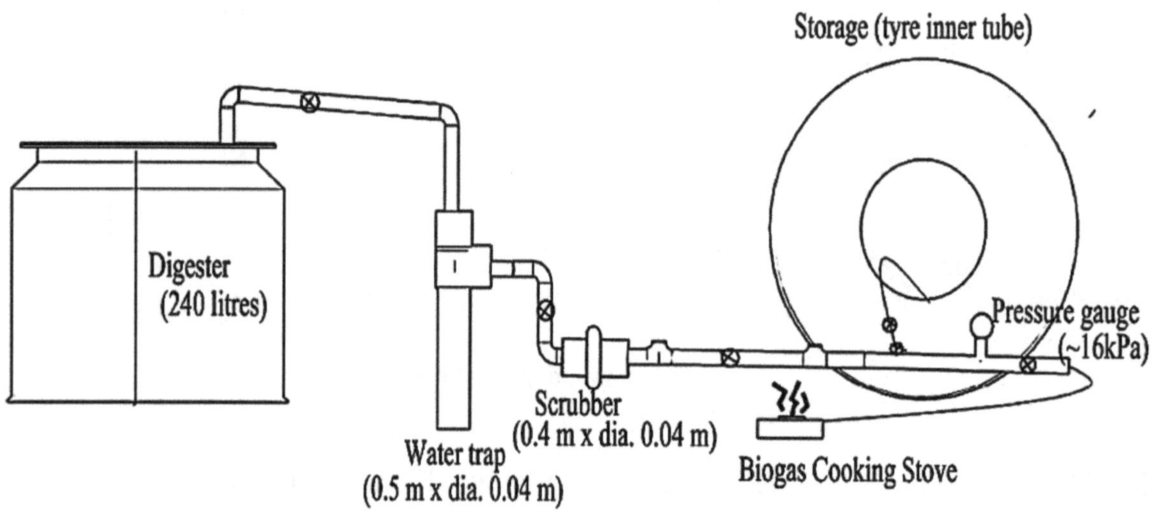

The Formulation and Demonstration of Digester Operating-Status Indicators

The authors reviewed the literature related to the assessment of the working condition of the biogas plants to collect information regarding the challenges of the monitoring process of the plant indicators applied. In addition, they considered the application of biomass-fed digesters in the semi-arid areas of least developing countries. The review outputs resulted in selection variables representing the indicators for demonstration using the assembled ABFAD system. The selected variables are piping system and digester temperatures, system pressure, methane content, pH and dissolved oxygen (DO).

Furthermore, the authors demonstrated the application of the ABFAD system by conducting experimental testing involving data collection. The testing involved using biomass waste, which is cow dung. The cow dung had the following properties: dry matter (DM) and Carbon Nitrogen ratio (C/N ratio) of 228.93 g/kg of feed (standard deviation of 5.21 g/kg) and 19.30 (standard deviation of 0.56), respectively.

Instrumentation and Measurements for ABFAD System Testing

A list of essential instruments and tools used during this study include the following:

1. UGT Humidity and temperature (thermocouple Type-K) sensor with Relative Humidity (RH) measuring range of up to 100%, humidity accuracy of ± 2% and temperature measuring range of -30.00 - +70.00°C with the accuracy of 0.01°C. The sensor connects, picks signals, and transfers them to the UGT DL 386 data logger. The authors set the sensor to sample ambient air temperature values at the one-hour interval.
2. Micro Footprint Dissolved Oxygen (DO) monitoring subsystem equipped with the DO circuit of Version 4.0 for hardware and firmware (*Atlas-Scientific.com*), with a full range of DO readings from 0.01 to +35.99 mg/L and the accuracy of ± 0.2, set to sample data after 15 minutes interval,
3. The device applied to measure the acidity or alkalinity of the ageing substrate in the digester was the Micro Footprint pH-monitoring subsystem. The device is equipped with a pH circuit Version 4.0 for hardware and firmware (*Atlas-Scientific.com*), with full range pH reading from 0.01 to 14.00 and accuracy within two significant figures (XX.XX); set to sample data after 15 minutes interval.
4. Ultrasonic biogas flow meter Version BF-2000 with a flow range of 0.025 – 4.000 m^3/h, methane (CH_4) content percentage to the total volume of the biogas measurement range of 30-100% (accuracy is 0.01%), temperature measurement ranges from -20°C to 60°C (accuracy being 0.01°C), the working pressure of less than 100 kPa and explosion-proof grade of Ex ibiiat4 Gb; CH_4 content was measured at three-hour interval, and
5. Simgas pressure gauge encased in plastic material with a measuring range up to 16 kPa with an accuracy of 0.5 kPa.

A Solar PV module with a battery bank powers all electronic instruments. Furthermore, the authors calibrated all the instruments for measuring pH, DO, digester temperature, room temperature, ambient air temperature, CH_4 content, pressure and piping system temperature and fixed them to the ABFAD system. In addition, the authors inserted probes for measuring pH, DO, and the digester temperature deep to 0.7 m from the top of the digester and axially 0.1 m away from the central core. The probe insertion was followed by feeding the digester with cow dung—Figure 11 illustrates the experimental setup.

Figure 11. Experimental set up of adapted batch anaerobic digestion system

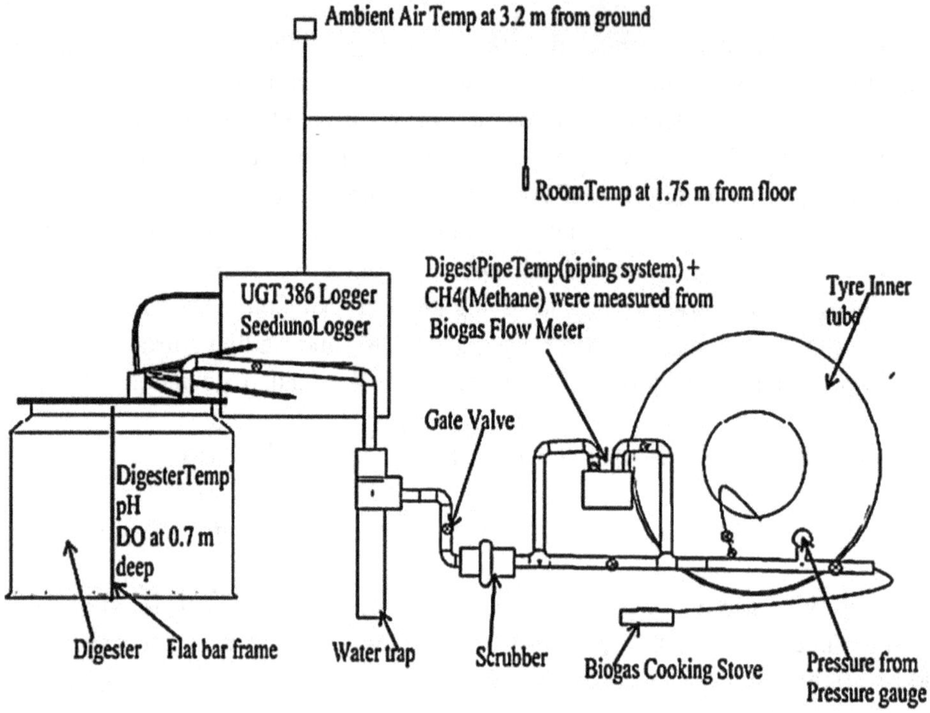

A room with a wall thickness of 0.3 m made of muddy materials (sandy clay soil) and lined with cement mortar accommodated the experiment. The inside, front and part of the outside left sides' room surface were lined. The room was slant roofed by corrugated iron sheets (gauge 30) with low and high heights of 2.5 m and 3.1 m, respectively. Its floor plan had 3.7 m by 5.1 m dimensions. The room had a screening floor (0.15 m thickness). Furthermore, the room had a window (0.6 m by 0.5 m dimensions) made of wooden frames and fitted with glass (2 mm thick). The room door (1.8 m by 0.8 m in dimensions) was made of timber and a corrugated iron sheet (gauge 30).

Furthermore, the collected data was analysed to get the average temperature, standard deviations, and rate of change of temperature (°C/h) values. The rate of temperature change was determined by dividing the standard deviation temperature by the mean of a given factor by 24 hours of a day.

Experimental Test Results and Discussion

The results include the independent variables: ambient air temperature, temperature response variables of room, digester and piping system. In addition, the results included the CH_4 content in the biogas, pH, DO inside the digester and pressure of the whole system. These parameters are essential for the technical evaluation and running of the system.

Temperature and CH_4 Content of the ABFAD System

The similar trends of the proportional relationship among temperature of the piping system, room, the digester and ambient air are shown clearly in Figure 12. The figure shows an analysis of the temperature data on an hourly basis within a day. Moreover, the figure shows that the piping-system temperature curve is not smooth compared to the other curves. The piping-system temperature curve indicates that temperature oscillates as it follows the trend of room temperature. The oscillation may be due to the intermittent flow of varying amounts of biogas at a low-temperature level through the piping system from the digester. Usually, the digester has a lower temperature compared to the piping system most of the time.

Figure 12. Hourly average temperature variations in ABFADS (Jun 16 - Oct 1/ 2017)

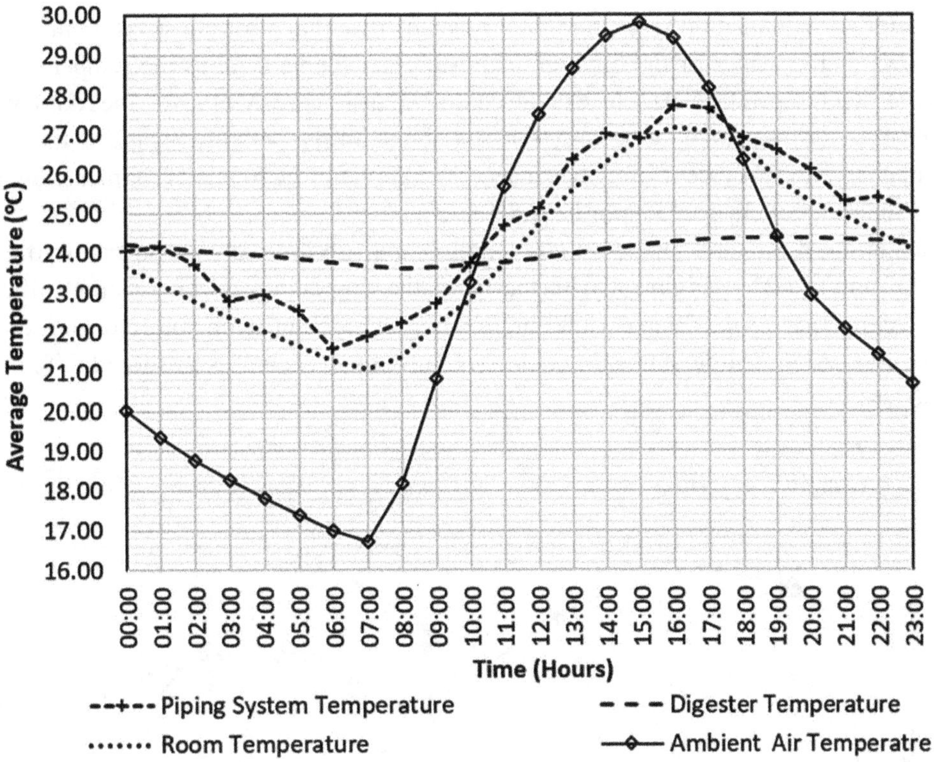

The authors computed the temperature change rate based on standard deviation from the mean temperature of the piping system, room, digester and ambient air. Table 5 shows the values of factors obtained for each of the ABFADS parameters. It shows that the temperature change rate inside the digester was the lowest with the value of 0.01°C/h, while ambient air was highest at 0.18°C/h. This revelation indicates that any ADS installed outside must be away from ambient air temperature variation, which has been the case for all biogas plants installed outside. The high-temperature variation affects biogas microbes adversely.

Table 5. Temperature rate of change and other factors of ABFADS's parameters

Factor (unit)	Parameter			
	Piping System	Digester Temperature	Room Temperature	Ambient Air Temperature
Mean temperature (°C)	24.71	24.04	24.05	22.67
Standard deviation (°C)	1.86	0.26	1.95	4.38
Temperature rate of change (°C/h)	0.08	0.01	0.08	0.18

Additionally, the average digester temperature of 24.04°C shown in Table 5 indicates that the ABFADS was operating in the mesophilic phase of temperature, i.e. 20-39°C of anaerobic digestion. The optimum temperature for mesophilic microbes' performance is 35°C.

The digester's low temperature change rate indicates that the substrate mass exhibited insulation behaviour towards the center of the digester. This situation seems true, as the temperature probe for measuring digester temperature was close to the central digester core. Moreover, the room walls and air-insulated the digester from an extreme variation of ambient air temperature regarding the experimental setup. The phenomenon is evident from the closeness of temperature trends of room and digester piping systems and their rate of change (Figure 12 and Table 5).

The ambient air, room, and piping system temperatures are inversely proportional to CH_4 content between 00:00 and 15:00 hrs (Figure 13). Furthermore, the temperatures decrease at the different slopes with CH_4 content between 15:00 and 18:00 hrs before the ambient temperature decreases fast than a room and piping system temperatures. In contrast, CH_4 content increases fast to 52.65% in a range of 18:00-00:00 hrs.

Figure 13. CH_4 content and piping system, room and ambient air temperatures (Jun 25- Oct 1/ 2017)

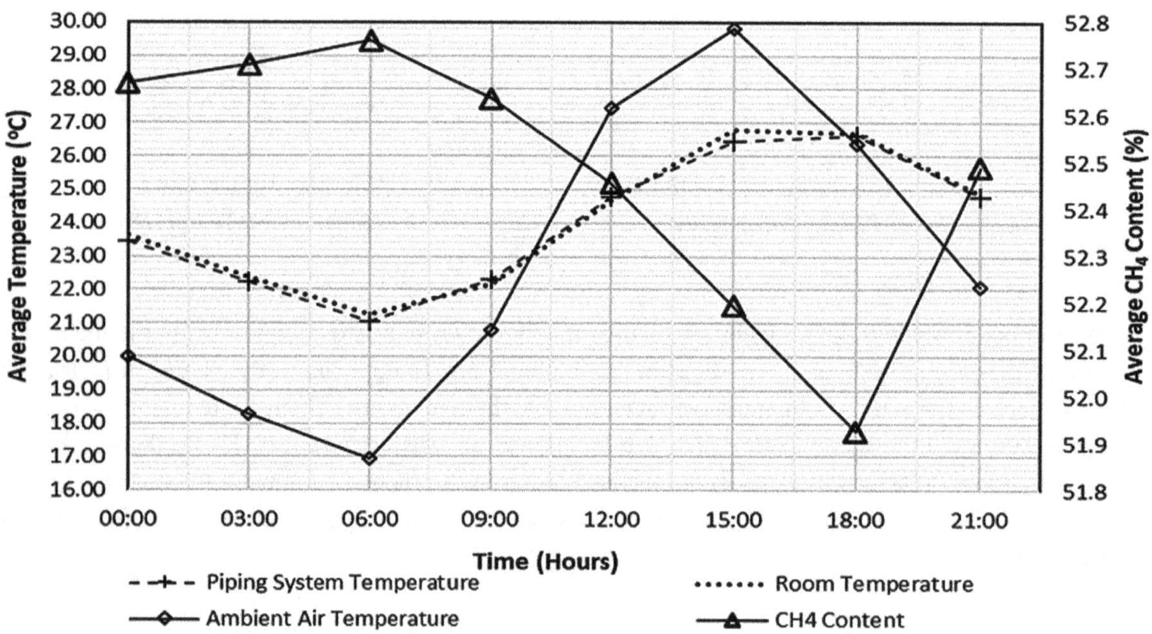

The digester temperature is directly proportional to CH_4 content in the 00:00 - 3:00 hrs and 6:00 -9:00 hrs. Between 18:00 and 00:00 hrs, the digester temperature slowly descends to 24.15°C while CH_4 content quickly ascends to 52.65%. In the remaining time ranges in Figure 14, digester temperature and CH_4 content are inversely proportional.

Figure 14. CH_4 content and digester temperature relationship (Jun 25- Oct 1/ 2017)

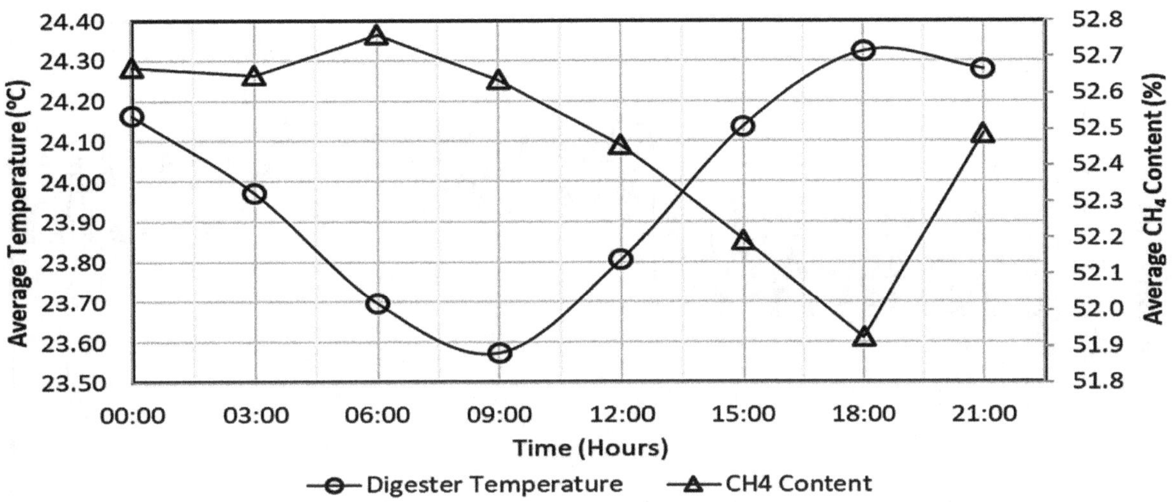

The average CH_4 content in generated biogas from 25/06/2017 to 01/10/2017 was 52.4%, with a standard deviation of 2.9%. Figure 14 shows that the room temperature and digester piping system increased as the average methane content decreased and vice versa.

CH_4 content was higher (averagely 52.8%, with a standard deviation of 0.3%) in the morning when the temperature was very low. Then CH_4 content decreased to 51.9% at 18:00 hours when the temperature inside the digester and the digester piping system was in the high range. This phenomenon may be due to the high solubility of carbon dioxide (CO_2), one of the principal constituents of biogas in the water at low temperatures. The solubility is given as moles of solute, i.e. CO_2 per 100 grams of solvent, e.g. water. Wu *et al.* (2006) reported that CO_2 has half of the solubility at 60°C compared to the same quantity of CO_2 when the temperature was 35°C and found this phenomenon after a laboratory-scale experiment. The aim was to assess the influence of temperature fluctuation on thermophilic anaerobic digestion of municipal organic solid waste. Therefore, this study's relationship between CH_4 content and temperature is supported scientifically.

Pressure, DO, and pH in the ABFAD System

There are various indicators of the functionality of the anaerobic digestion system, including pressure inside the digester. Pressure can indicate whether the system is leaking or not, as it was described by Oloche *et al.* (2017) on testing the underground ADS. In addition, pressure indicates the amount of biogas in the system. Figure 15 shows pressure variations of the ABFADS when the system leaks due to the failure of some pipe fittings. Pressure rose to maximum levels between 15:00 and 18:00 hours and then decreased afterwards before rising again from 9:00 hours. This phenomenon is indicated by the peak and trough of the pressure curve in Figure 15. Besides, the phenomenon is shown clearly in the expanded view in Figure 16. The curves in both figures (Figures 15 and 16) indicate leakage of biogas from the ABFADS, calling for immediate remedial measures.

Figure 15. Pressure variation of ABFADS (08/07/2017 (21:00 hours) - 10/08/2017)

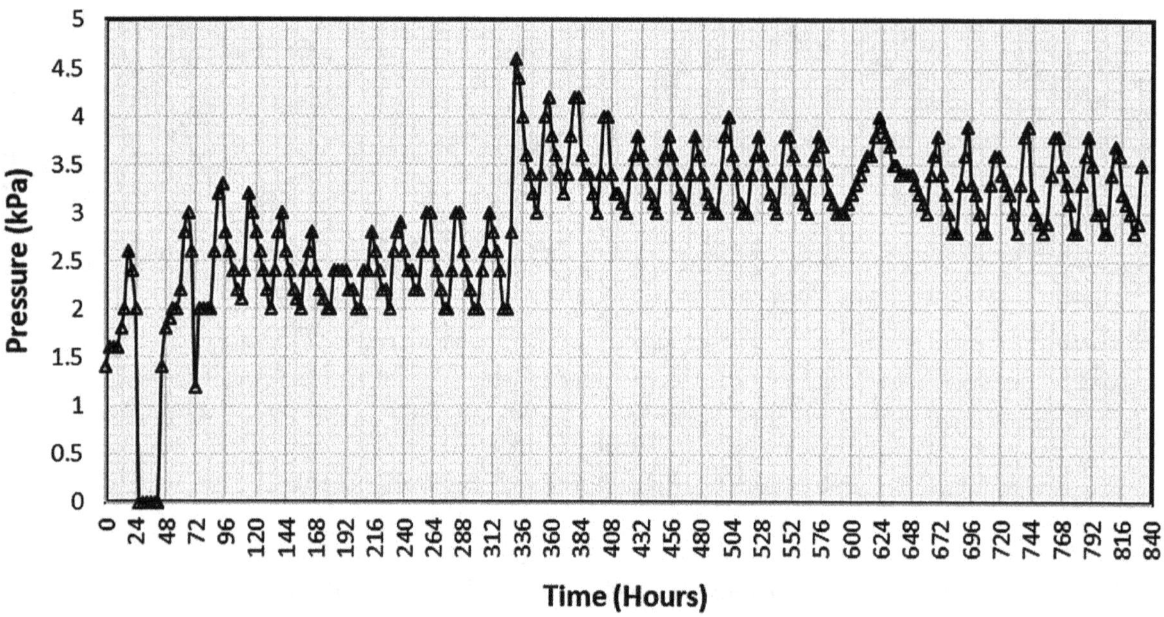

Furthermore, the rising pressure in the daytime at high temperatures indicates more biogas generation than biogas leakage from the system. This condition may also imply that biogas generation microbes are very active, producing biogas at high temperatures, as shown in Figure 16. As temperature decreases, the generation of biogas declines. It implies the reduced activity of microbes to generate biogas, while the rate of biogas leakage is higher than the rate of biogas generation; hence, the pressure in the system drops. In the phenomenon of system leakage, the pressure was hardly above 4.5 kPa.

Figure 16. Pressure variation of ABFADS (12/07/2017 (00:00 hours) - 16/07/2017)

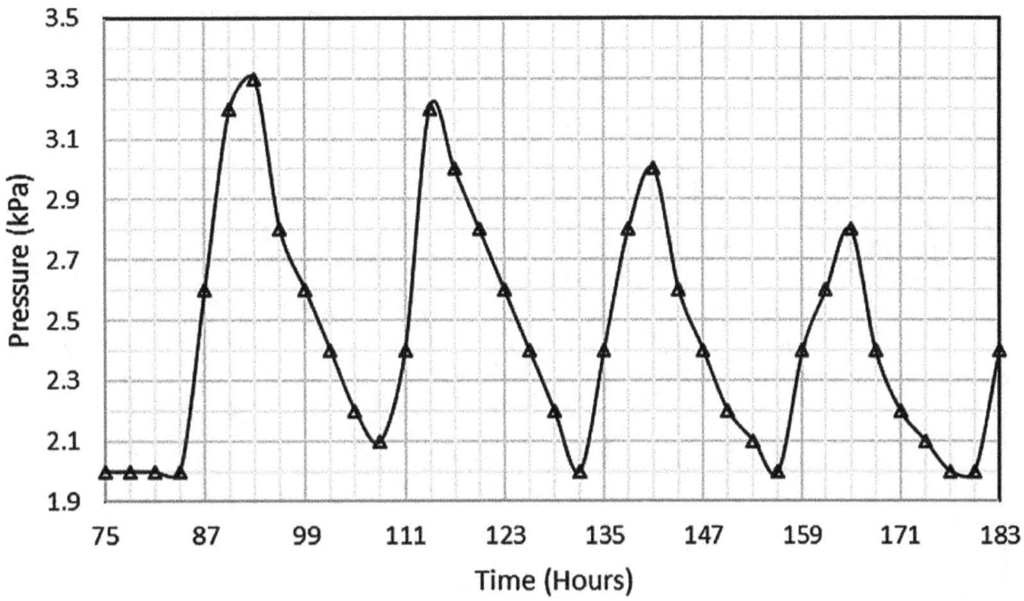

Window 24-48 hours, 75-81 hours, 321-327 hours, and 588-594 hours in Figure 15 indicate the time of remedying the fault pipefittings identified after troubleshooting. The pressure rise after window 321-327 hours indicates the improvement of sealing the biogas leakage. However, rising and falling pressure levels continued afterwards. Finally, at 831 hours, the pressure recording was stopped, and the piping system of ABFADS was disassembled and reassembled to replace all of its faulty pipe fittings. Generally, both pressure curves in Figures 15 and 16 have a shape similar to a sinusoidal curve presented by Kuhi *et al.* (2019).

Figures 17 and 18 illustrate the pressure variation in the ADS after fixing the faults causing leakage. The leakage occurred on the top lid. The rubber gasket of 2 mm thickness was replaced with that of 4 mm thick. In this case, a pressure drop did not occur without releasing biogas for the boiling test and replacing the biogas storage. The sudden drop in pressure seen in Figure 17 indicates the event of releasing generated biogas from the storage or replacing the biogas storage. In Figure 17, the pressure drops within windows of 12-24 hours and 228-240 hours were due to the change of biogas storage with the empty biogas storage. At the same time, the rest occurred due to boiling tests. However, the pressure rising rate decreased as the pressure level increased in the ABFADS (Figure 18). The situation might have been caused by the biogas compressibility, digester's biogas generating rate, and or physical elasticity property of the biogas storage. The biogas storage, i.e. inner tyre tube, was proven to safely withstand a pressure of up to 10.8 kPa (Figure 17 at 228 hours). Generally, the shape of increasing pressure curve before the pressure drop in the system is similar to that of the logarithmic curve (Figures 17 and 18). Grossman-Ponemon *et al.* (2019) have described the logarithmic curve.

Figure 17. Pressure variation of ABFADS (12/09/2017(03:00 hours) - 01/10/2017)

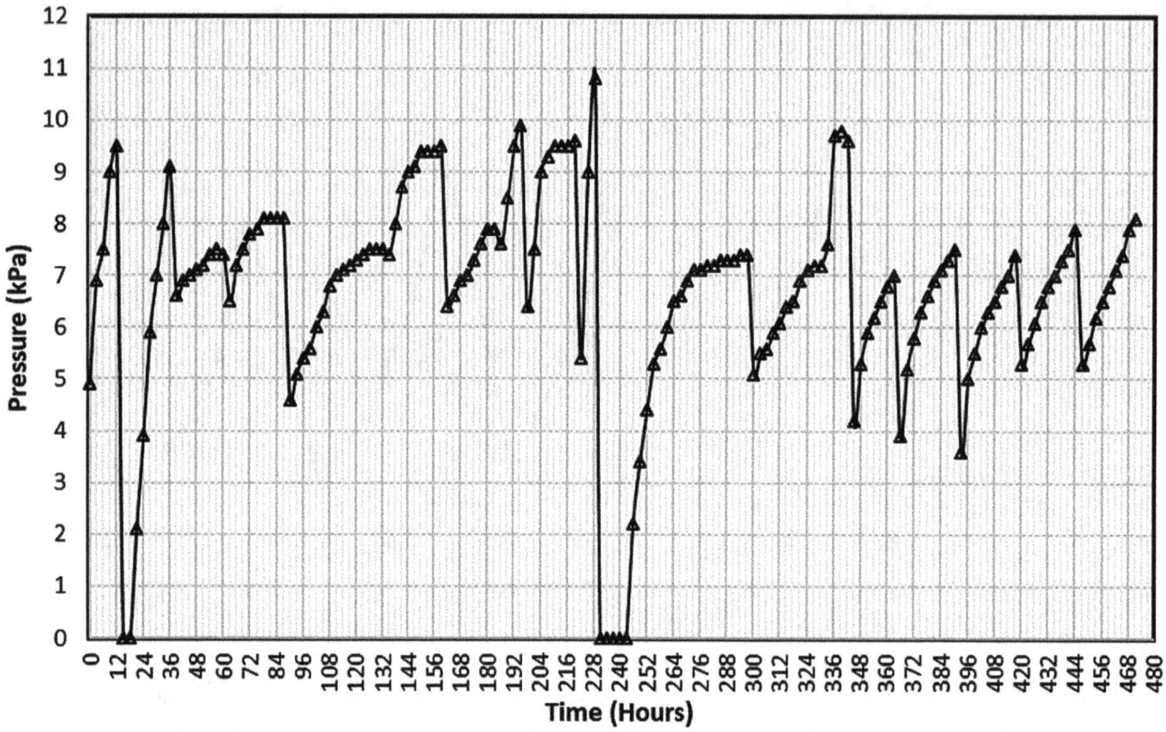

Figure 18. Pressure variation of ABFADS (22/09/2017(06:00 hours) - 24/09/2017)

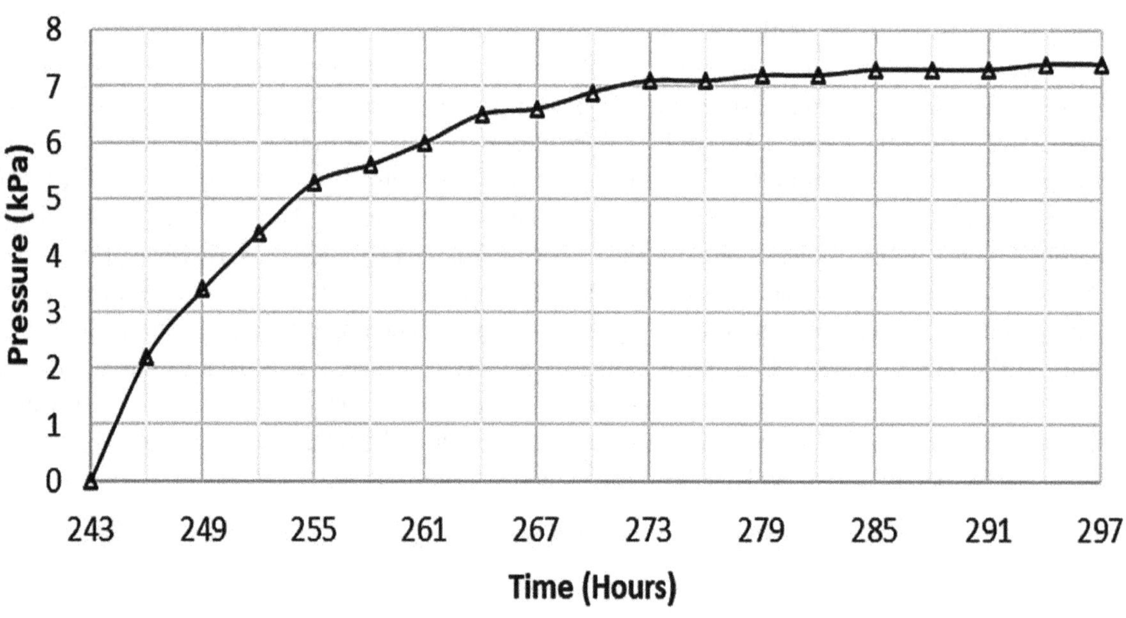

Generally, Figure 19 shows that pH dropped when the temperature was high in the afternoon. This condition is shown in windows 2144-2150 hours, 2168-2174 hours, 2192-2198 hours, and 2216-2222 hours, indicated with bolded vertical lines in the same figure. However, DO was not constantly dropping in the afternoon as pH did. This phenomenon is shown in windows 2192-2198 hours and 2216-2222 hours of Figure 19. Referring to the same figure and Figure 19, both DO and pH values were high when the digester temperature was low, i.e. from 19:00 to 11:00 hours. The digester's pH variation appeared to be influenced more by the digester temperature than the temperature did to DO. Generally, DO in the digester during testing was between 8.8 and 9.0 mg/L while pH was 5.11-5.15 range.

Figure 19. DO & pH versa time of ABFADS (13/09/2017 (00:00 hours) - 17/09/2017)

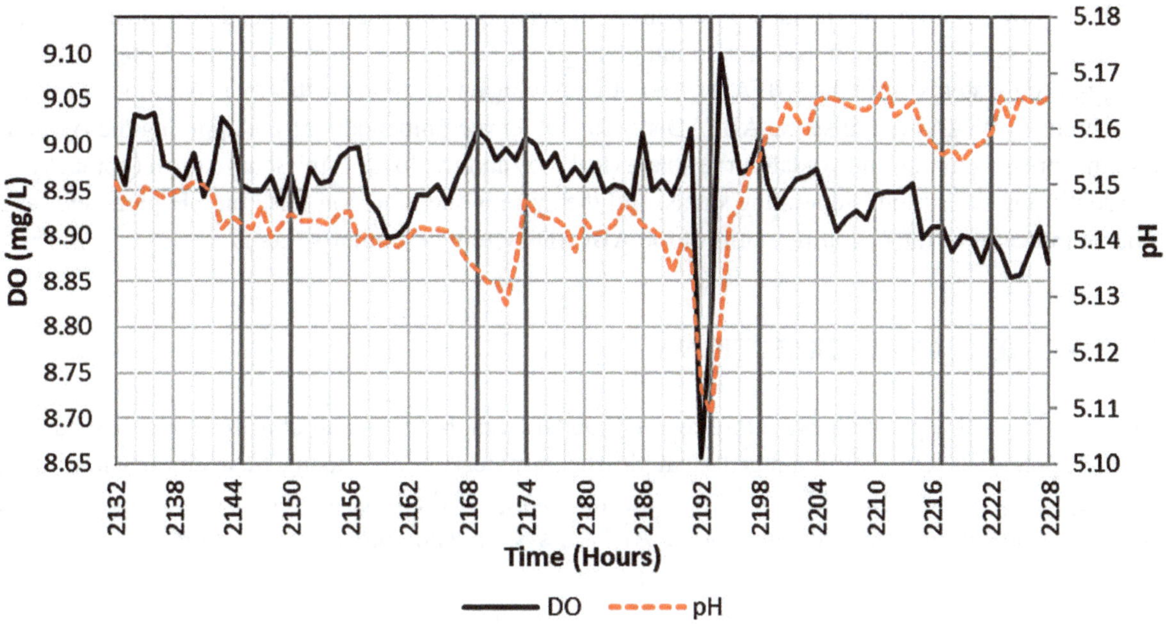

Although not tested, the durability of the ABFADS was envisaged to be more than 15 years if kept away from ultraviolet radiation and physical damage. However, the replacement of a valve made of iron-based material, located between the water trap and digester of ABFADS, is expected to occur after at least seven years (Sun and Nesic, 2007). The replacement is needed because biogas contains a trace of Hydrogen sulphide (H_2S) gas, which corrodes iron materials. The corrosion is caused by the reaction of iron with Sulphur found in H_2S.

SOLUTIONS AND RECOMMENDATIONS

The formulated approach of adapting the biomass-fed digestion system enables the development of the small-sized (with a digester of 0.24 m^3) ABFAD system. The system addresses designs and operational constraints when applying conventional biogas generation systems in RSAAs. Additionally, the system has the following features: no clogging of system biogas flow; easy handling of its auxiliary components like replacing of iron wool and portable biogas storage; withstanding appreciable pressure (10.8 kPa), and self-squeezing capability for easy release of quality biogas with a methane content of 52% ready for use.

The biogas leakage detection in the system using the pressure curves is reliable. The sinusoidal curve shape indicates system leakage, while the logarithmic curve form indicates no system leakage. The pressure values can be measured instantly using pressure gauge tools. The online pressure data collection and curve generation are done with the simple Arduino integrating circuit available in the market. Dissolved Oxygen and pH are indicators of the working condition of the ABFAD system's digester. However, pH is a more reliable indicator of the microbial activities in the system than DO, as the pH. Nevertheless, the DO-value tool can be applied in the absence of the pH-value indicator tool. Simple tools/devices are available for instant online measurement of pH and DO values inside the digester. Therefore, the pressure curves and pH values are appropriate for indicating the working condition of the biogas plant.

The following are recommended to system designers: applying the formulated approach of adapting the biomass-fed digestion system and integrating other renewable energy systems (solar and wind energy-based systems) with the ABFAD system for cooking energy and electricity generation. Also, the households in RSAAs to utilise the ABFAD system. Other recommendations are applying the ABFAD system's pressure in the sinusoidal-curve shape (leaking) and the logarithmic-curve form (not leaking) for identifying the system's operating status, and the system operators to utilise the pH parameter as a reliable indicator of the operating status of the ABFAD system or similar systems

FUTURE RESEARCH DIRECTIONS

The focus of this chapter unlocks future research potential on biomass and climate change issues. The research potentials are investigating the simple-cheap materials, optimising their amount fed in the auxiliary parts of the ABAFAD for improving the quality of the biogas generated, and studying the established operating status indicators of the ABFAD system using other types of biomass.

CONCLUSION

The challenges related to the existing design of the biomass-fed digester and the digester operating status indication are considered in this chapter. In addition, the solutions to address the challenges are devised and demonstrated. The solutions are the formulated explicit conceptual design approach for adapting the biomass-fed digester and the appropriate indicators of the operating condition of the digester for immediate fault detection. The solutions support the theme of Biomass and Bioenergy Solutions for Climate Change Mitigation and Sustainability. The solutions are recommended to the designers and operators of the digester, similar to the ABFAD system.

ACKNOWLEDGMENT

The German Federal Ministry of Education and Research (BMBF) and the Federal Ministry for Economic Cooperation and Development (BMZ) supported this research through the Trans-SEC project, with Project ID No.: 031A249 A-G. The above are ushered with gratitude for this vital support. However, the views expressed in this chapter are purely those of the author and, in any circumstances, may not be regarded as stating an official position of BMBF or BMZ.

REFERENCES

Abu-Taha, R., & Daim, T. (2011). Multi-criteria applications in renewable energy analysis: A literature review. Technology Management in the Energy Smart World Proceedings of Portland International Centre for Management of Engineering and Technology (PICMET) '11: Technology Management in the Energy Smart World, 1-8.

Achinas, S., Achinas, V., & Euverink, G. J. W. (2017). A technological overview of biogas production from biowaste. *Engineering*, *3*(3), 299–307. doi:10.1016/J.ENG.2017.03.002

Adekunle, K. F., & Okolie, J. A. (2015). A review of biochemical process of anaerobic digestion. *Advances in Bioscience and Biotechnology*, *6*(3), 205–212. doi:10.4236/abb.2015.63020

Al, R., Behera, C. R., Gernaey, K. V., & Sin, G. (2019). Towards development of a decision support tool for conceptual design of wastewater treatment plants using stochastic simulation optimisation. *Computer-Aided Chemical Engineering*, *46*, 325–330. doi:10.1016/B978-0-12-818634-3.50055-2

Andreasen, M. M., Hansen, C. T., & Cash, P. (2015). *Conceptual design*. Springer International Publishing. doi:10.1007/978-3-319-19839-2

Awe, O. W., Zhao, Y., Nzihou, A., Minh, D. P., & Lyczko, N. (2017). A review of biogas utilisation, purification and upgrading technologies. *Waste and Biomass Valorization*, *8*(2), 267–283. doi:10.100712649-016-9826-4

Diamond, L. W., & Akinfiev, N. N. (2003). Solubility of CO_2 in water from − 1.5 to 100 C and from 0.1 to 100 MPa: Evaluation of literature data and thermodynamic modelling. *Fluid Phase Equilibria*, *208*(1-2), 265–290. doi:10.1016/S0378-3812(03)00041-4

Divya, D., Gopinath, L. R., & Christy, P. M. (2015). A review on current aspects and diverse prospects for enhancing biogas production in sustainable means. *Renewable & Sustainable Energy Reviews*, *42*, 690–699. doi:10.1016/j.rser.2014.10.055

Garfi, M., Martí-Herrero, J., Garwood, A., & Ferrer, I. (2016). Household anaerobic digesters for biogas production in Latin America: A review. *Renewable & Sustainable Energy Reviews*, *60*, 599–614. doi:10.1016/j.rser.2016.01.071

Grossman-Ponemon, B. E., Heimisson, E., Lew, A. J., & Segall, P. (2019). *Logarithmic growth of dikes from a depressurising magma chamber*. doi:10.31223/osf.io/xekm6

Iswanto, A. M. A., Kebenaran, B., & Megantoro, P. (2021). Design of gas concentration measurement and monitoring system for biogas power plant. *Indonesian Journal of Electrical Engineering and Computer Science*, *22*(2), 726–732. doi:10.11591/ijeecs.v22.i2.pp726-732

Karellas, S., Boukis, I., & Kontopoulos, G. (2010). Development of an investment decision tool for biogas production from agricultural waste. *Renewable & Sustainable Energy Reviews*, *14*(4), 1273–1282. doi:10.1016/j.rser.2009.12.002

Kranert, M., Kusch, S., Huang, J., & Fischer, K. (2012). Anaerobic Digestion of Waste. In *Waste to Energy*. Springer. doi:10.1007/978-1-4471-2306-4_5

Kuhi, H. D., Hossein-Zadeh, N. G., López, S., Falahi, S., & And France, J. (2019). Sinusoidal function to describe the growth curve of dairy heifers. *Animal Production Science*, *59*(6), 1039–1047. doi:10.1071/AN18104

Kulkarni, M. B., & Ghanegaonkar, P. M. (2019). Hydrogen sulfide removal from biogas using chemical absorption technique in packed column reactors. *Global Journal of Environmental Science Management*, *5*(1), 155–166.

Kusekwa, M. A. (2011*). A review on the renewable energy resources for rural application in Tanzania, Renewable energy - Trends and applications*. http://www.intechopen.com/books/renewable-energy-trends-and-applications/areview-on-the-renewable-energy-resources-for-rural-application-in-tanzania

Li, Y., Park, S. Y., & Zhu, J. (2011). Solid-state anaerobic digestion for methane production from organic waste. *Renewable & Sustainable Energy Reviews*, *15*(1), 821–826. doi:10.1016/j.rser.2010.07.042

Mardani, A., Jusoh, A., Nor, K. M. D., Khalifah, Z., Zakwan, N., & Valipour, A. (2015). Multiple criteria decision-making techniques and their applications–A review of the literature from 2000 to 2014. *Economic Research-Ekonomska Istraživanja*, *28*(1), 516–571. doi:10.1080/1331677X.2015.1075139

Mrema, G., Gumbe, L. O., Chepete, H. J., & Agullo, J. (2011). Rural Energy. In *Rural Structures in the Tropics: Design and Development*. Technical Centre for Agricultural and Rural Cooperation.

Muñoz, R., Meier, L., Diaz, I., & Jeison, D. (2015). A review on the state-of-the-art of physical/chemical and biological technologies for biogas upgrading. *Reviews in Environmental Science and Biotechnology*, *14*(4), 727–759. doi:10.100711157-015-9379-1

Olabanji, O. M., & Mpofu, K. (2014). Comparison of weighted decision matrix and analytical hierarchy process for CAD design of reconfigurable assembly fixture. *Procedia Centre d'Insertion et de Réinsertion Professionnelle*, *23*, 264–269. doi:10.1016/j.procir.2014.10.088

Oloche, B., Ozigis, I., Adeyemi, K., & Ikpe, E. (2017). Construction and leakage detection of a dome-type biogas digester in a village at Abuja, Nigeria, Federal. *Journal of Engineering Technology*, *2*(1), 27–31.

Ordway, E. M., Asner, G. P., & Lambin, E. F. (2017). Deforestation risk due to commodity crop expansion in sub-Saharan Africa. *Environmental Research Letters*, *12*(4), 044015–044029. doi:10.1088/1748-9326/aa6509

Poulsen, T. G., & Adelard, L. (2016). Improving biogas quality and methane yield via co-digestion of agricultural and urban biomass wastes. *Waste Management (New York, N.Y.)*, *54*, 118–125. doi:10.1016/j.wasman.2016.05.020 PMID:27256782

Rebernik, M., Bradač, B., Rebernik, M., & Bradač, B. (2008). *Idea Evaluation Methods and Techniques*. Institute for Entrepreneurship and Small Business Management, University of Maribor.

Rowse, L. E. (2011). *Design of Small Scale Anaerobic Digesters for Application in Rural Developing Countries. In Graduate Theses and Dissertations. Civil and Environmental Engineering*. University of South Florida.

Sánchez, Ó. J., & Cardona, C. A. (2012). Conceptual design of cost-effective and environmentally friendly configurations for fuel ethanol production from sugarcane by knowledge-based process synthesis. *Bioresource Technology*, *104*, 305–314. doi:10.1016/j.biortech.2011.08.125 PMID:22137752

Siefers, A., Wang, N., Sindt, A., Dunn, J., McElvogue, J., Evans, E., & Ellis, T. (2010). A novel and cost-effective hydrogen sulfide removal technology using tire derived rubber particles. *Proceedings of the Water Environment Federation*, *12*(12), 4597–4622. doi:10.2175/193864710798182682

SimGas (2018). *The SimGas biogas system is an integrated farm solution for rural households in Africa and Asia*. https://simgas.org/biogas-system

Su, H., Liu, L., Wang, S., Wang, Q., Jiang, Y., Hou, X., & Tan, T. (2015). Semi-continuous anaerobic digestion for biogas production: Influence of ammonium acetate supplement and structure of the microbial community. *Biotechnology for Biofuels*, *8*(1), 1–11. doi:10.118613068-015-0197-z PMID:25705255

Sun, W., & Nesic, S. (2007). A mechanistic model of H2S corrosion of mild steel. *Corrosion-March 19, 2007, NACE International Conference and Expo*, Paper No. 07655. http://www.icmt.ohio.edu/nesic/papers/FullText/conference-91.pdf

Surendra, K. C., Takara, D., Hashimoto, A. G., & Khanal, S. K. (2014). Biogas as a sustainable energy source for developing countries: Opportunities and challenges. *Renewable & Sustainable Energy Reviews*, *31*, 846–859. doi:10.1016/j.rser.2013.12.015

TDA, Time and Date AS. (2018). *All year climate and weather averages in Dodoma, Annual weather averages in Dodoma based on weather reports collected during 1985–2015*. https://www.timeanddate.com/weather/tanzania/dodoma/climate

Theuerl, S., Kohrs, F., Benndorf, D., Maus, I., Wibberg, D., Schlüter, A., Kausmann, R., Heiermann, M., Rapp, E., Reichl, U., Pühler, A., & Klocke, M. (2015). Community Shifts in a Well-Operating Agricultural Biogas Plant: How Process Variations Are Handled by the Microbiome. *Applied Microbiology and Biotechnology*, *99*(18), 7791–7803. doi:10.100700253-015-6627-9 PMID:25998656

Tubecon. (2018). *Mechanical properties of steel for flat bars and sections*. Steel Properties, Tubecon Africa (PTY) Ltd. http://www.tubecon. co.za/en/technical-info/tubecon-wiki/properties-flat-steel-bars-and-sections. html

Wawa, A. I. (2012). *The challenges of promoting and adopting biogas technology in semi-arid areas of Tanzania: The Case of Kongwa and Bahi Districts in Dodoma Region* [PhD thesis]. Open University of Tanzania.

World Bank. (2015). *Promoting Biogas as Sustainable Clean Cooking Fuel for Rural Households in Kenya Project*. http://documents.worldbank.org/curated/en/ 350771468272038585/pdf/SFG1441-EA-P153493-ESMF-Box393229B-PUBLIC-Disclosed-10-20-2015.pdf

Wu, B. (2013). Advances in the use of CFD to characterise, design and optimise bioenergy systems. *Computers and Electronics in Agriculture*, *93*, 195–208. doi:10.1016/j.compag.2012.05.008

Wu, M. C., Sun, K. W., & Zhang, Y. (2006). Influence of temperature fluctuation on thermophilic anaerobic digestion of municipal organic solid waste. *Journal of Zhejiang University. Science*, *7*(3), 180–185. doi:10.1631/jzus.2006.B0180 PMID:16502503

Wu, Y., Kovalovszki, A., Pan, J., Lin, C., Liu, H., Duan, N., & Angelidaki, I. (2019). Early warning indicators for mesophilic anaerobic digestion of corn stalk: A combined experimental and simulation approach. *Biotechnology for Biofuels*, *12*(1), 1–15. doi:10.118613068-019-1442-7 PMID:31073330

Yang, L., Xu, F., Ge, X., & Li, Y. (2015). Challenges and strategies for solid-state anaerobic digestion of lignocellulosic biomass. *Renewable & Sustainable Energy Reviews*, *44*, 824–834. doi:10.1016/j.rser.2015.01.002

Zalm, J. E. (2017). *Performance of domestic biodigesters in Kenya towards the development of a comparative method for measuring the performance of different biodigester types in Sub-Saharan Africa* [MSc dissertation]. Utrecht University, Utrecht, Netherlands.

ADDITIONAL READING

Demirbas, A. (2005). Potential applications of renewable energy sources, biomass combustion problems in boiler power systems and combustion-related environmental issues. *Progress in Energy and Combustion Science*, *31*(2), 171–192. doi:10.1016/j.pecs.2005.02.002

Gielen, D., Boshell, F., Saygin, D., Bazilian, M. D., Wagner, N., & Gorini, R. (2019). The role of renewable energy in the global energy transformation. *Energy Strategy Reviews*, *24*, 38–50. doi:10.1016/j.esr.2019.01.006

Jiang, Y., Heaven, S., & Banks, C. J. (2012). Strategies for stable anaerobic digestion of vegetable waste. *Renewable Energy*, *44*, 206–214. doi:10.1016/j.renene.2012.01.012

Kuleape, R., Cobbina, S. J., Dampare, S. B., Duwiejuah, A. B., Amoako, E. E., & Asare, W. (2014). Assessment of the energy recovery potentials of solid waste generated in Akosombo, Ghana. *African Journal of Environmental Science and Technology*, *8*(5), 297–305. doi:10.5897/AJEST2014.1663

Panwar, N. L., Kaushik, S. C., & Kothari, S. (2011). Role of renewable energy sources in environmental protection: A review. *Renewable & Sustainable Energy Reviews*, *15*(3), 1513–1524. doi:10.1016/j.rser.2010.11.037

Sohni, S., Norulaini, N. N., Hashim, R., Khan, S. B., Fadhullah, W., & Omar, A. M. (2018). Physicochemical characterisation of Malaysian crop and agro-industrial biomass residues as renewable energy resources. *Industrial Crops and Products*, *111*, 642–650. doi:10.1016/j.indcrop.2017.11.031

Spataru, C. (2021). Business Models and Policies to Support Renewable Energy Community in Islands. In Research Anthology on Clean Energy Management and Solutions (pp. 1526-1543). IGI Global. doi:10.4018/978-1-7998-9152-9.ch065

Twidell, J., & Weir, T. (2015). *Renewable energy resources*. Routledge. doi:10.4324/9781315766416

KEY TERMS AND DEFINITIONS

Adaptation: The reference to a process of varying the system or structure design to get its improved design with high performance for the betterment of the user or community.

Conceptual Designing: An act of extensive outlining of purposes and forms of a system or structure to be established for facilitating the conception of a clear user interface to understand and interpret the system or structure.

Detailed Designing: The process in which the generated system design is refined where plans, specifications and estimates are created and ready for manufacturing or construction.

Rural Semi-Arid Areas: In the sub-Saharan Africa context, a land of marginal crop production potential, quick maturing cereals, and carrying natural vegetation of Acacia-Themeda association but including dry Brachystegia woodland. It is a potentially productive grazing area, limited by bush encroachment, leached soils, and inadequate water.

Chapter 18
Pretreatment of Lignocellulosic Biomass and 2G Ethanol

Hitesha J. Panchal
Parul University, Vadodara, India

Krishan Kumar
Parul University, Vadodara, India

ABSTRACT

Rapid depletion of fossil fuel-based energy sources increased the demand for alternate energy sources. Lignocellulosics-based 2G ethanol can be used as an alternative sustainable source that presents in ample amount. Sources of lignocellulose biomass are wood, food-agriculture wastes, and forest residues. Cellulose, hemicellulose, and lignin are the core components of lignocellulosic biomass. Cellulosic and hemicellulosic biomass are enzymatically hydrolyzed to produce the monomer sugar (such as glucose or xylose) which is further converted into ethanol using fermentation process. The presence of lignin provides physical barrier that limit the access of enzymes required for saccharification. Pretreatment helps in removing the lignin from biomass and reducing recalcitrance. Pretreatment can be done by conventional methods, which are chemical, physical, and biological. This study covers the different methods of pretreatment including their disadvantages and benefits along with saccharification and fermentation processes.

INTRODUCTION

Globally, the rapid use of energy causes the rapid reduction of fossil fuel-based energy sources. Also, the chief concern with its frequent use is that it releases harmful (unfavourable) gases like methane, CO, CO_2, nitrous oxide, etc. which are known as greenhouse gases (Houghton, 2005). Since these energy sources are vanishing, we need to emphasis on finding its sustainable alternative source. The goal should be to develop a long-term sustainable energy source that emphasizes "recovery and reuse" of bio-derived raw materials.

DOI: 10.4018/978-1-6684-5269-1.ch018

Lignocellulosic biomass (LCB) has acknowledged a lot of interest as a renewable, sustainable energy source. LCB has the potential for meeting future energy requirements while posing no environmental risks (Ahmad *et al.*,2020). Wheat straws, rice husk, sugarcane bagasse, forest and agricultural wastes, industrial and urban trash might all be good sources of lignocellulosic biomass (LB) (bioethanol) (Zheng *et al.*, 2014; Amin *et al.*, 2017).

Traditionally, sugar and starch-based feedstocks including sugarcane juice and molasses, as well as corn, have been used to make bioethanol (Balan *et al.*, 2013). Bioethanol is currently being generated on a commercial scale using lignocellulosic feedstocks (Balan *et al.*, 2013). LBM is the most abundant raw material found in nature which includes sources like hardwood, grasses, softwood, agricultural wastes, etc. Newsprint, office paper, municipal solid wastes, and other raw materials might be of interest as well. Starch is a simple substrate, whereas lignocellulosic biomass is more complicated. Second generation (2G) ethanol is less trending when compared to the first generation ethanol, and it is still less economical due to its high cost (Furlan *et al.*, 2013; Macrelli *et al.*, 2014). Several research of sugarcane bagasse-based 2G ethanol production techniques have been conducted. In one study, enzymatic hydrolysis of sugarcane bagasses after diluted acid pretreatment produced fermentable sugar for high yield bioethanol production (Benjamin *et al.*, 2013; Albuquerque *et al.*, 2013).

Lignocellulosic biomass is made of three key components which are cellulose, hemicellulose, and lignin. In which cellulose is the polysaccharide present in great amount therefore known as the most abundant LBM. Cellulose is the linear chain of glucose linked via β-1,4 glycosidic linkages. Hemicellulose is branched heteropolymer whereas coniferyl alcohol, sinaphyl alcohol and p-coumaryl alcohol make form the three-dimensional macromolecule lignin. Lignin causes the recalcitrance in the LBM (Alonso *et al.*,2012).

Since this recalcitrance of lignocellulose biomass is being the hurdle in bioethanol production, so, it should be reduced to get efficient yield of bioethanol. The recalcitrance of biomass can be reduced by using numerous pretreatment methods like chemical pretreatment, biological and physical pretreatment. By doing pretreatment we can increase the enzyme accessibility and hydrolysis of cellulose and hemicellulose polysaccharides, which further leads to higher bioethanol yield.

LIGNOCELLULOSIC BIOMASS

Lignocellulosic biomass is ubiquitous and sustainable resource which is composed of polysaccharides (cellulose and hemicelluloses) and an aromatic polymer (lignin) (Fengel *et al.*, 1984). The percentage of cellulose, hemicellulose and lignin varies with respect to sources (Table 1).

COMPONENTS OF LIGNOCELLULOSIC BIOMASS

Lignin, hemicellulose and cellulose together makes the lignocellulose biomass. The main chain of cellulose is made up of D-glucose subunits joined by β-1,4 glycosidic linkages (Fengel *et al.*, 1984). Cellulose is present in the cell wall of the plants and is the most abundant biomass. Cellulose is 'bundled' into cellulose fibrils or cellulose bundles (fig 1) and are interlinked to each other through van der Waals and hydrogen bonds (Fengel *et al.*, 1984). These cellulose fibrils are generally self-contained and have just a weak hydrogen bonding connection (Fengel *et al.*, 1984).

Figure 1. Structure of Cellulose

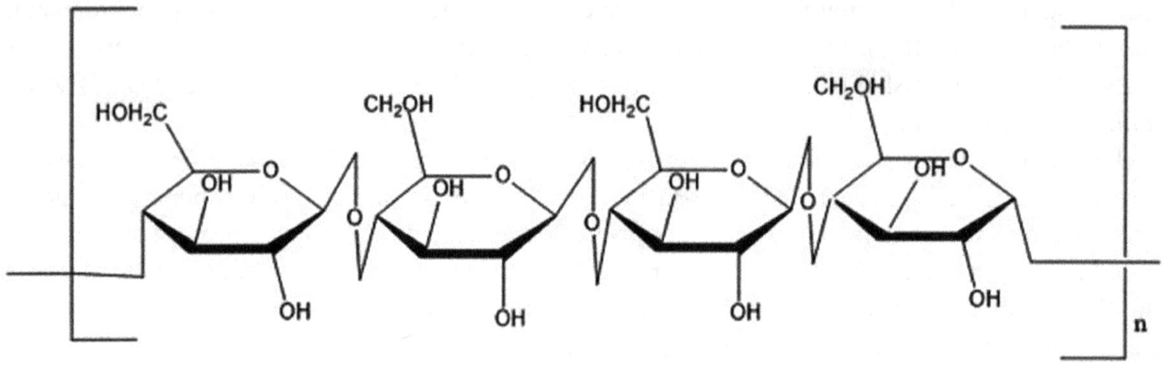

Xylose, arabinose, and other pentoses, as well as hexoses like mannose, glucose, and galactose, along with sugar acids make form the complex carbohydrate structure known as hemicellulose (fig. 2) (Grabber, 2005). In hardwood and agricultural plants like grasses and straw, xylan is the most prevalent kind of hemicellulose., whereas glucomannan is the most abundant component in softwood (Grabber, 2005). Lignin and cellulose fibers are connected with hemicellulose and this linkage increases the stiffness of the entire cellulose–hemicellulose–lignin complex (Levan et al., 1990; Winandy, 1995). Hemicelluloses are the more thermally reactive compared to cellulose and lignin, because, It's solubility increases as the temperature rises. (Levan et al., 1990; Winandy, 1995).

Figure 2. Structure of Hemicellulose a) Xyloglucan b) Glucomannan

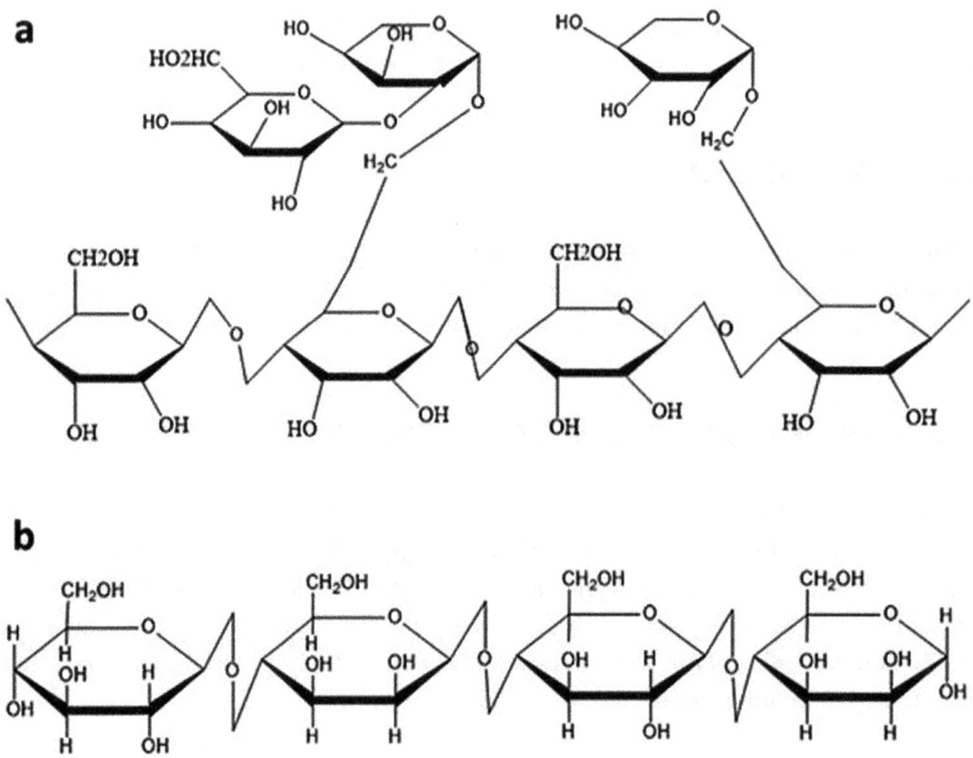

Plant cell walls include lignin, which is one of the plentiful polymer in nature (fig.3), along with hemicellulose and cellulose. Three different phenylpropane units (p-coumaryl, coniferyl, and sinapyl alcohol) together forms the amorphous heteropolymer, which are joined together by various connections (Fengel *et al.*, 1984). The key role of lignin in plant cell wall is to provide support to maintain the structure of cell walls and to maintain the impermeability and oxidative stress also, it gives resistance to microbial assault (Grabber, 2005). Under neutral circumstances, lignin dissolves into water at 180^{0}C (Bobleter,1994). But the lignin's precursor decides whether it dissolves in an acidic, neutral, or alkaline environment (p-coumaryl, coniferyl, sinapyl alcohol, or mixtures thereof).

Figure 3. Structure of Lignin

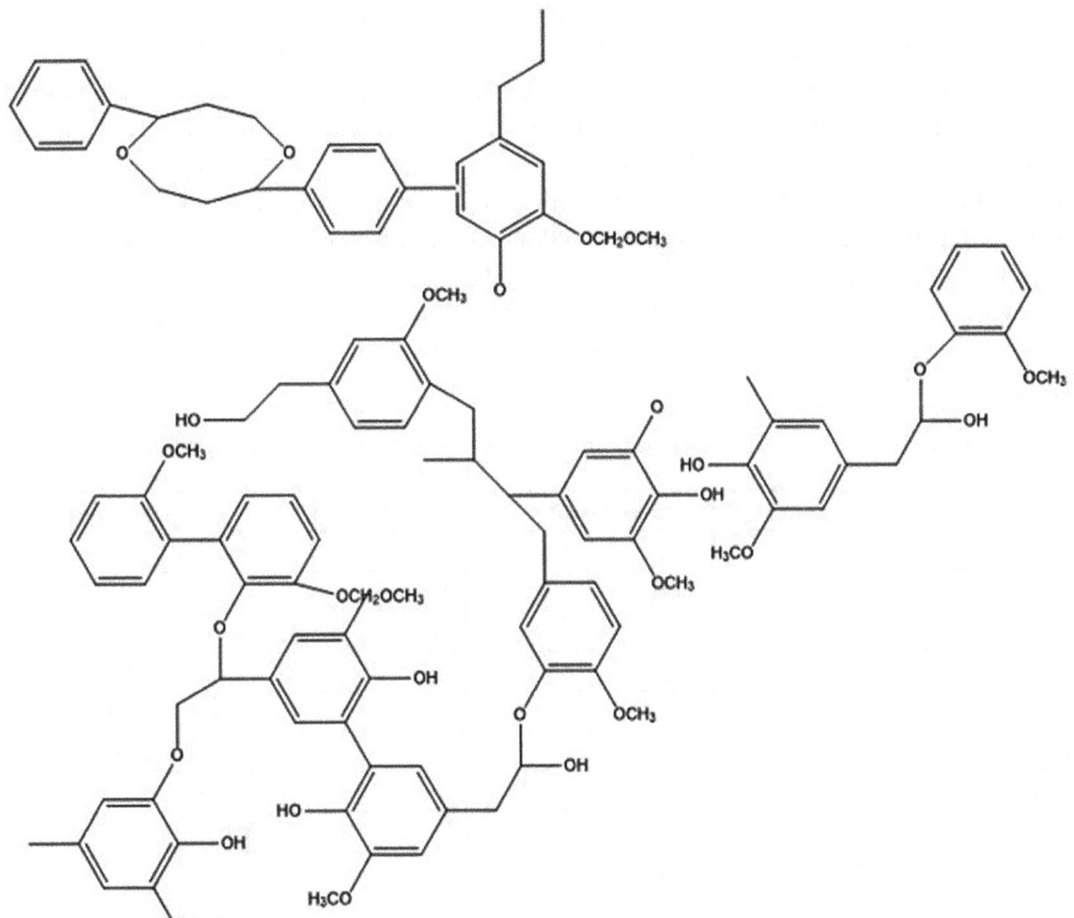

SOURCES OF LIGNOCELLULOSIC BIOMASS

To date, the majority of ethanol produced as a biofuel has come from edible crops, either through the fermentation of sugars or the conversion of starch from corn (Hill et al. 2006; Chandel and Singh 2011). The most accessible renewable resource on earth is the cell wall of a plant. A long-term option for the production of biofuels is therefore thought to be the manufacture of ethanol from non-crop plant cell walls or non-edible plant parts. Forages, in particular cultivated grasses, can be considered as significant prospects for the generation of cellulosic biomass, and they look to be very promising as a future biomass sources among the numerous kinds of biomass accessible for ethanol synthesis (Hill et al. 2006; Chandel and Singh 2011). The predominant herbaceous energy plants that have been chosen for the production of bioethanol (*Pennisetum purpureum*) include switchgrass (*Panicum virgatum*), miscanthus (*Miscanthus spp. Anderss.*), canary grass (*Phalaris arundinacea*), giant reed (*Arundo donax L.*), alfalfa (*Medicago sativa L.*), and Napier grass (Hill et al. 2006; Chandel and Singh 2011). Willow, eucalyptus, silver maple, eastern cottonwood, green ash, black walnut, sweetgum, and sycamore are a few examples of fast-growing, short-rotation forest trees that can be utilised as a feedstock for biofuels (Hadar et al., 2013). Sources of woody materials include natural forest wastes, forestry wastes including sawdust from sawmills, wood chips, and branches from dead trees, as well as short rotation energy forest plantations that use a variety of fast-growing tree species (Hadar et al., 2013). Municipal Solid Waste (MSW), which consists of paper and cardboard, food scraps, and garden waste, can be converted into biofuel, but it is not the best feedstock because of the variety of MSW components, as well as the incomplete source separation and contamination (Li et al., 2007).

Rice straw, wheat straw, maize stover, and sugarcane bagasse are the agricultural wastes that are most readily available in terms of biomass (Kim *et al.*, 2004). One of the most significant agricultural crops in countries with tropical climates is sugarcane. Bagasse, the byproduct of sugarcane obtained after crushing the sugarcane to get the juice, is required for the production of sugar and ethanol. It has an advantage over other crop leftovers since it saves the time and resources (Chandel *et al.*, 2012).

Table 1. Percentage of hemicellulose, lignin and cellulose in different LCB sources.

Sources	Lignin %	Hemicellulose%	Cellulose %	References
Sugar cane	20–42	19–25	42–48	Kim *et al.*,2011
Rice straw	12–14	23–28	28–36	Saini *et al.*,2015
Corn fiber	8.4	16.8	14.28	Mosier *et al.*,2005
Switch grass	12	31.4	45	Sun *et al.*,2005
Cattle manure	2.7–5.7	1.6–1.33	16–4.7	Sánchez *et al.*,2009
Wheat straw	17–19	26–32	33–38	Saini *et al.*,2015
Hardwood	18-30	25-40	40-50	Sun *et al.*,2005
Alfalfa stalks	16.6	6.5	48.5	Chan *et al.*,2007
General MSW	10–14	49.9–16	33	Li *et al.*,2012
Napier grass	9	20	32	Saxena *et al.*,2009
Food waste	11.4	7.2	55.4	Schmitt *et al.*,2012
Miscanthus	24.1-25	18-24.3	38.2-40	Brosse *et al.*,2010
Newspaper	18-30	25-40	40-55	Howard *et al.*,2003
Corn stover	7-19	24-26	38-40	Zhu *et al.*,2005
General grasses	10-30	25-50	24-40	Saini *et al.*,2015

PRETREATMENT

Hydrolysis of polysaccharide components to soluble sugars, which may then be fermented to produce desired end products, is used to produce fuels from renewable lignocellulosic sources (Singh *et al.*, 2015). Because such materials are only partially digestible in their natural state, some sort of pretreatment is usually necessary to make the carbohydrate component amenable to enzymatic hydrolysis (Singh *et al.*, 2015). Degradation of lignocellulosic biomass is difficult due to its intrinsic recalcitrance. Pretreatment disrupts the lignocellulosic biomass's tightly packed structure and reveals the fibres made of cellulose and hemicellulose (Singh *et al.,* 2015). Using a mix of chemical and structural changes to the lignin and carbohydrates, the lignocellulosic material is processed in order to overcome recalcitrance (Singh *et al.,* 2015). The diverse composition and refractory nature of the cellulosic fraction of the biomass provide major challenges in the conversion strategy. Pretreatment is used to render cellulosic and hemicellulosic biomass susceptible to enzyme activity (Bharat *et al.*, 2013). The main factors influencing accessibility and digestibility are cellulose crystallinity, hemicellulose disruption, accessible surface area (porosity), lignin protection, and cellulose-hemicellulose-lignin linkage.

Hemicellulose and lignin are ubiquitous polymers, while cellulose is the primary contributor to the crystalline component. Lignin serves as a barrier to prevent the deteriaration of cellulose and hemicellulose. Because lignin and hemicellulose are chemically related through covalent bonding, removing lignin will result in the elimination of hemicellulose as well. To improve the specific surface area and crystallinity of cellulose, it requires pretreatment to eliminate these amorphous polymers (Y.P *et al.*,2013).

Fractionation and delignification are two popular pretreatment methods for this lignocellulosic material. Fractionation is a process for separating the components of lignocellulosic biomass by disturbing the biopolymer pattern to enhance access to polysaccharides in lignocellulosic biomass. Delignification is used to remove lignin, although it can also separate hemicellulose fractions. To extract lignin for exposing cellulose to enzymatic hydrolysis, delignification is usually incorporated in the fractionation process. Physical, chemical, biological and physicochemical methods are some of the pretreatment methods (fig 4). Physical pretreatment method includes size reduction, microwave irradiation, etc. Chemical pretreatment method includes alkaline, acid and organic solvent methods. Physicochemical pretreatment methods are hot water, steam explosion whereas biological pretreatment methods deal with microorganisms. Advantages and disadvantages of different types pretreatment methods are shown in table 2.

Figure 4. Types of pretreatment methods

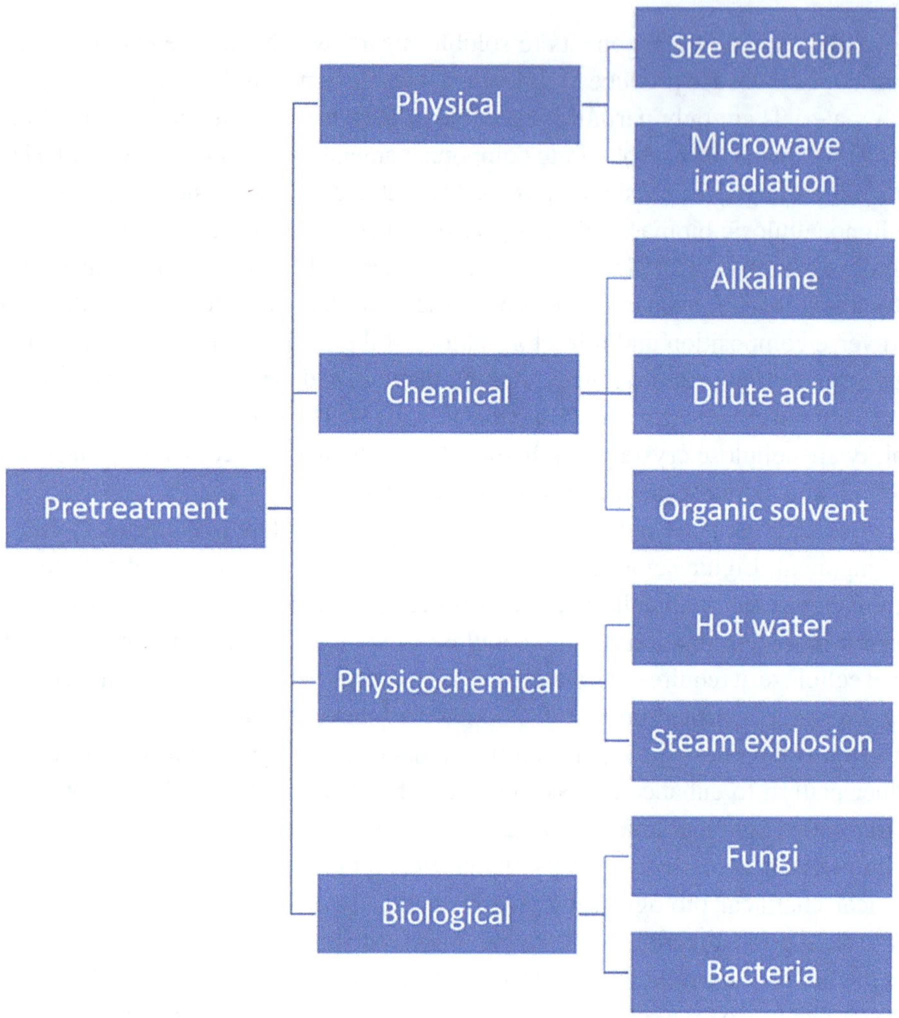

1. Physical Pretreatment Methods

These techniques are employed to change the size of the biomass and break the crystal structure of LBM.

a) Size Reduction Pretreatment Method

Size reduction is a common approach used for pretreatment of lignocellulose biomass through which polymerization and cellulose crystallinity is lowered. The size of biomass can be reduced by employing various physical size reduction methods like milling (using a ball or hammer and vibratory rod/ball milling), shredding, chipping and grinding (J.Y *et al.*, 2013). The amount of power required for mechanical size reduction is determined on the basis of the moisture content and the size of the biomass. As a result, particle size has an impact on specific energy consumption. Because of the need for high mass and heat transfer efficiency, the design of handling, transportation, and conversion facilities will certainly be determined by the selection of an appropriate particle size of the biomass (Barakat *et al.*, 2013; J.Y *et al.*, 2013).

b) Microwave Irradiation Pretreatment Method

Microwave irradiation for lignocellulosic biomass is another type of physical preparation. The remarkable heating efficiency and simplicity of this pretreatment method, which has been improved over time, are well known. Physical, chemical, and biological processes speed up as a result of heat and many collisions brought on by polar molecule oscillation and ion movement. The lignocellulosic material's performance is influenced by its dielectric properties. The ability of a material to convert electromagnetic energy into heat is measured by the dielectric loss factor, as opposed to a material's capability to store electromagnetic energy (H. Ma *et al.*, 2009).

2. Chemical Pretreatment Methods

The recalcitrant structure of lignocellulose materials can also be changed by treating with chemicals. These pretreatment methods majorly include alkaline method, dilute acid method and organic solvent method to reduce the crystallinity.

a) Acid Pretreatment Method

Acid pretreatment is used primary to remove hemicellulose from lignocellulose biomass. Enzymes are more easily accessible to cellulose after pretreatment of biomass using an acidic method (Velmurugan *et al.,* 2011). Sulfuric acid, acetic acid, and phosphoric acid are the three acids that are used for pretreatment the most frequently (Rezania *et al.,*2020). Acid pretreatment can be carried out in two distinct ways: concentrated acid pretreatment at comparatively lower temperatures, and diluted acid pretreatment at higher temperatures (Rezania *et al.,*2020). By-products of acid pretreatment include 5-hydroxymethyfurfural and 2-furfuraldehyde. Peracetic acid pretreatment can also successfully remove lignin and produce hemicellulose breakdown, revealing cellulose. Acid pretreatment eliminates almost all carboxylic acid changes, including acetyl groups and uronic acids. (Park *et al.*, 2012; Zhao *et al.*,2008). It removes the hemicellulose component and promotes the crystallinity of the biomass (Zhao *et al.*,2008). The residence time might last anywhere between a few seconds and minutes, depending on the pretreatment's temperature. In a stainless-steel reactor, an aqueous substrate solution is heated to the necessary temperature and then processed with warmed sulfuric acid.

b) Alkaline Pretreatment Method

For the pretreatment of biomass, hydroxides of potassium (K), sodium (Na), ammonium (NH_4), and calcium (Ca) are used in alkaline pretreatment method. This pretreatment technique causes the biomass to swell, increasing its internal surface area while decreasing its level of polymerization and cellulose crystallinity (Chandra *et al.,* 2007). Alkaline pretreatment alters the structure of lignin and removes the bonds that bind it to the other carbohydrate fractions, making the heteromatrix's carbohydrates more accessible. As the lignin is eliminated, the efficiency of polysaccharide enzyme hydrolysis rises (Galbe *et al.,* 2007). Alkali pretreatments also get rid of acetyl and other uronic acid substitutions on hemicellulose that make the surface of the cellulose less accessible to enzymes (Mosier *et al.*,2005). When, compared to acid pretreatment with HCl and H_2SO4, alkaline pretreatment with NaOH generally removes more lignin (Zhao *et al.*,2008). There are no by-products of the alkaline treatments. It results in increased lignin removal, biomass porosity, and cellulose swelling. Almost all carboxylic acid substitutions will be eliminated by alkaline pretreatment (Garlock *et al.*, 2011; Park *et al.*, 2012; Zhao *et al.*,2008).

c) Organic Solvent Pretreatment Method

The organosolv procedure is another appealing chemical pretreatment. This method has been widely used to extract high-quality lignin. Greater biomass conversion results from the removal of lignin from the biomass because this makes the cellulose fibres available to enzymes for hydrolysis and limits the amount of lignin that cellulolytic enzymes can digest. (Wormeyer *et al.*,2011). In this process mixture of aqueous organic solvent (ethanol, methanol, ethylene glycol, and acetone) is added with or without use of catalyst (Acid, base, or salt) along with high temperature. By doing so, linakages of hemiceelulose and lignin are cleaved due to solubilization which improves the surface area overall for enzymatic hydrolysis of cellulose (Hage *et al.*,2009 ; Wormeyer *et al.*,2011).

3. Physicochemical Pretreatment Methods

a) Liquid Hot Water Pretreatment Method

This pretreatment method uses H_2O at high temperature to reduce the crystallinity of lignocellulosic biomass. Using this technique, lignin is removed and the hemicellulose is hydrolyzed along with the cellulose making them easily accessible to the enzymes. The hemiacetal linkages are broken down followed by acid liberation during the hydrolysis of biomass by hot water (Antal,1996). Because lower temperatures are used, there is less chance of degradation products forming. And since, the pretreatment solvent in this case is water, there is no need for a final washing or neutralising step. For large-scale applications, the solvent's low cost is also a benefit.

b) Steam Explosion Pretreatment Method

In Steam explosion process, before being physically treated with high-pressure saturated steam in the steam explosion process, biomass is chipped, crushed, or just left raw. Because of the high temperature, the hemicellulose breaks down and lignin is transformed, enhancing the possibility for cellulose hydrolysis (Agbor *et al.*, 2011). Poplar plant chips processed by steam explosion showed enzymatic hydrolysis efficiency of 90% in 24 hours, while untreated chips showed only 15% hydrolysis efficiency (Grous *et al.*,1986). The increase in enzymatic efficiency gave the clear idea about the role of steam explosion in the LBM's pretreatment.

4. Biological Pretreatment Methods

Biological pretreatment methods include the use of Fungi and bacteria. Fungi like White-rot fungi (*Dichmitus squalens, Phlebia radiata, Rigidosporus lignosus, Phanerochaete chrysosporium,* and *Jungua separabilima*), brown-rot fungi (*Serpula lacrymans, sulfur shelf, Fomitopsis pinicola and Phae-*

olus schweinitzii), soft-rot fungi (*Chaetomium, Ceratocystis,* and *Kretzschmaria deusta*), and bacteria (*Clostridium thermocellum, Caldicellulosiruptor saccharolyticus, Caldicellulosiruptor bescii* (Putro *et al., 2016)* are used to change chemical composition of LBM to make it more enzymatically digestible. Fungi degrade lignocellulosic biomass in different ways. White-rot fungus breakdown lignin more aggressively, whereas brown and soft rots fungi target cellulose and showed less effect on lignin (Putro *et al., 2016*). Degradation of lignocellulosic materials has been explored using many microorganisms with thermostable or thermophilic activity such microorganisms are called thermophilic microorganisms. These microorganisms provide a number of benefits, including a faster hydrolysis time, great tolerance to low and high pH, reduced contamination risk, and cheap energy costs.

Table 2. Advantages and Disadvantages of various pretreatment methods.

Pretreatment Methods	Advantages	Disadvantages	Reference
Size reduction	Disrupt biomass crystallinity, reduces the degree of polymerization	Larger the particle size lowers the crystallinity.	Barakat *et al.*,2013; J.Y *et al.*, 2013.
Microwave irradiation	Facilitate the disruption of recalcitrant structures	High cost	H. Ma *et al.*, 2009
Alkaline method	Lignin structure is disrupted	Highly costly and used for small scale only.	Park *et al.*, 2012; Zhao *et al.*,2008
Organic solvent method	Hydrolyses hemicellulose and lignin.	High cost	Hage *et al.*,2009
Acid method	High yield of end product (sugars)	Degradation products are generated.	Zhao *et al.*,2008
Hot water method	No need to reduce size of biomass, chemicals	Toxic compounds are formed.	Bobleter, 1994
Steam explosion method	Alters lignin and solubilize hemicellulose	Partially degrade hemicellulose	Mes-Hartree *et al.*, 1988
Biological method	No requirement of chemicals. Causes delignification	Process is slow and not very efficient.	Putro *et al., 2016;* Maurya *et al.*,2015

2G ETHANOL

The biofuel with the greatest potential to replace oil-based fuels is second-generation (2G) ethanol made from lignocellulosic biomass (Seabra *et al.*,2010). The stages involved in producing ethanol from lignocellulosic biomass are as follows: 1) pretreatment of lignocellulosic biomass 2) saccharification 3) fermentation as shown in fig 5.

Because LBM's high resistant structure restricts the accessibility of cellulose and hemicelluloses to hydrolytic enzymes and prevents the release of sugars, it hinders the enzymatic hydrolysis process. To, achieve effective and quick enzymatic hydrolysis of carbohydrates, biomass must be pretreated to change its structure (Mosier *et al.*, 2005). Lignocellulosic ethanol production involves the enzymatic degradation of cellulose and hemicelluloses into fermentable sugars (Pan *et al.*, 2005). The composition and structure of the feedstocks, the pretreatment technique utilised, and the kind of enzymes used are all factors that affect this stage. Enzyme costs and catalytic efficiency are a key stumbling block to enhancing the bioethanol industry's profitability. Optimizing enzyme mixtures for enzymatic hydrolysis is crucial for the economic viability of lignocellulosic ethanol production. Now in case of saccharification and fermentation, it can be done separately or simultaneously.

Figure 5. Production of Bioethanol

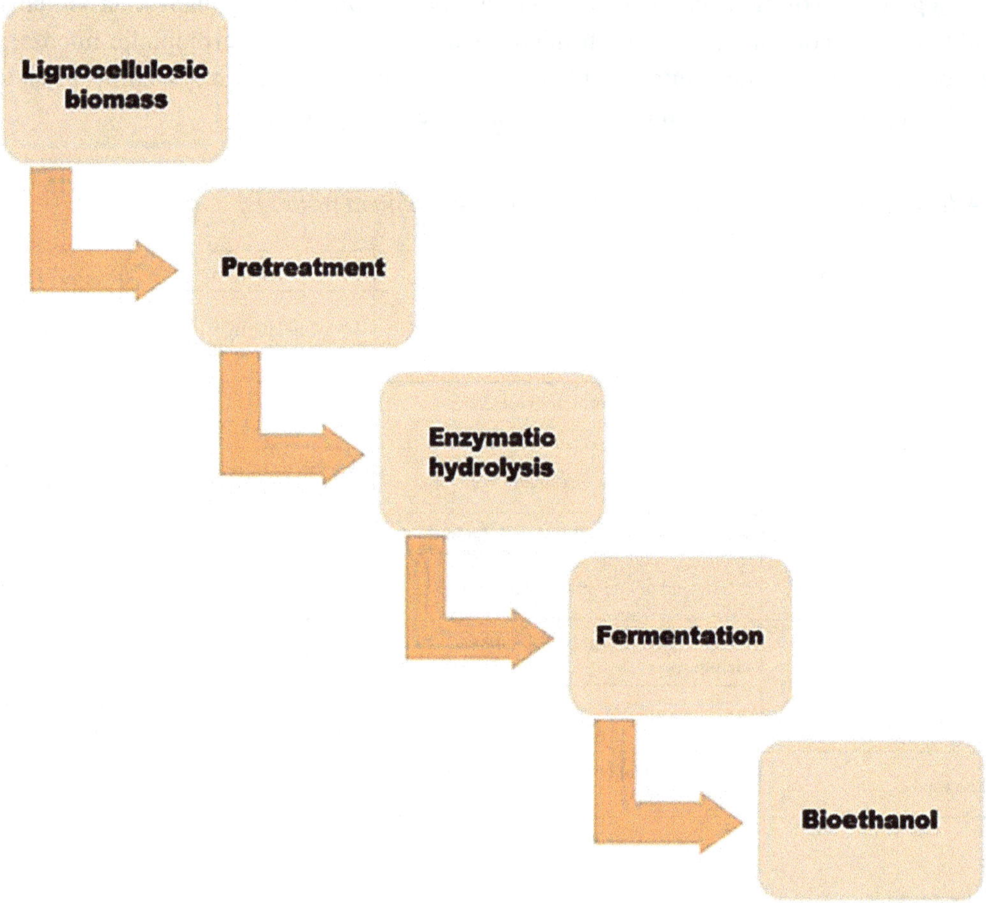

a) Hydrolysis of LBM

Fermentable sugars from hemicellulose and cellulose is obtained through the hydrolysis process. Hydrolysis process is of two types namely acidic and enzymatic. The process in which acid is used is called acid hydrolysis and enzymatic hydrolysis involves enzyme (Jatoi *etal.,*2021). In order to acquire a higher yield of fermentable sugars, which are then turned into ethanol, hydrolysis is one of the important steps in the manufacturing of ethanol.

i) Enzymatic Hydrolysis

Either commercial enzyme systems, which are more feasible and often used on an industrial scale, or microorganisms that release enzymes into the media can produce enzymatic hydrolysis. (Klein *et al.,*

2012). But for industrial utilization the only drawback is its higher enzyme cost (Klein *et al.,*2015). Hemicelluloses, have complex structure in comparison to cellulose. While cellulose is not that complex structure but still there is need of several enzymes for complete hydrolysis. Hemicelluloses include xylans, mannans, xyloglucans, and glucomannans, as well as β 1-3,1-4 glucans which are hydrolysed by their respective enzymes like xylanase, mannanase and β-glucnase etc., whereas cellulose is hydrolyzed by cellulase enzyme, which includes endoglucanase, exoglucanase, and glucosidses.

ii) Acid Hydrolysis

Acid hydrolysis is called the chemical hydrolysis. Acid hydrolysis method is advantageous in comparison to enzymatic hydrolysis since reaction rate achieved is higher than enzymatic hydrolysis (Cheah *et al.,*2020). Dilute acid hydrolysis is commonly used to break down hemicellulose and to prepare cellulose for enzyme digestion. This hydrolysis can be achieved by two ways 1) concentrated acids, low temperature, long reaction time 2) high temperature, dilute acids, short reaction time (Balat *et al.,*2011). The major drawback with concentrated acids is its toxicity and corrosive properties that can reduce the fermentation yield (Balat, 2007). While dilute acid provided by low temperature is favorable for depolymerization of hemicellulose and cellulose at high temperature (Rezania *et al.,*2018). Acids like hydrogen sulphate, phosphoric acids with appropriate concentrations and optimum required temperatures are often used in this acid hydrolysis method.

b) Fermentation of LBM

The hydrolysis step releases the simple sugars such as glucose, xylose and mannose etc. in the optimized condition. This simple sugar can be converted to bioethanol using fermentation techniques. Yeast strains such as *Sacchromyces cerevisiae*, *Pichia stipitis*, *Kluyveromyces fagilis* were used for bioethanol production from different types of monomer sugars. *S. cerevisiae*, one of these yeast varieties, has been widely employed in industry sector for the manufacture of bioethanol since it can withstand a wide pH range. (Azhar et al., 2017).

There are various ways through which fermentation is done namely continuous, batch, and fed-batch fermentation processing (Jatoi *et al.,*2021). Type of fermentation process is dependent or can say determined by feedstock origin and by nature and type of microorganisms used (Huang *et al.,*2018).

Advantage of Fed-batch fermentation is that if critical process parameters are maintained then maximum viable cell concentration is achieved as a result (Balat *et al.,*2011). One downside of this method is that Feed-rate limits productivity, which is restricted by concentration of cell mass. Continuous fermentation process gives more output in comparison to batch processing.

The two fermentation techniques are submerged fermentation and solid state fermentation. Water to lignocellulose ratios of 1:1 to 10:1 are used to wet the biomass by depositing a thin layer of water on its surface. (Lee *et al.,*1997). Fermentation process can be done separately or can be done simultaneously with saccharification. Separate fermentation is time consuming while simultaneous saccharification and fermentation is less time consuming. Still the research is going on this topic, through which we can reduce time and cost of process.

CONCLUSION

Lignocellulosic biomass is available at an affordable price, has a high carbohydrate content, is adaptable, and is not competitive with food and feed. The scientific and technological aspects of producing lignocellulosic ethanol have been extensively studied in recent years. However, the economic applicability of LBM is still constrained by the exceedingly challenging ethanol conversion process. Several parameters, including cellulose crystallinity, accessible surface area, and lignin concentration, restrict the biodegradability of lignocellulosic biomass. To, increase ethanol production, numerous pretreatment procedures for lignocellulosic biomass have been described. The hydrolysis of cellulose and hemicellulose is restricted by the existence of lignin and the crystallinity of cellulose. Lignin removal and crystallinity reduction are accomplished using a variety of pretreatment techniques, including physical, chemical, and biological ones. Pretreatment method is followed by saccharification and fermentation, these both processes can be done simultaneously or separately. saccharification is done by using different enzymes such as cellulases, xylanase, mannanase and β-glucnase etc. These enzymes depolymerize the ccomplex sugars and convert it into simple sugar. This simple sugar then subjected to fermentation to get ethanol as final product. *Sacchromyces cerevisiae*, *Pichia stipitis*, *Kluyveromyces fagilis* are the yeast used for conversion of sugar to ethanol. Thus, by using these techniques (pretreatment, saccharification and fermentation) lignocellulose biomass can be converted into bioethanol that can overcome the energy requirement and can a best alternative of fossil fuel-based energy source.

REFERENCES

Agbor, V. B., Cicek, N., Sparling, R., Berlin, A., & Levin, D. B. (2011). Biomass pretreatment: Fundamentals toward application. *Biotechnology Advances*, *29*(6), 675–685. doi:10.1016/j.biotechadv.2011.05.005 PMID:21624451

Ahmad, E., Khan, T. S., Alam, M. I., Pant, K. K., & Haider, M. A. (2020). Understanding reaction kinetics, deprotonation and solvation of brønsted acidic protons in heteropolyacid catalyzed synthesis of biorenewable alkyl levulinates. *Chemical Engineering Journal*, *400*, 125916. doi:10.1016/j.cej.2020.125916

Alonso, D. M., Wettstein, S. G., & Dumesic, J. A. (2012). Bimetallic catalysts for upgrading of biomass to fuels and chemicals. *Chemical Society Reviews*, *41*(24), 8075–8098. doi:10.1039/c2cs35188a PMID:22872312

Amin, F. R., Khalid, H., Zhang, H., Rahman, S., Zhang, R., Liu, G., & Chen, C. (2017). Pretreatment methods of lignocellulosic biomass for anaerobic digestion. *AMB Express*, *7*(1), 72. doi:10.118613568-017-0375-4 PMID:28353158

Antal, M. J., Jr. (1996). Water: A traditional solvent pregnant with new applications. In Physical Chemistry of Aqueous Systems: Meeting the Needs of Industry. Begell House.

Azhar, S. H. M., Abdulla, R., Jambo, S. A., Marbawi, H., Gansau, J. A., Faik, A. A. M., & Rodrigues, K. F. (2017). Yeasts in sustainable bioethanol production: A review. *Biochemistry and Biophysics Reports*, *10*, 52–61. doi:10.1016/j.bbrep.2017.03.003 PMID:29114570

Bajpai, P. (2016). Pretreatment of lignocellulosic biomass. In *Pretreatment of lignocellulosic biomass for Biofuel Production* (pp. 17–70). Springer. doi:10.1007/978-981-10-0687-6_4

Balan, V., Chiaramonti, D., & Kumar, S. (2013). Review of US and EU initiatives toward development, demonstration, and commercialization of lignocellulosic biofuels. *Biofuels, Bioproducts & Biorefining*, *7*(6), 732–759. doi:10.1002/bbb.1436

Balat, M. (2007). An overview of biofuels and policies in the European Union. Energy Sources. *Part B*, *2*(2), 167–181.

Balat, M. (2011). Production of bioethanol from lignocellulosic materials via the biochemical pathway: A review. *Energy Conversion and Management*, *52*(2), 858–875. doi:10.1016/j.enconman.2010.08.013

Barakat, A., de Vries, H., & Rouau, X. (2013). Dry fractionation process as an important step in current and future lignocellulose biorefineries: A review. *Bioresource Technology*, *134*, 362–373. doi:10.1016/j.biortech.2013.01.169 PMID:23499177

Benjamin, Y., Cheng, H., & Görgens, J. F. (2013). Evaluation of bagasse from different varieties of sugarcane by dilute acid pretreatment and enzymatic hydrolysis. *Industrial Crops and Products*, *51*, 7–18. doi:10.1016/j.indcrop.2013.08.067

Bobleter, O. (1994). Hydrothermal degradation of polymers derived from plants. *Progress in Polymer Science*, *19*(5), 797–841. doi:10.1016/0079-6700(94)90033-7

Brosse, N., El Hage, R., Sannigrahi, P., & Ragauskas, A. (2010). Dilute sulphuric acid and ethanol organosolv pretreatment of miscanthus x giganteus. *Cellulose Chemistry and Technology*, *44*(1-3), 71–78.

Ch, A. K., Chan, E. S., Rudravaram, R., Narasu, M. L., Rao, L. V., & Ravindra, P. (2007). Economics and environmental impact of bioethanol production technologies: An appraisal. *Biotechnology and Molecular Biology Reviews*, *2*(1), 14–32.

Chandel, A. K., da Silva, S. S., Carvalho, W., & Singh, O. V. (2012). Sugarcane bagasse and leaves: Foreseeable biomass of biofuel and bio-products. *Journal of Chemical Technology and Biotechnology (Oxford, Oxfordshire)*, *87*(1), 11–20. doi:10.1002/jctb.2742

Chandel, A. K., & Singh, O. V. (2011). Weedy lignocellulosic feedstock and microbial metabolic engineering: Advancing the generation of 'Biofuel'. *Applied Microbiology and Biotechnology*, *89*(5), 1289–1303. doi:10.100700253-010-3057-6 PMID:21181146

Chandra, R. P., Bura, R., Mabee, W. E., Berlin, D. A., Pan, X., & Saddler, J. N. (2007). Substrate pretreatment: The key to effective enzymatic hydrolysis of lignocellulosics. *Biofuels*, •••, 67–93. PMID:17530205

Cheah, W. Y., Sankaran, R., Show, P. L., Ibrahim, T. N. B. T., Chew, K. W., Culaba, A., & Jo-Shu, C. (2020). Pretreatment methods for lignocellulosic biofuels production: Current advances, challenges and future prospects. *Biofuel Research Journal*, *7*(1), 1115. doi:10.18331/BRJ2020.7.1.4

de Albuquerque Wanderley, M. C., Martín, C., de Moraes Rocha, G. J., & Gouveia, E. R. (2013). Increase in ethanol production from sugarcane bagasse based on combined pretreatments and fed-batch enzymatic hydrolysis. *Bioresource Technology*, *128*, 448–453. doi:10.1016/j.biortech.2012.10.131 PMID:23201527

Duan, X., Zhang, C., Ju, X., Li, Q., Chen, S., Wang, J., & Liu, Z. (2013). Effect of lignocellulosic composition and structure on the bioethanol production from different poplar lines. *Bioresource Technology*, *140*, 363–367. doi:10.1016/j.biortech.2013.04.101 PMID:23708852

El Hage, R., Brosse, N., Chrusciel, L., Sanchez, C., Sannigrahi, P., & Ragauskas, A. (2009). Characterization of milled wood lignin and ethanol organosolv lignin from miscanthus. *Polymer Degradation & Stability*, *94*(10), 1632–1638. doi:10.1016/j.polymdegradstab.2009.07.007

Faraco, V., & Hadar, Y. (2011). The potential of lignocellulosic ethanol production in the Mediterranean Basin. *Renewable & Sustainable Energy Reviews*, *15*(1), 252–266. doi:10.1016/j.rser.2010.09.050

Fengel, D., & Wegener, G. (1984). Lignin. *Wood: Chemistry, ultrastructure, reactions*, 132-191.

Furlan, F. F., Pinto, F. H., Costa, C. B., Cruz, A. J., Giordano, R. L., & Giordano, R. C. (2013). Bioelectricity versus bioethanol from sugarcane bagasse: Is it worth being flexible? *Biotechnology for Biofuels*, *6*(1), 1–12. doi:10.1186/1754-6834-6-142 PMID:24088415

Galbe, M., & Zacchi, G. (2007). Pretreatment of lignocellulosic materials for efficient bioethanol production. *Biofuels*, 41–65. PMID:17646946

Grabber, J. H. (2005). How do lignin composition, structure, and cross-linking affect degradability? A review of cell wall model studies. *Crop Science*, *45*(3), 820–831. doi:10.2135/cropsci2004.0191

Grous, W. R., Converse, A. O., & Grethlein, H. E. (1986). Effect of steam explosion pretreatment on pore size and enzymatic hydrolysis of poplar. *Enzyme and Microbial Technology*, *8*(5), 274–280. doi:10.1016/0141-0229(86)90021-9

Hadar, Y. (2013). Sources for lignocellulosic raw materials for the production of ethanol. In *Lignocellulose conversion* (pp. 21–38). Springer. doi:10.1007/978-3-642-37861-4_2

Hill, J., Nelson, E., Tilman, D., Polasky, S., & Tiffany, D. (2006). Environmental, economic, and energetic costs and benefits of biodiesel and ethanol biofuels. *Proceedings of the National Academy of Sciences of the United States of America*, *103*(30), 11206–11210. doi:10.1073/pnas.0604600103 PMID:16837571

Houghton, J. (2005). Global warming. *Reports on Progress in Physics*, *68*(6), 1343–1403. doi:10.1088/0034-4885/68/6/R02

Howard, R. L., Abotsi, E. L. J. R., Van Rensburg, E. J., & Howard, S. (2003). Lignocellulose biotechnology: Issues of bioconversion and enzyme production. *African Journal of Biotechnology*, *2*(12), 602–619. doi:10.5897/AJB2003.000-1115

Huang, X., Dong, W., Wang, H., & Feng, Y. (2018). Role of acid/alkali-treatment in primary sludge anaerobic fermentation: Insights into microbial community structure, functional shifts and metabolic output by high-throughput sequencing. *Bioresource Technology*, *249*, 943–952. doi:10.1016/j.biortech.2017.10.104 PMID:29145121

Jatoi, A. S., Abbasi, S. A., Hashmi, Z., Shah, A. K., Alam, M. S., Bhatti, Z. A., Maitlo, G., Hussain, S., Khandro, G. A., Usto, M. A., & Iqbal, A. (2021). Recent trends and future perspectives of lignocellulose biomass for biofuel production: A comprehensive review. *Biomass Conversion and Biorefinery*, •••, 1–13. doi:10.100713399-021-01853-8

Kim, M., & Day, D. F. (2011). Composition of sugar cane, energy cane, and sweet sorghum suitable for ethanol production at Louisiana sugar mills. *Journal of Industrial Microbiology & Biotechnology*, *38*(7), 803–807. doi:10.100710295-010-0812-8 PMID:20803247

Kim, S., & Dale, B. E. (2004). Global potential bioethanol production from wasted crops and crop residues. *Biomass and Bioenergy*, *26*(4), 361–375. doi:10.1016/j.biombioe.2003.08.002

Klein-Marcuschamer, D., & Blanch, H. W. (2015). Renewable fuels from biomass: Technical hurdles and economic assessment of biological routes. *AIChE Journal. American Institute of Chemical Engineers*, *61*(9), 2689–2701. doi:10.1002/aic.14755

Klein-Marcuschamer, D., Oleskowicz-Popiel, P., Simmons, B. A., & Blanch, H. W. (2012). The challenge of enzyme cost in the production of lignocellulosic biofuels. *Biotechnology and Bioengineering*, *109*(4), 1083–1087. doi:10.1002/bit.24370 PMID:22095526

Lee, J. (1997). Biological conversion of lignocellulosic biomass to ethanol. *Journal of Biotechnology*, *56*(1), 1–24. doi:10.1016/S0168-1656(97)00073-4 PMID:9246788

Lee, J. Y., Ryu, H. J., & Oh, K. K. (2013). Acid-catalyzed hydrothermal severity on the fractionation of agricultural residues for xylose-rich hydrolyzates. *Bioresource Technology*, *132*, 84–90. doi:10.1016/j.biortech.2013.01.019 PMID:23395759

LeVan, S., Ross, R. J., & Winandy, J. E. (1990). *Effects of fire retardant chemicals on the bending properties of wood at elevated temperatures* (Vol. 498). US Department of Agriculture, Forest Service, Forest Products Laboratory.

Li, A., Antizar-Ladislao, B., & Khraisheh, M. (2007). Bioconversion of municipal solid waste to glucose for bio-ethanol production. *Bioprocess and Biosystems Engineering*, *30*(3), 189–196. doi:10.100700449-007-0114-3 PMID:17458580

Li, S., Zhang, X., & Andresen, J. M. (2012). Production of fermentable sugars from enzymatic hydrolysis of pretreated municipal solid waste after autoclave process. *Fuel*, *92*(1), 84–88. doi:10.1016/j.fuel.2011.07.012

Ma, H., Liu, W. W., Chen, X., Wu, Y. J., & Yu, Z. L. (2009). Enhanced enzymatic saccharification of rice straw by microwave pretreatment. *Bioresource Technology*, *100*(3), 1279–1284. doi:10.1016/j.biortech.2008.08.045 PMID:18930389

Macrelli, S., Galbe, M., & Wallberg, O. (2014). Effects of production and market factors on ethanol profitability for an integrated first and second generation ethanol plant using the whole sugarcane as feedstock. *Biotechnology for Biofuels*, *7*(1), 1–16. doi:10.1186/1754-6834-7-26 PMID:24559312

Mankar, A. R., Pandey, A., Modak, A., & Pant, K. K. (2021). Pretreatment of lignocellulosic biomass: A review on recent advances. *Bioresource Technology*, *334*, 125235. doi:10.1016/j.biortech.2021.125235 PMID:33957458

Maurya, D. P., Singla, A., & Negi, S. (2015). An overview of key pretreatment processes for biological conversion of lignocellulosic biomass to bioethanol. *3 Biotech*, *5*(5), 597-609.

Mes-Hartree, M., Dale, B. E., & Craig, W. K. (1988). Comparison of steam and ammonia pretreatment for enzymatic hydrolysis of cellulose. *Applied Microbiology and Biotechnology*, *29*(5), 462–468. doi:10.1007/BF00269069

Mosier, N., Wyman, C., Dale, B., Elander, R., Lee, Y. Y., Holtzapple, M., & Ladisch, M. (2005). Features of promising technologies for pretreatment of lignocellulosic biomass. *Bioresource Technology*, *96*(6), 673–686. doi:10.1016/j.biortech.2004.06.025 PMID:15588770

Pan, X., Xie, D., Gilkes, N., Gregg, D. J., & Saddler, J. N. (2005). Strategies to enhance the enzymatic hydrolysis of pretreated softwood with high residual lignin content. In *Twenty-Sixth Symposium on Biotechnology for Fuels and Chemicals* (pp. 1069-1079). Humana Press. 10.1007/978-1-59259-991-2_90

Park, Y. C., & Kim, J. S. (2012). Comparison of various alkaline pretreatment methods of lignocellulosic biomass. *Energy*, *47*(1), 31–35. doi:10.1016/j.energy.2012.08.010

Pereira, S. C., Maehara, L., Machado, C. M. M., & Farinas, C. S. (2015). 2G ethanol from the whole sugarcane lignocellulosic biomass. *Biotechnology for Biofuels*, *8*(1), 1–16. doi:10.118613068-015-0224-0 PMID:25774217

Putro, J. N., Soetaredjo, F. E., Lin, S. Y., Ju, Y. H., & Ismadji, S. (2016). Pretreatment and conversion of lignocellulose biomass into valuable chemicals. *RSC Advances*, *6*(52), 46834–46852. doi:10.1039/C6RA09851G

Rezania, S., Din, M. F. M., Taib, S. M., Mohamad, S. E., Dahalan, F. A., Kamyab, H., Darajeh, N., & Ebrahimi, S. S. (2018). Ethanol production from water hyacinth (Eichhornia crassipes) using various types of enhancers based on the consumable sugars. *Waste and Biomass Valorization*, *9*(6), 939–946. doi:10.100712649-017-9883-3

Rezania, S., Oryani, B., Cho, J., Talaiekhozani, A., Sabbagh, F., Hashemi, B., Rupani, P. F., & Mohammadi, A. A. (2020). Different pretreatment technologies of lignocellulosic biomass for bioethanol production: An overview. *Energy*, *199*, 117457. doi:10.1016/j.energy.2020.117457

Saini, J. K., Saini, R., & Tewari, L. (2015). Lignocellulosic agriculture wastes as biomass feedstocks for second-generation bioethanol production: Concepts and recent developments. *3 Biotech*, *5*(4), 337-353.

Sánchez, C. (2009). Lignocellulosic residues: Biodegradation and bioconversion by fungi. *Biotechnology Advances*, *27*(2), 185–194. doi:10.1016/j.biotechadv.2008.11.001 PMID:19100826

Saxena, R. C., Adhikari, D. K., & Goyal, H. B. (2009). Biomass-based energy fuel through biochemical routes: A review. *Renewable & Sustainable Energy Reviews*, *13*(1), 167–178. doi:10.1016/j.rser.2007.07.011

Schmitt, E., Bura, R., Gustafson, R., Cooper, J., & Vajzovic, A. (2012). Converting lignocellulosic solid waste into ethanol for the State of Washington: An investigation of treatment technologies and environmental impacts. *Bioresource Technology*, *104*, 400–409. doi:10.1016/j.biortech.2011.10.094 PMID:22119432

Seabra, J. E., Tao, L., Chum, H. L., & Macedo, I. C. (2010). A techno-economic evaluation of the effects of centralized cellulosic ethanol and co-products refinery options with sugarcane mill clustering. *Biomass and Bioenergy*, *34*(8), 1065–1078. doi:10.1016/j.biombioe.2010.01.042

Singh, S., Cheng, G., Sathitsuksanoh, N., Wu, D., Varanasi, P., George, A., Balan, V., Gao, X., Kumar, R., Dale, B. E., Wyman, C. E., & Simmons, B. A. (2015). Comparison of different biomass pretreatment techniques and their impact on chemistry and structure. *Frontiers in Energy Research*, *2*, 62. doi:10.3389/fenrg.2014.00062

Sun, Y., & Cheng, J. J. (2005). Dilute acid pretreatment of rye straw and bermudagrass for ethanol production. *Bioresource Technology*, *96*(14), 1599–1606. doi:10.1016/j.biortech.2004.12.022 PMID:15978993

Timilsena, Y. P., Abeywickrama, C. J., Rakshit, S. K., & Brosse, N. (2013). Effect of different pretreatments on delignification pattern and enzymatic hydrolysability of miscanthus, oil palm biomass and typha grass. *Bioresource Technology*, *135*, 82–88. doi:10.1016/j.biortech.2012.09.010 PMID:23069607

Valdés, G., Mendonça, R. T., & Aggelis, G. (2020). Lignocellulosic biomass as a substrate for oleaginous microorganisms: A review. *Applied Sciences (Basel, Switzerland)*, *10*(21), 7698. doi:10.3390/app10217698

Velmurugan, R., & Muthukumar, K. (2011). Utilization of sugarcane bagasse for bioethanol production: Sono-assisted acid hydrolysis approach. *Bioresource Technology*, *102*(14), 7119–7123. doi:10.1016/j.biortech.2011.04.045 PMID:21570831

Wang, R., Wang, K., Zhou, M., Xu, J., & Jiang, J. (2021). Efficient fractionation of moso bamboo by synergistic hydrothermal-deep eutectic solvents pretreatment. *Bioresource Technology*, *328*, 124873. doi:10.1016/j.biortech.2021.124873 PMID:33639413

Winandy, J. E. (1995). *Effects of fire retardant treatments after 18 months of exposure at 150 F (66 C)* (Vol. 264). US Department of Agriculture, Forest Service, Forest Products Laboratory.

Xu, F., Shi, Y. C., & Wang, D. (2013). X-ray scattering studies of lignocellulosic biomass: A review. *Carbohydrate Polymers*, *94*(2), 904–917. doi:10.1016/j.carbpol.2013.02.008 PMID:23544649

Zabed, H., Sahu, J. N., Boyce, A. N., & Faruq, G. (2016). Fuel ethanol production from lignocellulosic biomass: An overview on feedstocks and technological approaches. *Renewable & Sustainable Energy Reviews*, *66*, 751–774. doi:10.1016/j.rser.2016.08.038

Zhang, Q., Zhang, P., Pei, Z. J., & Wang, D. (2013). Relationships between cellulosic biomass particle size and enzymatic hydrolysis sugar yield: Analysis of inconsistent reports in the literature. *Renewable Energy*, *60*, 127–136. doi:10.1016/j.renene.2013.04.012

Zhao, X., Zhang, L., & Liu, D. (2008). Comparative study on chemical pretreatment methods for improving enzymatic digestibility of crofton weed stem. *Bioresource Technology*, *99*(9), 3729–3736. doi:10.1016/j.biortech.2007.07.016 PMID:17709243

Zheng, Y., Zhao, J., Xu, F., & Li, Y. (2014). Pretreatment of lignocellulosic biomass for enhanced biogas production. *Progress in Energy and Combustion Science*, *42*, 35–53. doi:10.1016/j.pecs.2014.01.001

Zhu, Y., Lee, Y. Y., & Elander, R. T. (2005). Optimization of dilute-acid pretreatment of corn stover using a high-solids percolation reactor. *Applied Biochemistry and Biotechnology*, *124*(1), 1045–1054. doi:10.1385/ABAB:124:1-3:1045 PMID:15930580

Compilation of References

Ozawa, H., Ohmura, A., Lorenz, R. D., & Pujol, T. (2003). The second law of thermodynamics 30 and the global climate system, A review of the maximum entropy production principle. *Reviews of Geophysics*, *41*, 1018.

Fraedrich, K. (2001). Simple Climate Models. In Progress in Probability 49. Birkh¨auser.

Fraedrich, K., Jansen, H., Kirk, E., Luksch, U., & Lunkeit, F. (2005). The planet simulator: Towards a user friendly model. *Met. Zeit.*, *14*, 299–304.

Fraedrich, K., & Lunkeit, F. (2008). Diagnosing the entropy budget of a climate model. *Tellus. Series A, Dynamic Meterology and Oceanography*, *60*, 921–931.

Kunz, T., Fraedrich, K., & Kirk, E. (2008). Optimization of simplified GCMs using circulation indices and maximum entropy production. *Climate Dynamics*, *30*, 803–813.

Lorenz, E. N. (1955). Available potential energy and the maintenance of the general circulation. *Tellus*, *7*, 157–167.

Dones, R., Bauer, C., Bolliger, R., Burger, B., Faist, M., Frischknecht, R., Heck, T. N., Jungbluth, A., & Röder. (2016). Life Cycle Inventories of Energy Systems: Results for Current Systems in Switzerland and other UCTE Countries. Academic Press.

Villigen & Swiss. (2003). *Centre for Life Cycle Inventories*. www.ecoinvent.ch

Frischknecht, R., Jungbluth, N., Althaus, H. J., Doka, G., Dones, R., Hellweg, S., Hischier, R., Nemecek, T., Rebitzer, G., & Spielmann, M. (2003). *Overview and Methodology — Data v1.01*. Swiss Centre for Life Cycle Inventories. www.ecoinvent.ch

Inter-Governmental Panel on Climate Change (IPCC). (2001). *The Scientific Basis. Contribution of Working Group I to the Third Assessment Report*. Author.

Hondo. (2003). Life cycle GHG emission analysis of power generation systems: Japanese case. *Energy*.

Spath, Mann, & Kerr. (1999). Life-cycle assessment of coal-fired power production. National Renewable Energy Laboratory, NREL/TP-570-25119.

Dones, Zhou, & Tian. (n.d.). Life cycle assessment. In *Integrated assessment of sustainable energy systems in China — The China Energy Technology Program.* Academic Press.

Lee, Y. (1997). Comparison of energy sources in terms of their full energy chain emission factors of greenhouse gases. *Energy Policy*, *25*(1), 1–6.

Abdelrahman, M., Yassin, A. A., Hussein, I. H., Mirghani, M. E. S., & Karama, A. B. (2020). Biodiesel Production from Vegetable Oil: A Review. *J Marine Biol Environ Sci*, *2*(1), 1–11.

Abels, C., Carstensen, F., & Wessling, M. (2013). Membrane processes in biorefinery applications. *Journal of Membrane Science*, *444*, 285–317. doi:10.1016/j.memsci.2013.05.030

Abu-Taha, R., & Daim, T. (2011). Multi-criteria applications in renewable energy analysis: A literature review. Technology Management in the Energy Smart World Proceedings of Portland International Centre for Management of Engineering and Technology (PICMET) '11: Technology Management in the Energy Smart World, 1-8.

Acharya, S., & Chaudhary, A. (2012). Bioprospecting thermophiles for cellulase production: A review. *Brazilian Journal of Microbiology*, *43*(3), 844–856. doi:10.1590/S1517-83822012000300001 PMID:24031898

Achinas, S., Achinas, V., & Euverink, G. J. W. (2017). A technological overview of biogas production from biowaste. *Engineering*, *3*(3), 299–307. doi:10.1016/J.ENG.2017.03.002

Adams, J. M., Gallagher, J. A., & Donnison, I. S. (2009). Fermentation study on Saccharina latissima for bioethanol production considering variable pre-treatments. *Journal of Applied Phycology*, *21*(5), 569–574. doi:10.100710811-008-9384-7

Adarme, C., Lim, D. K. Y., Timmins, M., Vernen, F., Li, Y., & Schenk, P. M. (2012). Microalgal biofactories: A promising approach towards sustainable omega-3 fatty acid production. *Microbial Cell Factories*, *11*(1), 96. doi:10.1186/1475-2859-11-96 PMID:22830315

Adegboye, M. F., Ojuederie, O. B., Talia, P. M., & Babalola, O. O. (2021). Bioprospecting of microbial strains for biofuel production: Metabolic engineering, applications, and challenges. *Biotechnology for Biofuels*, *14*(1), 5. Advance online publication. doi:10.118613068-020-01853-2 PMID:33407786

Adekunle, K. F., & Okolie, J. A. (2015). A review of biochemical process of anaerobic digestion. *Advances in Bioscience and Biotechnology*, *6*(3), 205–212. doi:10.4236/abb.2015.63020

Adler, P. R., Sanderson, M. A., Boateng, A. A., Weimer, P. J., & Jung, H. J. G. (2006). Biomass yield and biofuel quality of switchgrass harvested in fall or spring. *Agronomy Journal*, *98*(6), 1518–1525. doi:10.2134/agronj2005.0351

Afify, A. E. M. R., Shalaby, E. A., & Shanab, S. M. M. (2010). Enhancement of biodiesel production from different species of algae. *Grasas y Aceites*, *61*(4), 416–422. doi:10.3989/gya.021610

Agbogbo, F. K., Haagensen, F. D., Milam, D., & Wenger, K. S. (2008). Fermentation of acid-pretreated corn stover to ethanol without detoxification using Pichia stipitis. *Applied Biochemistry and Biotechnology*, *145*(1–3), 53–58. doi:10.100712010-007-8056-4 PMID:18425611

Agbor, V. B., Cicek, N., Sparling, R., Berlin, A., & Levin, D. B. (2011). Biomass pretreatment: Fundamentals toward application. *Biotechnology Advances*, *29*(6), 675–685. doi:10.1016/j.biotechadv.2011.05.005 PMID:21624451

Ahmad, E., Khan, T. S., Alam, M. I., Pant, K. K., & Haider, M. A. (2020). Understanding reaction kinetics, deprotonation and solvation of brønsted acidic protons in heteropolyacid catalyzed synthesis of biorenewable alkyl levulinates. *Chemical Engineering Journal*, *400*, 125916. doi:10.1016/j.cej.2020.125916

Ahmad, M., Zafar, M., Khan, M. A., & Sultana, S. (2009). Biodiesel from pongamia pinnata l. oil: A promising alternative bioenergy source. *Energy Sources*, *31*(16), 1436–1442. doi:10.1080/15567030802092882

Ahmed, S. F., Liu, G., Mofijur, M., Azad, A. K., Hazrat, M. A., & Chu, Y. M. (2021). Physical and hybrid modelling techniques for earth-air heat exchangers in reducing building energy consumption: Performance, applications, progress, and challenges. *Solar Energy*, *216*(November), 274–294. doi:10.1016/j.solener.2021.01.022

Ahmed, M., Khan, M. A., Zafar, M., Sultana, S., & Gulzar, S. (2007). Indigenous plant-based biodiesel resources in Pakistan. *J. Ethnobotany Leaflets*, *11*, 224–230.

Alam, F., Mobin, S., & Chowdhury, H. (2015). Third generation biofuel from Algae. *Procedia Engineering, 105*, 763–768. doi:10.1016/j.proeng.2015.05.068

Alam, F., Date, A., Rasjidin, R., Mobin, S., Moria, H., & Baqui, A. (2012). Biofuel from algae-Is it a viable alternative? *Procedia Engineering*, *49*(December), 221–227. doi:10.1016/j.proeng.2012.10.131

Ali, M., Saleem, M., Khan, Z., & Watson, I. A. (2019). The use of crop residues for biofuel production. In Biomass, Biopolymer-Based Materials, and Bioenergy: Construction, Biomedical, and other Industrial Applications. doi:10.1016/B978-0-08-102426-3.00016-3

Alia, E. N., & Taya, C. I. (2013). Characterization of biodiesel produced from palm oil via base catalyzed transesterification. *Procedia Engineering*, *53*, 7–20. doi:10.1016/j.proeng.2013.02.002

Allen, O. N., & Allen, E. K. (1981). *The Leguminosae*. University of Wisconsin Press. doi:10.1007/978-1-349-06142-6

Allouache, A., Majda, A., Toudert, A. Z., Amrane, A., & Ballesteros, M. (2021). Cellulosic bioethanol production from *Ulva lactuca* macroalgae. *Cellulose Chemistry and Technology*, *55*(5-6), 629–635. doi:10.35812/CelluloseChemTechnol.2021.55.51

Alonso, D. M., Wettstein, S. G., & Dumesic, J. A. (2012). Bimetallic catalysts for upgrading of biomass to fuels and chemicals. *Chemical Society Reviews*, *41*(24), 8075–8098. doi:10.1039/c2cs35188a PMID:22872312

Al, R., Behera, C. R., Gernaey, K. V., & Sin, G. (2019). Towards development of a decision support tool for conceptual design of wastewater treatment plants using stochastic simulation optimisation. *Computer-Aided Chemical Engineering*, *46*, 325–330. doi:10.1016/B978-0-12-818634-3.50055-2

Altinkaynak, C., Tavlasoglu, S., Özdemir, N., & Ocsoy, I. (2016). A new generation approach in enzyme immobilization: Organic-inorganic hybrid nanoflowers with enhanced catalytic activity and stability. *Enzyme and Microbial Technology*, *93–94*, 105–112. doi:10.1016/j.enzmictec.2016.06.011 PMID:27702469

Alvira, P., Tomás-Pejó, E., Ballesteros, M., & Negro, M. J. (2010). Pretreatment technologies for an efficient bioethanol production process based on enzymatic hydrolysis: A review. *Bioresource Technology*, *101*(13), 4851–4861. doi:10.1016/j.biortech.2009.11.093 PMID:20042329

Amano, H., Kakinuma, M., Coury, D. A., Ohno, H., & Hara, T. (2005). Effect of a seaweed mixture on serum lipid level and platelet aggregation in rats. *Fisheries Science*, *71*(5), 1160–1166. doi:10.1111/j.1444-2906.2005.01076.x

Ameka, G. K., Doamekpor, L. K., Amadu, A. A., & Amamoo, A. P. (2019). Production of biodiesel from marine macroalgae occurring in the Gulf of Guinea, off the coast of Ghana. *Ghana Journal of Science*, *60*(1), 50–58. doi:10.4314/gjs.v60i1.5

Amin, F. R., Khalid, H., Zhang, H., Rahman, S., Zhang, R., Liu, G., & Chen, C. (2017). Pretreatment methods of lignocellulosic biomass for anaerobic digestion. *AMB Express*, *7*(1), 72. doi:10.118613568-017-0375-4 PMID:28353158

Amiri, H., Karimi, K., & Zilouei, H. (2014). Organosolv pretreatment of rice straw for efficient acetone, butanol, and ethanol production. *Bioresource Technology*, *152*, 450–456. doi:10.1016/j.biortech.2013.11.038 PMID:24321608

Amit Aradhey. (2010). *India-Biofuels Annual, Global Agricultural Information Network Report Number, IN1058*. USDA.

Amit Arora, S. S. (2015). Techniques for Exploitation of Gas Hydrate (Clathrates) an Untapped Resource of Methane Gas. *Journal of Microbial & Biochemical Technology*, *7*(2), 108–111. doi:10.4172/1948-5948.1000190

Andreasen, M. M., Hansen, C. T., & Cash, P. (2015). *Conceptual design*. Springer International Publishing. doi:10.1007/978-3-319-19839-2

Antal, M. J., Jr. (1996). Water: A traditional solvent pregnant with new applications. In Physical Chemistry of Aqueous Systems: Meeting the Needs of Industry. Begell House.

Anya, A. U., Chioma, N. N., & Obinna, O. (2012). Optimized reduction of free fatty acid content on neem seed oil for biodiesel production. *J. Basic. Appl. Chem.*, *2*, 21–28.

Aradhey, A. (2011). India Biofuels Annual. 2011. USDA Foreign Agricultural Service. Global Agricultural Information Network (GAIN). GAIN report number: IN 1159.

Araque, E., Parra, C., Freer, J., Contreras, D., Rodríguez, J., Mendonça, R., & Baeza, J. (2008). Evaluation of organosolv pretreatment for the conversion of Pinus radiata D. Don to ethanol. *Enzyme and Microbial Technology*, *43*(2), 214–219. doi:10.1016/j.enzmictec.2007.08.006

Aravantinou, A. F., Andreou, F., & Manariotis, I. D. (2020). Long-term toxicity of ZnO nanoparticles on Scenedesmus rubescens cultivated in semi-batch mode. *Nanomaterials (Basel, Switzerland)*, *10*(11), 2262.

Aresta, M., Dibenedetto, A., & Barberio, G. (2005). Utilization of macro-algae for enhanced CO_2 fixation and biofuels production: Development of a computing software for an LCA study. *Fuel Processing Technology*, *86*(14-15), 1679–1693. doi:10.1016/j.fuproc.2005.01.016

Arora, R., Sharma, N. K., Kumar, S., & Sani, R. K. (2019). Lignocellulosic Ethanol: Feedstocks and Bioprocessing. Bioethanol Production from Food Crops. doi:10.1016/B978-0-12-813766-6.00009-6

Arora, A., Cameotra Ss, S., & Balomajumder, C. (2015a). Natural Gas Hydrate as an Upcoming Resource of Energy. *Journal of Petroleum & Environmental Biotechnology*, *6*(1), 1–6. doi:10.4172/2157-7463.1000199

Arora, A., Chandrajit, B., & Cameotra, S. S. (2015b). Natural Gas Hydrate (Clathrates) as an Untapped Resource of Natural Gas. *Journal of Petroleum & Environmental Biotechnology*, •••, 6.

Arunprasad, J., Sener, A. S., Thirugnanasambantham, R., Elango, T., & Bothichandar, T. (2022). Experimental Investigation of Spirulina Microalgae Biodiesel with Metal Nanoadditive on Single-Cylinder Diesel Engine. *Journal of Nanomaterials*.

Arya, I., Poona, A., Dikshit, P. K., Pandit, S., Kumar, J., Singh, H. N., ... Kumar, S. (2021). Current Trends and Future Prospects of Nanotechnology in Biofuel Production. *Catalysts*, *11*(11), 1308.

Atabani, A. E., Silitonga, A. S., Badruddin, I. A., Mahlia, T. M. I., Masjuki, H. H., & Mekhilef, S. (2012). A comprehensive review on biodiesel as an alternative energy resource and its characteristics. *Renewable & Sustainable Energy Reviews*, *16*(4), 2070–2093. doi:10.1016/j.rser.2012.01.003

Awalgaonkar, N., Tibdewal, S., Singal, V., Mathew, J., & Karthikeyan, A. K. (2015). Biodiesel Production Status: Are the Present Policies Good Enough for the Growth of Biodiesel Sector in India? In *Engineering Asset Management - Systems, Professional Practices and Certification, Lecture Notes in Mechanical Engineering* (pp. 1199–1211). Springer.

Awe, O. W., Zhao, Y., Nzihou, A., Minh, D. P., & Lyczko, N. (2017). A review of biogas utilisation, purification and upgrading technologies. *Waste and Biomass Valorization*, *8*(2), 267–283. doi:10.100712649-016-9826-4

Awogbemi, O., Von Kallon, D. V., Onuh, E. I., & Aigbodion, V. S. (2021). An overview of the classification, production and utilization of biofuels for internal combustion engine applications. *Energies*, *14*(18), 1–42. doi:10.3390/en14185687

Axelsson, L., Franzén, M., Ostwald, M., Berndes, G., Lakshmi, G., & Ravindranath, N. H. (2012). Perspective: Jatropha cultivation in southern India: Assessing farmers' experiences. *Biofuels, Bioproducts & Biorefining*, *6*(3), 246–256. doi:10.1002/bbb.1324

Ayodele, B. V., Alsaffar, M. A., & Mustapa, S. I. (2020). An overview of integration opportunities for sustainable bioethanol production from first- and second-generation sugar-based feedstocks. *Journal of Cleaner Production, 245*, 118857. Advance online publication. doi:10.1016/j.jclepro.2019.118857

Azbar, N., Çetinkaya Dokgöz, F. T., Keskin, T., Korkmaz, K. S., & Syed, H. M. (2008). 2009. Continuous fermentative hydrogen production from cheese whey wastewater under thermophilic anaerobic conditions. *Int. J. Hydrogen Energy, IWBT, 34*, 7441–7447.

Azhar, S. H. M., Abdulla, R., Jambo, S. A., Marbawi, H., Gansau, J. A., Faik, A. A. M., & Rodrigues, K. F. (2017). Yeasts in sustainable bioethanol production: A review. *Biochemistry and Biophysics Reports*, *10*, 52–61. doi:10.1016/j.bbrep.2017.03.003 PMID:29114570

Azman, S. (2016). *Anaerobic Digestion of Cellulose and Hemiceullose in the Presence of Humic Acids*. Wageningen University.

Bajpai, P. (2016). Pretreatment of lignocellulosic biomass. In *Pretreatment of lignocellulosic biomass for Biofuel Production* (pp. 17–70). Springer. doi:10.1007/978-981-10-0687-6_4

Baka, J. (2014). What wastelands? A critique of biofuel policy discourse in South India. *Geoforum*, *54*, 315–323.

Balan, V., Chiaramonti, D., & Kumar, S. (2013). Review of US and EU initiatives toward development, demonstration, and commercialization of lignocellulosic biofuels. *Biofuels, Bioproducts & Biorefining*, *7*(6), 732–759. doi:10.1002/bbb.1436

Balat, H., & Kirtay, E. (2010). Hydrogen from biomass - Present scenario and future prospects. *International Journal of Hydrogen Energy*, *35*(14), 7416–7426. doi:10.1016/j.ijhydene.2010.04.137

Balat, M. (2007). An overview of biofuels and policies in the European Union. Energy Sources. *Part B*, *2*(2), 167–181.

Balat, M. (2007). Global Bio-Fuel Processing and Production Trends. *Energy Exploration & Exploitation*, *25*(3), 195–218. doi:10.1260/014459807782009204

Balat, M. (2011). Production of bioethanol from lignocellulosic materials via the biochemical pathway: A review. *Energy Conversion and Management*, *52*(2), 858–875. doi:10.1016/j.enconman.2010.08.013

Balat, M., Balat, H., & Öz, C. (2008). Progress in bioethanol processing. *Progress in Energy and Combustion Science*, *34*(5), 551–573. doi:10.1016/j.pecs.2007.11.001

Balat, M., & Balat, M. (2009). Political, economic and environmental impacts of biomass-based hydrogen. *International Journal of Hydrogen Energy*, *34*(9), 3589–3603. doi:10.1016/j.ijhydene.2009.02.067

Bandyopadhyay, K. R., & Das, K. (2014). Biofuels in South Asia: Connecting the Dots. *Renewable Energy Law and Policy Review*, *5*, 145–161.

Barakat, A., de Vries, H., & Rouau, X. (2013). Dry fractionation process as an important step in current and future lignocellulose biorefineries: A review. *Bioresource Technology*, *134*, 362–373. doi:10.1016/j.biortech.2013.01.169 PMID:23499177

Baral, S. S., Singh, K., & Sharma, P. (2015). The potential of sustainable algal biofuel production using CO2 from thermal power plant in India. *Renewable & Sustainable Energy Reviews*, *49*, 1061–1074. doi:10.1016/j.rser.2015.04.181

Baras, J., Gacesa, S., & Pejin, D. (2002). Ethanol is a strategic raw material. *Hemijska Industrija*, *56*(3), 89–104. doi:10.2298/HEMIND0203089B

Barcelos, C. A., Rocha, V. A., Groposo, C., de Castro, A. M., & Pereira, N. (2015). Enzymes and accessory proteins involved in the hydrolysis of lignocellulosic biomass for bioethanol production. In R. N. Silva (Ed.), *Mycology: Current and future developments* (pp. 23–56). Bentham Books. doi:10.2174/9781681080741115010005

Barisa, A., Romagnoli, F., Blumberga, A., & Blumberga, D. (2015). Future biodiesel policy designs and consumption patterns in Latvia: A system dynamics model. *Journal of Cleaner Production*, *88*, 71–82.

Barrak, H., Saied, T., Chevallier, P., Laroche, G., M'nif, A., & Hamzaoui, A. H. (2019). Synthesis, characterization, and functionalization of ZnO nanoparticles by N-(trimethoxysilylpropyl) ethylenediamine triacetic acid (TMSEDTA): Investigation of the interactions between Phloroglucinol and ZnO@ TMSEDTA. *Arabian Journal of Chemistry*, *12*(8), 4340–4347.

Basavaraj, G., Rao, P. P., Ravinder Reddy, C., Kumar, A. A., Srinivasa Rao, P., & Reddy, B. V. S. (2012). A Review of national biofuel policy in India: A critique- need for promotion of alternative feedstocks. *J Biofuels*, *3*, 65–78.

Bastos Lima, M. (2012). *An institutional analysis of biofuel policies and their social implications lessons from Brazil, India and Indonesia*. UNRISD Occasional Paper: Social Dimensions of Green Economy and Sustainable Development.

Bastos Lima, M. G., & Gupta, J. (2014). The extraterritorial dimensions of biofuel policies and the politics of scale: Live and let die? *Third World Quarterly*, *35*(3), 392–410.

Batista, A. P., Raymundo, A., Sousa, I., & Empis, J. (2006). Rheological characterization of coloured oil in-water food emulsions with lutein and fycocyanin added to the oil and aqueous phases. *Food Hydrocolloids*, *20*(1), 44–52. doi:10.1016/j.foodhyd.2005.02.009

Becker, E. W. (1994). *Microalgae: Biotechnology and microbiology*. Cambridge University Press.

Behera, S., Mohanty, R. C., & Ray, R. C. (2012). Ethanol fermentation of sugarcane molasses by Zymomonas mobilis MTCC 92 immobilized in Luffa cylindrica L. sponge discs and ca-alginate matrices. *Brazilian Journal of Microbiology*, *43*(4), 1499–1507. doi:10.1590/S1517-83822012000400034

Behera, S., Singh, R., Arora, R., Sharma, N. K., Shukla, M., & Kumar, S. (2015). Scope of algae as third generation biofuels. *Frontiers in Bioengineering and Biotechnology*, *2*, 90. doi:10.3389/fbioe.2014.00090 PMID:25717470

Belay, N., Sparling, R., & Daniels, L. (1986). Relationship of formate to growth and methanogenesis by Methanococcus thermolithotrophicus. *Applied and Environmental Microbiology*, *52*(5), 1080–1085. doi:10.1128/aem.52.5.1080-1085.1986 PMID:3098165

Bell, M. L., Zanobetti, A., & Dominici, F. (2014). Who is more affected by ozone pollution? A systematic review and meta-analysis. *American Journal of Epidemiology*, *180*, 15–28. PMID:24872350

Benemann, J. R., Tillett, D. M., & Weissman, J. C. (1987). Microalgae biotechnology. *Trends in Biotechnology*, *5*(2), 47–53. doi:10.1016/0167-7799(87)90037-0

Benjamin, Y., Cheng, H., & Görgens, J. F. (2013). Evaluation of bagasse from different varieties of sugarcane by dilute acid pretreatment and enzymatic hydrolysis. *Industrial Crops and Products*, *51*, 7–18. doi:10.1016/j.indcrop.2013.08.067

Benson, D., Kerry, K., & Malin, G. (2014). Algal biofuels: Impact significance and implications for EU multi-level governance. *Journal of Cleaner Production*, *72*, 4–13.

Berchamans, H. J., Morishita, K., & Takarada, T. (2010). Kinetic study of hydroxide-catalyzed methanolysis of Jatropha curcas-waste food oil mixture for biodiesel production. *Fuel*, *104*, 46–52. doi:10.1016/j.fuel.2010.01.017

Berłowska, J., Pielech-Przybylska, K., Balcerek, M., Dziekońska-Kubczak, U., Patelski, P., Dziugan, P., & Kręgiel, D. (2016). Simultaneous Saccharification and Fermentation of Sugar Beet Pulp for Efficient Bioethanol Production. *BioMed Research International, 2016*. Advance online publication. doi:10.1155/2016/3154929

Bertrand, E., Vandenberghe, L. P. S., Soccol, C. R., Sigoillot, J. C., & Faulds, C. (2016). First generation bioethanol. In Green Energy and Technology (Vol. 0). https://doi.org/ doi:10.1007/978-3-319-30205-8_8

Bertzky, M., Kapos, V., & Scharlemann, J.P.W. (2011). *Indirect land use change from biofuels production: Implications for biodiversity*. JNCC Report, No 456.

Bhardwaj, N., Kumar, B., Agrawal, K., & Verma, P. (2021). Current perspective on production and applications of microbial cellulases: A review. *Bioresources and Bioprocessing*, *8*(1), 95. Advance online publication. doi:10.118640643-021-00447-6

Bian, R., Ma, B., Zhu, X., Wang, W., Li, L., Joseph, S., ... Pan, G. (2016). Pyrolysis of crop residues in a mobile bench-scale pyrolyser: Product characterization and environmental performance. *Journal of Analytical and Applied Pyrolysis*, *119*, 52–59. doi:10.1016/j.jaap.2016.03.018

Bildirici, M. E. (2017). The effects of militarization on biofuel consumption and CO2 emission. *Journal of Cleaner Production*, *152*, 420–428.

Biswal, D. (2021). Impact of climate change on the livestock sector: An overview. *Uttar Pradesh Journal of Zoology*, *42*(22), 117–132.

Biswal, D. (2022). Soil nematodes as the silent sufferers of climate-induced toxicity: Analysing the outcomes of their interactions with climatic stress factors on land cover and agricultural production. *Applied Biochemistry and Biotechnology*. Advance online publication. doi:10.100712010-022-03965-x PMID:35593954

Blanchard, R. (2015). A review of biofuels in In-dia: challenges and opportunities. In *World Energy Engineering Congress*. Association of Energy Engineers.

Bobade, S. N., & Khyade, V. B. (2012a). Detail study on the properties of pongamia pinnata (karanja) for the production of biofuel. *Research Journal of Chemical Sciences*, *2*(7), 16–20.

Bobade, S. N., & Khyade, V. B. (2012b). Preparation of methyl ester (biodiesel) from Karanja (pongamia pinnata) oil. *Research Journal of Chemical Sciences*, *2*(8), 43–50.

Bobleter, O. (1994). Hydrothermal degradation of polymers derived from plants. *Progress in Polymer Science*, *19*(5), 797–841. doi:10.1016/0079-6700(94)90033-7

Boerjan, W., Ralph, J., & Baucher, M. (2003). Lignin Biosynthesis. *Annual Review of Plant Biology*, *54*, 519–546. doi:10.1146/annurev.arplant.54.031902.134938

Borines, M. G., de Leon, R. L., & Cuello, J. L. (2013). Bioethanol production from the macroalgae *Sargassum* spp. *Bioresource Technology*, *138*, 22–29. doi:10.1016/j.biortech.2013.03.108 PMID:23612158

Bošnjaković, M., & Sinaga, N. (2020). The perspective of large-scale production of algae biodiesel. *Applied Sciences (Basel, Switzerland)*, *10*(22), 8181. doi:10.3390/app10228181

Brennan, L., & Owende, P. (2010). Biofuels from microalgae—A review of technologies for production, processing, and extractions of biofuels and co-products. *Renewable & Sustainable Energy Reviews*, *14*(2), 557–577.

Brosse, N., El Hage, R., Sannigrahi, P., & Ragauskas, A. (2010). Dilute sulphuric acid and ethanol organosolv pretreatment of miscanthus x giganteus. *Cellulose Chemistry and Technology*, *44*(1-3), 71–78.

Brown, M. R., Jeffrey, S. W., Volkman, J. K., & Dunstan, G. A. (1997). Nutritional properties of microalgae for mariculture. *Aquaculture (Amsterdam, Netherlands)*, *151*(1–4), 315–331. doi:10.1016/S0044-8486(96)01501-3

Bucholc, K., Szymczak-Żyła, M., Lubecki, L., Zamojska, A., Hapter, P., Tjernström, E., & Kowalewska, G. (2014). Nutrient content in macrophyta collected from southern Baltic Sea beaches in relation to eutrophication and biogas production. *The Science of the Total Environment*, *473-474*, 298–307. doi:10.1016/j.scitotenv.2013.12.044 PMID:24374591

Budarin, V. L., Zhao, Y., Gronnow, M. J., Shuttleworth, P. S., Breeden, S. W., Macquarriea, D. J., & Clark, J. H. (2011). Microwave-mediated pyrolysis of macro-algae. *Green Chemistry*, *13*(9), 2330–2333. doi:10.1039/c1gc15560a

Bulawayo, B., Bvochora, J. M., Muzondo, M. I., & Zvauya, R. (1996). Ethanol production by fermentation of sweet-stem sorghum juice using various yeast strains. *World Journal of Microbiology & Biotechnology*, *12*(4), 357–360. doi:10.1007/BF00340211 PMID:24415310

Bušić, A., Mardetko, N., Kundas, S., Morzak, G., Belskaya, H., Šantek, M. I., Komes, D., Novak, S., & Šantek, B. (2018). Bioethanol production from renewable raw materials and its separation and purification: A review. *Food Technology and Biotechnology*, *56*(3), 289–311. doi:10.17113/ftb.56.03.18.5546 PMID:30510474

Business Line. (2015). *Centre mulls revising National Policy on Biofuels*. Business Line.

Canilha, L., Chandel, A. K., Suzane Dos Santos Milessi, T., Antunes, F. A. F., Luiz Da Costa Freitas, W., Das Graças Almeida Felipe, M., & Da Silva, S. S. (2012). Bioconversion of sugarcane biomass into ethanol: An overview about composition, pretreatment methods, detoxification of hydrolysates, enzymatic saccharification, and ethanol fermentation. *Journal of Biomedicine & Biotechnology*, *2012*, 1–15. Advance online publication. doi:10.1155/2012/989572 PMID:23251086

Canilha, L., Santos, V. T. O., Rocha, G. J. M., Almeida, E., Silva, J. B., Giulietti, M., Silva, S. S., ... Carvalho, W. (2011). A study on the pretreatment of a sugarcane bagasse sample with dilute sulfuric acid. *Journal of Industrial Microbiology & Biotechnology*, *38*(9), 1467–1475. doi:10.100710295-010-0931-2 PMID:21210180

Capelli, B., & Cysewski, G. (2010). Potential health benefits of spirulina microalgae. *NUTRAfoods: International Journal of Science and Marketing for Nutraceutical Actives, Raw Materials, Finish Products*, *9*(2), 19–26. Advance online publication. doi:10.1007/BF03223332

Cardona, E., Rios, J., Peña, J., Peñuela, M., & Rios, L. (2016). King Grass: A very promising material for the production of second generation ethanol in tropical countries. *Biomass and Bioenergy*, *95*, 206–213. doi:10.1016/j.biombioe.2016.10.008

Carrero, A., Vicente, G., Rodríguez, R., Linares, M., & Del Peso, G. L. (2011). Hierarchical zeolites as catalysts for biodiesel production from Nannochloropsis microalga oil. *Catalysis Today*, *167*(1), 148–153.

Carriquiry, M. A., Du, X., & Timilsina, G. R. (2011). Second generation biofuels: Economics and policies. *Energy Policy*, *39*(7), 4222–4234. doi:10.1016/j.enpol.2011.04.036

Carvalho, A. P., Meireles, L. A., & Malcata, F. X. (2006). Microalgal reactors: A review of enclosed system designs and performances. *Biotechnology Progress*, *22*(6), 1490–1506. doi:10.1002/bp060065r PMID:17137294

Castaldi, M. J., Zhou, Y., & Yegulalp, T. M. (2007). Down-hole combustion method for gas production from methane hydrates. *Journal of Petroleum Science Engineering*, *56*(1-3), 176–185. doi:10.1016/j.petrol.2006.03.031

Ch, A. K., Chan, E. S., Rudravaram, R., Narasu, M. L., Rao, L. V., & Ravindra, P. (2007). Economics and environmental impact of bioethanol production technologies: An appraisal. *Biotechnology and Molecular Biology Reviews*, *2*(1), 14–32.

Chandel, A. K., Bhatia, L., Garlapati, V. K., Roy, L., & Arora, A. (2017). Biofuel Policy in Indian Perspective: Socio-economic Indicators and Sustainable Rural Development. In *Sustainable Biofuels Development in India*. Springer.

Chandel, A. K., da Silva, S. S., Carvalho, W., & Singh, O. V. (2012). Sugarcane bagasse and leaves: Foreseeable biomass of biofuel and bio-products. *Journal of Chemical Technology and Biotechnology (Oxford, Oxfordshire)*, *87*(1), 11–20. doi:10.1002/jctb.2742

Chandel, A. K., & Singh, O. V. (2011). Weedy lignocellulosic feedstock and microbial metabolic engineering: Advancing the generation of 'Biofuel'. *Applied Microbiology and Biotechnology*, *89*(5), 1289–1303. doi:10.100700253-010-3057-6 PMID:21181146

Chandra, R. P., Bura, R., Mabee, W. E., Berlin, D. A., Pan, X., & Saddler, J. N. (2007). Substrate pretreatment: The key to effective enzymatic hydrolysis of lignocellulosics. *Biofuels*, •••, 67–93. PMID:17530205

Chang, H. N., Yang, J. W., Park, Y. S., Kim, D. J., & Han, K. C. (1992). Extractive ethanol production in a membrane cell recycle bioreactor. *Journal of Biotechnology*, *24*(3), 329–343. doi:10.1016/0168-1656(92)90041-7 PMID:1367988

Chatterji, J., & Griffith, J.E. (1998). *Methods of decomposing gas hydrates*. US Pat No 5,713,416.

Chauhan, N. M., Hajare, S. T., Mamo, B., & Madebo, A. A. (2021). Bioethanol Production from Stalk Residues of Chiquere and Gebabe Varieties of Sweet Sorghum. *International Journal of Microbiology*, *2021*, 1–16. Advance online publication. doi:10.1155/2021/6696254 PMID:33679985

Chávez, R., Bull, P., & Eyzaguirre, J. (2006). The xylanolytic enzyme system from the genus Penicillium. *Journal of Biotechnology*, *123*(4), 413–433. doi:10.1016/j.jbiotec.2005.12.036 PMID:16569456

Cheah, W. Y., Sankaran, R., Show, P. L., Ibrahim, T. N. B. T., Chew, K. W., Culaba, A., & Jo-Shu, C. (2020). Pretreatment methods for lignocellulosic biofuels production: Current advances, challenges and future prospects. *Biofuel Research Journal*, *7*(1), 1115. doi:10.18331/BRJ2020.7.1.4

Chen, H., Zhou, D., Luo, G., Zhang, S., & Chen, J. (2015). Macroalgae for biofuels production: Progress and perspectives. *Renewable and Sustainable Energy Reviews, 47*, 427-437. doi:10.1016/j.rser.2015.03.086

Chen. (2015). Prospects for Bioethanol Production from Macroalgae. *Trends in Renewable Energy, 1*(3), 185-197. Available at: http://www.futureenergysp.com/index.php/tre/article/view/16

Chen, F., Qiao, Z., Fan, Z., Zheng, Q., Wu, Y., Zhang, M., Cui, Y., Deng, Y., Luo, B., Zhang, W., Ji, K., Qiao, X., Zhao, X., & Li, X. (2017). The effects of Sulphur dioxide on acute mortality and years of life lost are modified by temperature in Chengdu, China. *The Science of the Total Environment*, *576*, 775–784. PMID:27810762

Cheng, D., & He, Q. (2014). Assessment of Environmental Stresses for Enhanced Microalgal Biofuel Production – An Overview. *Frontiers in Energy Research, 2*, 1–8.

Chen, H. (2014). *Chemical composition and structure of natural lignocellulose. Biotechnology of Lignocellulose*. Springer. doi:10.1007/978-94-007-6898-7

Chen, H., Zhou, D., Luo, G., Zhang, S., & Chen, J. (2015). Macroalgae for biofuels production: Progress and perspectives. *Renewable & Sustainable Energy Reviews*, *47*, 427–437.

Chen, S. Y., Mochizuki, T., Abe, Y., Toba, M., & Yoshimura, Y. (2014). Ti-incorporated SBA-15 mesoporous silica as an efficient and robust Lewis solid acid catalyst for the production of high-quality biodiesel fuels. *Applied Catalysis B: Environmental*, *148*, 344–356.

Chen, W. C., Lin, Y. C., Ciou, Y. L., Chu, I. M., Tsai, S. L., Lan, J. C. W., Chang, Y.-K., & Wei, Y. H. (2017). Producing bioethanol from pretreated-wood dust by simultaneous saccharification and co-fermentation process. *Journal of the Taiwan Institute of Chemical Engineers*, *79*, 43–48. doi:10.1016/j.jtice.2017.04.025

Chen, W. H., Chen, Y.-C., & Lin, J.-G. (2013). Evaluation of biobutanol production from non-pretreated rice straw hydrolysate under non-sterile environmental conditions. *Bioresource Technology*, *135*, 262–268. doi:10.1016/j.biortech.2012.10.140 PMID:23186674

Cherubini, F., Bird, N. D., Cowie, A., Jungmeier, G., Schlamadinger, B., & Woess-Gallasch, S. (2009). Energy- and greenhouse gas-based LCA of biofuel and bioenergy systems: Key issues, ranges and recommendations. *Resources, Conservation and Recycling*, *53*(8), 434–447. doi:10.1016/j.resconrec.2009.03.013

Chintagunta, A.D., Zuccaro, G., Kumar, M., Kumar, S.P.J., Garlapati, V.K., Postemsky, P.D., Kumar, N.S.S., Chandel, A.K., & Simal-Gandara, J. (2021). Biodiesel production from lignocellulosic biomass using oleaginous microbes: prospects for integrated biofuel production. *Front. Microbiol.*, *12*, 658284. doi:.2021.658284 doi:10.3389/fmicb

Chisti, Y. (2007). Biodiesel from microalgae. *Biotechnology Advances*, *25*(3), 294–306. PMID:17350212

Chisti, Y. (2008). Biodiesel from microalgae beats bioethanol. *Trends in Biotechnology*, *26*(3), 126–131. doi:10.1016/j.tibtech.2007.12.002 PMID:18221809

Choi, W., Kim, G., & Lee, K. (2012). Influence of the CO2 absorbent monoethanolamine on growth and carbon fixation by the green alga Scenedesmus sp. *Bioresource Technology*, *120*, 295–299.

Chong, Z. R., Yang, S. H. B., Babu, P., Linga, P., & Li, X. S. (2016). Review of natural gas hydrates as an energy resource: Prospects and challenges. *Applied Energy*, *162*, 1633–1652. doi:10.1016/j.apenergy.2014.12.061

Choudhury, H. A., Malani, R. S., & Moholkar, V. S. (2013). Acid catalysed biodiesel synthesis from Jatropha oil: Mechanistic aspects of ultrasonic intensification. *Chemical Engineering Journal*, *231*, 262–272. doi:10.1016/j.cej.2013.06.107

Christenson, L., & Sims, R. (2011). Production and harvesting of microalgae for wastewater treatment, biofuels, and bioproducts. *Biotechnology Advances*, *29*(6), 686–702. doi:10.1016/j.biotechadv.2011.05.015 PMID:21664266

Christie, H., Fredriksen, S., & Rinde, E. (1998). Regrowth of kelp and colonization of epiphyte and fauna community after kelp trawling at the coast of Norway. *Hydrobiologia*, *375*, 49–58. doi:10.1023/A:1017021325189

Churasia, A., Singh, J., & Kumar, A. (2016). Production of biodiesel from soybean oil biomass as renewable energy source. *Journal of Environmental Biology*, *37*, 1303–1307. PMID:29261252

Cirne, D. G., Paloumet, X., Björnsson, L., Alves, M. M., & Mattiasson, B. (2007). Anaerobic digestion of lipid-rich waste—Effects of lipid concentration. *Renewable Energy*, *32*(6), 965–975. doi:10.1016/j.renene.2006.04.003

Cleveland, C. J., Kaufmann, R. K., & Stern, D. I. (2000). Aggregation and the role of energy in the economy. *Ecological Economics*, *32*, 301–317.

Cleveland, C. J., & Morris, C. G. (2013). *Handbook of Energy: Diagrams, Charts, and Tables: 1* (1st ed.). Elsevier Science.

Collett, T. S. (2002). Energy Resource Potential of Natural Gas Hydrates. *The American Association of Petroleum Geologists Bulletin*, *86*, 1971–1992.

Contreras, F., Pramanik, S., Rozhkova, A. M., Zorov, I. N., Korotkova, O., Sinitsyn, A. P., Schwaneberg, U., & Davari, M. D. (2020). Engineering robust cellulases for tailored lignocellulosic degradation cocktails. *International Journal of Molecular Sciences*, *21*(5), 1589. doi:10.3390/ijms21051589 PMID:32111065

Correa, D. F., Beyer, H. L., Possingham, H. P., Thomas-Hall, S. R., & Schenk, P. M. (2017). Biodiversity impacts of bioenergy production: Microalgae vs. first generation biofuels. *Renewable & Sustainable Energy Reviews*, *74*, 1131–1146.

Coyle, W. (2007). The Future of Biofuels: A Global Perspective. *Amber Waves : The Economics of Food, Farming, Natural Resources, and Rural America*, *5*(5), 365–372. doi:10.22004/ag.econ.125366

Crago, C. L., Khanna, M., Barton, J., Giuliani, E., & Amaral, W. (2010). Competitiveness of Brazilian sugarcane ethanol compared to US corn ethanol. *Energy Policy*, *38*(11), 7404–7415. doi:10.1016/j.enpol.2010.08.016

Cranganu, C., & Nitzov, B. (2009) Natural Gas Production from Gas Hydrates–An Economic Perspective. *Analele Ştiinţifice Ale Univ "AL I CUZA" IAŞI Geol Tomul LV, 1*, 299–312.

Cranganu, C. (2009). In-situ thermal stimulation of gas hydrates. *Journal of Petroleum Science Engineering*, *65*(1-2), 76–80. doi:10.1016/j.petrol.2008.12.028

Cromey, C. J., Nickell, T. D., Treasurer, J., Black, K. D., & Inall, M. (2009). Modelling the impact of cod (*Gadus morhua* L) farming in the marine environment-CODMOD. *Aquaculture (Amsterdam, Netherlands)*, *289*(1-2), 42–53. doi:10.1016/j.aquaculture.2008.12.020

Cuishus. F., & Butler, C. (1992). *A Review of the Potential of Bio-Diesel as a Trampon Fut ETSU-R-71*. Academic Press.

Cultivating Entrepreneurship | University of Maryland Extension. (n.d.). Retrieved January 20, 2022, from https://extension.umd.edu/resource/cultivating-entrepreneurship

Czernik, S., & Bridgwater, A. V. (2004). Overview of applications of biomass fast pyrolysis oil. *Energy & Fuels*, *18*(2), 590–598. doi:10.1021/ef034067u

Czyrnek-Delêtre, M. M., Smyth, B. M., & Murphy, J. D. (2017). Beyond carbon and energy: The challenge in setting guidelines for life cycle assessment of biofuel systems. *Renewable Energy*, *105*, 436–448.

da Silva Vaz, B., Costa, J. A. V., & de Morais, M. G. (2020). Physical and biological fixation of CO2 with polymeric nanofibers in outdoor cultivations of Chlorella fusca LEB 111. *International Journal of Biological Macromolecules*, *151*, 1332–1339.

da Silva Vaz, B., da Silveira Mastrantonio, D. J., Costa, J. A. V., & de Morais, M. G. (2019). Green alga cultivation with nanofibers as physical adsorbents of carbon dioxide: Evaluation of gas biofixation and macromolecule production. *Bioresource Technology*, *287*, 121406.

Da Silva, A. S. A., Inoue, H., Endo, T., Yano, S., & Bon, E. P. S. (2010). Milling pretreatment of sugarcane bagasse and straw for enzymatic hydrolysis and ethanol fermentation. *Bioresource Technology*, *101*(19), 7402–7409. doi:10.1016/j.biortech.2010.05.008 PMID:20576565

Dadwal, A., Sharma, S., & Satyanarayana, T. (2020). Progress in ameliorating beneficial characteristics of microbial cellulases by genetic engineering approaches for cellulose saccharification. *Frontiers in Microbiology*, *11*, 1387. doi:10.3389/fmicb.2020.01387 PMID:32670240

Dahiya, A. (2014). *Bioenergy: Biomass to Biofuels*. Academic Press.

Dahman, Y., Dignan, C., Fiayaz, A., & Chaudhry, A. (2019). An introduction to biofuels, foods, livestock, and the environment. Biomass, Biopolymer-Based Materials, and Bioenergy: Construction, Biomedical, and other Industrial Applications. doi:10.1016/B978-0-08-102426-3.00013-8

Dahman, Y., Syed, K., Begum, S., Roy, P., & Mohtasebi, B. (2019). Biofuels: Their characteristics and analysis. Biomass, Biopolymer-Based Materials, and Bioenergy: Construction, Biomedical, and other Industrial Applications. doi:10.1016/B978-0-08-102426-3.00014-X

Dalemans, F., Muys, B., Verwimp, A., Van den Broeck, G., Bohra, B., Sharma, N., Gowda, B., Tollens, E., & Maertens, M. (2018). Redesigning oilseed tree biofuel systems in India. *Energy Policy*, *115*, 631–643.

Damanik, N., Ong, H. C., Tong, C. W., Mahlia, T. M. I., & Silitonga, A. S. (2018). A review on the engine performance and exhaust emission characteristics of diesel engines fueled with biodiesel blends. *Environmental Science and Pollution Research International*, *25*(16), 15307–15325.

Daniel-Gromke, J., Rensberg, N., Denysenko, V., Stinner, W., Schmalfuß, T., Scheftelowitz, M., Nelles, M., & Liebetrau, J. (2017). Current Developments in Production and Utilization of Biogas and Biomethane in Germany. *Chemieingenieurtechnik (Weinheim)*, *90*(1-2), 17–35. doi:10.1002/cite.201700077

Daroch, M., Geng, S., & Wang, G. (2013). Recent advances in liquid biofuel production from algal feedstocks. *Applied Energy*, *102*, 1371–1381. doi:10.1016/j.apenergy.2012.07.031

Darzins, A., Pienkos, P., & Edye, L. (2010). Current Status and Potential for Algal Biofuels Production Al Darzins (NREL) Philip Pienkos (NREL) Les Edye (BioIndustry Partners). *NREL*. http://task39.org/files/2013/05/IEA-Task-39-Current-Status-and-Potential-of-Algal-biofuels0.pdf

Database of State Incentives for Renewables & Efficiency®—DSIRE. (n.d.). Retrieved January 20, 2022, from https://www.dsireusa.org/

Datta, S., Christena, L. R., & Rajaram, Y. R. S. (2013). Enzyme immobilization: An overview on techniques and support materials. *3 Biotech*, *3*(1), 1–9. doi:10.1007/s13205-012-0071-7

Datta, A., Hossain, A., & Roy, S. (2019). An overview on biofuels and their advantages and disadvantages. *Asian Journal of Chemistry*, *31*(8), 1851–1858. doi:10.14233/ajchem.2019.22098

de Albuquerque Wanderley, M. C., Martín, C., de Moraes Rocha, G. J., & Gouveia, E. R. (2013). Increase in ethanol production from sugarcane bagasse based on combined pretreatments and fed-batch enzymatic hydrolysis. *Bioresource Technology*, *128*, 448–453. doi:10.1016/j.biortech.2012.10.131 PMID:23201527

De Oliveira, J. S., Leite, P. M., de Souza, L. B., Mello, V. M., Silva, E. C., Rubim, J. C., Meneghetti, S. M. P., & Suarez, P. A. Z. (2009). Characteristics and composition of Jatropha gossypiifolia and Jatropha curcas L. oils and application for biodiesel production. *Biomass and Bioenergy*, *33*(3), 449–453. doi:10.1016/j.biombioe.2008.08.006

Degfie, T. A., Mamo, T. T., & Mekonnen, Y. S. (2019). Optimized biodiesel production from waste cooking oil (WCO) using calcium oxide (CaO) nano-catalyst. *Scientific Reports*, *9*(1), 18982. doi:10.103841598-019-55403-4 PMID:31831823

Demirbas, A. (2008). Biofuels sources, biofuel policy, biofuel economy and global biofuel projections. *Energy Conversion and Management*, *49*(8), 2106–2116. doi:10.1016/j.enconman.2008.02.020

Demirbas, A. (2008). Production of Biodiesel from Tall Oil. *Energy Sources. Part A, Recovery, Utilization, and Environmental Effects*, *30*, 1896–1902.

Demirbas, A. (2008a). Biohydrogen generation from organic waste. *Energy Sources. Part A, Recovery, Utilization, and Environmental Effects*, *30*(5), 475–482. doi:10.1080/15567030600828909

Demirbas, A. (2008b). Products from lignocellulosic materials via degradation processes. *Energy Sources. Part A, Recovery, Utilization, and Environmental Effects*, *30*(1), 27–37. doi:10.1080/00908310600626705

Demirbas, A. (2009). Political, economic and environmental impacts of biofuels: A review. *Applied Energy*, *86*(Suppl. 1), S108–S117. doi:10.1016/j.apenergy.2009.04.036

Demirbas, A. (2009). Progress and recent trends in biodiesel fuels. *Energy Conversion and Management*, *50*(1), 14–34. doi:10.1016/j.enconman.2008.09.001

Demirbas, A. (2010). *Methane gas hydrate: as a natural gas source*. Springer London. doi:10.1007/978-1-84882-872-8

Demirbas, A. (2010). Use of algae as biofuel sources. *Energy Conversion and Management*, *51*(12), 2738–2749. doi:10.1016/j.enconman.2010.06.010

Demirbas, A. (2010a). Methane from gas hydrates in the Black Sea. *Energy Sources. Part A, Recovery, Utilization, and Environmental Effects*, *32*(2), 165–171. doi:10.1080/15567030802463885

Devalia, J. L., Rusznak, C., Herdman, M. J., Trigg, C. J., Tarraf, H., & Davies, R. J. (1994). Effect of nitrogen dioxide and sulphur dioxide on airway response of mild asthmatic patients to allergen inhalation. *Lancet*, *344*, 1668–1671. PMID:7996960

Devi, A., Bajar, S., Kour, H., Kothari, R., Pant, D., & Singh, A. (2022). Lignocellulosic biomass valorization for bioethanol production: A circular bioeconomy approach. *BioEnergy Research*. Advance online publication. doi:10.100712155-022-10401-9 PMID:35154558

Dewey, L. (1916). *Hemp Hards as Papermaking Material*. USDA Bulletin No. 404.

Dey, P., Pal, P., Kevin, J. D., & Das, D. B. (2020). Lignocellulosic bioethanol production: Prospects of emerging membrane technologies to improve the process - A critical review. *Reviews in Chemical Engineering*, *36*(3), 333–367. doi:10.1515/revce-2018-0014

Dhinesh, B., & Annamalai, M. (2018). A study on performance, combustion and emission behaviour of diesel engine powered by novel nano nerium oleander biofuel. *Journal of Cleaner Production*, *196*, 74–83.

Diamond, L. W., & Akinfiev, N. N. (2003). Solubility of CO_2 in water from − 1.5 to 100 C and from 0.1 to 100 MPa: Evaluation of literature data and thermodynamic modelling. *Fluid Phase Equilibria*, *208*(1-2), 265–290. doi:10.1016/S0378-3812(03)00041-4

Dinopoulou, G., Rudd, T., & Lester, J. N. (1988). Anaerobic acidogenesis of complex wastewater: I. The influence of operational parameters on reactor performance. *Biotechnology and Bioengineering*, *31*(9), 958–968. doi:10.1002/bit.260310908 PMID:18584703

Divya, D., Gopinath, L. R., & Christy, P. M. (2015). A review on current aspects and diverse prospects for enhancing biogas production in sustainable means. *Renewable & Sustainable Energy Reviews*, *42*, 690–699. doi:10.1016/j.rser.2014.10.055

Dodić, S., Popov, S., Dodić, J., Ranković, J., Zavargo, Z., & Jevtić Mučibabić, R. (2009). Bioethanol production from thick juice as intermediate of sugar beet processing. *Biomass and Bioenergy*, *33*(5), 822–827. doi:10.1016/j.biombioe.2009.01.002

Doshi, A., Pascoe, S., Coglan, L., & Rainey, T. J. (2016). Economic and policy issues in the production of algae-based biofuels: A review. *Renewable & Sustainable Energy Reviews*, *64*, 329–337.

Dragone, G., Fernandes, B., Vicente, A., & Teixeira, J. (2010). Third generation biofuels from microalgae. *Current Research, Technology and Education Topics in Applied Microbiology and Microbial Biotechnology,* 1355–1366. http://repositorium.sdum.uminho.pt/handle/1822/16807

Dresselhaus, M. S., & Thomas, I. L. (2001). Alternative energy technologies. *Nature*, *414*, 332–337. PMID:11713539

Duan, X., Zhang, C., Ju, X., Li, Q., Chen, S., Wang, J., & Liu, Z. (2013). Effect of lignocellulosic composition and structure on the bioethanol production from different poplar lines. *Bioresource Technology*, *140*, 363–367. doi:10.1016/j.biortech.2013.04.101 PMID:23708852

Dürre, P. (2007). Biobutanol: An attractive biofuel. *Biotechnology Journal*, *2*(12), 1525–1534. doi:10.1002/biot.200700168

Duvernay, W. H., Chinn, M. S., & Yencho, G. C. (2013). Hydrolysis and fermentation of sweetpotatoes for production of fermentable sugars and ethanol. *Industrial Crops and Products*, *42*(1), 527–537. doi:10.1016/j.indcrop.2012.06.028

Dwivedi, G., Sharma, M. P., & Kumar, M. (2014). Status and Policy of Biodiesel Development in India. *Int. J. Renew. Energy Res*, *4*, 246–254.

Economic Times. (2017). *Make ethanol from surplus food grains: Draft policy*. https://goo.gl/Af6zy7

Ejaz, U., Sohail, M., & Ghanemi, A. (2021). Cellulases: From bioactivity to a variety of industrial applications. *Biomimetics*, *6*(3), 44. doi:10.3390/biomimetics6030044 PMID:34287227

Ejikeme, P. M., Anyaogu, I. D., Ejikeme, C. L., Nwafor, N. P., Egbuonu, C. A. C., Ukogu, K., & Ibemesi, J. A. (2010). Catalysis in biodiesel production by transesterification processes-an insight. *E-Journal of Chemistry*, *7*(4), 1120–1132. doi:10.1155/2010/689051

El Hage, R., Brosse, N., Chrusciel, L., Sanchez, C., Sannigrahi, P., & Ragauskas, A. (2009). Characterization of milled wood lignin and ethanol organosolv lignin from miscanthus. *Polymer Degradation & Stability*, *94*(10), 1632–1638. doi:10.1016/j.polymdegradstab.2009.07.007

Elemike, E. E., Oseghale, O. C., & Okoye, A. C. (2015). Utilization of cellulosic cassava waste for bio-ethanol production. *Journal of Environmental Chemical Engineering*, *3*(4), 2797–2800. doi:10.1016/j.jece.2015.10.021

Elkadi, J. M., Pillay, A. E., Manuel, J., Stephen, S., & Khan, M. Z. (2013). Kinetic study of neem biodiesel producton. *British Biotechnology Journal*, *3*(4), 500–508. doi:10.9734/BBJ/2013/4434

Elkady, M. F., Zaatout, A., & Balbaa, O. (2015). Production of biodiesel from waste vegetable oil via KM micromixer. *Journal of Chemistry*, *2015*, 1–9. doi:10.1155/2015/630168

Encinar, J. M., Pardal, A., Sanchez, N., & Nogales, S. (2018). Biodiesel by transesterification of rapeseed oil using ultrasound: A kinetic study of base-catalysed reactions. *Energies*, *11*(2229), 1–13. doi:10.3390/en11092229

Eriksen, N. T., Poulsen, B. R., & Iversen, J. J. L. (1998). Dual sparging laboratory-scale photobioreactor for continuous production of microalgae. *Journal of Applied Phycology*, *10*(4), 377–382. doi:10.1023/A:1008029521111

Escaramboni, B., Fernández Núñez, E. G., Carvalho, A. F. A., & de Oliva Neto, P. (2018). Ethanol biosynthesis by fast hydrolysis of cassava bagasse using fungal amylases produced in optimized conditions. *Industrial Crops and Products*, *112*(December), 368–377. doi:10.1016/j.indcrop.2017.12.004

Espinoza, D. N., & Santamarina, J. C. (2011). P-wave monitoring of hydrate-bearing sand during CH4-CO2replacement. *International Journal of Greenhouse Gas Control*, *5*(4), 1031–1038. doi:10.1016/j.ijggc.2011.02.006

Ethanol Impacts | Agricultural Economics. (n.d.). Retrieved January 20, 2022, from https://agecon.unl.edu/ethanolimpacts

Europian Tech. and Innovation Platform. (2022). *Trans esterification to biodiesel*. http://www. etipbioenergy. eu/value-chains/conversion-technologies/convention al-technologies/ transesterifi cation-t o-biodiesel 22

Evets, L. (1994). Means to normalise the levels of immunoglobulin E, using the food supplement Spirulina. Grodenski State Medical University Russian Federation Committee of Patents and Trade. Patent (19) RU (11) 2005486.

Fabris, M., Abbriano, R. M., Pernice, M., Sutherland, D. L., Commault, A. S., Hall, C. C., Labeeuw, L., McCauley, J. I., Kuzhiuparambil, U., Ray, P., Kahlke, T., & Ralph, P. J. (2020). Emerging Technologies in Algal Biotechnology: Toward the Establishment of a Sustainable, Algae-Based Bioeconomy. *Frontiers in Plant Science*, *11*, 279. doi:10.3389/fpls.2020.00279 PMID:32256509

Falk, H. M., & Benz, H. C. (2011). Monitoring the Anaerobic Digestion Process. Information Resource Center der Jacobs University.

Fan, S., Zhang, Y., Tian, G., Liang, D., & Li, D. (2006). Natural gas hydrate dissociation by presence of ethylene glycol. *Energy & Fuels*, *20*(1), 324–326. doi:10.1021/ef0502204

FAO. (2008). *State of Food and Agriculture. Biofuels: prospects, risks, and opportunities. Food and Agricultural Organisation*. FAO.

Faraco, V., & Hadar, Y. (2011). The potential of lignocellulosic ethanol production in the Mediterranean Basin. *Renewable & Sustainable Energy Reviews*, *15*(1), 252–266. doi:10.1016/j.rser.2010.09.050

Fargione, J., Hill, J., Tilman, D., Polasky, S., & Hawthorne, P. (2008). Land clearing and the biofuel carbon debt. *Science*, *2008*(319), 1235–1238. PMID:18258862

Fatykhov, M. A., & Bagautdinov, N. Y. (2005). Experimental investigations of decomposition of gas hydrate in a pipe under the impact of a microwave electromagnetic field. *High Temperature*, *43*(4), 614–619. doi:10.100710740-005-0103-7

Fengel, D., & Wegener, G. (1984). Lignin. *Wood: Chemistry, ultrastructure, reactions*, 132-191.

Fernandes, B. S., Peixoto, G., Albrecht, F. R., Saavedra del Aguila, N. K., & Zaiat, M. (2010). Potential to produce biohydrogen from various wastewaters. *Energy for Sustainable Development*, *14*, 143–148.

Ferrell, J., & Sarisky-Reed, V. (2010). *National Algal Biofuels Technology Roadmap (No. DOE/EE—0332)*. EERE Publication and Product Library. doi:10.2172/1218560

Ferry, J. G. (2010). The chemical biology of methanogenesis. *Planetary and Space Science*, *58*(14-15), 1775–1783. doi:10.1016/j.pss.2010.08.014

Fitzgerald, G. C., & Castaldi, M. J. (2013). Thermal stimulation based methane production from hydrate bearing quartz sediment. *Industrial & Engineering Chemistry Research*, *52*(19), 6571–6581. doi:10.1021/ie400025f

Flach, B., Lieberz, S., & Bolla, S. (2019). *EU biofuels annual 2019*. Global Agriculturan Information Network. https://apps.fas.usda.gov/newgainapi/api/report/downloadreportbyfilename?filename=Biofuels Annual_The Hague_EU-28_7-15-2019.pdf

Foon, C. S., May, C. Y., Ngan, M. A., & Hock, C. C. (2004). Kinetics study on transesterification of palm oil. *Journal of Oil Palm Research*, *16*(2), 19–29.

Fournier, G. P., & Gogarten, J. P. (2008). Evolution of Acetoclastic Methanogenesis in Methanosarcina via Horizontal Gene Transfer from Cellulolytic Clostridia. *Journal of Bacteriology*, *190*(3), 1124–1127. doi:10.1128/JB.01382-07 PMID:18055595

Frank, E. D., Han, J., Palou-Rivera, I., Elgowainy, A., & Wang, M. Q. (2012). Methane and nitrous oxide emissions affect the life-cycle analysis of algal biofuels. *Environmental Research Letters*, *7*(1), 014030. doi:10.1088/1748-9326/7/1/014030

Freedman, B., Butterfield, R. O., & Pryde, E. H. (1986). Transesterification kinetics of soybean oil. *Journal of the American Oil Chemists' Society*, *63*(10), 1375–1380. doi:10.1007/BF02679606

Friedlingstein, P., Andrew, R. M., Rogelj, J., Peters, G. P., Canadell, J. G., Knutti, R., Luderer, G., Raupach, M. R., Schaeffer, M., van Vuuren, D. P., & Le Quéré, C. (2014). Persistent growth of CO2 emissions and implications for reaching climate targets. *Nature Geoscience*, *7*, 709–715.

Furlan, F. F., Pinto, F. H., Costa, C. B., Cruz, A. J., Giordano, R. L., & Giordano, R. C. (2013). Bioelectricity versus bioethanol from sugarcane bagasse: Is it worth being flexible? *Biotechnology for Biofuels*, *6*(1), 1–12. doi:10.1186/1754-6834-6-142 PMID:24088415

Gabitto, J. F., & Tsouris, C. (2010). Physical Properties of Gas Hydrates: A Review. *Journal of Thermodynamics*, *2010*, 1–12. doi:10.1155/2010/271291

Gajraj, R. S., Singh, G. P., & Kumar, A. (2018). Third-Generation Biofuel: Algal Biofuels as a Sustainable Energy Source. In A. Kumar, S. Ogita & Y.-Y. Yau (Eds.), Biofuels: Greenhouse Gas Mitigation and Global Warming (pp. 307-325). Springer (India) Pvt. Ltd.

Galbe, M., & Zacchi, G. (2007). Pretreatment of lignocellulosic materials for efficient bioethanol production. *Biofuels*, 41–65. PMID:17646946

Gallego-Schmid, A., Zepon, T., & Ricardo, R. (2019). Life cycle assessment of wastewater treatment in developing countries: A review. *Water Research*, *153*, 63–79. doi:10.1016/j.watres.2019.01.010 PMID:30690219

Ganesh, M., Hemalatha, P., Peng, M. M., & Jang, H. T. (2017). One pot synthesized Li, Zr doped porous silica nanoparticle for low temperature CO2 adsorption. *Arabian Journal of Chemistry*, *10*, S1501–S1505.

Ganzler, K., Salgó, A., & Valkó, K. (2012). Microwave extraction A novel sample preparation method for chromatography. *Comprehensive Sampling and Sample Preparation*, *2*, 135–149. doi:10.1016/B978-0-12-381373-2.00048-X

Gao, G., Burgess, J., Wu, M., Wang, S., & Gao, K. (2020). Using macroalgae as biofuel: Current opportunities and challenges. *Botanica Marina*, *63*(4), 355–370. doi:10.1515/bot-2019-0065

García, A., González Alriols, M., & Labidi, J. (2014). Evaluation of different lignocellulosic raw materials as potential alternative feedstocks in biorefinery processes. *Industrial Crops and Products*, *53*, 102–110. doi:10.1016/j.indcrop.2013.12.019

Garfi, M., Martí-Herrero, J., Garwood, A., & Ferrer, I. (2016). Household anaerobic digesters for biogas production in Latin America: A review. *Renewable & Sustainable Energy Reviews*, *60*, 599–614. doi:10.1016/j.rser.2016.01.071

Garrido-Cardenas, J. A., Manzano-Agugliaro, F., Acien-Fernandez, F. G., & Molina-Grima, E. (2018). Microalgae research worldwide. *Algal Research*, *35*, 50–60. doi:10.1016/j.algal.2018.08.005

Gaykawad, S. S., Zha, Y., Punt, P. J., van Groenestijn, J. W., van der Wielen, L. A. M., & Straathof, A. J. J. (2013). Pervaporation of ethanol from lignocellulosic fermentation broth. *Bioresource Technology*, *129*, 469–476. doi:10.1016/j.biortech.2012.11.104 PMID:23266848

Gemmeke, B., Rieger, C., Weiland, P., & Schröder, J. (2009). *Biogasanlagen im Vergleich*. Fachagentur Nachwachsende Rohstoffe E.V.

Gendy, T. S., & El-Temtamy, S. A. (2013). Commercialization potential aspects of microalgae for biofuel production: An overview. *Egyptian Journal of Petroleum*, *22*(1), 43–51.

Georgianna, D. R. G., & Mayfield, S. P. (2012). Exploring diversity and synthetic biology for the production of algal biofuels. *Nature*, *448*(7411), 329–335. doi:10.1038/nature11479 PMID:22895338

Ge, X., Xu, F., & Li, Y. (2016). Solid-state anaerobic digestion of lignocellulosic biomass: Recent progress and perspectives. *Bioresource Technology*, *205*(January), 239–249. doi:10.1016/j.biortech.2016.01.050 PMID:26832395

Ghadiryanfar, M., Rosentrater, K. A., Keyhani, A., & Omid, M. (2016). A review of macroalgae production, with potential applications in biofuels and bioenergy. *Renewable and Sustainable Energy Reviews, 54*, 473-481. doi:10.1016/j.rser.2015.10.022

Ghaffar, S. H., & Fan, M. (2013). Structural analysis for lignin characteristics in biomass straw. *Biomass and Bioenergy*, *57*, 264–279. doi:10.1016/j.biombioe.2013.07.015

Ghasemi, Y., Rasoul-Amini, S., Naseri, A. T., Montazeri-Najafabady, N., Mobasher, M. A., & Dabbagh, F. (2012). Microalgae biofuel potentials [Review]. *Applied Biochemistry and Microbiology*, *48*(2), 126–144. doi:10.1134/S0003683812020068 PMID:22586908

Ginsburg, G. D., Guseynov, R. A., Dadashev, A. A., Ivanova, G. A., Kazantsev, S. A., Solov'yev, V. A., Telepnev, E. V., Askeri-Nasirov, R. Y., Yesikov, A. D., Mal'tseva, V. I., Mashirov, Y. G., & Shabayeva, Y. (1992). Gas Hydrates of the Southern Caspian. *International Geology Review*, *34*(8), 765–782. doi:10.1080/00206819209465635

Ginsburg, G. D., & Soloviev, V. A. (1994). Mud Volcano Gas hydrates in the Caspian Sea. *Bulletin of the Geological Society of Denmark*, *41*, 95–100. doi:10.37570/bgsd-1995-41-10

Giugni, D., & Giugni, V. (2010). Intellectual Property: A powerful tool to develop biotech research. *Microbial Biotechnology*, *3*(5), 493–506. doi:10.1111/j.1751-7915.2010.00172.x PMID:21255349

Glaser, B., Haumaier, L., Guggenberger, G., & Zech, W. (2001). The "Terra Preta" phenomenon: A model for sustainable agriculture in the humid tropics. *Naturwissenschaften*, *88*(1), 37–41. doi:10.1007001140000193

Gnansounou, E., Dauriat, A., Villegas, J., & Panichelli, L. (2009). Life-cycle assessment of biofuels: Energy and greenhouse gas balances. *Bioresource Technology*, *100*(21), 4919–4930. PMID:19553106

Godvin Sharmila, Dinesh Kumar, Pugazhendi, Bajhaiya, Gugulothu, & Banu. (2021). Biofuel production from Macroalgae: Present scenario and future scope. *Bioengineered, 12*(2), 9216-9238. doi:10.1080/21655979.2021.1996019

Goel, N., Wiggins, M., & Shah, S. (2001). Analytical modeling of gas recovery from in situ hydrates dissociation. *Journal of Petroleum Science and Engineering, 29*(2), 115 – 127.

Gómez, D., Ramos-Suárez, J. L., Fernández, B., Muñoz, E., Tey, L., Romero-Güiza, M., & Hansen, F. (2019). Development of a Modified Plug-Flow Anaerobic Digester for Biogas Production from Animal Manures. *Energies*, *12*(13), 2628. doi:10.3390/en12132628

Gopinathan, M. C., & Sudhakaran, R. (2009). Biofuels: Opportunities and challenges in India. *In Vitro Cellular & Developmental Biology. Plant*, *45*(3), 350–371. doi:10.100711627-009-9217-7

Goran, B. (2010). Bioenergy land use change and climate change mitigation. IEA Technology Network.

Goswami, K., & Choudhury, H. K. (2015). To grow or not to grow? Factors influencing the adoption of and continuation with *Jatropha* in North East India. *Renewable Energy*, *81*, 627–638.

Goyal, H. B., Seal, D., & Saxena, R. C. (2008). Bio-fuels from thermochemical conversion of renewable resources: A review. *Renewable & Sustainable Energy Reviews*, *12*(2), 504–517. doi:10.1016/j.rser.2006.07.014

Grabber, J. H. (2005). How do lignin composition, structure, and cross-linking affect degradability? A review of cell wall model studies. *Crop Science*, *45*(3), 820–831. doi:10.2135/cropsci2004.0191

Grala, A., Zieliński, M., Dębowski, M., & Dudek, M. (2012). Effects of hydrothermal depolymerisation and enzymatic hydrolysis of algae biomass on yield of methane fermentation process. *Polish Journal of Environmental Studies*, *21*(2), 363–368.

Graue, A., Kvamme, B., Zornes, D.R., Stevens, J.C., Howard, J.J., & Baldwin, B.A. (2007). *Production of free gas by gas hydrate conversion.* US Pat No 7,222,673. doi:10.1126/science.Liquids

Greetham, D., Adams, J. M., & Du, C. (2020). The utilization of seawater for the hydrolysis of macroalgae and subsequent bioethanol fermentation. *Scientific Reports*, *10*(1), 9728. doi:10.103841598-020-66610-9 PMID:32546695

Griffiths, M.J., Dicks, R.G., Richardson, C., & Harrison, S.T.L. (2011). *Advantages and Challenges of Microalgae as a Source of Oil for Biodiesel*. doi:10.5772/30085

Grippi, D., Clemente, R., & Bernal, M. P. (2020). Chemical and bioenergetic characterization of biofuels from plant biomass: Perspectives for southern Europe. *Applied Sciences (Basel, Switzerland)*, *10*(10), 3571. doi:10.3390/app10103571

Grossman-Ponemon, B. E., Heimisson, E., Lew, A. J., & Segall, P. (2019). *Logarithmic growth of dikes from a depressurising magma chamber*. doi:10.31223/osf.io/xekm6

Grous, W. R., Converse, A. O., & Grethlein, H. E. (1986). Effect of steam explosion pretreatment on pore size and enzymatic hydrolysis of poplar. *Enzyme and Microbial Technology*, *8*(5), 274–280. doi:10.1016/0141-0229(86)90021-9

Guinée, J. B., Heijungs, R., & van der Voet, E. (2009). A greenhouse gas indicator for bioenergy: Some theoretical issues with practical implications. *Int. J. LCA*, *14*, 328–339.

Gujer, W., & Zehnder, A. J. B. (1983). Conversion processes in anaerobic digestion. *Water Science and Technology*, *15*(8-9), 127–167. doi:10.2166/wst.1983.0164

Gullapalli, I., Jones, E., & Moridis, G. (2009). *Method for gas production from gas hydrate reservoirs*. US Pat No 7,537,058.

Guo, B., Bretz, R.E., & Lee, R.L. (1995). *Method and apparatus for generating, transporting and dissociating gas hydrates*. US Pat No 5,473,904.

Guo, M., Song, W., & Buhain, J. (2015). Bioenergy and biofuels: History, status, and perspective. *Renewable & Sustainable Energy Reviews*, *42*, 712–725. doi:10.1016/j.rser.2014.10.013

Haas, M. J., Michalski, P. J., Runyon, S., Nunez, A., & Scott, K. M. (2003). Production of FAME from acid oil, a by-product of vegetable oil refining. *Journal of the American Oil Chemists' Society*, *80*, 97–102.

Hadar, Y. (2013). Sources for lignocellulosic raw materials for the production of ethanol. In *Lignocellulose conversion* (pp. 21–38). Springer. doi:10.1007/978-3-642-37861-4_2

Hafting, J. T., Craigie, J. S., Stengel, D. B., Loureiro, R. R., Buschmann, A. H., Yarish, C., Edwards, M. D., & Critchley, A. T. (2015). Prospects and challenges for industrial production of seaweed bioactives. *Journal of Phycology*, *51*(5), 821–837. doi:10.1111/jpy.12326 PMID:26986880

Haghighi Mood, S., Hossein Golfeshan, A., Tabatabaei, M., Salehi Jouzani, G., Najafi, G. H., Gholami, M., & Ardjmand, M. (2013). Lignocellulosic biomass to bioethanol, a comprehensive review with a focus on pretreatment. *Renewable & Sustainable Energy Reviews*, *27*, 77–93. doi:10.1016/j.rser.2013.06.033

Haligva, C., Linga, P., Ripmeester, J. A., & Englezos, P. (2010). Recovery of methane from a variable-volume bed of silica sand/hydrate by depressurization. *Energy & Fuels*, *24*(5), 2947–2955. doi:10.1021/ef901220m

Hall, C. A. S. (2016). Energy Return on Investment: A Unifying Principle for Biology, Economics, and Sustainability. Springer.

Hamouda, H. I., Nassar, H. N., Madian, H. R., Abu Amr, S. S., & El-Gendy, N. S. (2015). Response Surface Optimization of Bioethanol Production from Sugarcane Molasses by Pichia veronae Strain HSC-22. *Biotechnology Research International*, *2015*, 1–10. doi:10.1155/2015/905792

Han, H.-S. (n.d.). *Getting Biomass to Market: Harvesting Methods and Costs*. Academic Press.

Hancock, S.H., Carle, D., Weatherill, B., Dallimore, S.R., Collett, T.S., Satoh, T., & Inoue, T. (2005) Overview of pressure-drawdown production-test results for the JAPEX/JNOC/GSC et al. Mallik 5L-38 gas hydrate production research well. *Bull Surv Canada 585*:134

Handa, Y. P. (1986). Calorimetric determinations of the compositions, enthalpies of dissociation, and heat capacities in the range 85 to 270 K for clathrate hydrates of xenon and krypton. *The Journal of Chemical Thermodynamics*, *18*(9), 891–902. doi:10.1016/0021-9614(86)90124-2

Han, G. B., Jang, J. H., Ahn, M. H., & Jung, B. H. (2019). Recent application of bio-alcohol: Bio-jet fuel. In Y. Yun (Ed.), *Alcohol fuels - Current technologies and future prospect*. IntechOpen.

Han, K.-H., Ko, J.-H., & Yang, S. H. (2007). Optimizing lignocellulosic feedstock for improved biofuel productivity and processing. *Biofuels, Bioproducts & Biorefining*, *1*(2), 135–146. doi:10.1002/bbb.14

Hannon, M., Gimpel, J., Tran, M., Rasala, B., & Mayfield, S. (2010). Biofuels from algae: Challenges and potential Importance & challenges of algal biofuels. *Biofuels*, *1*(5), 763–784. doi:10.4155/bfs.10.44 PMID:21833344

Hansen, C. L., & Cheong, D. Y. (2013). Agricultural Waste Management in Food Processing. In *Handbook of Farm, Dairy, and Food Machinery Engineering*. Academic Press. doi:10.1016/B978-0-12-385881-8.00023-9

Harun, R., & Danquah, M. K. (2011). Influence of acid pre-treatment on microalgal biomass for bioethanol production. *Process Biochemistry*, *46*(1), 304–309.

Harun, R., Danquah, M. K., & Forde, G. M. (2010). Microalgal biomass as a fermentation feedstock for bioethanol production. *Journal of Chemical Technology and Biotechnology (Oxford, Oxfordshire)*, *85*(2), 199–203.

Harun, R., Singh, M., Forde, G. M., & Danquah, M. K. (2010). Bioprocess engineering of microalgae to produce a variety of consumer products. *Renewable & Sustainable Energy Reviews*, *14*(3), 1037–1047. doi:10.1016/j.rser.2009.11.004

Hasannuddin, A. K., Yahya, W. J., Sarah, S., Ithnin, A. M., Syahrullail, S., Sidik, N. A. C., ... Ramlan, N. A. (2018). Nano-additives incorporated water in diesel emulsion fuel: Fuel properties, performance and emission characteristics assessment. *Energy Conversion and Management*, *169*, 291–314.

Hashim, H., Narayanasamy, M., Yunus, N. A., Shiun, L. J., Muis, Z. A., & Ho, W. S. (2017). A cleaner and greener fuel: Biofuel blend formulation and emission assessment. *Journal of Cleaner Production*, *146*, 208–217.

Hassaan, M. A., El Nemr, A., Elkatory, M. R., Eleryan, A., Ragab, S., El Sikaily, A., & Pantaleo, A. (2021). Enhancement of biogas production from macroalgae *Ulva latuca* via ozonation pretreatment. *Energies*, *14*(6), 1703. doi:10.3390/en14061703

Hasunuma, T., & Kondo, A. (2012). Consolidated bioprocessing and simultaneous saccharification and fermentation of lignocellulose to ethanol with thermotolerant yeast strains. *Process Biochemistry*, *47*(9), 1287–1294. doi:10.1016/j.procbio.2012.05.004

Hauschild, M. Z., Rosenbaum, R. K., & Olsen, S. I. (2017). Life Cycle Assessment: Theory and Practice. doi:10.1007/978-3-319-56475-3

Hayyan, A., Alam, M. Z., Mirghani, M. E. S., Kabbashi, A. A., Hakimi, N. I. N. M., Siran, Y. M., & Tahiruddin, S. (2010). Production of biodiesel from sludge palm oil by esterification process. *Journal for Energy and Power Engineering*, *4*(1), 11–17.

Heinemann, R.F., Huang, D.D-T., Long, J., & Saeger, R. (2001). *Method for recovering gas from hydrates.* US Pat No 6,214,175. doi:10.1016/S0141

Hemaiswarya, S., Raja, R., Carvalho, I.S., Ravikumar, R., Zambare, V., & Barh, D. (2012). *An Indian scenario on renewable and sustainable energy sources with emphasis on algae*. Academic Press.

Hendriks, A. T. W. M., & Zeeman, G. (2009). Pretreatments to enhance the digestibility of lignocellulosic biomass. *Bioresource Technology*, *100*(1), 10–18. doi:10.1016/j.biortech.2008.05.027

Herbert, R. A. (1999). Nitrogen cycling in coastal marine ecosystems. *FEMS Microbiology Reviews*, *23*(5), 563–590. doi:10.1111/j.1574-6976.1999.tb00414.x PMID:10525167

He, S., Liang, D., Li, D., & Ma, L. (2011). Experimental investigation on the dissociation behavior of methane gas hydrate in an unconsolidated sediment by microwave stimulation. *Energy & Fuels*, *25*(1), 33–41. doi:10.1021/ef1011116

Hideno, A., Inoue, H., Tsukahara, K., Fujimoto, S., Minowa, T., Inoue, S., ... Sawayama, S. (2009). Wet disk milling pretreatment without sulfuric acid for enzymatic hydrolysis of rice straw. *Bioresource Technology*, *100*(10), 2706–2711. doi:10.1016/j.biortech.2008.12.057

Hill, J., Nelson, E., Tilman, D., Polasky, S., & Tiffany, D. (2006). Environmental, economic, and energetic costs and benefits of biodiesel and ethanol biofuels. *PNAS, 103*, 11206–11210.

Hill, J., Nelson, E., Tilman, D., Polasky, S., & Tiffany, D. (2006). Environmental, economic, and energetic costs and benefits of biodiesel and ethanol biofuels. *Proceedings of the National Academy of Sciences of the United States of America*, *103*(30), 11206–11210. doi:10.1073/pnas.0604600103 PMID:16837571

Hill, J., Tajibaeva, L., & Polasky, S. (2016). Climate consequences of low-carbon fuels: The United States Renewable Fuel Standard. *Energy Policy*, *97*, 351–353. doi:10.1016/j.enpol.2016.07.035

Hiloidhari, M., Bhuyan, N., Gogoi, N., Seth, D., Garg, A., & Singh, A. … Kataki, R. (2019). Agroindustry wastes: Biofuels and biomaterials feedstocks for sustainable rural development. In Refining Biomass Residues for Sustainable Energy and Bioproducts: Technology, Advances, Life Cycle Assessment, and Economics. https://doi.org/ doi:10.1016/B978-0-12-818996-2.00016-8

Himmel, M. E., Ding, S. Y., Johnson, D. K., Adney, W. S., Nimlos, M. R., Brady, J. W., & Foust, T. D. (2007). Biomass recalcitrance: Engineering plants and enzymes for biofuels production. *Science*, *315*(5813), 804–807. doi:10.1126cience.1137016 PMID:17289988

Hirani, A. H., Javed, N., Asif, M., Basu, S. K., & Kumar, A. (2018). A review on first- and second- generation biofuel productions. *Biofuels: Greenhouse Gas Mitigation and Global Warming: Next Generation Biofuels and Role of Biotechnology*, 141–154. doi:10.1007/978-81-322-3763-1_8

Hirano, A., Ueda, R., Hirayama, S., & Ogushi, Y. (1997). CO2 fixation and ethanol production with microalgal photosynthesis and intracellular anaerobic fermentation. *Energy*, *22*(2-3), 137–142. doi:10.1016/S0360-5442(96)00123-5

Hirayama, S., Ueda, R., Ogushi, Y., Hirano, A., Samejima, Y., Hon-Nami, K., & Kunito, S. (1998). Ethanol production from carbon dioxide by fermentative microalgae. *Studies in Surface Science and Catalysis*, 657–660.

Ho, D. P., Ngo, H. H., & Guo, W. (2014). A mini review on renewable sources for biofuel. *Bioresource Technology*, *169*(July), 742–749. doi:10.1016/j.biortech.2014.07.022 PMID:25115598

Hoekman, S. K., Broch, A., Robbins, C., Ceniceros, E., & Natarajan, M. (2012). Review of biodiesel composition, properties, and specifications. *Renewable & Sustainable Energy Reviews*, *16*(1), 143–169. doi:10.1016/j.rser.2011.07.143

Hoffmann, J. P. (1998). Wastewater treatment with suspended and nonsuspended algae. *Journal of Phycology*, *34*(5), 757–763. doi:10.1046/j.1529-8817.1998.340757.x

Hoffman, S. M., Alvarez, M., Alfassi, G., Rein, D. M., Garcia-Echauri, S., Cohen, Y., & Avalos, J. L. (2021). Cellulosic biofuel production using emulsified simultaneous saccharification and fermentation (eSSF) with conventional and thermotolerant yeasts. *Biotechnology for Biofuels*, *14*(1), 157. doi:10.118613068-021-02008-7 PMID:34274018

Hombach, L. E., Cambero, C., Sowlati, T., & Walther, G. (2016). Optimal design of supply chains for second generation biofuels incorporating European biofuel regulations. *Journal of Cleaner Production*, *133*, 565–575.

Hong, B., Xue, G., Weng, L., & Guo, X. (2012). Pretreatment of moso bamboo with dilute phosphoric acid. *BioResources*, *7*(4), 4902–4913. doi:10.15376/biores.7.4.4902-4913

Hon-Nami, K. (2006). A unique feature of hydrogen recovery in endogenous starch-to-alcohol fermentation of the marine microalga, *Chlamydomonas perigranulata*. *Applied Biochemistry and Biotechnology*, *131*, 808–828. PMID:18563656

Horn, S. J. (2000). *Bioenergy from brown seaweeds* [Ph.D. thesis]. NTNU.

Horn, H. J., Aasen, I. M., & Østgaard, M. (2000). Production of ethanol from mannitol by *Zymobacter palmae*. *Journal of Industrial Microbiology & Biotechnology*, *24*(1), 51–57. doi:10.1038j.jim.2900771

Horn, S. J., Vaaje-Kolstad, G., Westereng, B., & Eijsink, V. G. (2012). Novel enzymes for the degradation of cellulose. Biotechnol Biofuels. *Biotechnology for Biofuels*, *5*(1), 45. doi:10.1186/1754-6834-5-45 PMID:22747961

Ho, S. H., Chen, W. M., & Chang, J. S. (2010). Scenedesmus obliquus CNW-N as a potential candidate for CO2 mitigation and biodiesel production. *Bioresource Technology*, *101*(22), 8725–8730.

Hossain, A. B. M. S., & Aishah Salleh, A. (2008). Biodiesel fuel production from algae as renewable energy. *American Journal of Biochemistry and Biotechnology*, *4*(3), 250–254. doi:10.3844/ajbbsp.2008.250.254

Hossain, A. B. M. S., & Al-saif, A. M. (2010). Biodiesel fuel production from soybean oil waste as agricultural bioresource. *Australian Journal of Crop Science*, *4*(7), 538–542.

Hossain, A. B. M. S., & Mekhled, M. A. (2010). Biodiesel fuel production from waste canola cooking oil as sustainable energy and environmental recycling process. *Australian Journal of Crop Science*, *4*(7), 543–549.

Hossain, A. B. M. S., Saleh, A. A., Aishah, S., Boyce, A. N., Chowdhury, P. P., & Naqiuddin, M. (2008). Bioethanol production from agricultural waste biomass as a renewable bioenergy resource in biomaterials. In N. A. A. Osman, F. Ibrahim, W. A. B. W. Abas, H. S. A. Rahman, & H. N. Ting (Eds.), *Biomed 2008* (Vol. 21, pp. 300–305). Proceedings. doi:10.1007/978-3-540-69139-6_77

Hossain, N., Mahlia, T. M. I., & Saidur, R. (2019). Latest development in microalgae-biofuel production with nano-additives. *Biotechnology for Biofuels*, *12*(1), 1–16.

Hossain, N., Zaini, J., Mahlia, T. M. I., & Azad, A. K. (2019). Elemental, morphological and thermal analysis of mixed microalgae species from drain water. *Renewable Energy*, *131*, 617–624.

Houghton, J. (2005). Global warming. *Reports on Progress in Physics*, *68*(6), 1343–1403. doi:10.1088/0034-4885/68/6/R02

Howard, R. L., Abotsi, E. L. J. R., Van Rensburg, E. J., & Howard, S. (2003). Lignocellulose biotechnology: Issues of bioconversion and enzyme production. *African Journal of Biotechnology*, *2*(12), 602–619. doi:10.5897/AJB2003.000-1115

Hribernik, A., & Kegl, B. (2007). The influence of biodiesel on combustion and emission characteristicsof a diesel engine. *Jixie Gongcheng Xuebao*, *53*, 683–695.

Huang, D., Zhou, H., & Lin, L. (2012). Biodiesel: An Alternative to Conventional Fuel. *Energy Procedia*, *16*, 1874–1885.

Huang, C. H., & Tan, C. S. (2014). A review: CO2 utilization. *Aerosol and Air Quality Research*, *14*(2), 480–499.

Huang, W., Wang, Z., Zhou, Y., & Ng, W. J. (2015). The role of hydrogenotrophic methanogens in an acidogenic reactor. *Chemosphere*, *140*, 40–46. doi:10.1016/j.chemosphere.2014.10.047 PMID:25466820

Huang, X., Dong, W., Wang, H., & Feng, Y. (2018). Role of acid/alkali-treatment in primary sludge anaerobic fermentation: Insights into microbial community structure, functional shifts and metabolic output by high-throughput sequencing. *Bioresource Technology*, *249*, 943–952. doi:10.1016/j.biortech.2017.10.104 PMID:29145121

Huang, Y. J., Wu, C. Y., & Hsieh, B. Z. (2016). Effect of Fluid Saturation on Gas Recovery from Class-3 Hydrate Accumulations Using Depressurization: A Case Study of the Yuan-an Ridge Site in Southwestern Offshore Taiwan. *Energy Procedia*, *97*, 310–317. doi:10.1016/j.egypro.2016.10.006

Hughes, A. D., Kelly, M. S., Black, K. D., & Stanley, M. S. (2012). Biogas from Macroalgae: Is it time to revisit the idea? *Biotechnology for Biofuels and Bioproducts*, *5*(1), 86. doi:10.1186/1754-6834-5-86 PMID:23186536

Husain, Q. (2017). Nanomaterials immobilized cellulolytic enzymes and their industrial applications : A literature review. *JSM Biochemistry and Molecular Biology*, *4*(3), 1029.

Hussain, F., Soudagar, M. E. M., Afzal, A., Mujtaba, M. A., Fattah, I. M., Naik, B., ... Rahman, S. M. (2020). Enhancement in combustion, performance, and emission characteristics of a diesel engine fueled with Ce-ZnO nanoparticle additive added to soybean biodiesel blends. *Energies*, *13*(17), 4578.

Hu, X., Zhang, Y., Ding, Z., Wang, T., Lian, H., Sun, Y., & Wu, J. (2012). Bioaccessibility and health risk of arsenic and heavy metals (Cd, Co, Cr, Cu, Ni, Pb, Zn and Mn) in TSP and PM2.5 in Nanjing, China. *Atmospheric Environment*, *57*, 146–152.

Hu, Z., & Wen, Z. (2008). Enhancing enzymatic digestibility of switchgrass by microwave-assisted alkali pretreatment. *Biochemical Engineering Journal*, *38*(3), 369–378. doi:10.1016/j.bej.2007.08.001

IAE. (2017). *International Energy Efficiency*. https://www.iea.org/reports/world-energy-outlook-2017

Ibeto, C. N., Ofoefule, A. U., & Agbo, K. E. (2011). A global overview of biomass potentials for bioethanol production: A renewable alternative fuel. *Trends in Applied Sciences Research*, *6*(5), 410–425. doi:10.3923/tasr.2011.410.425

Ibrahim, I. S., Abdullahi, I. T., & Muhammad, F. Y. (2021). Comparative analyses of biodiesel produced from neem and jatropha seed oil. *Bayero Journal of Pure and Applied Sciences*, *12*(2), 141–143. doi:10.4314/bajopas.v12i2.20

IEA Bioenergy Annual Report. (2020). Under the IEA Framework for International Energy Technology Cooperation, the Executive Committee of each Technology Collaboration Programme (TCP) must produce an Annual Report for IEA Headquarters.

Iens, J. O. H. N. W., & Argione, J. O. F. (2011). Biofuels and biodiversity notifications. *Environmental Impact of Biofuels*, *21*(June), 1085–1095.

Ige, E., Olanrewaju, A., Duffy, J., & Obiora, C. (2021). A review of the effectiveness of Life Cycle Assessment for gauging environmental impacts from cement production. *Journal of Cleaner Production*, *324*, 129213.

Indian Power Alcohol (Repeal) Act 2000 | National Portal of India [WWW Document]. https://www.india.gov.in//indian-power-alcohol-repeal-act-2000

Inoue, H., Yano, S., Endo, T., Sakaki, T., & Sawayama, S. (2008). Combining hot-compressed water and ball milling pretreatments to improve the efficiency of the enzymatic hydrolysis of eucalyptus. *Biotechnology for Biofuels*, *1*(1), 1–9. doi:10.1186/1754-6834-1-2 PMID:18471309

Intelligence, M. (2016). *Global bio-ethanol market 2022-2027.* https://www.mordorintelligence.com/industry-reports/bio-ethanol-market

International—U.S. Energy Information Administration (EIA). (n.d.). Retrieved January 20, 2022, from https://www.eia.gov/international/data/world

Islam, M. R. (1994). A new recovery technique for gas production from Alaskan gas hydrates. *Journal of Petroleum Science Engineering, 11*(4), 267–281. doi:10.1016/0920-4105(94)90046-9

Issariyakul, T., Kulkarni, M. G., Dalai, A. K., & Bakhshi, N. N. (2007). Production of biodiesel from waste fryer grease using mixed methanol/ethanol system. *Fuel Processing Technology, 88*, 429–436.

Iswanto, A. M. A., Kebenaran, B., & Megantoro, P. (2021). Design of gas concentration measurement and monitoring system for biogas power plant. *Indonesian Journal of Electrical Engineering and Computer Science, 22*(2), 726–732. doi:10.11591/ijeecs.v22.i2.pp726-732

Izmirlioglu, G., & Demirci, A. (2012). Ethanol production from waste potato mash by using Saccharomyces cerevisiae. *Applied Sciences (Switzerland), 2*(4), 738–753. doi:10.3390/app2040738

Jabasingh, S. A. (2011). Optimization and kinetics of cellulase immobilization on modified chitin using response surface methodology. *Adsorption Science and Technology, 29*(9), 897–916. doi:10.1260/0263-6174.29.9.897

Jain, S. K., Prakash, V., & Kapoor, S. (2004). Flue gas treatment alternatives or enhancing ESP collection efficiency: NTPC experience. *Workshop on ESP performance: Role of Fly Ash Resistivity. IIT-DandNTPC*, 1–5.

James, G. A., & James, G. (1999). *Ethical Perspectives on Environmental Issues in India* (1st ed.). APH Publishing Corporation.

Jamil, F., Kumar, P.S.M., Al-Haj, L., Myint, M.T.Z., & Al-Muhtaseb, A.H. (2021). Heterogeneous carbon-based catalyst modified by alkaline earth metal oxides for biodiesel production: Parametric and kinetic study. *Energy Convers. Manag. X 2020, 10*, 10047. doi:10.1016/j.ecmx.2020.100047

Järvinen, P. (2005). *Action research as an approach in design science.* Academic Press.

Jatoi, A. S., Abbasi, S. A., Hashmi, Z., Shah, A. K., Alam, M. S., Bhatti, Z. A., Maitlo, G., Hussain, S., Khandro, G. A., Usto, M. A., & Iqbal, A. (2021). Recent trends and future perspectives of lignocellulose biomass for biofuel production: A comprehensive review. *Biomass Conversion and Biorefinery*, •••, 1–13. doi:10.100713399-021-01853-8

Jayus, Nurhayati, Mayzuhroh, A., Arindhani, S., & Caroenchai, C. (2016). Studies on Bioethanol Production of Commercial Baker's and Alcohol Yeast under Aerated Culture Using Sugarcane Molasses as the Media. *Agriculture and Agricultural Science Procedia, 9*, 493–499. doi:10.1016/j.aaspro.2016.02.168

Jeong, T. S., Choi, C. H., Lee, J. Y., & Oh, K. K. (2012). Behaviors of glucose decomposition during acid-catalyzed hydrothermal hydrolysis of pretreated *Gelidium amansii. Bioresource Technology, 116*, 435–440. doi:10.1016/j.biortech.2012.03.104 PMID:22522017

Ji, C., Ahmadi, G., & Smith, D. H. (2001). Natural Gas Production from Hydrate Decomposition by Depressurization. *Chemical Engineering Science, 56*(20), 5801–5814. doi:10.1016/S0009-2509(01)00265-2

Jin, B. B., Duan, P. G., Xu, Y. P., Wang, F., & Fan, Y. C. (2013). Co-liquefaction of micro- and macroalgae in subcritical water. *Bioresource Technology, 149*, 103–110. doi:10.1016/j.biortech.2013.09.045 PMID:24096026

John, R. P., Anisha, G. S., Nampoothiri, K. M., & Pandey, A. (2011). Micro and macro algal biomass: A renewable source for bioethanol. *Bioresource Technology, 102*(1), 186–193. doi:10.1016/j.biortech.2010.06.139 PMID:20663661

Johnravindar, D., Murugesan, K., Wong, J. W. C., & Elangovan, N. (2017). Waste-to-biofuel: Production of biobutanol from sago waste residues. *Environmental Technology (United Kingdom)*, *38*(13–14), 1725–1734. doi:10.1080/09593330.2017.1283362

Johnson, P., & Shubert, E. (1986). Availability of iron to rats from spirulina, a Bluegreen algae. *Nutrition Research (New York, N.Y.)*, *6*(1), 85–94. doi:10.1016/S0271-5317(86)80202-0

Joniver, C. F. H., Photiades, A., Moore, P. J., Winters, A. L., Woolmer, A., & Adams, J. M. M. (2021). The global problem of nuisance macroalgal blooms and pathways to its use in the circular economy. *Algal Research*, *58*, 102407. doi:10.1016/j.algal.2021.102407

Jordão, T. (2011). The strategic planning for renewable energy sources deployment in the Czech Republic with the support of balanced scorecard. *International Journal of Energy and Environment*, *5*, 364–376.

Joshi, S. M., Waghmare, J. S., Sonawane, K. D., & Waghmare, S. R. (2015). Bio-ethanol and bio-butanol production from orange peel waste. *Biofuels*, *6*(1–2), 55–61. doi:10.1080/17597269.2015.1045276

Juneja, A., Ceballos, R. M., & Murthy, G. S. (2013). Effects of Environmental Factors and Nutrient Availability on the Biochemical Composition of Algae for Biofuels Production: A Review. *Energies*, *6*, 4607–4638.

Jung, A. K., Lim, S. R., Kim, Y., & Park, J. M. (2013). Potentials of macroalgae as feedstocks for biorefinery. *Bioresource Technology*, *135*, 182–190. doi:10.1016/j.biortech.2012.10.025 PMID:23186669

Jung, J. W., Espinoza, D. N., & Santamarina, J. C. (2010). Properties and phenomena relevant to CH4-CO2replacement in hydrate-bearing sediments. *Journal of Geophysical Research. Solid Earth*, *115*(B10), 1–16. doi:10.1029/2009JB000812

Jung, K. W., Kim, D. H., & Shin, H. S. (2011). Fermentative hydrogen production from *Laminaria japonica* and optimization of thermal pretreatment conditions. *Bioresource Technology*, *102*(3), 2745–2750. doi:10.1016/j.biortech.2010.11.042 PMID:21123054

Kadar, E., Rooks, P., Lakey, C., & White, D. A. (2012). The effect of engineered iron nanoparticles on growth and metabolic status of marine microalgae cultures. *The Science of the Total Environment*, *439*, 8–17.

Käkölä, T. (2003). *Software business models and contexts for software innovation: Key areas software business research.* doi:10.1109/HICSS.2003.1174425

Kaliaperumal, N., Ramalingam, J. R., Kalimuthu, S., & Ezhil Valavan, R. (2002). Seasonal changes in growth, biochemical constituents and phycocolloid of some marine algae of mandapam coast. *Seaweed Research and Utilisation*, *24*(1), 73–77.

Kambo, H. S., & Dutta, A. (2015). A comparative review of biochar and hydrochar in terms of production, physico-chemical properties and applications. *Renewable & Sustainable Energy Reviews*, *45*, 359–378. doi:10.1016/j.rser.2015.01.050

Kampa, M., & Castanas, E. (2008). Human health effects of air pollution. *Environmental Pollution*, *151*, 362–367. PMID:17646040

Karellas, S., Boukis, I., & Kontopoulos, G. (2010). Development of an investment decision tool for biogas production from agricultural waste. *Renewable & Sustainable Energy Reviews*, *14*(4), 1273–1282. doi:10.1016/j.rser.2009.12.002

Karisathan Sundararajan, N., & Ammal, A. R. B. (2018). Improvement studies on emission and combustion characteristics of DICI engine fuelled with colloidal emulsion of diesel distillate of plastic oil, TiO2 nanoparticles and water. *Environmental Science and Pollution Research International*, *25*(12), 11595–11613.

Karthikeyan, S., & Dharma Prabhakaran, T. (2018). Emission analysis of Botryococcus braunii algal biofuel using Ni-Doped ZnO nano additives for IC engines. *Energy Sources. Part A, Recovery, Utilization, and Environmental Effects*, *40*(9), 1060–1067.

Karthikeyan, S., Elango, A., Prathima, A., & Raja, K. (2014). Environmental effects of nano additive Co3O4 in grape seed oil biofuel fuelled in CI engine. *Research Journal of Chemistry and Environment*, *18*, 5.

Karthikeyan, S., & Prathima, A. (2017). Analysis of emissions from use of an algae biofuel with nano-ZrO2. *Energy Sources. Part A, Recovery, Utilization, and Environmental Effects*, *39*(5), 473–479.

Karthikeyan, S., & Prathima, A. (2017). Microalgae biofuel with CeO2 nano additives as an eco-friendly fuel for CI engine. *Energy Sources. Part A, Recovery, Utilization, and Environmental Effects*, *39*(13), 1332–1338.

Katinas, V., Marčiukaitis, M., Perednis, E., Dzenajavičienė, E. F. (2019). Analysis of biodegradable waste use for energy generation in Lithuania. *Renewable and Sustainable Energy Reviews*, *101*, 559– 567. 5.

Kawai, S., & Murata, K. (2016). Biofuel Production Based on Carbohydrates from Both Brown and Red Macroalgae: Recent Developments in Key Biotechnologies. *International Journal of Molecular Sciences*, *17*(2), 145. doi:10.3390/ijms17020145 PMID:26861307

Kawamura, T., Ohtake, M., Sakamoto, Y., Yamamoto, Y., Haneda, H., Komai, T., & Higuchi, S. (2007). Experimental study on steam injection method using methane hydrate core samples. *7th Int Offshore Polar Eng Conf (ISOPE), Ocean Min Symp*, 83–86.

Kawamura, T., Sakamoto, Y., Ohtake, M., Yamamoto, Y., Haneda, H., & Komai, T. (2006). Dissociation Behavior of Hydrate Core Sample Using Thermodynamic Inhibitor — Part 2 : Experimental Investigation Using Long Core Samples. *International Journal of Offshore and Polar Engineering*, *16*, 5–9.

Kawamura, T., Yamamoto, Y., Yoon, J. H., Haned, H., Ohga, K., & Higuchi, K. (2003) Dissociation behavior of methane gas hydrate in ethylene glycol and silicone oil. *Proc ISOPE Ocean Min Symp*, *2*, 181–184

Kawentara, W. A., & Budiman, A. (2013). Synthesis of biodiesel from second-used cooking oil. *Energy Procedia*, *32*, 190–199. doi:10.1016/j.egypro.2013.05.025

Kemper, J. (2015). Biomass and carbon dioxide capture and storage: A review. *International Journal of Greenhouse Gas Control*, (40), 401–430.

Kendall, A., & Chang, B. (2009). Estimating life cycle greenhouse gas emissions from corn-ethanol: A critical review of current US practices. *Journal of Cleaner Production*, *17*(13), 1175–1182.

Khan, M. I., Shin, J. H., & Kim, J. D. (2018). The promising future of microalgae: Current status, challenges, and optimization of a sustainable and renewable industry for biofuels, feed, and other products. *Microbial Cell Factories*, *17*(1), 36. doi:10.118612934-018-0879-x PMID:29506528

Khan, N., Park, S. H., Kadima, L., Bourdeau, C., Calina, E., Edmunds, C. W., & Pursell, D. P. (2021). Locally sustainable biodiesel production from waste cooking oil and grease using a deep eutectic solvent: Characterization, thermal properties, and blend performance. *ACS Omega*, *6*(13), 9204–9212. doi:10.1021/acsomega.1c00556 PMID:33842789

Khanna, M., Önal, H., Crago, C. L., & Mino, K. (2013). Can India Meet Biofuel Policy Targets? Implications for Food and Fuel Prices. *American Journal of Agricultural Economics*, *95*, 296–302.

Khan, Z., Bhadouria, P., & Bisen, P. S. (2005). Nutritional and therapeutic potential of Spirulina. *Current Pharmaceutical Biotechnology*, *6*(5), 373–379. doi:10.2174/138920105774370607 PMID:16248810

Khattab, S. M. R., & Watanabe, T. (2019). Bioethanol From Sugarcane Bagasse: Status and Perspectives. In Bioethanol Production from Food Crops. https://doi.org/ doi:10.1016/B978-0-12-813766-6/00010-2

Khedkar, M. A., Nimbalkar, P. R., Chavan, P. V., Chendake, Y. J., & Bankar, S. B. (2017). Cauliflower waste utilization for sustainable biobutanol production: Revelation of drying kinetics and bioprocess development. *Bioprocess and Biosystems Engineering*, *40*(10), 1493–1506. doi:10.100700449-017-1806-y

Khlebnikov, V. N., Antonov, S. V., Mishin, A. S., Bakulin, D. A., Khamidullina, I. V., Liang, M., Vinokurov, V. A., & Gushchin, P. A. (2016). A new method for the replacement of CH 4 with CO 2 in natural gas hydrate production. *Nat Gas Ind B*, *3*(5), 445–451. doi:10.1016/j.ngib.2017.02.005

Kiener, A., & Leisinger, T. (1983). Oxygen sensitivity of methanogenic bacteria. *Systematic and Applied Microbiology*, *4*(3), 305–312. doi:10.1016/S0723-2020(83)80017-4 PMID:23194731

Kim, I., Seo, Y. H., Kim, G.-Y., & Han, J.-I. (2015). Co-production of bioethanol and biodiesel from corn stover pretreated with nitric acid. *Fuel*, *143*, 285–289.

Kim, J., Jia, H., & Wang, P. (2006). Challenges in biocatalysis for enzyme-based biofuel cells. *Biotechnology Advances*, *24*(3), 296–308.

Kim, M., & Day, D. F. (2011). Composition of sugar cane, energy cane, and sweet sorghum suitable for ethanol production at Louisiana sugar mills. *Journal of Industrial Microbiology & Biotechnology*, *38*(7), 803–807. doi:10.100710295-010-0812-8 PMID:20803247

Kim, M.-S., Baek, J.-S., Yun, Y.-S., Sim, S. J., Park, S., & Kim, S.-C. (2006). Hydrogen production from Chlamydomonas reinhardtii biomass using a two-step conversion process: Anaerobic conversion and photosynthetic fermentation. *International Journal of Hydrogen Energy*, *31*(6), 812–816. doi:10.1016/j.ijhydene.2005.06.009

Kim, S., & Dale, B. E. (2004). Global potential bioethanol production from wasted crops and crop residues. *Biomass and Bioenergy*, *26*(4), 361–375. doi:10.1016/j.biombioe.2003.08.002

Kim, S.-K., & Lee, C.-G. (2015). *Marine bioenergy: Trends and developments*. CRC Press.

Klein-Marcuschamer, D., & Blanch, H. W. (2015). Renewable fuels from biomass: Technical hurdles and economic assessment of biological routes. *AIChE Journal. American Institute of Chemical Engineers*, *61*(9), 2689–2701. doi:10.1002/aic.14755

Klein-Marcuschamer, D., Oleskowicz-Popiel, P., Simmons, B. A., & Blanch, H. W. (2012). The challenge of enzyme cost in the production of lignocellulosic biofuels. *Biotechnology and Bioengineering*, *109*(4), 1083–1087. doi:10.1002/bit.24370 PMID:22095526

Klinke, H. B., Thomsen, A. B., & Ahring, B. K. (2004). Inhibition of ethanol-producing yeast and bacteria by degradation products produced during pre-treatment of biomass. *Applied Microbiology and Biotechnology*, *66*(1), 10–26. doi:10.100700253-004-1642-2 PMID:15300416

Koçar, G., & Civaş, N. (2013). An overview of biofuels from energy crops: Current status and future prospects. *Renewable & Sustainable Energy Reviews*, *28*, 900–916.

Koh, C. A., Sum, A. K., & Sloan, E. D. (2009). Gas hydrates: Unlocking the energy from icy cages. *Journal of Applied Physics*, *106*(6), 061101. Advance online publication. doi:10.1063/1.3216463

Kolesinska, B., Fraczyk, J., Binczarski, M., Modelska, M., Berlowska, J., Dziugan, P., ... Kregiel, D. (2019). Butanol synthesis routes for biofuel production: Trends and perspectives. *Materials (Basel)*, *12*(3). Advance online publication. doi:10.3390/ma12030350

Kono, H. O., Narasimhan, S., Song, F., & Smith, D. H. (2002). Synthesis of Methane Gas Hydrate in Porous Sediments and its Dissociation by Depressurizing. *Powder Technology*, *122*(2-3), 239–246. doi:10.1016/S0032-5910(01)00420-X

Koradiya, M., Duggirala, S., Tipre, D., & Dave, S. (2016). Pretreatment optimization of Sorghum pioneer biomass for bioethanol production and its scale-up. *Bioresource Technology*, *199*, 142–147. doi:10.1016/j.biortech.2015.08.156 PMID:26384087

Koshijima, T., & Watanabe, T. (2003). *Preparation and Characterization of Lignin-Carbohydrate Complexes*. doi:10.1007/978-3-662-05191-7_1

Kothari, R., Vashishtha, A., Singh, H. M., Pathak, V. V., Tyagi, V. V., Yadav, B. C., Ashokkumar, V., & Singh, D. P. (2020). Assessment of Indian bioenergy policy for sustainable environment and its impact for rural India: Strategic implementation and challenges. *Environmental Technology and Innovation*, *20*, 101078. doi:10.1016/j.eti.2020.101078

Kovács, E., Wirth, R., Maróti, G., Bagi, Z., Rákhely, G., & Kovács, K. L. (2013). Biogas production from protein-rich biomass: Fed-batch anaerobic fermentation of casein and of pig blood and associated changes in microbial community composition. *PLoS One*, *8*(10), 8. doi:10.1371/journal.pone.0077265 PMID:24146974

Kraan, S. (2013). Mass-cultivation of carbohydrate rich macroalgae, a possible solution for sustainable biofuel production. *Mitigation and Adaptation Strategies for Global Change*, *18*(1), 27–46. doi:10.100711027-010-9275-5

Kranert, M., Kusch, S., Huang, J., & Fischer, K. (2012). Anaerobic Digestion of Waste. In *Waste to Energy*. Springer. doi:10.1007/978-1-4471-2306-4_5

Kricka, W., Fitzpatrick, J., & Bond, U. (2015). Challenges for the production of bioethanol from biomass using recombinant yeasts. *Advances in Applied Microbiology*, *92*, 89–125. doi:10.1016/bs.aambs.2015.02.003 PMID:26003934

Kuhi, H. D., Hossein-Zadeh, N. G., López, S., Falahi, S., & And France, J. (2019). Sinusoidal function to describe the growth curve of dairy heifers. *Animal Production Science*, *59*(6), 1039–1047. doi:10.1071/AN18104

Kulkarni, M. B., & Ghanegaonkar, P. M. (2019). Hydrogen sulfide removal from biogas using chemical absorption technique in packed column reactors. *Global Journal of Environmental Science Management*, *5*(1), 155–166.

Kumar Biswas, P., & Pohit, S. (2013). What ails India's biodiesel programme? *Energy Policy*, *52*, 789–796.

Kumar Biswas, P., Pohit, S., & Kumar, R. (2010). Biodiesel from jatropha: Can India meet the 20% blending target? *Energy Policy*, *38*, 1477–1484.

Kumar Tiwari, A., Kumar, A., & Raheman, H. (2007). Biodiesel production from jatropha oil (Jatropha curcas) with high free fatty acids: An optimized process. *Biomass and Bioenergy*, *31*, 569–575.

Kumar, A., & Kumar, P. (1984). Improving the productivity of the petrocrops in Rajasthan. *Proc. Bioenergy Soc., India, First convention and symposium,* 125.

Kumar, D., Eswari, A. P., Park, J.-H., Adishkumar, S., & Banu, J. R. (2019). Biohydrogen generation from macroalgal biomass, *Chaetomorpha antennina* through surfactant aided microwave disintegration. *Frontiers in Energy Research*, *7*, 78. doi:10.3389/fenrg.2019.00078

Kumari, A., Khan, S. H., Misra, A. K., Majumder, C. B., & Arora, A. (2020). Hydrates of Binary Guest Mixtures: Fugacity Model Development and Experimental Validation. *Journal of Non-Equilibrium Thermodynamics*, *45*(1), 39–58. doi:10.1515/jnet-2019-0062

Kumari, V., Shah, S., & Gupta, M. N. (2007). Preparation of Biodiesel by Lipase-Catalyzed Transesterification of High Free Fatty Acid Containing Oil from Madhuca indica. *Energy & Fuels*, *21*, 368–372.

Kumar, K., Ghosh, S., Angelidaki, I., Holdt, S. L., Karakashev, D. B., Morales, M. A., & Das, D. (2016). Recent developments on biofuels production from microalgae and macroalgae. *Renewable & Sustainable Energy Reviews*, *65*, 235–249. doi:10.1016/j.rser.2016.06.055

Kumar, L., & Bharadvaja, N. (2020). A review on microalgae biofuel and biorefinery: Challenges and way forward. *Energy Sources. Part A, Recovery, Utilization, and Environmental Effects*, *00*(00), 1–24. doi:10.1080/15567036.2020.1836084

Kumar, M. D., Tamilarasan, K., Kaliappan, S., Banu, J. R., Rajkumar, M., & Kim, S. H. (2018). Surfactant assisted disperser pretreatment on the liquefaction of *Ulva reticulata* and evaluation of biodegradability for energy efficient biofuel production through nonlinear regression modelling. *Bioresource Technology*, *255*, 116–122. doi:10.1016/j.biortech.2018.01.116 PMID:29414156

Kumar, M., Sun, Y., Rathour, R., Pandey, A., Thakur, I. S., & Tsang, D. C. W. (2020). Algae as potential feedstock for the production of biofuels and value-added products: Opportunities and challenges. *The Science of the Total Environment*, *716*, 137116. doi:10.1016/j.scitotenv.2020.137116 PMID:32059310

Kumar, R. S., Manimaran, R., & Gopalakrishnan, V. (2013). Performance and emission analysis using pongamia oil biodiesel fuel with an artificial neural network. *Advanced Engineering and Applied Sciences*, *3*(1), 17–20.

Kumar, S., Chaube, A., & Jain, S. K. (2012). Critical review of Jatropha biodiesel promotion policies in India. *Energy Policy*, *41*, 775–781.

Kumar, S., Shrestha, P., & Salam, P. A. (2013). A review of biofuel policies in the major biofuel producing countries of ASEAN: Production, targets, policy drivers and impacts. *Renewable & Sustainable Energy Reviews*, *26*, 822–836.

Kunjapur, A. M., & Eldridge, R. B. (2010). Photobioreactor design for commercial biofuel production from microalgae. *Industrial & Engineering Chemistry Research*, *49*(8), 3516–3526. doi:10.1021/ie901459u

Kusekwa, M. A. (2011*). A review on the renewable energy resources for rural application in Tanzania, Renewable energy - Trends and applications*. http://www.intechopen.com/books/renewable-energy-trends-and-applications/areview-on-the-renewable-energy-resources-for-rural-application-in-tanzania

Kvenvolden, K.A. (1998). A primer on the geological occurrence of gas hydrate. *Geol Soc London, 137*(9).

Kvenvolden, K. A., & Rogers, B. W. (2005). Gaia's breath - Global methane exhalations. *Marine and Petroleum Geology*, *22*(4), 579–590. doi:10.1016/j.marpetgeo.2004.08.004

Kwapinski, W., Byrne, C. M. P., Kryachko, E., Wolfram, P., Adley, C., Leahy, J. J., Novotny, E. H., & Hayes, M. H. B. (2010). Biochar from biomass and waste. *Waste and Biomass Valorization*, *1*(2), 177–189. doi:10.100712649-010-9024-8

Kywe, T. T., & Oo, M. M. (2009). Production of biodiesel from jatropha oil (jatropha curcas) in pilot plant. *Proceedings of World Academy of Science, Engineering and Technology*, *38*, 481–489.

Kywe, T. T., & Oo, M. M. (2009). Production of biodiesel from Jatropha Oil (Jatropha curcas) in pilot plant. *Proceedings of World Academy of Science, Engineering, and Technology*, *38*, 481–489.

Lali, A. (2016). Biofuels for India: What, when and how. *Current Science*, *110*, 552–555.

Lal, R. (2008). Crop residues as soil amendments and feedstock for bioethanol production. *Waste Management (New York, N.Y.)*, *28*(4), 747–758. doi:10.1016/j.wasman.2007.09.023 PMID:18053700

Lamers, P., Hamelinck, C., Junginger, M., & Faaij, A. (2011). International bioenergy trade - A review of past developments in the liquid biofuel market. *Renewable & Sustainable Energy Reviews*, *15*(6), 2655–2676. doi:10.1016/j.rser.2011.01.022

Lamichhane, G., Acharya, A., Poudel, D. K., Aryal, B., Gyawali, N., Niraula, P., Phuyal, S. R., Budhathoki, P., Bk, G., & Parajuli, N. (2021). Recent advances in bioethanol production from lignocellulosic biomass. *International Journal of Green Energy*, *18*(7), 731–744. doi:10.1080/15435075.2021.1880910

Lam, M. K., & Lee, K. T. (2019). Scale-Up and Commercialization of Algal Cultivation and Biofuel Production. In A. Pandey, D.-J. Lee, Y. Chisti, & C. R. Soccol (Eds.), *Biofuels from Algae* (pp. 261–286). doi:10.1016/B978-0-444-64192-2.00019-6

Lamnatou, C., Ezcurra-Ciaurriz, X., Chemisana, D., & Plà-Aragonés, L.M. (2022). Life Cycle Assessment (LCA) of a food-production system in Spain: Iberian ham based on an extensive system. *The Science of the Total Environment*, *808*, 151900. PMID:34838553

Larson, E. D. (2006). A review of life-cycle analysis studies on liquid biofuel systems for the transport sector. *Energy for Sustainable Development*, *2*, 109–126.

Lau, M. W., & Dale, B. E. (2009). Cellulosic ethanol production from AFEX-treated corn stover using Saccharomyces cerevisiae 424A(LNH-ST). *Proceedings of the National Academy of Sciences of the United States of America*, *106*(5), 1368–1373. doi:10.1073/pnas.0812364106 PMID:19164763

Lecksiwilai, N., Gheewala, S. H., Silalertruksa, T., & Mungkalasiri, J. (2017). LCA of biofuels in Thailand using Thai Ecological Scarcity method. *Journal of Cleaner Production*, *142*, 1183–1191.

Lee, B. R., Koh, C. A., & Sum, A. K. (2014). Quantitative measurement and mechanisms for CH4production from hydrates with the injection of liquid CO2. *Physical Chemistry Chemical Physics*, *16*(28), 14922–14927. doi:10.1039/C4CP01780C PMID:24931508

Lee, H., Seo, Y., Seo, Y. T., Moudrakovski, I. L., & Ripmeester, J. A. (2003). Recovering Methane from Solid Methane Hydrate with Carbon Dioxide. *Angewandte Chemie International Edition*, *42*(41), 5048–5051. doi:10.1002/anie.200351489 PMID:14595627

Lee, J. (1997). Biological conversion of lignocellulosic biomass to ethanol. *Journal of Biotechnology*, *56*(1), 1–24. doi:10.1016/S0168-1656(97)00073-4 PMID:9246788

Lee, J. (2010). Experimental study on the dissociation behavior and productivity of gas hydrate by brine injection scheme in porous rock. *Energy & Fuels*, *24*(1), 456–463. doi:10.1021/ef900791r

Lee, J. E., Lee, N., Kim, T., Kim, J., & Hyeon, T. (2011). Multifunctional mesoporous silica nanocomposite nanoparticles for theranostic applications. *Accounts of Chemical Research*, *44*(10), 893–902.

Lee, J. Y., Ryu, H. J., & Oh, K. K. (2013). Acid-catalyzed hydrothermal severity on the fractionation of agricultural residues for xylose-rich hydrolyzates. *Bioresource Technology*, *132*, 84–90. doi:10.1016/j.biortech.2013.01.019 PMID:23395759

Lee, J., Park, S., & Sung, W. (2010). An experimental study on the productivity of dissociated gas from gas hydrate by depressurization scheme. *Energy Conversion and Management*, *51*(12), 2510–2515. doi:10.1016/j.enconman.2010.05.015

Lee, K. C., Arai, T., Ibrahim, D., Kosugi, A., Prawitwong, P., Lan, D., Murata, Y., & Mori, Y. (2014). Purification and characterization of a thermostable laminarinase from *Penicillium rolfsii* c3-2(1) IBRL. *Bioresource*, *9*(1), 1072–1084. doi:10.15376/biores.9.1.1072-1084

Lee, R. A., & Lavoie, J. M. (2013). From first- to third-generation biofuels: Challenges of producing a commodity from a biomass of increasing complexity. *Animal Frontiers*, *3*(2), 6–11. doi:10.2527/af.2013-0010

Lee, Y., Kim, Y., Lee, J., Lee, H., & Seo, Y. (2015). CH4 recovery and CO2 sequestration using flue gas in natural gas hydrates as revealed by a micro-differential scanning calorimeter. *Applied Energy*, *150*, 120–127. doi:10.1016/j.apenergy.2015.04.012

LeVan, S., Ross, R. J., & Winandy, J. E. (1990). *Effects of fire retardant chemicals on the bending properties of wood at elevated temperatures* (Vol. 498). US Department of Agriculture, Forest Service, Forest Products Laboratory.

Levine, R. B., Pinnarat, T., & Savage, P. E. (2010). Biodiesel production from wet algal biomass through in situ lipid hydrolysis and supercritical transesterification. *Energy & Fuels*, *24*(9), 5235–5243. doi:10.1021/ef1008314

Li, X.S., Wan, L.H., Li, G., Li, Q.P., Chen, Z.Y., & Yan, K.F. (2008). *Experimental Investigation into the Production Behavior of Methane Hydrate in Porous Sediment with Hot Brine Stimulation*. doi:10.1021/acs.energyfuels.7b00464

Li, A., Antizar-Ladislao, B., & Khraisheh, M. (2007). Bioconversion of municipal solid waste to glucose for bio-ethanol production. *Bioprocess and Biosystems Engineering*, *30*(3), 189–196. doi:10.100700449-007-0114-3 PMID:17458580

Liang, D., He, S., & Li, D. (2009). Effect of microwave on formation/decomposition of natural gas hydrate. *Chinese Science Bulletin*, *54*(6), 965–971. doi:10.100711434-009-0116-4

Liang, L., Zhang, Y. P., Zhang, L., Zhu, M. J., Liang, S. Z., & Huang, Y. N. (2008). Study of sugarcane pieces as yeast supports for ethanol production from sugarcane juice and molasses. *Journal of Industrial Microbiology & Biotechnology*, *35*(12), 1605–1613. doi:10.100710295-008-0404-z PMID:18685877

Liao, Y. C., Chang, C. C., Nagarajan, D., Chen, C. Y., & Chang, J. S. (2021). Algae-derived hydrocolloids in foods: Applications and health-related issues. *Bioengineered*, *12*(1), 3787–3801. doi:10.1080/21655979.2021.1946359 PMID:34281484

Liew, W. H., Hassim, M. H., & Ng, D. K. S. (2014). Review of evolution, technology and sustainability assessments of biofuel production. *Journal of Cleaner Production*, *71*, 11–29. doi:10.1016/j.jclepro.2014.01.006

Li, G., Li, X.-S., Tang, L.-G., & Zhang, Y. (2007). Experimental investigation of production behavior of methane hydrate under ethylene glycol injection in unconsolidated sediment. *Energy & Fuels*, *21*(6), 3388–3393. doi:10.1021/ef060644d

Ling, X., Guo, J., Liu, X., Zhang, X., Wang, N., Lu, Y., & Ng, I. S. (2015). Impact of carbon and nitrogen feeding strategy on high production of biomass and docosahexaenoic acid (DHA) by Schizochytrium sp. LU310. *Bioresource Technology*, *184*, 139–147.

Lin, L., Yan, R., Liu, Y., & Jiang, W. (2010). In-depth investigation of enzymatic hydrolysis of biomass wastes based on three major components: Cellulose, hemicellulose and lignin. *Bioresource Technology*, *101*(21), 8217–8223. doi:10.1016/j.biortech.2010.05.084 PMID:20639116

Lin, S. P., Kuo, T. C., Wang, H. T., Ting, Y., Hsieh, C. W., Chen, Y. K., Hsu, H.-Y., & Cheng, K. C. (2020). Enhanced bioethanol production using atmospheric cold plasma-assisted detoxification of sugarcane bagasse hydrolysate. *Bioresource Technology*, *313*, 123704. doi:10.1016/j.biortech.2020.123704 PMID:32590306

Li, S., Zhang, X., & Andresen, J. M. (2012). Production of fermentable sugars from enzymatic hydrolysis of pretreated municipal solid waste after autoclave process. *Fuel*, *92*(1), 84–88. doi:10.1016/j.fuel.2011.07.012

Liu, G., Liao, Y., Wu, Y., & Ma, X. (2018). Synthesis gas production from microalgae gasification in the presence of Fe2O3 oxygen carrier and CaO additive. *Applied Energy*, *212*, 955–965.

Liu, J., Yan, Y., Xu, J., Li, S., Chen, G., & Zhang, J. (2016). Replacement micro-mechanism of CH 4 hydrate by N 2 /CO 2 mixture revealed by ab initio studies. *Computational Materials Science*, *123*, 106–110. doi:10.1016/j.commatsci.2016.06.025

Liu, L., Cheng, S. Y., Li, J. B., & Huang, Y. F. (2007). Mitigating Environmental Pollution and Impacts from Fossil Fuels: The Role of Alternative Fuels. *Energy Sources. Part A, Recovery, Utilization, and Environmental Effects*, *29*, 1069–1080.

Liu, W., Dongxue, H., Tie, G., Zicheng, P., Zishen, M., & Maochuan, L. (2022). Comparative life cycle assessment of cement, sintered bricks and non-sintered bricks manufacturing using water-based drilling cuttings from shale gas production in the Sichuan Basin, China. *Journal of Environmental Management*, *314*, 115135. PMID:35487056

Liu, Y., Strumendo, M., & Arastoopour, H. (2009). Simulation of methane production from hydrates by depressurization and thermal stimulation. *Industrial & Engineering Chemistry Research*, *48*(5), 2451–2464. doi:10.1021/ie8005275

Liu, Z., Ho, S. H., Sasaki, K., Den Haan, R., Inokuma, K., Ogino, C., Van Zyl, W. H., Hasunuma, T., & Kondo, A. (2016). Engineering of a novel cellulose-adherent cellulolytic Saccharomyces cerevisiae for cellulosic biofuel production. *Scientific Reports*, *6*(1), 24550. doi:10.1038rep24550 PMID:27079382

Li, W. W., & Yu, H. Q. (2011). From wastewater to bioenergy and biochemicals via two-stage bioconversion processes: A future paradigm. *Biotechnology Advances*, *29*(6), 972–982. doi:10.1016/j.biotechadv.2011.08.012 PMID:21884782

Li, X. S., Wang, Y., Li, G., Zhang, Y., & Chen, Z. Y. (2011). Experimental investigation into methane hydrate decomposition during three-dimensional thermal huff and puff. *Energy & Fuels*, *25*(4), 1650–1658. doi:10.1021/ef101731g

Li, X. S., Yang, B., Duan, L. P., Li, G., Huang, N. S., & Zhang, Y. (2013). Experimental study on gas production from methane hydrate in porous media by SAGD method. *Applied Energy*, *112*, 1233–1240. doi:10.1016/j.apenergy.2013.02.007

Li, Y., Park, S. Y., & Zhu, J. (2011). Solid-state anaerobic digestion for methane production from organic waste. *Renewable & Sustainable Energy Reviews*, *15*(1), 821–826. doi:10.1016/j.rser.2010.07.042

Lo Leggio, L., & Larsen, S. (2002). The 1.62 Å structure of Thermoascus aurantiacus endoglucanase: Completing the structural picture of subfamilies in glycoside hydrolase family 5. *FEBS Letters*, *523*(1–3), 103–108. doi:10.1016/S0014-5793(02)02954-X PMID:12123813

Lorentsen, S.-H., Sjøtun, K., & Grémillet, D. (2010). Multi-trophic consequences of kelp harvest. *Biological Conservation*, *143*(9), 2054–2062. doi:10.1016/j.biocon.2010.05.013

Lovley, D. R., & Klug, M. J. (1983). Methanogenesis from Methanol and Methylamines and Acetogenesis from Hydrogen and Carbon Dioxide in the Sediments of a Eutrophic Lake. *Applied and Environmental Microbiology*, *45*(4), 1310–1315. doi:10.1128/aem.45.4.1310-1315.1983 PMID:16346271

Low, S. S., Bong, K. X., Mubashir, M., Cheng, C. K., Lam, M. K., Lim, J. W., ... Show, P. L. (2021). Microalgae cultivation in palm oil mill effluent (POME) treatment and biofuel production. *Sustainability*, *13*(6), 3247.

Lu, H., Lorenson, T. D., Moudrakovski, I. L., Ripmeester, J. A., Collett, T. S., Hunter, R. B., & Ratcliffe, C. I. (2011). The characteristics of gas hydrates recovered from the Mount Elbert Gas Hydrate Stratigraphic Test Well, Alaska North Slope. *Marine and Petroleum Geology*, *28*(2), 411–418. doi:10.1016/j.marpetgeo.2010.01.002

Lundquist, T. J., Woertz, I. C., Quinn, N. W. T., & Benemann, J. R. (2010). *A Realistic Technology and Engineering Assessment of Algae Biofuel*. Academic Press.

Lüning, K., & Pang, S. (2003). Mass cultivation of seaweeds: Current aspects and approaches. *Journal of Applied Phycology*, *15*(2/3), 115–119. doi:10.1023/A:1023807503255

Luo, K., Yang, Q., Li, X., Yang, G., Liu, Y., Wang, D., Zheng, W., & Zeng, G. (2012). Hydrolysis kinetics in anaerobic digestion of waste activated sludge enhanced by α-amylase. *Biochemical Engineering Journal*, *62*, 17–21. doi:10.1016/j.bej.2011.12.009

Lupoi, J. S., & Smith, E. A. (2011). Evaluation of nanoparticle-immobilized cellulase for improved ethanol yield in simultaneous saccharification and fermentation reactions. *Biotechnology and Bioengineering*, *108*(12), 2835–2843. doi:10.1002/bit.23246 PMID:21702028

Luque, R. (2010). Algal biofuels: The eternal promise? *Energy & Environmental Science*, *3*, 254–257.

Mabee, W. E., Gregg, D. J., & Saddler, J. N. (2005). Assessing the emerging biorefinery sector in Canada. *Applied Biochemistry and Biotechnology*, *123*(1-3), 765–778. doi:10.1385/ABAB:123:1-3:0765 PMID:15930556

Maceiras, R., Rodríguez, M., Cancela, A., Urréjola, S., & Sánchez, A. (2011). Macroalgae: Raw material for biodiesel production. *Applied Energy*, *88*(10), 3318–3323. doi:10.1016/j.apenergy.2010.11.027

Mackenzie, A. (2013). Synthetic biology and the technicity of biofuels. Studies in History and Philosophy of Science Part C: Studies in History and Philosophy of Biological and Biomedical Sciences. *Philosophical Perspectives on Synthetic Biology*, *44*, 190–198.

Mackie, K. L., Brownell, H. H., West, K. L., & Saddler, J. N. (1985). Effect of sulphur dioxide and sulphuric acid on steam explosion of aspenwood. *Journal of Wood Chemistry and Technology*, *5*(3), 405–425. doi:10.1080/02773818508085202

Macrelli, S., Galbe, M., & Wallberg, O. (2014). Effects of production and market factors on ethanol profitability for an integrated first and second generation ethanol plant using the whole sugarcane as feedstock. *Biotechnology for Biofuels*, *7*(1), 1–16. doi:10.1186/1754-6834-7-26 PMID:24559312

Mączyńska, J., Krzywonos, M., Kupczyk, A., Tucki, K., Sikora, M., Pińkowska, H., … Wielewska, I. (2019). Production and use of biofuels for transport in Poland and Brazil – The case of bioethanol. *Fuel, 241*(September), 989–996. doi:10.1016/j.fuel.2018.12.116

Ma, F., & Hanna, M. A. (1999). Biodiesel production: A review. *Bioresource Technology*, *70*(1), 1–15. doi:10.1016/S0960-8524(99)00025-5

Ma, H., Liu, W. W., Chen, X., Wu, Y. J., & Yu, Z. L. (2009). Enhanced enzymatic saccharification of rice straw by microwave pretreatment. *Bioresource Technology*, *100*(3), 1279–1284. doi:10.1016/j.biortech.2008.08.045 PMID:18930389

Maity, J. P., Bundschuh, J., Chen, C. Y., & Bhattacharya, P. (2014). Microalgae for third generation biofuel production, mitigation of greenhouse gas emissions and wastewater treatment: Present and future perspectives - A mini-review. *Energy*, *78*, 104–113. doi:10.1016/j.energy.2014.04.003

Maity, J. P., Bundschuh, J., Chen, C. Y., & Bhattacharya, P. (2014). Microalgae for third generation biofuel production, mitigation of greenhouse gas emissions and wastewater treatment: Present and future perspectives–A mini review. *Energy*, *78*, 104–113.

Ma, J., Frear, C., Wang, Z., Yu, L., Zhao, Q., Li, X., & Chen, S. (2013). A simple methodology for rate-limiting step determination for anaerobic digestion of complex substrates and effect of microbial community ratio. *Bioresource Technology*, *134*, 391–395. doi:10.1016/j.biortech.2013.02.014 PMID:23489573

Makogon, Y. (1997a). *Hydrates of Hydrocarbons*. Academic Press.

Makogon, Y. F. (1997b). *Hydrates of natural gas* (1st ed.). Penn Well Publ Co Tulsa.

Makogon, Y. F. (2010). Natural gas hydrates - A promising source of energy. *Journal of Natural Gas Science and Engineering*, *2*(1), 49–59. doi:10.1016/j.jngse.2009.12.004

Makogon, Y. F., Holditch, S. A., & Makogon, T. Y. (2007). Natural gas-hydrates - A potential energy source for the 21st Century. *Journal of Petroleum Science Engineering*, *56*(1-3), 14–31. doi:10.1016/j.petrol.2005.10.009

Mamilla, V. R., Mallikarjun, M. V., & Rao, G. L. (2012). Biodiesel production from palm oil by transesterification method. *International Journal of Current Research, 4*(8), 83–88.

Mamilla, V. R., Mallikarjun, M. V., & Rao, G. L. (2012). Biodiesel Production from Palm Oil by Transesterification Method. *International Journal of Current Research, 4*(8), 83–88.

Manasa, P., Saroj, P., & Korrapati, N. (2017). Immobilization of cellulase enzyme on zinc ferrite nanoparticles in increasing enzymatic hydrolysis on ultrasound-assisted alkaline pretreated Crotalaria juncea biomass. *Indian Journal of Science and Technology, 10*(24), 1–7. doi:10.17485/ijst/2017/v10i24/112798

Mankar, A. R., Pandey, A., Modak, A., & Pant, K. K. (2021). Pretreatment of lignocellulosic biomass: A review on recent advances. *Bioresource Technology, 334*, 125235. doi:10.1016/j.biortech.2021.125235 PMID:33957458

Mansha, M., Qurashi, A., Ullah, N., Bakare, F. O., Khan, I., & Yamani, Z. H. (2016). Synthesis of In2O3/graphene heterostructure and their hydrogen gas sensing properties. *Ceramics International, 42*(9), 11490–11495.

Mardani, A., Jusoh, A., Nor, K. M. D., Khalifah, Z., Zakwan, N., & Valipour, A. (2015). Multiple criteria decision-making techniques and their applications–A review of the literature from 2000 to 2014. *Economic Research-Ekonomska Istraživanja, 28*(1), 516–571. doi:10.1080/1331677X.2015.1075139

Margareta, W., Nagarajan, D., Chang, J.-S., & Lee, D.-J. (2020). Dark fermentative hydrogen production using macroalgae (*Ulva* sp.) as the renewable feedstock. *Applied Energy, 262*, 114574. doi:10.1016/j.apenergy.2020.114574

Mariod, A., Klupsch, S., Hussein, I. H., & Ondruschka, B. (2006). Synthesis of Alkyl Esters from Three Unconventional Sudanese Oils for Their Use as Biodiesel. *Energy & Fuels, 20*, 2249–2252.

Martinez, D. G., Feiden, A., Bariccatti, R., & Zara, K. R. de F. (2018). Ethanol production from waste of cassava processing. *Applied Sciences (Switzerland), 8*(11), 1–8. doi:10.3390/app8112158

Martínez-Gutiérrez, E. (2018). Biogas production from different lignocellulosic biomass sources: advances and perspectives. *3 Biotech, 8*(5), 1–18. doi:10.1007/s13205-018-1257-4

Martin, M. A. (2010). First generation biofuels compete. *New Biotechnology, 27*(5), 596–608. doi:10.1016/j.nbt.2010.06.010 PMID:20601265

Martinov, M., Djatkov, D., Golub, M., Viskovic, M., Bojic, S., & Krstic, J. (2015). *Plant for lignocellulosic bioethanol. Case study*. Department of Environmental Engineering, Faculty of Technical Sciences.

Marzo, C., Díaz, A. B., Caro, I., & Blandino, A. (2019). Status and Perspectives in Bioethanol Production From Sugar Beet. Bioethanol Production from Food Crops. doi:10.1016/B978-0-12-813766-6.00004-7

Mat Aron, N. S., Khoo, K. S., Chew, K. W., Show, P. L., Chen, W.-H., & Nguyen, T. H. P. (2020). Sustainability of the four generations of biofuels – A review. *International Journal of Energy Research, 44*(12), 9266–9282. doi:10.1002/er.5557

Mata, T. M., Martins, A. A., & Caetano, N. S. (2010). Microalgae for biodiesel production and other applications: A review. *Renewable & Sustainable Energy Reviews, 14*, 217–232.

Mathimani, T., Uma, L., & Prabaharan, D. (2017). Optimization of direct solvent lipid extraction kinetics on marine trebouxiophycean alga by central composite design – Bioenergy perspective. *Energy Conversion and Management, 142*, 334–346.

Mathushika, J., & Gomes, C. (2022). Development of microalgae-based biofuels as a viable green energy source: Challenges and future perspectives. *Biointerface Research in Applied Chemistry, 12*(3), 3849–3882. doi:10.33263/BRIAC123.38493882

Maurya, D. P., Singla, A., & Negi, S. (2015). An overview of key pretreatment processes for biological conversion of lignocellulosic biomass to bioethanol. *3 Biotech, 5*(5), 597-609.

Mautner, H. G. (1954). The chemistry of brown algae. *Economic Botany*, *8*(2), 174–192. doi:10.1007/BF02984737

Max, M. D., & Cruickshank, M. J. (1999) Extraction of methane from oceanic hydrate system deposits. *Proc Annu Offshore Technol Conf, 1*, 71–78. 10.4043/10727-MS

Max, M. D., & Johnson, A. H. (2018). *Exploration and production of oceanic natural gas hydrate: Critical factors for commercialization*. Springer.

McHugh, D. J. (2003). *A guide to the seaweed industry*. Food and Agriculture Organization of the United Nations.

McKendry, P. (2002). Energy production from biomass (part 1): Overview of biomass. *Bioresource Technology*, *83*(1), 37–46. doi:10.1016/S0960-8524(01)00118-3 PMID:12058829

Mckone, T. E., Nazaro, W. W., Berck, P., Au, M., Lipman, T., Torn, M. S., Masanet, E., Lobscheid, A., Santero, N., Mishra, U., Barrett, A., Bomberg, M., Fingerman, K., Scown, C., Strogen, B., & Horvath, A. (2011). *Grand Challenges for Life-Cycle Assessment of Biofuels*. Academic Press.

McMillan, J. D., Newman, M. M., Templeton, D. W., & Mohagheghi, A. (1999). Simultaneous saccharification and cofermentation of dilute-acid pretreated yellow poplar hardwood to ethanol using xylose-fermenting Zymomonas mobilis. *Applied Biochemistry and Biotechnology - Part A Enzyme Engineering and Biotechnology*, *77–79*(1-3), 649–665. doi:10.1385/ABAB:79:1-3:649 PMID:15304685

Meher, L. C., Naik, S. N., & Das, L. M. (2004). Metanalysis of pongamia pinnata (karanja) oil for production of biodiesel. *Journal of Scientific and Industrial Research*, *63*, 913–918.

Meher, L. C., Vidya Sagar, D., & Naik, S. N. (2006). Technical aspects of biodiesel production by transesterification – a review. *Renewable & Sustainable Energy Reviews*, *10*(3), 248–268. doi:10.1016/j.rser.2004.09.002

Mekala, N. K., Potumarthi, R., Rajubaadhe, R., & Gupta, V. K. (2014). Current Bioenergy Researches: Strengths And Future Challenges. *Bioenergy Research: Advances And Applications*, *1*, 1–21. doi:10.1016/B978-0-444-59561-4.00001-2

Mendes-Pinto, M. M., Raposo, M. F. J., Bowen, J., Young, A. J., & Morais, R. (2001). Evaluation of different cell disruption processes on encysted cells of Haematococcus pluvialis. *Journal of Applied Phycology*, *13*(1), 19–24. doi:10.1023/A:1008183429747

Menendez, E., Garcia-Fraile, P., & Rivas, R. (2015). Biotechnological applications of bacterial cellulases. *AIMS Bioengineering*, *2*(3), 163–182. doi:10.3934/bioeng.2015.3.163

Meng, X., Chen, G., & Wang, Y. (2008). Biodiesel production from waste cooking oil via alkali catalyst and its engine test. *Fuel Processing Technology*, *89*(9), 851–857. doi:10.1016/j.fuproc.2008.02.006

Meng, X., Yang, J., Xu, X., Zhang, L., Nie, Q., & Xian, M. (2009). Biodiesel production from oleaginous microorganisms. *Renewable Energy*, *34*, 1–5.

Meramo-Hurtado, S. I., Puello, P., & Cabarcas, A. (2020). Process analysis of hydrogen production via biomass gasification under computer-aided safety and environmental assessments. *ACS Omega*, *5*(31), 19667–19681. doi:10.1021/acsomega.0c02344 PMID:32803062

Mesa, L., González, E., Cara, C., González, M., Castro, E., & Mussatto, S. I. (2011). The effect of organosolv pretreatment variables on enzymatic hydrolysis of sugarcane bagasse. *Chemical Engineering Journal*, *168*(3), 1157–1162. doi:10.1016/j.cej.2011.02.003

Mes-Hartree, M., Dale, B. E., & Craig, W. K. (1988). Comparison of steam and ammonia pretreatment for enzymatic hydrolysis of cellulose. *Applied Microbiology and Biotechnology*, *29*(5), 462–468. doi:10.1007/BF00269069

Miao, X., & Wu, Q. (2006). Biodiesel production from heterotrophic microalgal oil. *Bioresource Technology*, *97*, 841–846. PMID:15936938

Mihailović, M., Trbojević-Ivić, J., Banjanac, K., Milosavić, N., Velicković, D., Carević, M., & Bezbradica, D. (2016). Immobilization of maltase from Saccharomyces cerevisiae on thiosulfonate supports. *Journal of the Serbian Chemical Society*, *81*(12), 1371–1382. doi:10.2298/JSC160730099M

Milledge, J. J., Smith, B., Dyer, P. W., & Harvey, P. (2014). Macroalgae-Derived Biofuel: A Review of Methods of Energy Extraction from Seaweed Biomass. *Energies*, *7*(11), 7194–7222. doi:10.3390/en7117194

Milosavić, N., Bogdanović Pristov, J., Veličković, D. V., Dimitrijević, A. S., Kalauzi, A., & Radotić, K. (2012). Study of the covalently immobilized amyloglucosidase on macroporous polymer by mathematical modeling of the pH optima. *Journal of Chemical Technology and Biotechnology (Oxford, Oxfordshire)*, *87*(10), 1450–1457. doi:10.1002/jctb.3768

Milosavic, N., Trbojevic Ivic, J., Veličković, D., Dimitrijević, A., & Bezbradica, D. (2016). Immobilized lipases: An old-fashioned twist for a new generation of industrial biocatalysts. *Journal of Microbial & Biochemical Technology*, *8*(6), 8–9. doi:10.4172/1948-5948.1000e127

Mishra, A., & Ghosh, S. (2017). Key Pretreatment Technologies for An Efficient Bioethanol Production from Lignocellulosics. *Advances in Biofeedstocks and Biofuels*, *1*, 55–83. doi:10.1002/9781119117322.ch3

Mishra, V. K., & Goswami, R. (2018). A review of production, properties and advantages of biodiesel. *Biofuels*, *9*(2), 273–289. doi:10.1080/17597269.2017.1336350

Misson, M., Zhang, H., & Jin, B. (2015). Nanobiocatalyst advancements and bioprocessing applications. *Journal of the Royal Society, Interface*, *12*(102), 20140891. Advance online publication. doi:10.1098/rsif.2014.0891 PMID:25392397

Miyafuji, H. (2015). Application of ionic liquids for effective use of woody biomass. *Journal of Wood Science*, *61*(4), 343–350. doi:10.100710086-015-1489-4

Mizik, T., Nagy, L., Gabnai, Z., & Bai, A. (2020). The major driving forces of the EU and US ethanol. *Energies*, *13*(21), 5614. doi:10.3390/en13215614

Mobin, S. M. A., & Alam, F. (2018). A review of microalgal biofuels, challenges and future directions. *Green Energy and Technology*, 83–108. doi:10.1007/978-981-10-0697-5_4

Mobin, S., & Alam, F. (2014). Biofuel production from algae utilizing wastewater. *Proceedings of the 19th Australasian Fluid Mechanics Conference, AFMC 2014.*

Mofijur, M., Masjuki, H., Kalam, M., Rahman, S. A., & Mahmudul, H. (2015). Energy scenario and biofuel policies and targets in ASEAN countries. *Renewable & Sustainable Energy Reviews*, *46*, 51–61.

Mohan, D., Pittman, C. U., & Philip, S. (2017). Pyrolysis of Wood/Biomass for Bio-oil: A Critical Review Dinesh. *Progress in Energy and Combustion Science*, *62*(4), 848–889. doi:10.1016/j.pecs.2017.05.004

Mohan, M. R., Phillippe, G. T., & Shiju, M. (2006). Biofuel laws in Asia: Instruments for energy access, security, environmental protection and rural empowerment. *Asian Biotechnology and Development Review*, *8*(2), 51–75.

Mohanram, S., Amat, D., Choudhary, J., Arora, A., & Nain, L. (2013). Novel perspectives for evolving enzyme cocktails for lignocellulose hydrolysis in biorefineries. *Sustainable Chemical Processes*, *1*(1), 15. doi:10.1186/2043-7129-1-15

Mohapatra, S., Ray, R. C., & Ramachandran, S. (2019). Bioethanol From Biorenewable Feedstocks: Technology, Economics, and Challenges. Bioethanol Production from Food Crops. doi:10.1016/B978-0-12-813766-6.00001-1

Molina Grima, E., Belarbi, E. H., Acién Fernández, F. G., Robles Medina, A., & Chisti, Y. (2003). Recovery of microalgal biomass and metabolites: Process options and economics. *Biotechnology Advances*, *20*(7–8), 491–515. doi:10.1016/S0734-9750(02)00050-2 PMID:14550018

Monari, C., Righi, S., & Olsen, S. I. (2016). Greenhouse gas emissions and energy balance of biodiesel production from microalgae cultivated in photobioreactors in Denmark: A lifecycle modeling. *Journal of Cleaner Production*, *112*, 4084–4092.

Moodley, P., & Gueguim Kana, E. B. (2019). Bioethanol production from sugarcane leaf waste: Effect of various optimized pretreatments and fermentation conditions on process kinetics. *Biotechnology Reports (Amsterdam, Netherlands)*, *22*, e00329. doi:10.1016/j.btre.2019.e00329 PMID:31008065

MoP & NG. (2006). *Bio-diesel Purchase Policy*. Government of India.

Moradi, F., Amiri, H., Soleimanian-Zad, S., Ehsani, M. R., & Karimi, K. (2013). Improvement of acetone, butanol and ethanol production from rice straw by acid and alkaline pretreatments. *Fuel*, *112*, 8–13. doi:10.1016/j.fuel.2013.05.011

Moridis, G. J., Kowalsky, M., & Pruess, K. (2005). Depressurization-Induced Gas Production From Class 1 Hydrate Deposits. *Lawrence Berkeley Natl Lab*, *10*, 458–481. doi:10.2118/97266-MS

Moridis, G. J., & Reagan, M. T. (2007). *Strategies for Gas Production From Oceanic Class 3 Hydrate Accumulations*. Offshore Technol Conf. doi:10.4043/18865-MS

Morone, P., & Cottoni, L. (2016). Biofuels: Technology, economics, and policy issues. Technology, economics, and policy issues. In Handbook of Biofuels Production: Processes and Technologies: Second Edition. doi:10.1016/B978-0-08-100455-5.00004-7

Moser, B. R., & Vaughn, S. F. (2012). Biodiesel from Corn Distillers Dried Grains with Solubles: Preparation, Evaluation, and Properties. *Bioenerg. Res.*, *5*, 439–449.

Moset, V., Poulsen, M., Wahid, R., Højberg, O., & Møller, H. B. (2015). Mesophilic versus thermophilic anaerobic digestion of cattle manure: Methane productivity and microbial ecology. *Microbial Biotechnology*, *8*(5), 787–800. doi:10.1111/1751-7915.12271 PMID:25737010

Mosier, N., Wyman, C., Dale, B., Elander, R., Lee, Y. Y., Holtzapple, M., & Ladisch, M. (2005). Features of promising technologies for pretreatment of lignocellulosic biomass. *Bioresource Technology*, *96*(6), 673–686. doi:10.1016/j.biortech.2004.06.025 PMID:15588770

Moulthrop, J. S., Swatloski, R. P., Moyna, G., & Rogers, R. D. (2005). High-resolution 13C NMR studies of cellulose and cellulose oligomers in ionic liquid solutions. *Chemical Communications*, *40*(12), 1557–1559. doi:10.1039/b417745b PMID:15770258

Mrema, G., Gumbe, L. O., Chepete, H. J., & Agullo, J. (2011). Rural Energy. In *Rural Structures in the Tropics: Design and Development*. Technical Centre for Agricultural and Rural Cooperation.

Munir, N., Sharif, N., Naz, S., Saleem, F., & Manzoor, F. (2013). Harvesting and processing of microalgae biomass fractions for biodiesel production (a review). *Science Technology and Development*, *32*(3), 235–243.

Muñoz, I. G., Ubhayasekera, W., Henriksson, H., Szabó, I., Pettersson, G., Johansson, G., Mowbray, S. L., & Ståhlberg, J. (2001). Family 7 cellobiohydrolases from Phanerochaete chrysosporium: Crystal structure of the catalytic module of Cel7D (CBH58) at 1.32 Å resolution and homology models of the isozymes. *Journal of Molecular Biology*, *314*(5), 1097–1111. doi:10.1006/jmbi.2000.5180 PMID:11743726

Muñoz, R., Meier, L., Diaz, I., & Jeison, D. (2015). A review on the state-of-the-art of physical/chemical and biological technologies for biogas upgrading. *Reviews in Environmental Science and Biotechnology*, *14*(4), 727–759. doi:10.100711157-015-9379-1

Murali, P., Hari, K., & Prathap, D. P. (2016). An Economic Analysis of Biofuel Production and Food Security in India. *Sugar Tech*, *18*, 447–456.

Mussatto, S. I., Dragone, G., Guimarães, P. M. R., Silva, J. P. A., Carneiro, L. M., Roberto, I. C., Vicente, A., Domingues, L., & Teixeira, J. A. (2010). Technological trends, global market, and challenges of bio-ethanol production. *Biotechnology Advances*, *28*(6), 817–830. doi:10.1016/j.biotechadv.2010.07.001 PMID:20630488

Mustafa, A. M., Poulsen, T. G., & Sheng, K. (2016). Fungal pretreatment of rice straw with Pleurotus ostreatus and Trichoderma reesei to enhance methane production under solid-state anaerobic digestion. *Applied Energy*, *180*, 661–671. doi:10.1016/j.apenergy.2016.07.135

Mutripah, S., Meinita, M. D. N., Kang, J.-Y., Jeong, G.-T., Susanto, A. B., Prabowo, R. E., & Hong, Y.-K. (2014). Bioethanol production from the hydrolysate of *Palmaria palmata* using sulfuric acid and fermentation with brewer's yeast. *Journal of Applied Phycology*, *26*(1), 687–693. doi:10.100710811-013-0068-6

Mykhaylenko, N. F., & Zolotareva, E. K. (2017). The effect of copper and selenium nanocarboxylates on biomass accumulation and photosynthetic energy transduction efficiency of the green algae *Chlorella vulgaris*. *Nanoscale Research Letters*, *12*(1), 1–8.

Naik, K. K., & Naik, T. P. (2021). Synthesis of biodiesel by using marine Macro-algae of Karwar Region, Karnataka, India. *Gorteria Journal*, *34*(1), 26–34.

Naik, S. N., Goud, V. V., Rout, P. K., & Dalai, A. K. (2010). Production of first and second generation biofuels: A comprehensive review. *Renewable & Sustainable Energy Reviews*, *14*(2), 578–597. doi:10.1016/j.rser.2009.10.003

Naito, S., Maeda, N., & Fujimoto, A. (2020). *U.S. patent application No. 16/627,888. 9*. US Patent Office.

Najafi, G., Ghobadian, B., & Yusaf, T. F. (2011). Algae as a sustainable energy source for biofuel production in Iran: A case study. *Renewable & Sustainable Energy Reviews*, *15*(8), 3870–3876. doi:10.1016/j.rser.2011.07.010

Nanaki, E. A., & Koroneos, C. J. (2011). Comparative LCA of the use of biodiesel, diesel and gasoline for transportation. *Journal of Cleaner Production*, *20*(1), 14–19.

Naqvi, M., & Yan, J. (2015). First-Generation Biofuels. Handbook of Clean Energy Systems, 1–18. doi:10.1002/9781118991978.hces207

Nasyrov, N. M., Nizaeva, I. G., & Sayakhov, F. L. (1997). Mathematical simulation of heat-and-mass transfer in gas hydrate deposits in a high-frequency electromagnetic field. *Journal of Applied Mechanics and Technical Physics*, *38*(6), 895–905. doi:10.1007/BF02467955

Natarajan, E., & Venkatesh, K.P. (2012). Production and characterization of biodiesel using pongamia oil by immobilized rhizopusoryzae. *Advanced Biotech*, *11*(7), 32–34.

National Policy on Biofuels. (2008). Ministry of New & Renewable Energy, Government of India. https://mnre.gov.in/file-manager/UserFiles/biofuel-policy.pdf

Nebel, B. A., & Mittelbach, M. (2006). Biodiesel from extracted fat out of meat and bone meal. *European Journal of Lipid Science and Technology*, *108*, 398–403.

Ngo, H. L., Zafiropoulos, N. A., Foglia, T. A., Samulski, E. T., & Lin, W. (2008). Efficient Two- Step Synthesis of Biodiesel from Greases. *Energy & Fuels*, *22*, 626–634.

Ngo, H. L., Zafiropoulos, N. A., Foglia, T. A., Samulski, E. T., & Lin, W. (2008). Efficient two-step synthesis of biodiesel from grease. *Energy & Fuels*, *22*(1), 626–634. doi:10.1021/ef700343b

Nikolov, Z. L., Meagher, M. M., & Reilly, P. J. (1989). Kinetics, equilibria, and modeling of the formation of oligo-saccharides from D-glucose with Aspergillus niger glucoamylases I and II. *Biotechnology and Bioengineering*, *34*(5), 694–704. doi:10.1002/bit.260340514 PMID:18588154

Ning, F., Wu, N., Jiang, G., Zhang, L., Guan, J. A., Yu, Y., & Tang, F. (2010). A method to use solar energy for the production of gas from marine hydrate-bearing sediments: A case study on the Shenhu Area. *Energies*, *3*(12), 1861–1879. doi:10.3390/en3121861

Nizami, A. S., & Rehan, M. (2018). Towards nanotechnology-based biofuel industry. *Biofuel Res. J.*, *5*, 798–799.

Njoku, S. I., Uellendahl, H., & Ahring, B. K. (2013). Comparing oxidative and dilute acid wet explosion pretreatment of cocksfoot grass at high dry matter concentration for cellulosic ethanol production. *Energy Science & Engineering*, *1*(2), 89–98. doi:10.1002/ese3.11

Notoya, M. (2010). Production of Biofuel by Macroalgae with Preservation of Marine Resources and Environment. In J. Seckbach, R. Einav, & A. Israel (Eds.), *Seaweeds and their Role in Globally Changing Environments. Cellular Origin, Life in Extreme Habitats and Astrobiology* (Vol. 15). Springer., doi:10.1007/978-90-481-8569-6_13

November & December 2008 Newsletter | Agricultural Marketing Resource Center. (n.d.). Retrieved January 20, 2022, from https://www.agmrc.org/renewable-energy/renewable-energy-climate-change-report/renewable-energy-climate-change-report/november--december-2008-newsletter

OECD. (2012). *Organization for Economic Co-operation and Development*. https://www.oecd.org/env/cc/Outlook%20to%202050_Climate%20Change%20Chapter_HIGLIGHTS-FINA-8pager-UPDATED%20NOV2012.pdf

Ogbonna, J. C., Mashima, H., & Tanaka, H. (2001). Scale up of fuel ethanol production from sugar beet juice using loofa sponge immobilized bioreactor. *Bioresource Technology*, *76*(1), 1–8. doi:10.1016/S0960-8524(00)00084-5 PMID:11315804

Ojala, P. (2008). Experiences of implementing a value-based approach. *WSEAS Transactions on Information Science and Applications*, *5*(4), 385–395.

Ojewumi, M. E., Job, A. I., Samson Taiwo, O., Mopelola Obanla, O., Ayoola, A. A., Ojewumi, E. O., & Oyeniyi, E. A. (2018). Bio-Conversion of Sweet Potato Peel Waste to Bio-Ethanol Using Saccharomyces Cerevisiae. *International Journal of Pharmaceutical and Phytopharmacological Research*, *8*(3), 46–54. www.eijppr.com

Olabanji, O. M., & Mpofu, K. (2014). Comparison of weighted decision matrix and analytical hierarchy process for CAD design of reconfigurable assembly fixture. *Procedia Centre d'Insertion et de Réinsertion Professionnelle*, *23*, 264–269. doi:10.1016/j.procir.2014.10.088

Olguin-Maciel, E., Singh, A., Chable-Villacis, R., Tapia-Tussell, R., & Ruiz, H. A. (2020). Consolidated bioprocessing, an innovative strategy towards sustainability for biofuels production from crop residues: An overview. *Agronomy (Basel)*, *10*(11), 1834. Advance online publication. doi:10.3390/agronomy10111834

Oloche, B., Ozigis, I., Adeyemi, K., & Ikpe, E. (2017). Construction and leakage detection of a dome-type biogas digester in a village at Abuja, Nigeria, Federal. *Journal of Engineering Technology*, *2*(1), 27–31.

Olofsson, K., Bertilsson, M., & Lidén, G. (2008). A short review on SSF - An interesting process option for ethanol production from lignocellulosic feedstocks. *Biotechnology for Biofuels*, *1*(1), 1–14. doi:10.1186/1754-6834-1-7 PMID:18471273

Olsson, J., Toth, G. B., & Albers, E. (2020). Biochemical composition of red, green and brown seaweeds on the Swedish west coast. *Journal of Applied Phycology*, *32*(5), 3305–3317. doi:10.100710811-020-02145-w

Ong, H. C., Milano, J., Silitonga, A. S., Hassan, M. H., Wang, C. T., Mahlia, T. M. I., ... Sutrisno, J. (2019). Biodiesel production from Calophyllum inophyllum-Ceiba pentandra oil mixture: Optimization and characterization. *Journal of Cleaner Production*, *219*, 183–198.

Ordway, E. M., Asner, G. P., & Lambin, E. F. (2017). Deforestation risk due to commodity crop expansion in sub-Saharan Africa. *Environmental Research Letters*, *12*(4), 044015–044029. doi:10.1088/1748-9326/aa6509

Ors, O., & Sinayuc, C. (2014). An experimental study on the CO2-CH4 swap process between gaseous CO2 and CH4 hydrate in porous media. *Journal of Petroleum Science Engineering*, *119*, 156–162. doi:10.1016/j.petrol.2014.05.003

Ota, M., Morohashi, K., Abe, Y., Watanabe, M., Smith, R. L. Jr, & Inomata, H. (2005). Replacement of CH4in the hydrate by use of liquid CO2. *Energy Conversion and Management*, *46*(11-12), 1680–1691. doi:10.1016/j.enconman.2004.10.002

Ove Arup and Partners. (1989). *Monitoring of a Commercial Demon stration of Harvesting and Combudion of Forestry Wastes*. ETSU.

Øverland, M., Mydland, L. T., & Skrede, A. (2019). Marine macroalgae as sources of protein and bioactive compounds in feed for monogastric animals. *Journal of the Science of Food and Agriculture*, *99*(1), 13–24. doi:10.1002/jsfa.9143 PMID:29797494

Padam, B. S., Tin, H. S., Chye, F. Y., Abdullah, M. I. (2014). Banana by–products: An under– utilized renewable food biomass with great potential. *Journal of Food Science and Technology, 51*(12), 3527–3545.

Padam, B. S., Tin, H. S., Chye, F. Y., & Abdullah, M. I. (2014). Banana by–products: An under–utilized renewable food biomass with great potential. *Journal of Food Science and Technology*, *51*(12), 3527–3545. doi:10.100713197-012-0861-2 PMID:25477622

Pádrová, K., Lukavský, J., Nedbalová, L., Čejková, A., Cajthaml, T., Sigler, K., ... Řezanka, T. (2015). Trace concentrations of iron nanoparticles cause overproduction of biomass and lipids during cultivation of cyanobacteria and microalgae. *Journal of Applied Phycology*, *27*(4), 1443–1451.

Palaniappan, K. (2017). An overview of applications of nanotechnology in biofuel production. *World Appl Sci J, 35*(8), 1305-1311.

Pandey, A., Soccol, C. R., Nigam, P., Soccol, V. T., Vandenberghe, L. P. S., & Mohan, R. (2000). Biotechnological potential of agro-industrial residues. II: Cassava bagasse. *Bioresource Technology*, *74*(1), 81–87. doi:10.1016/S0960-8524(99)00143-1

Pandey, V. C., Singh, K., Singh, J. S., Kumar, A., Singh, B., & Singh, R. P. (2012). Jatropha curcas: A potential biofuel plant for sustainable environmental development. *Renewable & Sustainable Energy Reviews*, *16*(5), 2870–2883. doi:10.1016/j.rser.2012.02.004

Panella, L. (2010). Sugar Beet as an Energy Crop. *Sugar Tech*, *12*(3–4), 288–293. doi:10.100712355-010-0041-5

Pang, W. X., Xu, W. Y., Sun, C. Y., Zhang, C. L., & Chen, G. J. (2009). Methane hydrate dissociation experiment in a middle-sized quiescent reactor using thermal method. *Fuel*, *88*(3), 497–503. doi:10.1016/j.fuel.2008.11.002

Pan, X., Xie, D., Gilkes, N., Gregg, D. J., & Saddler, J. N. (2005). Strategies to enhance the enzymatic hydrolysis of pretreated softwood with high residual lignin content. In *Twenty-Sixth Symposium on Biotechnology for Fuels and Chemicals* (pp. 1069-1079). Humana Press. 10.1007/978-1-59259-991-2_90

Pardilhó, S., Boaventura, R., Almeida, M., & Dias, J. M. (2022). Marine macroalgae waste: A potential feedstock for biogas production. *Journal of Environmental Management, 304*, 114309. doi:10.1016/j.jenvman.2021.114309 PMID:34933268

Park, J. H., Hong, J. Y., Jang, H. C., Oh, S. G., Kim, S. H., Yoon, J. J., & Kim, Y. J. (2012). Use of *Gelidium amansii* as a promising resource for bioethanol: A practical approach for continuous dilute-acid hydrolysis and fermentation. *Bioresource Technology, 108*, 83–88. doi:10.1016/j.biortech.2011.12.065 PMID:22261657

Park, J. H., Yoon, J. J., Park, H. D., Kim, Y. J., Lim, D. J., & Kim, S. H. (2011). Feasibility of biohydrogen production from *Gelidium amansii*. *International Journal of Hydrogen Energy, 36*(21), 13997–14003. doi:10.1016/j.ijhydene.2011.04.003

Park, J., Park, S., & Kim, M. (2014). Anaerobic degradation of amino acids generated from the hydrolysis of sewage sludge. *Environmental Technology, 35*(9), 1133–1139. doi:10.1080/09593330.2013.863951 PMID:24701908

Park, J.-Y., Kim, D.-K., Wang, Z.-M., Lu, P., Park, S.-C., & Lee, J.-S. (2008). Production and characterization of biodiesel from tung oil. *Applied Biochemistry and Biotechnology, 148*, 109–117. PMID:18418744

Park, Y. C., & Kim, J. S. (2012). Comparison of various alkaline pretreatment methods of lignocellulosic biomass. *Energy, 47*(1), 31–35. doi:10.1016/j.energy.2012.08.010

Pascaud M. (1993), The essential polyunsaturated fatty acids of spirulina and our immune response. *Bull. Inst. Océano, 12*, 49-57.

Pasha, M. K., Dai, L., Liu, D., Guo, M., & Du, W. (2021). An overview to process design, simulation and sustainability evaluation of biodiesel production. *Biotechnology for Biofuels, 14*(1), 129. Advance online publication. doi:10.118613068-021-01977-z PMID:34074344

Patel, K., & Shah, S. N. (2015). Biodiesel from plant oils. In *Food, Energy, and Water*. Elsevier. https: / /Doi. Org/10.1016/ B978-0-12-800211-7.00011-9. 23.

Pathak, V. V., Singh, D. P., Kothari, R., & Chopra, A. K. (2014). Phycoremediation of textile wastewater by unicellular microalga Chlorella pyrenoidosa. *Cellular and Molecular Biology, 60*(5), 35–40.

Patil, V., Tran, K. Q., & Giselrod, H. R. (2008). Towards sustainable production of biofuels. *International Journal of Molecular Sciences, 9*(7), 1188–1195. doi:10.3390/ijms9071188 PMID:19325798

Pattarkine, M. V., & Pattarkine, V. M. (2012). Nanotechnology for algal biofuels. In *The science of algal fuels* (pp. 147–163). Springer.

Paul Abishek, M., Patel, J., & Prem Rajan, A. (2014). Algae oil: A sustainable renewable fuel of future. *Biotechnology Research International.*

Pavlečić, M., Rezić, T., Šantek, M. I., Horvat, P., & Šantek, B. (2017). Bioethanol production from raw sugar beet cossettes in horizontal rotating tubular bioreactor. *Bioprocess and Biosystems Engineering, 40*(11), 1679–1688. doi:10.100700449-017-1823-x PMID:28770369

Pavlović, M., Dimitrijević, A., Trbojević, J., Milosavić, N., Gavrović-Jankulović, M., Bezbradica, D., & Veličković, D. (2013). A study of transglucosylation kinetic in an enzymatic synthesis of benzyl alcohol glucoside by α-glucosidase from S. cerevisiae. *Russian Journal of Physical Chemistry A, 87*(13), 2285–2288. doi:10.1134/S0036024413130207

Peng, F., Peng, P., Xu, F., & Sun, R. C. (2012). Fractional purification and bioconversion of hemicelluloses. *Biotechnology Advances*, *30*(4), 879–903. doi:10.1016/j.biotechadv.2012.01.018 PMID:22306329

Perego, P., Converti, A., Palazzi, E., Del Borghi, M., & Ferraiolo, G. (1990). Fermentation of hardwood hemicellulose hydrolysate by Pachysolen tannophilus, candida shehatae and Pichia stipitis. *Journal of Industrial Microbiology*, *6*(3), 157–164. doi:10.1007/BF01577690

Pereira, S. C., Maehara, L., Machado, C. M. M., & Farinas, C. S. (2015). 2G ethanol from the whole sugarcane lignocellulosic biomass. *Biotechnology for Biofuels*, *8*(1), 1–16. doi:10.118613068-015-0224-0 PMID:25774217

Perez-Verdin, G., Grebner, D. L., Sun, C., Munn, I. A., Schultz, E. B., & Matney, T. G. (2009). Woody biomass availability for bioethanol conversion in Mississippi. *Biomass and Bioenergy*, *33*(3), 492–503. doi:10.1016/j.biombioe.2008.08.021

Perwez, M., Ahmad, R., & Sardar, M. (2017). A reusable multipurpose magnetic nanobiocatalyst for industrial applications. *International Journal of Biological Macromolecules*, *103*, 16–24. doi:10.1016/j.ijbiomac.2017.05.029 PMID:28495621

Peter, A. P., Khoo, K. S., Chew, K. W., Ling, T. C., Ho, S.-H., Chang, J.-S., & Show, P. L. (2021). Microalgae for biofuels, wastewater treatment and environmental monitoring. *Environmental Chemistry Letters*, *19*(4), 2891–2904. doi:10.100710311-021-01219-6

Pham Quoc, K., Dubacq, J. P., & Demandre, C. (1994). Comparative effects of exogenous fatty acid supplementations on the lipids from the cyanobacterium Spirulina platensis. *Plant Physiology and Biochemistry*, *32*(4), 501–509.

Phirani, J., & Mohanty, K. K. (2009). Warm water flooding of confined gas hydrate reservoirs. *Chemical Engineering Science*, *64*(10), 2361–2369. doi:10.1016/j.ces.2009.02.019

Pikula, K., Zakharenko, A., Stratidakis, A., Razgonova, M., Nosyrev, A., Mezhuev, Y., Tsatsakis, A., & Golokhvast, K. (2020). The advances and limitations in biodiesel production: Feedstocks, oil extraction methods, production, and environmental life cycle assessment. *Green Chemistry Letters and Reviews*, *13*(4), 275–294. doi:10.1080/17518253.2020.1829099

Pimentel, D., & Patzek, T. W. (2005). Ethanol Production Using Corn, Switchgrass, and Wood; Biodiesel Production Using Soybean and Sunflower. *Natural Resources Research*, *14*, 65–76.

Piwowar, A. (2020). Agricultural Biogas—An Important Element in the Circular and Low-Carbon Development in Poland. *Energies*, *13*(7), 1733. doi:10.3390/en13071733

Planning Commission. (2003). *Report of the Committee on Development of Biofuels*. Government of India. http://mnre.gov.in/file-manager/UserFiles/biofuelpolicy. pdf

Pohit, S., Biswas, P. K., Kumar, R., & Goswami, A. (2010). Pricing model for biodiesel feedstock: A case study of Chhattisgarh in India. *Energy Policy*, *38*, 7487–7496.

Pohlman, J. W., Kaneko, M., Heuer, V. B., Coffin, R. B., & Whiticar, M. (2009). Methane sources and production in the northern Cascadia margin gas hydrate system. *Earth and Planetary Science Letters*, *287*(3-4), 504–512. doi:10.1016/j.epsl.2009.08.037

Pooladi-Darvish, M. (2004). Gas Production From Hydrate Reservoirs and Its Modeling. *Soc Pet Eng*, 90–97. doi:10.2118/13340-PA

Popoola, T. O. S., & Yangomodou, O. D. (2006). Extraction, properties and utilization potentials of cassava seed oil. *Biotechnology*, *5*(1), 38–41. doi:10.3923/biotech.2006.38.41

Potumarthi, R., Baadhe, R. R., Nayak, P., & Jetty, A. (2013). Simultaneous pretreatment and sacchariffication of rice husk by Phanerochete chrysosporium for improved production of reducing sugars. *Bioresource Technology*, *128*, 113–117. doi:10.1016/j.biortech.2012.10.030 PMID:23196230

Poulsen, T. G., & Adelard, L. (2016). Improving biogas quality and methane yield via co-digestion of agricultural and urban biomass wastes. *Waste Management (New York, N.Y.)*, *54*, 118–125. doi:10.1016/j.wasman.2016.05.020 PMID:27256782

Prabhu, M., Chemodanov, A., Gottlieb, R., Kazir, M., Nahor, O., Gozin, M., Israel, A., Livney, Y. D., & Golberg, A. (2019). Starch from the sea: The green macroalga Ulva ohnoi as a potential source for sustainable starch production in the marine biorefinery. *Algal Research, 37*, 215-227. doi:10.1016/j.algal.2018.11.007

Prabu, A. (2018). Nanoparticles as additive in biodiesel on the working characteristics of a DI diesel engine. *Ain Shams Engineering Journal*, *9*(4), 2343–2349.

Pradhan, S., & Ruysenaar, S. (2014). Burning Desires: Untangling and Interpreting "Pro-Poor" Biofuel Policy Processes in India and South Africa. *Environment & Planning A*, *46*, 299–317.

Pratto, B., dos Santos-Rocha, M. S. R., Longati, A. A., de Sousa Júnior, R., & Cruz, A. J. G. (2020). Experimental optimization and techno-economic analysis of bioethanol production by simultaneous saccharification and fermentation process using sugarcane straw. *Bioresource Technology*, *297*, 122494. doi:10.1016/j.biortech.2019.122494 PMID:31813817

Predojevic, Z., Skrbic, B., & Durisic-Mladenovicn, N. (2012). Transesterification of linoleic and oleic sunflower oils to biodiesel using Cao as a solid base catalyst. *Journal of the Serbian Chemical Society*, *77*(6), 815–832. doi:10.2298/JSC110824206P

Preethi, G., M, G., Kumar, G., Karthikeyan, O. P., Varjani, S., & J, R. B. (2021). Lignocellulosic biomass as an optimistic feedstock for the production of biofuels as valuable energy source: Techno-economic analysis, environmental impact analysis, breakthrough and perspectives. *Environmental Technology and Innovation*, *24*, 102080. doi:10.1016/j.eti.2021.102080

Prier, B. (1998). *Electricity from Bonon*. Financial Times Business Lid.

Procentese, A., Raganati, F., Olivieri, G., Elena Russo, M., & Marzocchella, A. (2017). Pre-treatment and enzymatic hydrolysis of lettuce residues as feedstock for bio-butanol production. *Biomass and Bioenergy*, *96*, 172–179. doi:10.1016/j.biombioe.2016.11.015

PTI. (2015). *Steps for Implementation of Biofuel Programme*. http://pib.nic.in/newsite/PrintRelease.aspx?relid=124076

Putro, J. N., Soetaredjo, F. E., Lin, S. Y., Ju, Y. H., & Ismadji, S. (2016). Pretreatment and conversion of lignocellulose biomass into valuable chemicals. *RSC Advances*, *6*(52), 46834–46852. doi:10.1039/C6RA09851G

Qari, H., Rehan, M., & Nizami, A. S. (2017). Key Issues in Microalgae Biofuels: A Short Review. *Energy Procedia*, *142*, 898–903. doi:10.1016/j.egypro.2017.12.144

Quintero, J. A., Montoya, M. I., Sánchez, O. J., Giraldo, O. H., & Cardona, C. A. (2008). Fuel ethanol production from sugarcane and corn: Comparative analysis for a Colombian case. *Energy*, *33*(3), 385–399. doi:10.1016/j.energy.2007.10.001

Qureshi, N., Saha, B. C., Hector, R. E., Dien, B., Hughes, S., Liu, S., Iten, L., Bowman, M. J., Sarath, G., & Cotta, M. A. (2010). Production of butanol (a biofuel) from agricultural residues: Part II - Use of corn stover and switchgrass hydrolysates. *Biomass and Bioenergy*, *34*(4), 566–571. doi:10.1016/j.biombioe.2009.12.023

R&D | Ministry of New and Renewable Energy, Government of India. (n.d.). Retrieved January 20, 2022, from https://mnre.gov.in/research-and-development/bio-energy

Radha, M., & Murugesan, A. G. (2017). Enhanced dark fermentative biohydrogen production from marine macroalgae *Padina tetrastromatica* by different pretreatment processes. *Biofuel Research Journal, 13*(1), 551–558. doi:10.18331/BRJ2017.4.1.5

Rafa, N., Ahmed, S. F., Badruddin, I. A., Mofijur, M., & Kamangar, S. (2021). Strategies to Produce Cost-Effective Third-Generation Biofuel From Microalgae. *Frontiers in Energy Research, 9*(September), 1–11. doi:10.3389/fenrg.2021.749968

Rafiqul, I. M., Hassan, A., Sulebele, G., Orosco, C. A., Roustaian, P., & Jalal, K. C. A. (2003). Salt stress culture of blue-green algae Spirulina fusiformis. *Pakistan Journal of Biological Sciences, 6*(7), 648–650. doi:10.3923/pjbs.2003.648.650

Raganati, F., Procentese, A., Olivieri, G., Russo, M. E., & Marzocchella, A. (2016). Butanol production by fermentation of fruit residues. *Chemical Engineering Transactions, 49*, 229–234. doi:10.3303/CET1649039

Rahim, I., Nomura, S., Mukasa, S., & Toyota, H. (2015). Decomposition of methane hydrate for hydrogen production using microwave and radio frequency in-liquid plasma methods. *Applied Thermal Engineering, 90*, 120–126. doi:10.1016/j.applthermaleng.2015.06.074

Rajagopal, D. (2008). Implications of India's biofuel policies for food, water and the poor. *Water Policy, 10*(S1), 95–106.

Raju, S. S., Parappurathu, S., Chand, R., Joshi, P. K., Kumar, P., & Msangi, S. (2012). *Biofuels in India: potential, policy and emerging paradigms*. Policy Paper - National Centre for Agricultural Economics and Policy Research.

Raju, S. S., Shinoj, P., & Joshi, P. K. (2009). Sustainable development of biofuels: Prospects and challenges. *Economic and Political Weekly, 26*, 65–72.

Ramacharyulu, P. V. R. K., Muhammad, R., Kumar, J. P., Prasad, G. K., & Mohanty, P. (2015). Iron phthalocyanine modified mesoporous titania nanoparticles for photocatalytic activity and CO 2 capture applications. *Physical Chemistry Chemical Physics, 17*(39), 26456–26462.

Ramadhas, A. S., Jayaraj, S., & Muraleedharan, C. (2005). Biodiesel production from high FFA rubber seed oil. *Fuel, 84*, 335–340.

Ramadhas, A. S., Jayraj, S., & Muralidharan, C. (2004). Use of vegetable oil as engine fuels: A review. *Renewable Energy, 29*(5), 727–742. doi:10.1016/j.renene.2003.09.008

Ramasubramanian, S., Manavalan, S., Gnanavel, C., & Balakrishnan, G. (2017). Processing of neem and jatropha methyl esters –alternative fuels from vegetable oil. *IOP Conf. Series: Materials Science and Engineering, 183*, 012036. 10.1088/1757-899X/183/1/012036

Ramesh, C., Ray, K. B. U., & Chandrasekaran Trilokesh, C. L. (2019). Sweet Sorghum for Bioethanol Production: Scope, Technology, and Economics. In Bioethanol Production from Food Crops (pp. 81–100). doi:10.1016/B978-0-12-813766-6/00005-9

Rana, M. S., Bhushan, S., Sudhakar, D. R., & Prajapati, S. K. (2020). Effect of iron oxide nanoparticles on growth and biofuel potential of Chlorella spp. *Algal Research, 49*, 101942.

Raqeeb, M. A., & Bhargavi, R. (2015). Biodiesel production from waste cooking oil. *Journal of Chemical and Pharmaceutical Research, 7*(12), 670–681.

Rashid, U., & Anwar, F. (2008). Production of biodiesel through optimized alkaline-catalyzed transesterification of rapeseed oil. *Fuel, 87*, 265–273.

Rasmey, A.-H., Hassan, H., Aboseidah, A., & Abdulwahid, O. (2018). Enhancing Bioethanol Productivity from Sugarcane Molasses by Saccharomyces cerevisiae Y17 KP096551. *Egyptian Journal of Botany*. doi:10.21608/ejbo.2018.1820.1126

Rasool, U., & Hemalatha, S. (2016). A review on bioenergy and biofuels: Sources and their production. *Brazilian Journal of Biological Sciences*, *3*(5), 3. doi:10.21472/bjbs.030501

Ratnavathi, C. V. (2017). Prospects of Sweet Sorghum as a Raw Material for Ethanol Production. *Recent Advances in Petrochemical Science*, *3*(3), 51–53. doi:10.19080/RAPSCI.2017.03.555613

Ratnavathi, C. V., Chakravarthy, S. K., Komala, V. V., Chavan, U. D., & Patil, J. V. (2011). Sweet Sorghum as Feedstock for Biofuel Production: A Review. *Sugar Tech*, *13*(4), 399–407. doi:10.100712355-011-0112-2

Raud, M., Kikas, T., Sippula, O., & Shurpali, N. J. (2019). Potentials and challenges in lignocellulosic biofuel production technology. *Renewable & Sustainable Energy Reviews*, *111*(May), 44–56. doi:10.1016/j.rser.2019.05.020

Raud, M., Olt, J., & Kikas, T. (2016). N2 explosive decompression pretreatment of biomass for lignocellulosic ethanol production. *Biomass and Bioenergy*, *90*, 1–6.

Ravindranath, N. H., Sita Lakshmi, C., Manuvie, R., & Balachandra, P. (2011). Biofuel production and implications for land use, food production and environment in India. *Energy Policy*, *39*, 5737–5745.

Ravindranath, N. H., Somashekar, H. I., Nagaraja, M. S., Sudha, P., Sangeetha, G., Bhattacharya, S. C., & Abdul Salam, P. (2005). Assessment of sustainable non-plantation biomass resources potential for energy in India. *Biomass and Bioenergy*, *29*(3), 178–190. doi:10.1016/j.biombioe.2005.03.005

Rawal, I., & Kaur, A. (2013). Synthesis of mesoporous polypyrrole nanowires/nanoparticles for ammonia gas sensing application. *Sensors and Actuators. A, Physical*, *203*, 92–102.

Ray, S., Miglani, S., & Goldar, A. (2011). *Ethanol blending policy in India: demand and supply issues.* ICRIER Policy Series.

Ray, N. H. S., Mohanty, M. K., & Mohanty, R. C. (2013). Anaerobic digestion of kitchen wastes: Biogas production and pre-treatment of wastes, a review. *International Journal of Scientific and Research Publications*, *3*(11), 1–6.

Ray, S., Goldar, A., & Miglani, S. (2012). The ethanol blending policy in India. *Economic and Political Weekly*, *7*, 23–25.

Razzak, S. A., Hossain, M. M., Lucky, R. A., Bassi, A. S., & de Lasa, H. (2013). Integrated CO2 capture, wastewater treatment and biofuel production by microalgae culturing—A review. *Renewable & Sustainable Energy Reviews*, *27*, 622–653.

Rebernik, M., Bradač, B., Rebernik, M., & Bradač, B. (2008). *Idea Evaluation Methods and Techniques*. Institute for Entrepreneurship and Small Business Management, University of Maribor.

Reid, W. V., Ali, M. K., & Field, C. B. (2020). The future of bioenergy. *Global Change Biology*, *26*(1), 274–286. doi:10.1111/gcb.14883 PMID:31642554

Reijnders, L., & Huijbregts, M. A. J. (2008). Palm oil and the emission of carbon-based greenhouse gases. *Journal of Cleaner Production*, *16*(4), 477–482.

Reinis, Z., Lojda, Z., Heyrovský, A., Horáková, D., & Reisenauer, R. (1975). Influence of beta-blocking agents on experimental atherosclerosis of cocks. *Casopis Lekaru Ceskych*, *114*(43), 1320–1325.

Renewable Fuels Association. (2021). *The economic impact of Covid-19 on the ethanol industry*. https://ethanolrfa.org/file/1753/COVID-19-Ethanol-Industry-Economic-Impact-July-2020-Update.pdf

Ren, H. Y., Dai, Y. Q., Kong, F., Xing, D., Zhao, L., Ren, N. Q., ... Liu, B. F. (2020). Enhanced microalgal growth and lipid accumulation by addition of different nanoparticles under xenon lamp illumination. *Bioresource Technology*, *297*, 122409.

Repérez, P. (2002). Mineral content of edible marine seaweeds. *Food Chemistry, 79*(1), 23–26. doi:10.1016/S0308-8146(02)00171-1

Reyimu, Z., & Özçimen, D. (2017). Batch cultivation of marine microalgae Nannochloropsis oculata and Tetraselmis suecica in treated municipal wastewater toward bioethanol production. *Journal of Cleaner Production, 150*, 40–46.

Rezania, S., Din, M. F. M., Taib, S. M., Mohamad, S. E., Dahalan, F. A., Kamyab, H., Darajeh, N., & Ebrahimi, S. S. (2018). Ethanol production from water hyacinth (Eichhornia crassipes) using various types of enhancers based on the consumable sugars. *Waste and Biomass Valorization, 9*(6), 939–946. doi:10.100712649-017-9883-3

Rezania, S., Oryani, B., Cho, J., Talaiekhozani, A., Sabbagh, F., Hashemi, B., Rupani, P. F., & Mohammadi, A. A. (2020). Different pretreatment technologies of lignocellulosic biomass for bioethanol production: An overview. *Energy, 199*, 117457. doi:10.1016/j.energy.2020.117457

Riayatsyah, T. M. I., Sebayang, A. H., Silitonga, A. S., Padli, Y., Fattah, I. M. R., Kusumo, F., Ong, H. C., & Mahlia, T. M. I. (2022). Current progress of jatropha curcascommoditisation as biodiesel feedstock: A comprehensive review. *Frontiers in Energy Research, 9*, 815416. doi:10.3389/fenrg.2021.815416

Ribeiro, L. A., Pereira da Silva, P., Ribeiro, L., & Dotti, F. L. (2017). Modelling the impacts of policies on advanced biofuel feedstocks diffusion. *Journal of Cleaner Production, 142*, 2471–2479.

Robak, K., & Balcerek, M. (2018). Review of second generation bioethanol production from residual biomass. *Food Technology and Biotechnology, 56*(2), 174–187. doi:10.17113/ftb.56.02.18.5428 PMID:30228792

Robak, K., & Balcerek, M. (2020). Current state-of-the-art in ethanol production from lignocellulosic feedstocks. *Microbiological Research, 240*, 126534. doi:10.1016/j.micres.2020.126534 PMID:32683278

Rodolfi, L., Chini Zittelli, G., Bassi, N., Padovani, G., Biondi, N., Bonini, G., & Tredici, M. R. (2009). Microalgae for oil: Strain selection, induction of lipid synthesis and outdoor mass cultivation in a low-cost photobioreactor. *Biotechnology and Bioengineering, 102*, 100–112. PMID:18683258

Rodríguez, L. A., Toro, M. E., Vazquez, F., Correa-Daneri, M. L., Gouiric, S. C., & Vallejo, M. D. (2010). Bioethanol production from grape and sugar beet pomaces by solid-state fermentation. *International Journal of Hydrogen Energy, 35*(11), 5914–5917. doi:10.1016/j.ijhydene.2009.12.112

Roesijadi, G. (2010). Macroalgae as a biomass feedstock: a preliminary analysis. U.S. Department of Energy under contract, Pacific Northwest National Laboratory. doi:10.2172/1006310

Rohstoffe, F. N. (2012). Guide to Biogas from Production to Use. Federal Ministry of Food, Agriculture and Consumer Protection, Fachagentur Nachwachsende Rohstoffe E.V.

Romagnoli, F., Blumberga, D., & Gigli, E. (2011). Biogas from marine macroalgae: A new environmental technology-life cycle inventory for a further LCA. *Environmental and Climate Technologies, 4*(1), 97–108.

Romagnoli, F., Pastare, L., Sabūnas, A., Bāliņa, K., & Blumberga, D. (2017). Effects of pre-treatment on Biochemical Methane Potential (BMP) testing using Baltic Sea *Fucus vesiculosus* feedstock. *Biomass and Bioenergy, 105*, 23–31. doi:10.1016/j.biombioe.2017.06.013

Romano, S.D., & Sorichetti, P.A. (2011). Dielectric spectroscopy in biodiesel production and characterization. *Green Energy and Technology*, 7-27. doi:10.1007/978-1-84996-519-4_2

Roopa, N. (n.d.). *Startups are helping North India's farmers to dispose of crop residue cleanly (and reduce pollution).* Retrieved January 20, 2022, from https://scroll.in/article/909699/startups-are-helping-north-indias-farmers-to-dispose-of-crop-residue-cleanly-and-reduce-pollution

Rosén, T., & Ödlund, L. (2019). System Perspective on Biogas Use for Transport and Electricity Production. *Energies*, *12*(21), 4159. doi:10.3390/en12214159

Rowse, L. E. (2011). *Design of Small Scale Anaerobic Digesters for Application in Rural Developing Countries. In Graduate Theses and Dissertations. Civil and Environmental Engineering*. University of South Florida.

Roy, M. M., & Chandra, A. (2019). Promoting biofuels: The case of ethanol blending initiative in India. *Clean Technologies and Environmental Policy*, *21*(5), 953–965. doi:10.100710098-019-01687-z

Roy, S., Kumar, K., Ghosh, S., & Das, D. (2014). Thermophilic biohydrogen production using pre-treated algal biomass as substrate. *Biomass and Bioenergy*, *61*, 157–166. doi:10.1016/j.biombioe.2013.12.006

Ruan, R., Zhang, Y., Chen, P., Liu, S., Fan, L., Zhou, N., . . . Li, B. (2019). Biofuels: Introduction. In Biomass, Biofuels, Biochemicals: Biofuels: Alternative Feedstocks and Conversion Processes for the Production of Liquid and Gaseous Biofuels. doi:10.1016/B978-0-12-816856-1.00001-4

Ruiz, J. A., Juárez, M. C., Morales, M. P., Muñoz, P., & Mendívil, M. A. (2013). Biomass gasification for electricity generation: Review of current technology barriers. *Renewable & Sustainable Energy Reviews*, *18*, 174–183. doi:10.1016/j.rser.2012.10.021

Rulli, M. C., Bellomi, D., Cazzoli, A., De Carolis, G., & D'Odorico, P. (2016). The water-land-food nexus of first-generation biofuels. *Scientific Reports*, *6*(March), 1–10. doi:10.1038/srep22521

Rupp, S. P., Bies, L., Glaser, A., Kowaleski, C., McCoy, T., Rentz, T., Riffell, S., Sibbing, J., Verschuyl, J., & Wigley, T. (2012). *Effects of bioenergy production on wildlife and wildlife habitat*. Wildlife Society Technical.

Safarik, I., Prochazkova, G., Pospiskova, K., & Branyik, T. (2016). Magnetically modified microalgae and their applications. *Critical Reviews in Biotechnology*, *36*(5), 931–941.

Saha, B. C. (2003). Hemicellulose bioconversion. *Journal of Industrial Microbiology & Biotechnology*, *30*(5), 279–291. doi:10.100710295-003-0049-x PMID:12698321

Sahoo, P. K., Das, L. M., Babu, M. K. G., & Naik, S. N. (2007). Biodiesel development from high acid value polanga seed oil and performance evaluation in a CI engine. *Fuel*, *86*, 448–454.

Saini, J. K., Saini, R., & Tewari, L. (2015). Lignocellulosic agriculture wastes as biomass feedstocks for second-generation bioethanol production: Concepts and recent developments. *3 Biotech*, *5*(4), 337-353.

Saini, J. K., Saini, R., & Tewari, L. (2015). Lignocellulosic agriculture wastes as biomass feedstocks for second-generation bioethanol production: concepts and recent developments. *3 Biotech*, *5*(4), 337–353. doi:10.1007/s13205-014-0246-5

Sakamoto, Y., Komai, T., Kawabe, Y., Tenma, N., & Yamaguchi, T. (2004) Gas Hydrate Extraction from Marine Sediments by Heat Stimulation Method. *Proc Fourteenth Int Offshore Polar Eng Conf*, 52.

Salapa, I., Katsimpouras, C., Topakas, E., & Sidiras, D. (2017). *Organosolv pretreatment of wheat straw for efficient ethanol production using various solvents*. Academic Press.

Sánchez, C. (2009). Lignocellulosic residues: Biodegradation and bioconversion by fungi. *Biotechnology Advances*, *27*(2), 185–194. doi:10.1016/j.biotechadv.2008.11.001 PMID:19100826

Sánchez, Ó. J., & Cardona, C. A. (2012). Conceptual design of cost-effective and environmentally friendly configurations for fuel ethanol production from sugarcane by knowledge-based process synthesis. *Bioresource Technology*, *104*, 305–314. doi:10.1016/j.biortech.2011.08.125 PMID:22137752

Santana, A., Jesus, S., Larrayoz, M. A., & Filho, R. M. (2012). Supercritical carbon dioxide extraction of algae lipids for the biodiesel production. *Procedia Engineering*, *42*, 1755–1761. doi:10.1016/j.proeng.2012.07.569

Šantek, B., Gwehenberger, G., Šantek, M. I., Narodoslawsky, M., & Horvat, P. (2010). Evaluation of energy demand and the sustainability of different bioethanol production processes from sugar beet. *Resources, Conservation and Recycling*, *54*(11), 872–877. doi:10.1016/j.resconrec.2010.01.006

Santelices, B. (2007). The discovery of kelp forests in deep-water habitats of tropical regions. *Proceedings of the National Academy of Sciences of the United States of America*, *104*(49), 19163–19164. doi:10.1073/pnas.0708963104 PMID:18042707

Saqib, A., Tabbssum, M. R., Rashid, U., Ibrahim, M., Gill, S. S., & Mehmood, M. A. (2013). Marine macroalgae *Ulva*: A potential feed-stock for bioethanol and biogas production. *Asian Journal of Agriculture and Biology*, *1*, 155–163.

Saran, S., Malaviya, A., & Chaubey, A. (2019). Introduction, Scope and Significance of Fermentation Technology. In High Value Fermentation Products (Vol. 1, pp. 1–25). Academic Press.

Saribiyik, O. Y., Ozcanli, M., Serin, H., Serin, S., & Aydin, K. (2010). Biodiesel production from ricinus communis oil and its blends with soybean biodiesel. *Jixie Gongcheng Xuebao*, *56*, 811–816.

Sarkar, R. D., Singh, H. B., & Kalita, M. C. (2021). Enhanced lipid accumulation in microalgae through nanoparticle-mediated approach, for biodiesel production: A mini-review. *Heliyon*, *7*(9), e08057.

Sarker, S., Lamb, J. J., Hjelme, D. R., & Lien, K. M. (2019). A Review of the Role of Critical Parameters in the Design and Operation of Biogas Production Plants. *Applied Sciences (Basel, Switzerland)*, *9*(9), 1915. doi:10.3390/app9091915

Sarma, S. J., Das, R. K., Brar, S. K., Le Bihan, Y., Buelna, G., Verma, M., & Soccol, C. R. (2014). Application of magnesium sulfate and its nanoparticles for enhanced lipid production by mixotrophic cultivation of algae using biodiesel waste. *Energy*, *78*, 16–22.

Sasaki, K., Ono, S., & Sugai, Y. (2007) Gas Production System From Methane Hydrates Layers by Hot Water Injection Using Dual Horizontal Wells. *Can Int Petroelum Conf*, *48*, 1–5. 10.2118/2007-195

Sasaki, K., Tsuge, Y., Sasaki, D., Kawaguchi, H., Sazuka, T., Ogino, C., & Kondo, A. (2015). Repeated ethanol production from sweet sorghum juice concentrated by membrane separation. *Bioresource Technology*, *186*, 351–355. doi:10.1016/j.biortech.2015.03.127 PMID:25857769

Sasaki, Y., Takagi, T., Motone, K., Shibata, T., Kuroda, K., & Ueda, M. (2018). Direct bioethanol production from brown macroalgae by co-culture of two engineered *Saccharomyces cerevisiae* strains. *Bioscience, Biotechnology, and Biochemistry*, *82*(8), 1459–1462. doi:10.1080/09168451.2018.1467262 PMID:29708475

Sathya, T., & Manivannan, A. (2013). Biodiesel production from neem oil using two step transesterification. *International Journal of Engineering Research and Applications*, *3*, 488–492.

Sawyer, W.K., Boyer, C.M., Frantz, J.H., & Yost, A.B. (2000). Comparative Assessment of Natural Gas Hydrate Production Models. *SPE/CERI Gas Technol Symp*, 1–9. doi:10.2118/62513-MS

Saxena, R. C., Adhikari, D. K., & Goyal, H. B. (2009). Biomass-based energy fuel through biochemical routes: A review. *Renewable & Sustainable Energy Reviews*, *13*(1), 167–178. doi:10.1016/j.rser.2007.07.011

Sayre, R. (2010). Microalgae: The Potential for Carbon Capture. *BioSci.*, *60*, 722–727.

Schenk, P. M., Thomas-Hall, S. R., Stephens, E., Marx, U. C., Mussgnug, J. H., Posten, C., Kruse, O., & Hankamer, B. (2008). Second Generation Biofuels: High-Efficiency Microalgae for Biodiesel Production. *Bioenerg. Res.*, *1*, 20–43.

Schmitt, E., Bura, R., Gustafson, R., Cooper, J., & Vajzovic, A. (2012). Converting lignocellulosic solid waste into ethanol for the State of Washington: An investigation of treatment technologies and environmental impacts. *Bioresource Technology*, *104*, 400–409. doi:10.1016/j.biortech.2011.10.094 PMID:22119432

Scott, S. A., Davey, M. P., Dennis, J. S., Horst, I., Howe, C. J., Lea-Smith, D. J., & Smith, A. G. (2010). Biodiesel from algae: Challenges and prospects. *Current Opinion in Biotechnology*, *21*(3), 277–286. doi:10.1016/j.copbio.2010.03.005 PMID:20399634

Seabra, J. E., Tao, L., Chum, H. L., & Macedo, I. C. (2010). A techno-economic evaluation of the effects of centralized cellulosic ethanol and co-products refinery options with sugarcane mill clustering. *Biomass and Bioenergy*, *34*(8), 1065–1078. doi:10.1016/j.biombioe.2010.01.042

Searchinger, T., Heimlich, R., Houghton, R. A., Dong, F., Elobeid, A., Fabiosa, J., Tokgoz, S., Hayes, D., & Yu, T. H. (2008). Use of U.S. croplands for biofuels increases greenhouse gases through emissions from land-use change. *Science*, *319*, 1238–1240. PMID:18258860

Sebastian, K., Jinqin, Z., Zhen, X., Laner, D., & Woon, K. (2022). A systematic review of life cycle assessment of solid waste management: Methodological trends and prospects. *The Science of the Total Environment*, *831*, 154903. PMID:35367543

Selvan, V. A. M., Anand, R. B., & Udayakumar, M. (2014). Effect of cerium oxide nanoparticles and carbon nanotubes as fuel-borne additives in diesterol blends on the performance, combustion and emission characteristics of a variable compression ratio engine. *Fuel*, *130*, 160–167.

Sengupta, M., & Poddar, A. (2013). National policy on biofuel under the scanner. *International Journal of Emerging Technology and Advanced Engineering*, *3*, 521–526.

Seo, Y. J., Park, S., Kang, H., Ahn, Y. H., Lim, D., Kim, S. J., Lee, J., Lee, J. Y., Ahn, T., Seo, Y., & Lee, H. (2016). Isostructural and cage-specific replacement occurring in sII hydrate with external CO 2 / N 2 gas and its implications for natural gas production and CO 2 storage. *Applied Energy*, *178*, 579–586. doi:10.1016/j.apenergy.2016.06.072

Seo, Y., Lee, S., & Lee, J. (2013). Experimental Verification of Methane Replacement in Gas Hydrates by Carbon Dioxide. *Ital Assoc Chem Eng*, *47*, 131184–13190.

Serrano, P. T., Ulrich, S., Szymcek, P., McCallum, S., & Phelps, T. (2009). Multiphase, Microdispersion Reactor for the Continuous Production of Methane Gas Hydrate. *Industrial & Engineering Chemistry Research*, *48*(13), 6448–6452. doi:10.1021/ie8019517

Sexton, S. E., & Zilberman, D. (2011). Biotechnology and biofuel. In Frontiers of Economics and Globalization (Vol. 10). doi:10.1108/S1574-8715(2011)0000010014

Shaah, M. A. H., Hossain, M. S., Allafi, F. A. S., Alsaedi, A., Ismail, N., Kadir, M. O. A., & Ahmad, M. I. (2021). A review on non-edible oil as a potential feedstock for biodiesel: Physicochemical properties and production technologies. *RSC Advances*, *11*(40), 25018–25037. doi:10.1039/D1RA04311K PMID:35481051

Shaalan, M., Saleh, M., El-Mahdy, M., & El-Matbouli, M. (2016). Recent progress in applications of nanoparticles in fish medicine: A review. *Nanomedicine; Nanotechnology, Biology, and Medicine*, *12*(3), 701–710.

Shah, M. A., Khan, M. N. S., & Kumar, V. (2018). Biomass residue characterization for their potential application as biofuels. *Journal of Thermal Analysis and Calorimetry*, *134*(3), 2137–2145. doi:10.100710973-018-7560-9

Shaishav, S., Singh, R. N., & Satyendra, T. (2013). Biohydrogen from algae: Fuel of the future. *International Research Journal of Environmental Sciences*, *2*(4), 44–47.

Shane, S., & Venkataraman, S. (2000). The Promise of Entrepreneurship as a Field of Research. *Academy of Management Review*, *25*(1), 217–226. https://doi.org/10.2307/259271

Sheldon, R. A. (2007). Enzyme immobilization: The quest for optimum performance. *Advanced Synthesis & Catalysis*, *349*(8–9), 1289–1307. doi:10.1002/adsc.200700082

Shilton, A., & Guieysse, B. (2010). Sustainable sunlight to biogas is via marginal organics. *Current Opinion in Biotechnology*, *21*(3), 287–291. doi:10.1016/j.copbio.2010.03.008 PMID:20378331

Shimada, W., Takeya, S., Kamata, Y., Uchida, T., Nagao, J., Ebinuma, T., & Narita, H. (2005). Texture change of ice on anomalously preserved methane clathrate hydrate. *The Journal of Physical Chemistry B*, *109*(12), 5802–5807. doi:10.1021/jp044624t PMID:16851632

Shin, W. K., Cho, J., Kannan, A. G., Lee, Y. S., & Kim, D. W. (2016). Cross-linked composite gel polymer electrolyte using mesoporous methacrylate-functionalized SiO2 nanoparticles for lithium-ion polymer batteries. *Scientific Reports*, *6*(1), 1–10.

Shishir, P. S. (2007). Effect of Particle Size Based Separation of Milled Corn Stover on AFEX Pretreatment and Enzymatic Digestibility. *Biotechnology and Bioengineering*, *96*(2), 219–231. doi:10.1002/bit.21132 PMID:16903002

Shi, X.-S., Dong, J.-J., Yu, J.-H., Yin, H., Hu, S.-M., Huang, S.-X., & Yuan, X.-Z. (2017). Effect of hydraulic retention time on anaerobic digestion of wheat straw in the semicontinuous continuous stirred-tank reactors. *BioMed Research International*, *2017*, 1–6. doi:10.1155/2017/2457805 PMID:28589134

Shklar, G., & Swartz, J. (1988). Tumor necrosis factor in experimental cancer regression with alpha- tocopherol, beta-carotene, canthaxanthin and algae extract. *Eur Cancer Clin Oncol*, *24*(5), 839–850. doi:10.1016/0277-5379(88)90192-7 PMID:3139418

Shrestha, P., Pometto, A. L. III, Khanal, S. K., & van Leeuwen, J. (2012). Second-Generation Biofuel Production from Corn-Ethanol Industry Residues. *Green Energy and Technology*, *62*, 71–87. doi:10.1007/978-1-4471-2324-8_5

Shruthi, M. S., & Ramani, D. (2021). Statistical analysis of impact of COVID 19 on India commodity markets. *Materials Today: Proceedings*, *37*(Part 2), 2306–2311. doi:10.1016/j.matpr.2020.07.729 PMID:32837924

Shukla, R., Kumar, M., Chakraborty, S., Gupta, R., Kumar, S., Sahoo, D., & Kuhad, R. C. (2016). Process development for the production of bioethanol from waste algal biomass of *Gracilaria verrucosa*. *Bioresource Technology*, *220*, 584–589. doi:10.1016/j.biortech.2016.08.096 PMID:27619709

Sibi, G., Ananda Kumar, D., Gopal, T., Harinath, K., Banupriya, S., & Chaitra, S. (2017). Metal Nanoparticle Triggered Growth and Lipid Production in Chlorella Vulgaris. *Int J Sci Res Environ Sci Toxicol*, *2*(1), 1–8.

Siciliano, A., Limonti, C., Mehariya, S., Molino, A., & Calabrò, V. (2019). Biofuel production and phosphorus recovery through an integrated treatment of agro-industrial waste. *Sustainability (Switzerland)*, *11*(1), 1–17. doi:10.3390u11010052

Siddiqua, S., Mamun, A. A., & Enayetul Babar, S. M. (2015). Production of biodiesel from coastal macroalgae (*Chara vulgaris*) and optimization of process parameters using Box-Behnken design. *SpringerPlus*, *4*(1), 720. doi:10.118640064-015-1518-1 PMID:26636008

Siefers, A., Wang, N., Sindt, A., Dunn, J., McElvogue, J., Evans, E., & Ellis, T. (2010). A novel and cost-effective hydrogen sulfide removal technology using tire derived rubber particles. *Proceedings of the Water Environment Federation*, *12*(12), 4597–4622. doi:10.2175/193864710798182682

Silalertruksa, T., & Gheewala, S. H. (2009). Environmental sustainability assessment of bio-ethanol production in Thailand. *Energy*, *34*(11), 1933–1946. doi:10.1016/j.energy.2009.08.002

Silitonga, A. S., Atabani, A. E., Mahlia, T. M. I., Masjuki, H. H., Badruddin, I. A., & Mekhilef, S. (2011). A review on prospect of *Jatropha curcas* for biodiesel in Indonesia. *Renewable & Sustainable Energy Reviews*, *15*(8), 3733–3756.

Silitonga, A. S., Masjuki, H. H., Ong, H. C., Sebayang, A. H., Dharma, S., Kusumo, F., ... Sugiyanto, B. (2018). Evaluation of the engine performance and exhaust emissions of biodiesel-bioethanol-diesel blends using kernel-based extreme learning machine. *Energy*, *159*, 1075–1087.

Sillu, D., & Agnihotri, S. (2020). Cellulase immobilization onto magnetic halloysite nanotubes: Enhanced enzyme activity and stability with high cellulose saccharification. *ACS Sustainable Chemistry & Engineering*, *8*(2), 900–913. doi:10.1021/acssuschemeng.9b05400

Silva, Â., Wilson, K., Lee, A. F., dos Santos, V. C., Bacilla, A. C. C., Mantovani, K. M., & Nakagaki, S. (2017). Nb2O5/SBA-15 catalyzed propanoic acid esterification. *Applied Catalysis B: Environmental*, *205*, 498–504.

SimGas (2018). *The SimGas biogas system is an integrated farm solution for rural households in Africa and Asia*. https://simgas.org/biogas-system

Simon, P., Lima, J. S., Valério, A., De Oliveira, D., Araújo, P. H. H., Sayer, C., De Souza, A. A. U., Guelli, U., & De Souza, S. M. A. (2018). Cellulase immobilization on poly(methyl methacrylate) nanoparticles by miniemulsion polymerization. *Brazilian Journal of Chemical Engineering*, *35*(2), 649–657. doi:10.1590/0104-6632.20180352s20160094

Sindhu, R., Binod, P., & Pandey, A. (2016). Biological pretreatment of lignocellulosic biomass - An overview. *Bioresource Technology*, *199*, 76–82. doi:10.1016/j.biortech.2015.08.030 PMID:26320388

Sindhu, R., Binod, P., Pandey, A., Ankaram, S., Duan, Y., & Awasthi, M. K. (2019). Biofuel production from biomass: Toward sustainable development. In *Current Developments in Biotechnology and Bioengineering*. Waste Treatment Processes for Energy Generation. doi:10.1016/B978-0-444-64083-3.00005-1

Singh, S. (2009). *India Biofuels Annual*. USDA Foreign Agricultural Service, GAIN Report Number IN9080.

Singh, A., Nigam, P. S., & Murphy, J. D. (2011). Renewable fuels from algae: An answer to debatable land based fuels. *Bioresource Technology*, *102*(1), 10–16. doi:10.1016/j.biortech.2010.06.032 PMID:20615690

Singh, A., & Olsen, S. I. (2011). A critical review of biochemical conversion, sustainability and life cycle assessment of algal biofuels. *Applied Energy*, *88*(10), 3548–3555. doi:10.1016/j.apenergy.2010.12.012

Singh, D. P., & Trivedi, R. K. (2013). Production of biofuel from algae: An economic and eco-friendly resource. *International Journal of Scientific Research*, *2*(11), 352–357.

Singhe, U.W. (2013). *Production of Methane from Abundant Hydrate Deposits*. US 2013/0341179 A1. doi:10.1037/t24245-000

Singh, S., Cheng, G., Sathitsuksanoh, N., Wu, D., Varanasi, P., George, A., Balan, V., Gao, X., Kumar, R., Dale, B. E., Wyman, C. E., & Simmons, B. A. (2015). Comparison of different biomass pretreatment techniques and their impact on chemistry and structure. *Frontiers in Energy Research*, *2*, 62. doi:10.3389/fenrg.2014.00062

Singh, S., Sharma, S., & Mohapatra, S. K. (2015). A Production of biodiesel from waste cotton seed oil and testing on small capacity diesel engine. *International Journal of Advance Research in Science and Engineering*, *4*(2), 172–176.

Singh, S., Sharma, S., & Mohapatra, S. K. (2015). A Production Of Biodiesel from Waste Cotton Seed Oil and Testing on Small Capacity Diesel Engine. *International Journal of Advance Research In Science And Engineering*, *4*(2), 172–176.

Singh, V. (2012). Effect of corn quality on bioethanol production. *Biocatalysis and Agricultural Biotechnology*, *1*(4), 353–355. doi:10.1016/j.bcab.2012.06.001

Sinha, V. K. (2017). Sharlin. A review approach on exhaust emission reduction by nanoparticle additives in diesel engine. *Int J Innov Res Sci Eng Technol*, *6*, 15046–15050.

Sira, J. H., Pat, H. S. L., & Kamath, V. (1990) Study of Hydrate Dissociation by Methanol and Glycol Injection. *PE Annual Technical Conference and Exhibition*, 977–984. 10.2118/20770-MS

Sivaramakrishnan, R., & Incharoensakdi, A. (2017). Microalgae as feedstock for biodiesel production under ultrasound treatment–A review. *Bioresource Technology*, *250*, 877–887. PMID:29221914

Slade, R., Bauen, A., & Shah, N. (2009). The greenhouse gas emissions performance of cellulosic ethanol supply chains in Europe. *Biotechnology for Biofuels*, *2*(15), 1–19. PMID:19682352

Slette, J., & Aradhey, A. (2014). *India Biofuels Annual*. USDA Gain Report: IN4055.

Slocombe, S.P., Zhang, Q., Ross, M., Anderson, A., Thomas, N.J., & Lapresa, Á., Rad-Menéndez, C., Campbell, C.N., Black, K.D., Stanley, M.S., & Day, J.G. (2015). Unlocking nature's treasure-chest: Screening for oleaginous algae. *Scientific Reports*, *5*, srep09844. PMID:26202369

Smith, S. R., Lang, N. L., Cheung, K. H. M., & Spanoudaki, K. (2005). Factors controlling pathogen destruction during anaerobic digestion of biowastes. *Waste Management (New York, N.Y.)*, *25*(4), 417–425. doi:10.1016/j.wasman.2005.02.010 PMID:15869985

Smyth, B., Gallachóir, B. Ó., Korres, N., & Murphy, J. (2010). Can we meet targets for biofuels and renewable energy in transport given the constraints imposed by policy in agriculture and energy? *Journal of Cleaner Production*, *18*(16-17), 1671–1685.

Soliman, R. M., Younis, S. A., El-Gendy, N. S., Mostafa, S., El-Temtamy, S. A., & Hashim, A. I. (2018). Batch bioethanol production via the biological and chemical saccharification of some Egyptian marine macroalgae. *Journal of Applied Microbiology*, *125*(2), 422–440. doi:10.1111/jam.13886 PMID:29675837

Somerville, C. (2007).. . *Biofuels*, *17*(4), 115–119.

Somerville, C., Youngs, H., Taylor, C., Davis, S. C., & Long, S. P. (2010). Feedstocks for lignocellulosic biofuels. *Science*, *329*, 790–792. PMID:20705851

Song, Y., Cheng, C., Zhao, J., Zhu, Z., Liu, W., Yang, M., & Xue, K. (2015). Evaluation of gas production from methane hydrates using depressurization, thermal stimulation and combined methods. *Applied Energy*, *145*, 265–277. doi:10.1016/j.apenergy.2015.02.040

Song, Y., Zhang, L., Lv, Q., Yang, M., Ling, Z., & Zhao, J. (2016). Assessment of gas production from natural gas hydrate using depressurization, thermal stimulation and combined methods. *RSC Advances*, *6*(53), 47357–47367. doi:10.1039/C6RA05526E

Sorda, G., Banse, M., & Kemfert, C. (2010). An overview of biofuel policies across the world. *Energy Policy*, *38*, 6977–6988.

Spolaore, P., Joannis-Cassan, C., Duran, E., & Isambert, A. (2006). Review: Commercial application of microalgae. *Journal of Bioscience and Bioengineering*, *101*(2), 87–96. doi:10.1263/jbb.101.87 PMID:16569602

Springer, Y. P., Hays, C. G., Carr, M. H., & Mackey, M. R. (2010). Toward ecosystem-based management of marine macroalgae-the bull kelp, *Nereocystis luetkeana. Oceanography and Marine Biology*, *48*, 1–41.

Sreenivas, C., Rao, A. B., & Patwardhan, A. (2018). *Critical Evaluation of Biodiesel Production Initiatives in India, Biofuels: Greenhouse Gas Mitigation and Global Warming*. Springer.

Srivastava, N., Srivastava, M., Mishra, P. K., Gupta, V. K., Molina, G., Rodriguez-Couto, S., Manikanta, A., & Ramteke, P. W. (2018). Applications of fungal cellulases in biofuel production: Advances and limitations. *Renewable & Sustainable Energy Reviews*, *82*, 2379–2386. doi:10.1016/j.rser.2017.08.074

Srivastava, N., Srivastava, M., Mishra, P. K., Singh, P., & Ramteke, P. W. (2015). Application of cellulases in biofuels industries: An overview. *Journal of Biofuels and Bioenergy*, *1*(1), 55–63. doi:10.5958/2454-8618.2015.00007.3

Stams, A. J. M., & Plugge, C. M. (2009). Electron transfer in syntrophic communities of anaerobic bacteria and archaea. *Nature Reviews. Microbiology*, *7*(8), 568–577. doi:10.1038/nrmicro2166 PMID:19609258

Standing Committee Report. (2015). *Ethanol Blended Petrol & Bio Diesel Policy*. Standing Committee on Petroleum and Natural Gas, MoPNG.

Steele, J. M., Grady, N. K., Nordlander, P., & Halas, N. J. (2007). Plasmon hybridization in complex nanostructures. In *Surface plasmon nanophotonics* (pp. 183–196). Springer.

Stephane Bungener, L. G. (2016). *Energy efficiency and integration in the refining and petrochemical industries* [PhD thesis].

Stern, D. I. (2011). The role of energy in economic growth. *Annals of the New York Academy of Sciences*, *1219*, 26–51. PMID:21332491

Stevens, C. V., & Verhe, R. (2004). *Renewable bioresources scope and modification for nonfood application*. John Wiley and Sons Ltd.

Stolarski, M. J., Krzyzaniak, M., Łuczyński, M., Załuski, D., Szczukowski, S., Tworkowski, J., & Gołaszewski, J. (2015). Lignocellulosic biomass from short rotation woody crops as a feedstock for second-generation bioethanol production. *Industrial Crops and Products*, *75*, 66–75. doi:10.1016/j.indcrop.2015.04.025

Stolarski, M. J., Szczukowski, S., Tworkowski, J., & Klasa, A. (2013). Yield, energy parameters and chemical composition of short-rotation willow biomass. *Industrial Crops and Products*, *46*, 60–65. doi:10.1016/j.indcrop.2013.01.012

Stolze, Y., Bremges, A., Maus, I., Pühler, A., Sczyrba, A., & Schlüter, A. (2017). Targeted in situ metatranscriptomics for selected taxa from mesophilic and thermophilic biogas plants. *Microbial Biotechnology*, *11*(4), 667–679. doi:10.1111/1751-7915.12982 PMID:29205917

Street, S. (2012). *Meeting U.S. Fuel Standards*. Academic Press.

Subba Rao, P. V., & Mantri, V. A. (2006). Indian seaweed resources and sustainable utilization: Scenario at the dawn of a new century. *Current Science*, *91*(2), 164–174.

Sudhakar, K., Mamat, R., Samykano, M., Azmi, W. H., & Ishak, W. F. W. (2018). An overview of marine macroalgae as bioresource. *Renewable and Sustainable Energy Reviews, 91*, 165-179. doi:10.1016/j.rser.2018.03.100

Sudhakar, K., & Premalatha, M. (2012). Energy Balance and Exergy analysis of large scale algal biomass production 1. Goal Definition 2. Life Cycle Inventory 4. *Interpretation (Tulsa)*, *1454*(June), 13–15.

Suganya, T., Nagendra Gandhi, N., & Renganathan, S. (2013). Production of algal biodiesel from marine macroalgae Enteromorpha compressa by two step process: Optimization and kinetic study. *Bioresource Technology*, *128*, 392–400. doi:10.1016/j.biortech.2012.10.068 PMID:23201520

Suganya, T., & Renganathan, S. (2014). Ultrasonic assisted acid base transesterification of algal oil from marine macroalgae *Caulerpa peltata*: Optimization and characterization studies. *Fuel*, *128*, 347–355. doi:10.1016/j.fuel.2014.03.037

Suganya, T., Varman, M., Masjuki, H. H., & Renganathan, S. (2016). Macroalgae and microalgae as a potential source for commercial applications along with biofuels production: A biorefinery approach. *Renewable & Sustainable Energy Reviews*, *55*, 909–941. doi:10.1016/j.rser.2015.11.026

Su, H., Liu, L., Wang, S., Wang, Q., Jiang, Y., Hou, X., & Tan, T. (2015). Semi-continuous anaerobic digestion for biogas production: Influence of ammonium acetate supplement and structure of the microbial community. *Biotechnology for Biofuels*, *8*(1), 1–11. doi:10.118613068-015-0197-z PMID:25705255

Sun, W., & Nesic, S. (2007). A mechanistic model of H2S corrosion of mild steel. *Corrosion-March 19, 2007, NACE International Conference and Expo*, Paper No. 07655. http://www.icmt.ohio.edu/nesic/papers/FullText/conference-91.pdf

Sun, F., & Chen, H. (2008). Organosolv pretreatment by crude glycerol from oleochemicals industry for enzymatic hydrolysis of wheat straw. *Bioresource Technology*, *99*(13), 5474–5479. doi:10.1016/j.biortech.2007.11.001 PMID:18077155

Sung, W., Lee, H., Lee, H., & Lee, C. (2002). Numerical Study for Production Performances of a Methane Hydrate Reservoir Stimulated by Inhibitor Injection. *Energy Sources*, *8312*(6), 37–41. doi:10.1080/00908310290086527

Sun, J. X., Sun, X. F., Sun, R. C., & Su, Y. Q. (2004). Fractional extraction and structural characterization of sugarcane bagasse hemicelluloses. *Carbohydrate Polymers*, *56*(2), 195–204. doi:10.1016/j.carbpol.2004.02.002

Sunna, A., & Antranikian, G. (1997). Xylanolytic enzymes from fungi and bacteria. *Critical Reviews in Biotechnology*, *17*(1), 39–67. doi:10.3109/07388559709146606 PMID:9118232

Sun, Y., & Cheng, J. J. (2005). Dilute acid pretreatment of rye straw and bermudagrass for ethanol production. *Bioresource Technology*, *96*(14), 1599–1606. doi:10.1016/j.biortech.2004.12.022 PMID:15978993

Surendhiran, D., & Vijay, M. (2012). Microalgal biodiesel-a comprehensive review on the potential and alternative biofuel. *Research Journal of Chemical Sciences*, *2*(11), 71–82.

Surendra, K. C., Takara, D., Hashimoto, A. G., & Khanal, S. K. (2014). Biogas as a sustainable energy source for developing countries: Opportunities and challenges. *Renewable & Sustainable Energy Reviews*, *31*, 846–859. doi:10.1016/j.rser.2013.12.015

Surriya, O., Saleem, S. S., Waqar, K., Kazi, A. G., & Öztürk, M. (2015). Bio-fuels: A Blessing in Disguise. In *Phytoremediation for Green Energy* (pp. 11–54). Springer.

Su, Y., Zhang, P., & Su, Y. (2015). An overview of biofuels policies and industrialization in the major biofuel producing countries. *Renewable & Sustainable Energy Reviews*, *50*, 991–1003.

Su, Z., He, Y., Wu, N., Zhang, K., & Moridis, G. J. (2012). Evaluation on gas production potential from laminar hydrate deposits in Shenhu Area of South China Sea through depressurization using vertical wells. *Journal of Petroleum Science Engineering*, *86–87*, 87–98. doi:10.1016/j.petrol.2012.03.008

Tai, A. P. K., Martin, M. V., & Heald, C. L. (2014). Threat to future global food security from climate change and ozone air pollution. *Nature Climate Change*, *4*, 817–821.

Takara, D., & Khanal, S. K. (2015). Characterizing compositional changes of Napier grass at different stages of growth for biofuel and biobased products potential. *Bioresource Technology*, *188*, 103–108. doi:10.1016/j.biortech.2015.01.114 PMID:25727997

Tan, I. S., Lam, M. K., Henry, C. Y. F., Lim, S., & Lee, K. T. (2020). Advances of macroalgae biomass for the third generation of bioethanol production. *Chinese Journal of Chemical Engineering*, *28*(2), 502-517. doi:10.1016/j.cjche.2019.05.012

Tang, L. G., Li, X. S., Feng, Z. P., Li, G., & Fan, S. S. (2007). Control mechanisms for gas hydrate production by depressurization in different scale hydrate reservoirs. *Energy & Fuels*, *21*(1), 227–233. doi:10.1021/ef0601869

Tan, I. S., Lam, M. K., & Lee, K. T. (2013). Hydrolysis of macroalgae using heterogeneous catalyst for bioethanol production. *Carbohydrate Polymers*, *94*(1), 561–566. doi:10.1016/j.carbpol.2013.01.042 PMID:23544575

TDA, Time and Date AS. (2018). *All year climate and weather averages in Dodoma, Annual weather averages in Dodoma based on weather reports collected during 1985–2015*. https://www.timeanddate.com/weather/tanzania/dodoma/climate

Teymouri, F., Laureano-Perez, L., Alizadeh, H., & Dale, B. E. (2005). Optimization of the ammonia fiber explosion (AFEX) treatment parameters for enzymatic hydrolysis of corn stover. *Bioresource Technology*, *96*(18), 2014–2018. doi:10.1016/j.biortech.2005.01.016

Thamsiriroj, T., & Murphy, J. D. (2010). Can rape seed biodiesel meet the European Union sustainability criteria for biofuels? *Energy & Fuels*, *24*, 1720–1730.

The Hindu. (2016). Flex fuel policy of government to curb pollution worry automakers (Sriram Lakshman). *The Hindu.*

Theuerl, S., Herrmann, C., Heiermann, M., Grundmann, P., Landwehr, N., Kreidenweis, U., & Prochnow, A. (2019). The Future Agricultural Biogas Plant in Germany: A Vision. *Energies*, *12*(3), 396. doi:10.3390/en12030396

Theuerl, S., Kohrs, F., Benndorf, D., Maus, I., Wibberg, D., Schlüter, A., Kausmann, R., Heiermann, M., Rapp, E., Reichl, U., Pühler, A., & Klocke, M. (2015). Community Shifts in a Well-Operating Agricultural Biogas Plant: How Process Variations Are Handled by the Microbiome. *Applied Microbiology and Biotechnology*, *99*(18), 7791–7803. doi:10.100700253-015-6627-9 PMID:25998656

Thirugnanasambantham, R., & Elango, T. (2020). Emission Analysis of Chlorella sp. Microalgae Biodiesel with Oxide Nano Additives in Diesel Engine. *Journal of Scientific and Industrial Research*, *79*(11), 1031–1034.

Thirumarimurugan, M., Sivakumar, V. M., Xavier, A. M., Prabhakaran, D., & Kannadasan, T. (2012). Preparation of biodiesel from sunflower oil by transesterification. *International Journal of Bioscience, Biochemistry, Bioinformatics*, *2*(6), 441–445. doi:10.7763/IJBBB.2012.V2.151

This woman entrepreneur's startup collects used cooking oil to convert into biodiesel. (2021, February 15). https://yourstory.com/herstory/2021/02/woman-entrepreneur-startup-cooking-oil-biodiesel/amp

Thompson, J. L., & Tyner, W. E. (2014). Corn stover for bioenergy production: Cost estimates and farmer supply response. *Biomass and Bioenergy*, *62*, 166–173. doi:10.1016/j.biombioe.2013.12.020

Timilsena, Y. P., Abeywickrama, C. J., Rakshit, S. K., & Brosse, N. (2013). Effect of different pretreatments on delignification pattern and enzymatic hydrolysability of miscanthus, oil palm biomass and typha grass. *Bioresource Technology*, *135*, 82–88. doi:10.1016/j.biortech.2012.09.010 PMID:23069607

Timilsina, G. R., Mevel, S., & Shrestha, A. (2011). Oil price, biofuels and food supply. *Energy Policy*, *39*(12), 8098–8105. doi:10.1016/j.enpol.2011.10.004

Tiwari, J. N., Tiwari, R. N., & Kim, K. S. (2012). Zero-dimensional, one-dimensional, two-dimensional and three-dimensional nanostructured materials for advanced electrochemical energy devices. *Progress in Materials Science*, *57*(4), 724–803.

Tiwari, R., Nain, L., Labrou, N. E., & Shukla, P. (2018). Bioprospecting of functional cellulases from metagenome for second generation biofuel production: A review. *Critical Reviews in Microbiology*, *44*(2), 244–257. doi:10.1080/1040841X.2017.1337713 PMID:28609211

Topare, N., Rout, S. J., Renge, V. C., Khedkar, S. V., Chavan, Y. P., & Bhagat, S. L. (2011). Extraction of oil from algae by solvent extraction and oil expeller method. *International Journal of Chemical Sciences*, *9*(4), 1746–1750.

Topi, D. (2020). Transforming waste vegetable oils to biodiesel, establishing of a waste oil management system in Albania. *SN Applied Sciences*, *2*(513), 1–7. doi:10.100742452-020-2268-4

Torkamani, S., Wani, S. N., Tang, Y. J., & Sureshkumar, R. (2010). Plasmon-enhanced microalgal growth in miniphotobioreactors. *Applied Physics Letters*, *97*(4), 043703.

Trbojevic Ivic, J., Dimitrijevic, A., Milosavic, N., Bezbradica, D., & Drakulic, B. J. (2016). Assessment of the interacting mechanism between Candida rugosa lipases and hydroxyapatite and identification of the hydroxyapatite-binding sequence through proteomics and molecular modelling. *RSC Advances*, *41*(41), 34818–34824. doi:10.1039/C6RA07521E

Trbojević Ivić, J., Veličković, D., Dimitrijević, A., Bezbradica, D., Dragačević, V., Gavrović Jankulović, M., & Milosavić, N. (2016). Design of biocompatible immobilized Candida rugosa lipase with potential application in food industry. *Journal of the Science of Food and Agriculture*, *96*(12), 4281–4287. doi:10.1002/jsfa.7641 PMID:26801832

Trbojević-Ivić, J., Milosavić, N., Dimitrijević, A., Gavrović-Jankulović, M., Bezbradica, D., Kolarski, D., & Veličković, D. (2017). Synthesis of medium-chain length capsinoids from coconut oil catalyzed by Candida rugosa lipases. *Food Chemistry*, *218*, 505–508. doi:10.1016/j.foodchem.2016.09.049 PMID:27719942

Trbojević-Ivić, J., Veličković, D., Dimitrijević, A., & Milosavić, N. (2018). Immobilization and stabilization of microbial lipases. In H. Broeken (Ed.), *Lipases* (pp. 55–107). Nova Science Publishers, Inc.

Trbojević, J. N., Dimitrijević, A. S., Veličković, D. V., Gavrović-Jankulović, M., & Milosavić, N. B. (2013). Isolation of Candida rugosa lipase isoforms. *Chemistry & Industry*, *67*(5), 703–706. doi:10.2298/HEMIND120828113T

Tredici, M. R., & Materassi, R. (1992). From open ponds to vertical alveolar panels: The Italian experience in the development of reactors for the mass cultivation of phototrophic microorganisms. *Journal of Applied Phycology*, *4*(3), 221–231. doi:10.1007/BF02161208

Trindade, S. C. (2011). *Nanotech biofuels and fuel additives. Biofuel's Engineering Process Technology*. InTech.

Trivedi, J., Aila, M., Bangwal, D. P., Kaul, S., & Garg, M. O. (2015). Algae based biorefinery—How to make sense? *Renewable & Sustainable Energy Reviews*, *47*, 295–307.

Tsolcha, O.N., Tekerlekopoulou, A.G., Akratos, C.S., Aggelis, G., & Genitsaris, S., Moustaka-Gouni, M., & Vayenas, D.V. (2017). Biotreatment of raisin and winery wastewaters and simultaneous biodiesel production using a Leptolyngbya-based microbial consortium. *Journal of Cleaner Production*, *148*, 185–193.

Tsubota, Y., & Ichiryu, K. (2005). Development of riding simulator. *Proceedings of the 6th International Conference on Fluid Power Transmission and Control, ICFP 2005, 6*, 686–689.

Tsypkin, G. G. (2005). Effect of decomposition of a gas hydrate on the gas recovery from a reservoir containing hydrate and gas in the free state. *Fluid Dynamics*, *40*(1), 117–125. doi:10.100710697-005-0050-3

Tubecon. (2018). *Mechanical properties of steel for flat bars and sections*. Steel Properties, Tubecon Africa (PTY) Ltd. http://www.tubecon. co.za/en/technical-info/tubecon-wiki/properties-flat-steel-bars-and-sections. html

Tutt, M., Kikas, T., & Olt, J. (2012). Influence of different pretreatment methods on bioethanol production from wheat straw. *Agronomy Research, 10*(1), 269–276.

Tye, Y. Y., Lee, K. T., Wan Abdullah, W. N., & Leh, C. P. (2016). The world availability of non-wood lignocellulosic biomass for the production of cellulosic ethanol and potential pretreatments for the enhancement of enzymatic saccharification. *Renewable & Sustainable Energy Reviews*, *60*, 155–172. doi:10.1016/j.rser.2016.01.072

u, T., Guan, G., Abudula, A., Yoshida, A., Wang, D., & Song, Y. (2019). Gas recovery enhancement from methane hydrate reservoir in the Nankai Trough using vertical wells. *Energy*, *166*, 834–844.

Ubando, A. T., Del Rosario, A. J. R., Chen, W. H., & Culaba, A. B. (2021, January 15). A state-of-the-art review of biowaste biorefinery. *Environ Pollut.*, *269*, 116149. doi:10.1016/j.envpol.2020.116149 PMID:33280912

UNDESA. (2006). *Indicators of Sustainable Development* (3rd ed.). Guidelines, and Methodologies.

UNDP & WHO. (2009). *The energy access situation in developing countries*. Author.

UNEP. (2019). *United Nation Environment Programme, Emmision Gap Report*. UNEP.

United Nations Development Programme. (2020). Human development report. United Nations Development Programme.

UPES Report. (2016). *UPES, CSTEP, and PLR Chambers, Fuel Blending in India: Learning and Way Forward (Report, 2016)*. Author.

USDA GAIN Report. (2010). *Amit Aradhey, India-Biofuels Annual*. Global Agricultural Information Network Report Number, IN1058.

Uwe, Klaus, Wiegeman, & Herrera. (2010). *FAO Environmental and Natural Resources Service Series, No. 46*. FAO.

V, G. S., M, D. K., Pugazhendi, A., Bajhaiya, A. K., Gugulothu, P., & J, R. B. (2021). Biofuel production from Macroalgae: Present scenario and future scope. *Bioengineered*, *12*(2), 9216–9238.

Valdés, G., Mendonça, R. T., & Aggelis, G. (2020). Lignocellulosic biomass as a substrate for oleaginous microorganisms: A review. *Applied Sciences (Basel, Switzerland)*, *10*(21), 7698. doi:10.3390/app10217698

Vamvuka, D. (2011). Bio-oil, solid and gaseous biofuels from biomass pyrolysis processes—An overview. *International Journal of Energy Research*, *35*, 835–862. doi:10.1002/er

Van Der Voet, E., Lifset, R. J., & Luo, L. (2010). Life-cycle assessment of biofuels, convergence and divergence. *Biofuels*, *1*(3), 435–449. doi:10.4155/bfs.10.19

van der Weijde, T., Alvim Kamei, C. L., Torres, A. F., Vermerris, W., Dolstra, O., Visser, R. G. F., & Trindade, L. M. (2013). The potential of C4 grasses for cellulosic biofuel production. *Frontiers in Plant Science*, *4*(MAY), 1–18. doi:10.3389/fpls.2013.00107 PMID:23653628

van Lier, J. B. (1995). *Thermophilic Anaerobic Wastewater Treatment: Temperature Aspects and Process Stability*. Available online: https://edepot.wur.nl/200091

Van Lier, J. B., Mahmoud, N., & Zeeman, G. (2008). Anaerobic Wastewater Treatment. In *Biological Wastewater Treatment: Principles, Modeling and Design* (pp. 401–442). International Water Association.

Vaughan, A. (2021). Will the IPCC report matter? *New Scientist*, *251*(3348), 8–9. doi:10.1016/S0262-4079(21)01442-1 PMID:34456414

Vaz, B. D. S., Costa, J. A. V., & Morais, M. G. D. (2018). Innovative nanofiber technology to improve carbon dioxide biofixation in microalgae cultivation. *Bioresource Technology*, *273*, 592–598.

Veeramuthu, A., & Ngamcharussrivichai, C. (2020). Potential of Microalgal Biodiesel: Challenges and Applications. *Potential of Microalgal Biodiesel: Challenges and Applications*, 1–14.

Veličković, D., Milosavić, N., Bezbradica, D., Bihelović, F., Segal, A. M., Šegan, D., Trbojević, J., & Dimitijević, A. (2014). The specificity of α-glucosidase from Saccharomyces cerevisiae differs depending on the type of reaction: Hydrolysis versus transglucosylation. *Applied Microbiology and Biotechnology*, *98*(14), 6317–6328. doi:10.100700253-014-5587-9 PMID:24682477

Veljković, V. B., Lakićević, S. H., Stamenković, O. S., Todorović, Z. B., & Lazić, M. L. (2006). Biodiesel production from tobacco (*Nicotiana tabacum* L.) seed oil with a high content of free fatty acids. *Fuel*, *85*, 2671–2675.

Velmurugan, R., & Muthukumar, K. (2011). Utilization of sugarcane bagasse for bioethanol production: Sono-assisted acid hydrolysis approach. *Bioresource Technology*, *102*(14), 7119–7123. doi:10.1016/j.biortech.2011.04.045 PMID:21570831

Vendruscolo, F. (2014). Biohydrogen Production from Starch Residues. *International Journal of Bioengineering and Life Sciences, 8*(12), 1399–1406. Retrieved from https://waset.org/publications/10000017

Venkataraman, N. S. (2012). Have we given a quiet burial to jatropha biofuel projects? *Deccan Herald*. https://www.deccanherald.com/content/246514/have-we-given-quiet-burial.html

Verma, N., & Kumar, V. (2021). Microbial conversion of waste biomass into bioethanol: Current challenges and future prospects. *Biomass Conversion and Biorefinery*. Advance online publication. doi:10.100713399-021-01824-z

Verma, S. (2002). *Anaerobic Digestion of Biodegradable Organics in Municipal Solid Wastes*. Columbia University.

Verma, S. K., & Hussain, R. (n.d.). Comparative analysis of growth kinetics. *Concentration of Sugar and Alcohol Tolerance of Various Yeast Strains Isolated From Five Different Sources*, *2*(8), 7.

Vignesh, P., Kumar, A. R. P., Ganesh, N. S., Jayaseelan, V., & Sudhakar, K. (2021). Biodiesel and green diesel generation: An overview. *Oil & Gas Science and Technology – Rev. IFP Energies Nouvelles*, *76*, 1–15.

Vimali, E., Gunaseelan, S., Devi, V. C., Mothil, S., Arumugam, M., Ashokkumar, B., Innasi, M. G. M., Pugazhendhi, A., & Varalakshmi, P. (2022). Comparative study of different catalysts mediated FAME conversion from macroalga Padina tetrastromatica biomass and hydrothermal liquefaction facilitated bio-oil production. *Chemosphere*, *292*, 133485. doi:10.1016/j.chemosphere.2021.133485 PMID:34979211

Vivekanand, V., Eijsink, V. G. H., & Horn, S. J. (2012). Biogas production from the brown seaweed *Saccharina latissima*: Thermal pretreatment and co-digestion with wheat straw. *Journal of Applied Phycology*, *24*(5), 1295–1301. doi:10.100710811-011-9779-8

Vohra, M., Manwar, J., Manmode, R., Padgilwar, S., & Patil, S. (2014). Bioethanol production: Feedstock and current technologies. *Journal of Environmental Chemical Engineering*, *2*(1), 573–584. doi:10.1016/j.jece.2013.10.013

Voloshin, R. A., Rodionova, M. V., Zharmukhamedov, S. K., Nejat Veziroglu, T., & Allakhverdiev, S. I. (2016). Review: Biofuel production from plant and algal biomass. *International Journal of Hydrogen Energy*, *41*, 17257–17273.

Vučurović, V. M., & Razmovski, R. N. (2012). Sugar beet pulp as support for Saccharomyces cerivisiae immobilization in bioethanol production. *Industrial Crops and Products*, *39*(1), 128–134. doi:10.1016/j.indcrop.2012.02.002

Wahid, R., Hjorth, M., Kristensen, S., & Møller, H. B. (2015). Extrusion as pretreatment for boosting methane production: Effect of screw configurations. *Energy & Fuels*, *29*(7), 4030–4037. doi:10.1021/acs.energyfuels.5b00191

Wang, B., Dong, H., Fan, Z., Zhao, J., & Song, Y. (2017a). Gas Production from Methane Hydrate Deposits Induced by Depressurization in Conjunction with Thermal Stimulation. *Energy Procedia*, *105*, 4713–4717. doi:10.1016/j.egypro.2017.03.1022

Wang, B., Li, Y., Wu, N., & Lan, C. Q. (2008). CO2 bio-mitigation using microalgae. *Applied Microbiology and Biotechnology*, *79*(5), 707–718. doi:10.100700253-008-1518-y PMID:18483734

Wang, N., Zhang, C., Zhu, W., & Weng, Y. (2020). Improving interfacial adhesion of PLA/Lignin composites byone–step solvent-free modification method. *Journal of Renewable Materials*, *8*(9), 1139–1147. doi:10.32604/jrm.2020.09961

Wang, R., Wang, K., Zhou, M., Xu, J., & Jiang, J. (2021). Efficient fractionation of moso bamboo by synergistic hydrothermal-deep eutectic solvents pretreatment. *Bioresource Technology*, *328*, 124873. doi:10.1016/j.biortech.2021.124873 PMID:33639413

Wang, X. H., Sun, Y. F., Wang, Y. F., Li, N., Sun, C. Y., Chen, G. J., Liu, B., & Yang, L. Y. (2017b). Gas Production from Hydrates by CH4-CO2_H2 Replacement. *Applied Energy*, *188*, 305–314. doi:10.1016/j.apenergy.2016.12.021

Wang, Y., Li, X. S., Li, G., Huang, N. S., & Feng, J. C. (2014). Experimental study on the hydrate dissociation in porous media by five-spot thermal huff and puff method. *Fuel*, *117*, 688–696. doi:10.1016/j.fuel.2013.09.088

Wang, Y., Ou, S., Liu, P., Xue, F., & Tang, S. (2006). Comparison of two different processes to synthesize biodiesel by waste cooking oil. *Journal of Molecular Catalysis A Chemical*, *252*(1-2), 107–112. doi:10.1016/j.molcata.2006.02.047

Wang, Z., Keshwani, D. R., Redding, A. P., & Cheng, J. J. (2010). Sodium hydroxide pretreatment and enzymatic hydrolysis of coastal Bermuda grass. *Bioresource Technology*, *101*(10), 3583–3585. doi:10.1016/j.biortech.2009.12.097 PMID:20089396

Wargacki, A. J., Leonard, E., Win, M. N., Regitsky, D. D., Santos, C. N., Kim, P. B., Cooper, S. R., Raisner, R. M., Herman, A., Sivitz, A. B., Lakshmanaswamy, A., Kashiyama, Y., Baker, D., & Yoshikuni, Y. (2012). An engineered microbial platform for direct biofuel production from brown macroalgae. *Science*, *335*(6066), 308–313. doi:10.1126cience.1214547 PMID:22267807

Warren, T. J., Poulter, R., & Partin, R. (1995). Converting biomass to electricity on a farm-sized scale using downdraft gasification and a spark ignon engine. *Bioresource Technology*, *52*(1), 95–98. doi:10.1016/0960-8524(95)00022-7

Wasserscheid, P., & Keim, W. (2010). Ionic Liquids - New Solutions for Transition Metal Catalysis. *Angewandte Chemie International Edition*, *39*(21), 3772–3789. doi:10.1002/1521-3773(20001103)39:21<3772::AID-ANIE3772>3.0.CO;2-5 PMID:11091453

Watanabe, F. (2007). Vitamin B12 sources and bioavailability. *Experimental Biology and Medicine (Maywood, N.J.)*, *232*(10), 1266–1274. doi:10.3181/0703-MR-67 PMID:17959839

Wawa, A. I. (2012). *The challenges of promoting and adopting biogas technology in semi-arid areas of Tanzania: The Case of Kongwa and Bahi Districts in Dodoma Region* [PhD thesis]. Open University of Tanzania.

Wei, N., Quarterman, J., & Jin, Y. S. (2013). Marine macroalgae: An untapped resource for producing fuels and chemicals. *Trends in Biotechnology*, *31*(2), 70–77. doi:10.1016/j.tibtech.2012.10.009 PMID:23245657

Wei, S. (2016). The application of biotechnology on the enhancing of biogas production from lignocellulosic waste. *Applied Microbiology and Biotechnology*, *100*(23), 9821–9836. doi:10.100700253-016-7926-5 PMID:27761635

Wijeyekoon, S. L. J., & Vaidya, A. A. (2021). Woody biomass as a potential feedstock for fermentative gaseous biofuel production. *World Journal of Microbiology & Biotechnology*, *37*(8), 1–15. doi:10.100711274-021-03102-6 PMID:34258684

Wilcox, H. A. (1977). The U.S. Navy's Ocean food and energy farm project. In *Ocean energy resources; Proceedings of the Energy Technology Conference, Houston, Texas* (pp. 83-104). American Society of Mechanical Engineers.

Wilkins, M. R., Suryawati, L., Maness, N. O., & Chrz, D. (2007). Ethanol production by Saccharomyces cerevisiae and Kluyveromyces marxianus in the presence of orange-peel oil. *World Journal of Microbiology & Biotechnology, 23*(8), 1161–1168. doi:10.100711274-007-9346-2

Williams, T. E., Millheim, K., & Liddell, B. (2005). *Methane hydrate production from Alaskan permafrost.* Maurer Technology Inc.

Winandy, J. E. (1995). *Effects of fire retardant treatments after 18 months of exposure at 150 F (66 C)* (Vol. 264). US Department of Agriculture, Forest Service, Forest Products Laboratory.

Wingren, A., Galbe, M., & Zacchi, G. (2003). Techno-economic evaluation of producing ethanol from softwood: Comparison of SSF and SHF and identification of bottlenecks. *Biotechnology Progress, 19*(4), 1109–1117. doi:10.1021/bp0340180 PMID:12892470

Woiciechowski, A. L., Medeiros, A. B. P., Rodrigues, C., de Souza Vandenberghe, L. P., de Andrade Tanobe, V. O., Dall'Agnol, A., Gazzoni, D. L., & Soccol, C. R. (2016). Feedstocks for biofuels. *Green Energy and Technology, 0*, 15–39. doi:10.1007/978-3-319-30205-8_2

World Bank. (2015). *Promoting Biogas as Sustainable Clean Cooking Fuel for Rural Households in Kenya Project.* http://documents.worldbank.org/curated/en/ 350771468272038585/pdf/SFG1441-EA-P153493-ESMF-Box393229B-PUBLIC-Disclosed-10-20-2015.pdf

Wortmann, C. S., Liska, A. J., Ferguson, R. B., Lyon, D. J., Klein, R. N., & Dweikat, I. (2010). Dryland performance of sweet sorghum and grain crops for biofuel in nebraska. *Agronomy Journal, 102*(1), 319–326. doi:10.2134/agronj2009.0271

Wu, B. (2013). Advances in the use of CFD to characterise, design and optimise bioenergy systems. *Computers and Electronics in Agriculture, 93*, 195–208. doi:10.1016/j.compag.2012.05.008

Wu, M. C., Sun, K. W., & Zhang, Y. (2006). Influence of temperature fluctuation on thermophilic anaerobic digestion of municipal organic solid waste. *Journal of Zhejiang University. Science, 7*(3), 180–185. doi:10.1631/jzus.2006.B0180 PMID:16502503

Wu, Q., Wang, Y., & Zhan, J. (2012). Effect of rapidly depressurizing and rising temperature on methane hydrate dissociation. *Journal of Natural Gas Chemistry, 21*(1), 91–97. doi:10.1016/S1003-9953(11)60338-1

Wu, X., Yao, W., & Zhu, J. (2010). Effect of pH on continuous biohydrogen production from liquid swine manure with glucose supplement using an anaerobic sequencing batch reactor. *International Journal of Hydrogen Energy, 35*(13), 6592–6599. doi:10.1016/j.ijhydene.2010.03.097

Wu, Y. N., Mattsson, M., Ding, M. W., Wu, M. T., Mei, J., & Shen, Y. L. (2019). Effects of different pretreatments on improving biogas production of macroalgae *Fucus vesiculosus* and *Fucus serratus* in Baltic Sea. *Energy & Fuels, 33*(3), 2278–2284. doi:10.1021/acs.energyfuels.8b04224

Wu, Y., Kovalovszki, A., Pan, J., Lin, C., Liu, H., Duan, N., & Angelidaki, I. (2019). Early warning indicators for mesophilic anaerobic digestion of corn stalk: A combined experimental and simulation approach. *Biotechnology for Biofuels, 12*(1), 1–15. doi:10.118613068-019-1442-7 PMID:31073330

Xiang, J. (2014). Research on Teaching Methods for Communication Engineering Students in Colleges. *Open Journal of Social Sciences, 2*(5), 9–12. doi:10.4236/jss.2014.25003

Xiu, S., & Shahbazi, A. (2012). Bio-oil production and upgrading research: A review. *Renewable & Sustainable Energy Reviews, 16*(7), 4406–4414. doi:10.1016/j.rser.2012.04.028

Xu, C. G., Cai, J., Lin, F. H., Chen, Z. Y., & Li, X. S. (2015). Raman analysis on methane production from natural gas hydrate by carbon dioxide–methane replacement. *Energy*, *79*, 111–116. doi:10.1016/j.energy.2014.10.068

Xu, C. G., Cai, J., Yu, Y. S., Yan, K. F., & Li, X. S. (2018). Effect of pressure on methane recovery from natural gas hydrates by methane-carbon dioxide replacement. *Applied Energy*, *217*, 527–536. doi:10.1016/j.apenergy.2018.02.109

Xue, C., & Cheng, C. (2019). Butanol production by Clostridium. In Advances in Bioenergy (1st ed., Vol. 4). doi:10.1016/bs.aibe.2018.12.001

Xu, F., Shi, Y. C., & Wang, D. (2013). X-ray scattering studies of lignocellulosic biomass: A review. *Carbohydrate Polymers*, *94*(2), 904–917. doi:10.1016/j.carbpol.2013.02.008 PMID:23544649

Xu, J., & Cheng, J. J. (2011). Pretreatment of switchgrass for sugar production with the combination of sodium hydroxide and lime. *Bioresource Technology*, *102*(4), 3861–3868. doi:10.1016/j.biortech.2010.12.038 PMID:21194931

Xu, X., Kim, J. Y., Oh, Y. R., & Park, J. M. (2014). Production of biodiesel from carbon sources of macroalgae, *Laminaria japonica*. *Bioresource Technology*, *169*, 455–461. PMID:25084043

Yacobucci, B. D., & Bracmort, K. (2010). *Calculation of lifecycle greenhouse gas emissions for the Renewable Fuel Standard (RFS). R40460*. Congressional Research Service.

Yahmed, N. B., Jmel, M. A., Alaya, M. B., Bouallagui, H., Marzouki, M. N., & Smaali, I. (2016). A biorefinery concept using the green macroalgae *Chaetomorpha linum* for the coproduction of bioethanol and biogas. *Energy Conversion and Management*, *119*, 257–265. doi:10.1016/j.enconman.2016.04.046

Yan, G., & Skutsch, M. (2011). *A global analysis of deforestation due to biofuel development*. Working Paper 68. CIFOR.

Yanagisawa, M., Nakamura, K., Ariga, O., & Nakasaki, K. (2011). Production of high concentrations of bioethanol from seaweeds that contain easily hydrolyzable polysaccharides. *Process Biochemistry*, *46*(11), 2111–2116. doi:10.1016/j.procbio.2011.08.001

Yan, F., Luo, S., Hu, Z., Xiao, B., & Cheng, G. (2010). Hydrogen-rich gas production by steam gasification of char from biomass fast pyrolysis in a fixed-bed reactor: Influence of temperature and steam on hydrogen yield and syngas composition. *Bioresource Technology*, *101*(14), 5633–5637. doi:10.1016/j.biortech.2010.02.025 PMID:20194019

Yang, J., Okwananke, A., Tohidi, B., Chuvilin, E., Maerle, K., Istomin, V., Bukhanov, B., & Cheremisin, A. (2017). Flue gas injection into gas hydrate reservoirs for methane recovery and carbon dioxide sequestration. *Energy Conversion and Management*, *136*, 431–438. doi:10.1016/j.enconman.2017.01.043

Yang, L., Xu, F., Ge, X., & Li, Y. (2015). Challenges and strategies for solid-state anaerobic digestion of lignocellulosic biomass. *Renewable & Sustainable Energy Reviews*, *44*, 824–834. doi:10.1016/j.rser.2015.01.002

Yang, X., Sun, C. Y., Yuan, Q., Ma, P. C., & Chen, G. J. (2010). Experimental study on gas production from methane hydrate-bearing sand by hot-water cyclic injection. *Energy & Fuels*, *24*(11), 5912–5920. doi:10.1021/ef100367a

Yanik, J., Stahl, R., Troeger, N., & Sinag, A. (2013). Pyrolysis of algal biomass. *Journal of Analytical and Applied Pyrolysis*, *103*, 134–141. doi:10.1016/j.jaap.2012.08.016

Yen, H. W., & Brune, D. E. (2007). Anaerobic co-digestion of algal sludge and waste paper to produce methane. *Bioresource Technology*, *98*(1), 130–134. doi:10.1016/j.biortech.2005.11.010 PMID:16386894

Yen, H.-W., Hu, I.-C., Chen, C.-Y., Ho, S.-H., Lee, D.-J., & Chang, J.-S. (2013). Microalgaebased biorefinery – From biofuels to natural products. *Bioresource Technology*, *135*, 166–174. PMID:23206809

Yin, Y., Hu, J., & Wang, J. (2019). Fermentative hydrogen production from macroalgae *Laminaria japonica* pretreated by microwave irradiation. *International Journal of Hydrogen Energy, 44*(21), 10398–10406. doi:10.1016/j.ijhydene.2019.03.034

Yin, Y., & Wang, J. (2019). Enhanced biohydrogen production from macroalgae by zero-valent iron nanoparticles: Insights into microbial and metabolites distribution. *Bioresource Technology, 282*, 110–117. doi:10.1016/j.biortech.2019.02.128 PMID:30852330

Yoon, J.-H., Kawamura, T., Yamamoto, Y., & Komai, T. (2004). Transformation of Methane Hydrate to Carbon Dioxide Hydrate: In Situ Raman Spectroscopic Observations. *The Journal of Physical Chemistry A, 108*(23), 5057–5059. doi:10.1021/jp0496831

Yoon, M., Choi, J. I., Lee, J. W., & Park, D. H. (2012). Improvement of saccharification process for bioethanol production from *Undaria* sp. by gamma irradiation. *Radiation Physics and Chemistry, 81*(8), 999–1002. doi:10.1016/j.radphyschem.2011.11.035

Yuan, Q., Sun, C. Y., Liu, B., Wang, X., Ma, Z. W., Ma, Q. L., Yang, L. Y., Chen, G. J., Li, Q. P., Li, S., & Zhang, K. (2013). Methane recovery from natural gas hydrate in porous sediment using pressurized liquid CO2. *Energy Conversion and Management, 67*, 257–264. https://doi.org/10.1016/j.enconman.2012.11.018

Yu, J., XuZhang, & Tan, T. (2008). Ethanol production by solid state fermentation of sweet sorghum using thermotolerant yeast strain. *Fuel Processing Technology, 89*(11), 1056–1059. doi:10.1016/j.fuproc.2008.04.008

Yu, J., Zhang, J., He, J., Liu, Z., & Yu, Z. (2009). Combinations of mild physical or chemical pretreatment with biological pretreatment for enzymatic hydrolysis of rice hull. *Bioresource Technology, 100*(2), 903–908. doi:10.1016/j.biortech.2008.07.025 PMID:18760920

Zabaniotou, A., Ioannidou, O., & Skoulou, V. (2008). Rapeseed residues utilization for energy and 2nd generation biofuels. *Fuel, 87*(8-9), 1492–1502. doi:10.1016/j.fuel.2007.09.003

Zabed, H., Sahu, J. N., Boyce, A. N., & Faruq, G. (2016). Fuel ethanol production from lignocellulosic biomass: An overview on feedstocks and technological approaches. *Renewable & Sustainable Energy Reviews, 66*, 751–774. doi:10.1016/j.rser.2016.08.038

Zah, R., & Gauch, M. (2007). *Life cycle assessment of energy products : Environmental impact assessment of biofuels.* Academic Press.

Zahran, H. H. (1999). Rhizobium -Legume Symbiosis and Nitrogen Fixation under Severe Conditions and in an Arid Climate. *Microbiology and Molecular Biology Reviews, 63*(4), 968–989. doi:10.1128/MMBR.63.4.968-989.1999 PMID:10585971

Zalm, J. E. (2017). *Performance of domestic biodigesters in Kenya towards the development of a comparative method for measuring the performance of different biodigester types in Sub-Saharan Africa* [MSc dissertation]. Utrecht University, Utrecht, Netherlands.

Zanuso, E., Ruiz, H. A., Domingues, L., & Teixeira, J. A. (2022). Magnetic nanoparticles as support for cellulase immobilization strategy for enzymatic hydrolysis using hydrothermally pretreated corn cob biomass. *BioEnergy Research.* Advance online publication. doi:10.100712155-021-10384-z

Zhang, H.-Q., Brill, T., & Sarica, C. (2009). *Extracting gas hydrates from marine sediments.* US Pat No 7,546,880. doi:10.1038/incomms1464

Zhang, J., & Zhang, X. (2019). The thermochemical conversion of biomass into biofuels. Biomass, Biopolymer-Based Materials, and Bioenergy: Construction, Biomedical, and other Industrial Applications. doi:10.1016/B978-0-08-102426-3.00015-1

Zhang, J., & Jiang, L. (2008). Acid-catalyzed esterification of *Zanthoxylum bungeanum* seed oil with high free fatty acids for biodiesel production. *Bioresource Technology*, *99*, 8995–8998. PMID:18562195

Zhang, L., Yang, L., Wang, J., Zhao, J., Dong, H., Yang, M., Liu, Y., & Song, Y. (2017). Enhanced CH4 recovery and CO2 storage via thermal stimulation in the CH4/CO2 replacement of methane hydrate. *Chemical Engineering Journal*, *308*, 40–49. https://doi.org/10.1016/j.cej.2016.09.047

Zhang, Q., Zhang, P., Pei, Z. J., & Wang, D. (2013). Relationships between cellulosic biomass particle size and enzymatic hydrolysis sugar yield: Analysis of inconsistent reports in the literature. *Renewable Energy*, *60*, 127–136. doi:10.1016/j.renene.2013.04.012

Zhang, Y., Zhao, Y., Gao, X., Li, B., & Huang, J. (2015). Energy and exergy analyses of syngas produced from rice husk gasification in an entrained flow reactor. *Journal of Cleaner Production*, *95*, 273–280. doi:10.1016/j.jclepro.2015.02.053

Zhao, J., Fan, Z., Wang, B., Dong, H., Liu, Y., & Song, Y. (2016). Simulation of microwave stimulation for the production of gas from methane hydrate sediment. *Applied Energy*, *168*, 25–37. https://doi.org/10.1016/j.apenergy.2016.01.091

Zhao, X., Peng, F., Cheng, K., & Liu, D. (2009). Enhancement of the enzymatic digestibility of sugarcane bagasse by alkali-peracetic acid pretreatment. *Enzyme and Microbial Technology*, *44*(1), 17–23. doi:10.1016/j.enzmictec.2008.07.011

Zhao, X., Zhang, L., & Liu, D. (2008). Comparative study on chemical pretreatment methods for improving enzymatic digestibility of crofton weed stem. *Bioresource Technology*, *99*(9), 3729–3736. doi:10.1016/j.biortech.2007.07.016 PMID:17709243

Zhao, X., Zhang, L., & Liu, D. (2012). Biomass recalcitrance. Part I: The chemical compositions and physical structures affecting the enzymatic hydrolysis of lignocellulose. *Biofuels, Bioproducts & Biorefining*, *6*(3), 246–256. doi:10.1002/bbb.1331

Zheng, Y., Zhao, J., Xu, F., & Li, Y. (2014). Pretreatment of lignocellulosic biomass for enhanced biogas production. *Progress in Energy and Combustion Science*, *42*, 35–53. doi:10.1016/j.pecs.2014.01.001

Zhou, A., & Thomson, E. (2009). The development of biofuels in Asia. *Applied Energy*, *86*, S11–S20.

Zhou, D., Zhang, L., Zhang, S., Fu, H., & Chen, J. (2010). Hydrothermal liquefaction of macroalgae *Enteromorpha prolifera* to bio-oil. *Energy & Fuels*, *24*(7), 4054–4061. doi:10.1021/ef100151h

Zhou, Y., Castaldi, M. J., & Yegulalp, T. M. (2009). Experimental Investigation of Methane Gas Production from Methane Hydrate. *Industrial & Engineering Chemistry Research*, *48*, 3142–3149. https://doi.org/10.1021/ie801004z

Zhu, Y., Lee, Y. Y., & Elander, R. T. (2005). Optimization of dilute-acid pretreatment of corn stover using a high-solids percolation reactor. *Applied Biochemistry and Biotechnology*, *124*(1), 1045–1054. doi:10.1385/ABAB:124:1-3:1045 PMID:15930580

Zimmerman, W. B., Hewakandamby, B. N., Tesař, V., Bandulasena, H. H., & Omotowa, O. A. (2009). On the design and simulation of an airlift loop bioreactor with microbubble generation by fluidic oscillation. *Food and Bioproducts Processing*, *87*(3), 215–227.

Zimmerman, W. B., Tesař, V., & Bandulasena, H. H. (2011). Towards energy efficient nanobubble generation with fluidic oscillation. *Current Opinion in Colloid & Interface Science*, *16*(4), 350–356.

Ziolkowska, J. R., & Simon, L. (2014). Recent developments and prospects for algae-based fuels in the US. *Renewable & Sustainable Energy Reviews*, *29*, 847–853.

Zoghlami, A., & Paës, G. (2019). Lignocellulosic biomass: Understanding recalcitrance and predicting H hydrolysis. *Frontiers in Chemistry*, *7*, 874. Advance online publication. doi:10.3389/fchem.2019.00874 PMID:31921787

Zulqarnain, A., Ayoub, M., Yusoff, M. H. M., Nazir, M. H., Zahid, I., Ameen, M., Sher, F., Floresyona, D., & Budi Nursanto, E. (2021). A comprehensive review on oil extraction and biodiesel production technologies. *Sustainability*, *13*(788), 1–28. doi:10.3390u13020788

Žuža, M., Milosavić, N., & Knežević-Jugović, Z. (2009). Immobilization of modified penicillin G acylase on Sepabeads carriers. *Chemical Papers, 63*(2), 117–124. doi:10.2478/s11696-009-0012-z

About the Contributors

Ashok Kumar Rathoure holds doctorate degree in Env. Biotechnology from Central University (HNBGU Srinagar UK India) and M.Tech. Biotechnology. Currently, he is serving as CEO, Akone Services. He served as Managing Director Biohm Consultare Pvt. Ltd. Surat India. Previously, he was associated with Eco Group of Companies Surat Gujarat; Vardan Environet Gurgaon; En-vision Group Surat (En-vision Environmental Services and En-vision Enviro Engineers Pvt. Ltd.) for EIA studies; Himachal Institute of Life Sciences Paonta and Beehive College of Ad. Studies Dehradun for teaching to Biotechnology, Microbiology, Biochemistry and other biosciences subjects. His area of research is environmental biotechnology and publication includes 100 full length research papers in international and national journals of repute, more than 50 books from reputed publishers in India & abroad and more than 25 Book chapters. He is also associated as Editor-in-Chief for Octa Journal of Environmental Research, Managing Editor for Octa Journal of Biosciences and Executive Editor for Scientific India Magazine. He also holds PG Diploma in Human Resource Management.

Shankar Mukundrao Khade earned his Ph.D. in Biochemical Engineering from IIT, Varanasi. He is having overall 4 years of experience in industry and academia post-PhD. He is currently working as Assistant Professor and Program Lead in Biotechnology at School of Engineering, Ajeenkya D Y Patil University, Pune, India. His research area involves Enzymology, Bioprocess technology, Bioenergy etc. He is also an Editorial Board Member of the journal American Journal of Chemical and Biochemical Engineering and Reviewer of Preparative Biochemistry and Biotechnology and Signa Vitae Journals.

* * *

Vaishali Arora is working as Ph.D. Scholar, Under guidance of Dr. Babita Khosla, in Environmental Science department, Maharshi Dayanand University, Rohtak, Haryana. She is keen interested in soil microbiology, agricultural implication management. She has contributed many book chapters in reputed books.

Sompalli Bhavna is pursuing PhD at CHRIST University Bangalore. Her area of Research is Industrial Biotechnology where she is working on "Production and purification of bioactive secondary metabolites from Mangrove sediment Microbiota and exploration of their biomedical applications". The research mainly focuses on screening of fibrinolytic enzymes and antimicrobial peptides producing Microbiota from mangrove soils and evaluating their effect invitro. Her research interest lies in the area of Industrial Biotechnology.

Debraj Biswal is currently employed as Assistant Professor of Zoology in a Government college in West Bengal, India. His teaching career spans for more than 13 years. His research activities for over 14 years mainly include areas in parasitology, helminthology, vector biology and microbiology. He has published a number of articles related to these topics. Additionally, he has keen interests in ecology and environmental biology with a few publications in these fields too.

Priyanka Bumbra is working as Ph.D. Scholar, Under guidance of Dr. Babita Khosla, in Environmental Science department, Maharshi Dayanand University, Rohtak, Haryana. She is keen interested in soil microbiology, agricultural implication management. She has contributed many research paper and book chapters in reputed books and journals.

Shital Giri is specialized in Food Technology and Biotechnology domain with award of Ph.D. in Food Engineering and Technology from Institute of Chemical Technology, Matunga, Mumbai. She received her M.Tech in Food Technology from LIT College, Nagpur University and B.Tech in Food Sciences from the CFT, Parbhani University. She joined School of Engineering, ADYPU in September 2020.

Ajaygiri K. Goswami has completed Masters and Doctorate in Chemical engineering and technology and presently working as Head Pharmaceutical Technology, KBC North Maharashtra University, Jalgaon. He has presented more than fifty research articles in International Conferences and also published research papers more than hundred in National and international reputed Journals. He is working in various academic bodies of the University like BOS,BOG and Coordinators for IIC, Startup, KCIL cell, etc. He has more than 25 years Teaching and research experience in Engineering College and University Departments and is research area are Pharmaceutical technology, Chemical engineering, Oil technology, Bio fuel, Nano catalyst, etc.

Kiran Grewal is a M.Sc. Environmental Science student in Environmental Science department, Maharshi Dayanand University, Rohtak, Haryana. She is keen interested in water pollution, its management.

Namdev Hadapad is very meticulous in his work and reliable in his duties. He has been inquisitive learner and strives diligently to achieve his goals. He can study efficiently even under any situations

Bela Khiratkar is currently pursuing graduation in Biotechnology Engineering from Ajeenkya D.Y. Patil University. With main fields of interest being microbiology and bioenergy. Bela hopes to pursue a career in microbiology particularly focusing on practical laboratory work.

Babita Khosla is working as an assistant professor in Environmental Science Department, Maharshi Dayanand University, Rohtak, Haryana, India since 2009. She is specialised in the field of Microbiology and bioremediation. Beside undertaking various responsibilities of environmental science dept. and university, she has completed various research projects. Now she's engaged as Principal Investigator in major research project on water quality assessment funded from Board of Research in Nuclear Sciences, Department of Atomic Energy, Mumbai.

Sakshi Kolpe currently lives in Pune, pursuing B.Tech in Food biotechnology at Ajeenkya D.Y. Patil University. Her interest in biofuels began while studying Biochemical engineering in the second year of her B.Tech program. This intrigued her to further understand the topic and initiate research in developing alternative fuels from bioresources for the future. Furthermore, she developed an interest in studying the process and decided to enroll in an internship program on fermentation technology.

Koushik Koujalagi is a Doctoral student at Christ University, working on formulation of probiotics for cardiovascular and liver diseases. He has three year of industry experience before he started his PhD work. His area of interest lies in Applied Microbiology and Fermentation.

Kondapalli Vamsi Krishna is a doctoral student working on the research topic - "Process development for fermentative production of cordycepin". He is a Master in Biotechnology with more than 8 years of research experience in Industry and academia. His research interest lies in the area of Industrial Biotechnology, Fermentation and Production of Bioactive molecules for food and feed applications.

Sunil Jayant Kulkarni has completed his PhD from Sant Gadge baba Amravati University and Masters in Chemical Engineering from Tatyasaheb Kore Institute of Engineering and Technology, Warananagar, Maharashtra, India. He is working as an Associate Professor in Chemical Engineering Department of Gharda Institute of Technology, Chiplun, India. He has 20 years of teaching and research experience. Dr. Sunil is involved in different academic activities across the universities in Maharashtra. He is the Board of Studies member of Chemical Engineering in Mumbai University since September 2019. His area of investigation in doctoral studies was phenol and heavy metal removal from wastewater. His area of research includes adsorption, clean technology and environmental biotechnology. He has also worked on kinetics and isotherms of various sorption operations. He is reviewer board member of many international journals and reviewed many international papers for Springer and Elsevier journals. He has published many international review and research papers in journals and presented 19 research papers in international conferences. He has published 11 books on various topics on chemical, environmental and biochemical engineering. He is also governing body member, promotional editor, and executive member of many reputed international journals.

Krishan Kumar has completed PhD in biotechnology from IIT Guwahati. He has Worked in cloning, expression, purification and biochemical characterization of new glycoside hydrolase enzymes, which have several biotechnological applications. Also learnt In silico homology modeling, structural characterization and molecular dynamics studies of proteins. Synthesized laminari-oligosaccharides by enzymatic hydrolysis of curdlan for prebiotic applications. He has also worked as senior project associate under project entitled "Design of a layered 3-D composite scaffold for articular cartilage tissue engineering" in IIT Kanpur. He has published 14 research publication under reputed journal like Carbohydrate Research, Bioresource Technology, Journal of Biomolecular Structure and Dynamics, International Journal of Biological Macromolecules etc. Currently, He is working as an assistant professor in Parul University, Vadodara.

Prashant Kumar is currently an Assistant Professor at the Department of Bioinformatics Kalinga University, Raipur, Chhattisgarh, India. He received his Bachelor of Dental Surgery (B.D.S.) from the M.J.P Rohilkhand University, Bareilly, India. Prashant Kumar received his M.Tech. in Bioinformatics from the Indian Institute of Information Technology, Allahabad, India. The Motilal Nehru National Institute of Technology in Allahabad, India, is where he received his PhD in bioinformatics. His research interests include computational drug design, molecular dynamic simulations, structural bioinformatics, biotechnology, and dentistry.

Anupama Kumari is a PhD Scholar at Indian Institute of Technology, Roorkee and currently working in the field of Gas Hydrate. Currently, she is working on the project entitled "Development of Thermo Catalytic Technique for Gas Production from Hydrate Bearing Sediments" since December 2017. This project is funded by the "Gas Hydrate Research and Technology Center" and "Oil and Natural Gas Corporation Limited". In this project, her research work is to perform the experiments on the formation and dissociation of gas hydrates in porous sediments. Her Ph.D thesis topic is "Studies on the Formation and Dissociation of Gas Hydrates in Porous Sediments''. She published four research papers in reputed journals as first author and another two research papers as second author. She has attended more than five international conferences during her Ph.D. She also worked on the simulation and modeling of gas hydrates formation and dissociation. She performed the thermodynamic modeling of gas hydrate formation to predict the phase equilibrium conditions of gas hydrates. For the thermodynamic modeling; She used the Fugacity model, Artificial Neural Network and Least Square Vector Machine through MATLAB. She also performed the characterization experiments of the gas hydrate bearing sediments of the 'National Gas Hydrate Program' received from GHRTC. For the characterization, she observed the TDS, pH, water content, BET, FESEM, ICPMS, FTIR, XRD and Raman Spectroscopy etc. The area of research during her M.Tech was waste water treatment in which she worked on the phytoremediation of phenol and cyanide through Water Hyacinth and Maize plants.

Jitender Singh Laura is working as a professor in Environmental Science Department, Maharshi Dayanand University, Rohtak, Haryana, India since 1992. He is specialised in the field of Environment monitoring and management. Beside undertaking various responsibilities of environmental science dept. and university, he has completed various research projects UGC, BRNS ETC. He has published more than 80 research articles, chapters in national and international journal of repute.

Mukund Madhaw is a self-employed person who wishes to pursue further. He is currently involved in the research work based on the gas hydrates. He studied the equilibrium conditions of gas hydrates in pure water, sea water and in the presence or absence of different inhibitors and promoters. He is currently working on the development of different thermodynamic model for the prediction of phase equilibrium data of gas hydrates. Currently, he developed the thermodynamic model for the gas hydrates using the Artificial Neural Network and Least Square Vector Machine through MATLAB. He submitted the book chapter based on Artificial Neural Network for the publication in IGI-Global and a research paper based on Least Square Vector Machine. He also compared the results obtained through these developed thermodynamic models with the results obtained by the available correlations for the prediction of phase equilibrium data of gas hydrates.

C B Majumder is a professor in the department of Chemical Engineering, Indian Instituteof Technology, Roorkee, Uttarakhand, India. His area of expertise are gas hydrates, waste water treatment, biofuels, soil remediation etc. He has published for than 300 research papers in various reputed international journals. He has been awarded many awards for his scientific work. He has given many planetary talks at various international conferences.

Alok Malaviya is an Associate Professor at the Department of Life Sciences, CHRIST (Deemed to be University), Bangalore, India. He received his PhD degree in Biochemical Engg and Biotechnology from Indian Institute of Technology Delhi, India. Later, he worked in various organisations such as Reliance Life Sciences - India, Korea Advanced Institute of Science and Technology - South Korea, International Centre for Genetic Engineering and Biotechnology - India, DuPont Knowledge Centre - India before joining CHRIST University. He has also consulted the Venture firm InvAscent as well as Slovenia based Biotechnology company Acies Bio for their R&D. He is also serving as associate editor for Springer Journal Applied Biochemistry and Biotechnology. He was awarded with the prestigious Marie Curie Award in 2012. He has published several papers in International Journals of repute and is also having a patent to his credit which is already licensed out to a biotech company. His research interests lies in Industrial Biotechnology and Fermentation Process Development, especially in the areas of Food and Nutrition, Probiotics, Industrial Enzymes and Biofuel production.

Shanthala Mallikarjunaiah obtained her Bachelor of Science and Master of Science from Kuvempu University and received a Doctor of philosophy in the field of Environmental Biology. She has 15 years of teaching experience working in various institutions and presently working as an Assistant professor in Bangalore University at Centre for Applied Genetics.

R. Nagarajan received his B.E. in Electrical and Electronics Engineering from Madurai Kamarajar University, Madurai, India, in 1997. He received his M.E. in Power Electronics and Drives from Anna University, Chennai, India, in 2008. He received his Ph.D in Electrical Engineering from Anna University, Chennai, India, in 2014. He has worked in the industry as an Electrical Engineer. He is currently working as Professor of Electrical and Electronics Engineering at Gnanamani College of Technology, Namakkal, Tamilnadu, India. He has published more than 90 papers in International Journals and Conferences and he has published more than 35 books. His research interest includes Power Electronics, Power System, Communication Engineering, Network Security, Soft Computing Techniques, Cloud Computing, Big Data Analysis and Renewable Energy Sources.

Marjorie Nyasani is a Graduate Student at Machakos University and Quality Assurance Manager at Somaliland Quality Control Commission.

Mahesh Pattabhiramaiah obtained his Bachelor, Master, and Doctoral degrees from Bangalore University. Having more than 15 years of PG teaching and Research experience at the Centre for Applied Genetics, Bangalore University, Bangalore, and at the PG Dept. Biotechnology and Biochemistry, SRN Adarsh College, affiliated College of Bangalore University, Bangalore, India. Guided more than 20 PG students. Published 50 scientific publications in peer-reviewed International Journals. Reviewer of various International Journals. Attended several International workshops and conferences in Berlin, Germany. BOS member and question paper setter for PG Biotechnology, Microbiology, and Biochemistry at CMR Institute. Awarded Postdoctoral fellowship (Guest Scientist) at the Institute for Advanced Studies, Germany. Worked as an International Visiting Research scholar at the University of Bremen, Bremen, Germany. During the Doctoral research investigation, work relating to the research problem has been carried out at Germany's prestigious research Institute, Max-Plank research institute for Evolutionary Biology, Ploen, Germany.

Hitesha Panchal has completed BSc and MSc in Biotechnology. Then, she has joined PhD program in biotechnology under the supervision of Dr. Krishan Kumar in Parul Institute of Applied Sciences, Parul University, Vadodara. Here, she is working on mesophilic cellulase enzyme required for hydrolysis of cellulose polysaccharides from lignocellulosic biomass. So that the hydrolysed cellulose can be used for bioethanol production. In her PhD work, she is working on cloning, expression, and biochemical characterization of cellulase enzymes along with structure characterization by in silico approach.

Dipanwita Sarkar (Paria) has been engaged in diverse activities in the fields of academic, training, extension and research activities towards protection of environment, uplift of rural economy including lab to land technology transfer programme. She is currently engaged as Assistant Professor of Zoology in a Government college in West Bengal and has been teaching for over twelve years. She has a vast experience of work on wetlands during her doctoral and post-doctoral research. She has already published a number of research papers on wetlands and its various aspects. She has also working experience with carbon sequestration and global warming effects on aquaculture and biodiversity. Her research career is more than 20 years long and includes toxicity effects of insecticides, heavy metals on nutrient cycling bacteria, microbial responses to excess nutrient contamination in fish ponds, long term residual effects of chemical and organic fertilizers and other similar topics.

Victor Shikuku holds a Ph.D. degree in Chemistry from Maseno University (Kenya) and postdoctoral research training at the University of Trier (Germany). He has authored over 25 peer-reviewed articles in referred journals and several book chapters. He has a keen interest in Physical, Materials, and Environmental Chemistry.

Kannadhasan Suriyan is working as an Assistant Professor in the department of Electronics and Communication Engineering in Cheran College of Engineering, karur, Tamilnadu, India. He is currently doing research in the field of Smart Antenna for Anna University. He is ten years of teaching and research experience. He obtained his B.E in ECE from Sethu Institute of Technology, Kariapatti in 2009 and M.E in Communication Systems from Velammal College of Engineering and Technology, Madurai in 2013. He obtained his M.B.A in Human Resources Management from Tamilnadu Open University, Chennai. He obtained his PGVLSI in Post Graduate diploma in VLSI design from Annamalai University, Chidambaram in 2011 and PGDCA in Post Graduate diploma in Computer Applications from Tamil University in 2014. He obtained his PGDRD in Post Graduate diploma in Rural Development from Indira Gandhi National Open University in 2016. He has published around 18 papers in the reputed indexed international journals and more than 125 papers presented/published in national, international journal and conferences. Besides he has contributed a book chapter also. He also serves as a board member, reviewer, speaker, session chair, advisory and technical committee of various colleges and conferences. He is also to attend the various workshop, seminar, conferences, faculty development programme, STTP and Online courses. His areas of interest are Smart Antennas, Digital Signal Processing, Wireless Communication, Wireless Networks, Embedded System, Network Security, Optical Communication, Microwave Antennas, Electromagnetic Compatability and Interference, Wireless Sensor Networks, Digital Image Processing, Satellite Communication, Cognitive Radio Design and Soft Computing techniques. He is Member of IEEE, ISTE, IEI, IETE, CSI, IAENG, SEEE, IEAE, INSC, IARDO, ISRPM, IACSIT, ICSES, SPG, SDIWC, IJSPR, and EAI Community.

Naresh Tanwer is working as Ph.D. Scholar, Under guidance of Dr. Babita Khosla, in Environmental Science department, Maharshi Dayanand University, Rohtak, Haryana. He is keen interested in environmental pollution and its management. He has done two projects, one is on waste water treatment and another one is on analysis of ambient air in Faridabad city, Haryana. He is also worked as Junior Research Fellow in major research project, entitled, "spatial distribution of Uranium and associated water quality parameters in seven districts of Haryana" funded from Board of Research in Nuclear sciences, Department of Atomic Energy, Mumbai. He has published many research articles in reputed journals.

Jovana Trbojević-Ivić received her Ph.D. in Biochemistry from the University of Belgrade - Faculty of Chemistry in 2019. From 2011 she was hired at Innovative Center of Faculty of Chemistry, Belgrade, Ltd., first as Junior Researcher Assistant, then from 2015 as Research Associate Assistant and from 2020 as Research Associate. From 2022 she is engaged as one of the lecturers on Department of Biochemistry at the University of Belgrade – Faculty of Chemistry on the subject Writing and Publishing Scientific Papers 1 for the PhD students. The main areas of her expertise and research focus are applied enzymology and enzyme tailoring by immobilization for an efficient practical use in synthetic reactions, biosensor development and development of clinical enzyme assays. Thus far she has published her results in several high-quality peer-reviewed journals: Food Chemistry, Jornal of the Science of Food and Agriculture, RSC Advances and Enzyme and Microbial Technology. She is also the first author of one book chapter, regarding immobilization and stabilization of microbial lipases.

Ghayas Usmani has completed Masters and Doctorate in Oil technology from HBTI, Kanpur and presently working as Head Oil Technology, UICT, KBC North Maharashtra University, Jalgaon. He has presented d published more than hundred research articles in International Conferences reputed Journals. He is working in various academic bodies of the University like BOS, BOG and Coordinators for KCIL cell, various Academic bodies etc. He has more than 32 years Teaching and research experience in Engineering College and University Departments and is research area are Chemical engineering and technology, Oil technology, Bio fuel, etc.

Sunil Kumar Verma is currently an Assistant Professor at the Department of Biotechnology, BN College of Engineering and Technology, Lucknow. He is UGC-NET, GATE qualified & Experienced with a demonstrated history of working in the higher education industry with Strong education professional with a Master of Technology (M.Tech.) focused in Biotechnology.

Yusto Mugisha Yustas received his PhD in Renewable Energy from the Sokoine University of Agriculture (SUA) in November 2021. He worked on developing a predictive model for integrated renewable energy system performance. In 2010, as the District Agriculture Engineer, he was employed by the Rufiji District Council-Coast region, Tanzania and is currently the academic member of staff in the School of Engineering and Technology at SUA. Yusto writes and presents widely on issues of renewable energy, upscaling strategies for ensuring food security, agricultural mechanisation and post-harvest processing.

Index

2G Ethanol 322-323, 331, 338

A

ABFAD 291, 306-307, 309, 311, 315-316
Adaptation 125, 163, 214, 291, 294-297, 321
Agriculture 3-4, 8, 11, 13-15, 21, 24, 51-53, 63, 65, 70-71, 73, 78, 81, 90, 107, 109, 115, 127, 151, 165-166, 215, 246, 255, 291, 296, 319, 337-339
Algae 2, 4, 16, 34, 48, 58-59, 62-63, 110, 129-132, 135-136, 140-141, 146-148, 151-152, 154, 156-158, 160-164, 166-167, 188, 190, 194, 203-204, 206-207, 209-210, 212-213, 215, 220, 222, 224, 227, 229-232, 289-290
Alternative Fuel 188, 233, 240, 252, 258, 274-275, 287
Aniline Point 258-262, 264-265, 274, 277, 279-280, 288

B

Biodiesel 4, 8, 20, 25-26, 29, 31, 37-38, 41-44, 47, 49, 51, 53-64, 77-80, 82, 89, 102, 104, 108, 110-111, 144, 146, 148, 150, 152, 157, 160-161, 164-168, 190-191, 193, 196-197, 199-206, 209, 211, 221, 223, 228, 230-232, 258-284, 286-290, 336
Bioenergy 1-21, 23, 30, 32-37, 39, 42, 57, 59-60, 78, 103-107, 109-110, 113-114, 116, 118-119, 141, 145, 147, 162, 166, 213-214, 231, 251, 254-256, 270-271, 288-289, 291-292, 296, 316, 319, 337-338
Bioethanol 4, 14, 26, 41, 47, 51-53, 55, 59, 62, 77, 79-80, 82-87, 89-90, 92, 95-96, 99-100, 102, 104-111, 113-117, 144, 146, 148, 150-152, 155-156, 158-163, 165-168, 190-191, 193, 202, 206, 208-216, 224, 233-234, 238, 240, 242-256, 323, 326, 331-339
Biofuel Policy 44, 53, 55-57, 59, 61, 77, 106
Biofuel Technology 52, 188
Biofuels 2, 4, 12-13, 16-17, 21-27, 29-32, 34, 36, 41-42, 47, 50-53, 55-64, 75-79, 81-83, 88-89, 93-96, 99-112, 114, 116-118, 141-142, 144-152, 154-155, 157-165, 167, 169, 188-191, 193, 196-197, 199, 201-207, 214-215, 217-222, 224-226, 228-234, 245, 250-252, 254, 256, 268, 271, 289-290, 319-320, 326, 335-338
Biogas 27, 33, 37, 39-40, 42, 77-78, 81-82, 91, 103, 111, 117, 120-121, 123-128, 141, 150-152, 154, 157, 160-162, 165-166, 168, 229, 272, 291-297, 299, 301-309, 311-313, 315-319, 339
Biogas Generation 125, 291-292, 295-296, 312, 315
Biomass 1-7, 9, 14, 16, 20-21, 23-24, 27-29, 32-36, 38, 40-42, 44, 47-48, 51, 55, 60, 63, 69, 72, 75-85, 88-92, 94-111, 113-118, 122, 124, 127, 129, 135, 137, 140-147, 149-152, 154-159, 162-166, 168, 188-190, 193-196, 198-199, 202-203, 207-216, 218-221, 223-225, 227-229, 232-236, 238, 240-246, 249-256, 270, 288, 291-293, 296, 307, 316-318, 320, 322-323, 325-331, 333-339
Biotransformation 233, 247
Biowastes 75, 128
Blue Bioeconomy 148-149
Boiling Point 258, 266-267, 281-282

C

Capital Investment 1, 8
Catalyst 5, 10, 97-98, 167, 196, 201, 216, 250, 258-259, 271, 274-276, 289, 330
Cellulase 99, 101, 121, 156, 233-234, 236-240, 244, 246, 248-250, 252-256, 333
Cellulose Hydrolysis 100, 233, 236-237, 246, 249, 330
Cetane Number 80, 198, 258-260, 266, 274, 286, 288
Challenges 13, 16, 22, 28-29, 31, 39, 54-55, 57-58, 61, 65, 73, 75, 94, 107, 109-112, 114, 149, 154, 157-160, 162-163, 172, 181, 203, 214, 217-218, 227-232, 250, 252, 255, 268, 271, 295, 307, 316, 319-320, 327, 335
Chemical Pretreatment 118, 322-323, 327, 329-330, 339

Clathrates 170, 180
Climate Change 1-2, 5, 11, 15, 21-24, 27-28, 35, 48, 63, 65-67, 69-71, 73-74, 102, 119, 150, 157, 159-161, 217-218, 291-292, 296, 316
CO2 Sequestration 170, 177, 183, 194, 198
Conceptual Designing 301, 305, 321
Conversion Technologies 75
Cow Dung 291, 296, 302, 307

D

Depressurization 170-171, 173-175, 177, 179, 182-184, 186
Detailed Designing 321
Development 2, 4, 8, 10-11, 15, 17, 20-23, 27-28, 30, 32, 34, 36-37, 39-40, 42, 44-45, 49-53, 55-58, 60-62, 64, 72-73, 77-79, 81, 91, 96, 102-103, 108-109, 115-116, 126-127, 136-137, 142, 148-149, 151, 159-160, 165-166, 174, 183, 191, 198, 201-202, 211, 213, 217-218, 223-224, 228-232, 242, 246, 249-250, 258-259, 271, 292, 315-318, 320, 335
Diesel Index 258-259, 261-266, 268, 274, 277, 279-280, 284, 286, 288
DMA 291, 300-301, 303-305

E

Effect on Environment 129
Energy Balance 19, 22, 60, 69, 85, 232
Energy Demand 13, 20, 44, 48, 76-78, 115, 120, 170, 180, 217-218
Entrepreneur 32, 34, 36, 42-43
Environment 2, 16-17, 19-20, 26, 28, 30-33, 37, 41-42, 53, 57, 59, 61-62, 66-67, 72-73, 103, 106, 109-110, 120-121, 129, 131, 141, 143, 150, 152, 158, 161, 172, 177, 190, 196, 202, 209-211, 213, 215, 218, 225, 227, 229, 232, 275, 291-292, 319, 325
Enzyme 85-86, 92, 97-99, 101, 104, 106, 111, 118, 154, 156, 195-196, 212-213, 234, 236-238, 242, 244, 247-248, 250-251, 253-257, 323, 327, 329, 331-333, 336-337
Enzyme Immobilization 247-248, 250-251, 254-256
Enzyme Reusability 234, 256
Ethanol 4, 25-27, 31-34, 36, 39, 42, 49, 51, 55-56, 58-59, 61-63, 75, 77-78, 80, 82-87, 90-92, 97, 100-118, 144, 147, 156, 162, 177, 206, 210-214, 224, 251-254, 319, 322-323, 325-326, 330-332, 334-339

F

Feedstocks 31, 33, 36-37, 42, 48, 52, 56, 62, 75, 79, 81-83, 86, 89-90, 93, 98, 102, 104, 107-108, 111-112, 114-115, 117, 144, 148, 151-152, 161, 163, 222, 254, 258, 268, 271, 323, 331, 338-339
Fermentable Sugar 242, 256, 323
Fermentation 59, 79-81, 84-87, 91, 93, 96-97, 100-101, 103, 105-108, 111, 113-115, 118, 121, 124, 127, 144-145, 149-151, 154-157, 159, 162, 165, 202, 206, 209-214, 224, 238, 242-244, 246, 249-250, 252, 322, 325, 331-334, 336

G

Gravity 72, 173, 258-260, 264-265, 278-279, 284-285, 288
Green House Effect 65
Green House Gases 65, 68
Greenhouse Gas Emissions 22-24, 26-27, 29-31, 34, 38, 45-46, 60, 64, 66, 68, 70-71, 102, 120, 144, 218, 231

H

Harvesting Techniques 217
Hydrolysis 86-87, 92, 94, 97-100, 104, 106-109, 111, 113, 116-118, 120-123, 125, 127, 154-157, 159, 162-165, 167, 211-214, 216, 233-234, 236-237, 241-242, 244, 246, 249-251, 253, 255-256, 323, 327, 329-339

I

Ignition Properties 258
India 1, 4, 12-14, 18, 32-33, 37, 40-41, 43-44, 46-47, 49, 51-63, 65, 75, 77-78, 82-84, 91-92, 103-104, 107, 109, 114, 119, 129, 143-145, 148, 162, 165, 170, 180, 188, 206, 217, 230, 244-245, 254, 258-259, 274-275, 289, 294, 306, 322

K

Kinematic Viscosity 258-261, 268, 274, 278, 288

L

Life Cycle Assessment 22-24, 29-31, 57, 73, 108, 167, 271
Lignocellulolosic Biomass 322
Lignocellulose Biomass 233-235, 238, 240-241, 244-246, 249-250, 256, 322-323, 328-329, 334, 336, 338
Lignocellulosic Biomass 79, 89-90, 94-96, 98, 100-

102, 107-108, 111, 115-116, 145, 209, 211-212, 233, 251-252, 254-255, 270, 320, 322-323, 325, 327, 329-331, 334-335, 337-339
Living of Standard Biofuel 1

M

Macroalgae 57, 64, 129, 135, 148, 150-154, 156-169, 206-207, 209-216
Marine Biofuel Resources 148
Metathesis Reaction 1, 5
Microalgae 44, 57-58, 60, 62, 79, 129, 135, 137, 141, 146, 149, 152, 154, 156-157, 160-161, 163-165, 167, 188-191, 193-196, 198-207, 209, 215, 217, 220-232, 268
Microalgae-Biofuels 188

N

Nano-Additives 188, 190, 193-199, 202
Nanobiocatalyst 248, 253, 257
Nanotechnology 188, 190, 200-201, 203-204, 249

O

Opportunities 3-4, 8, 11-12, 21-22, 28, 32, 35-39, 57, 92, 96, 104, 107, 110, 149, 152, 159, 162, 207, 214, 319

P

Physical Pretreatment 97, 322-323, 327-328
Policy 4, 22-23, 28, 30, 32-33, 44, 47, 50-59, 61-63, 74, 77-78, 102, 105-106, 109, 112, 114, 116, 119, 143
Policy Barriers 44
Potential 4, 20, 23, 28, 36-37, 40, 58, 61-62, 68, 70-74, 78, 84, 90-91, 93, 98, 107, 109-110, 113-114, 116-117, 129, 135-137, 141-143, 145-147, 165-167, 171, 173, 175, 178, 181, 184, 186, 202, 204, 213-215, 217-219, 228-230, 232, 249, 255-256, 258, 271-272, 274, 288, 300, 316, 320-321, 323, 331, 336-337
Pretreatment 76, 78, 81, 84, 86, 89-90, 92, 94-100, 104-109, 111-113, 115-118, 147, 155-157, 159, 162-164, 166, 168, 212, 241-242, 245, 249-250, 322-323, 327-331, 334-339
Price Stability 1, 9
Procedures 66, 71, 94-97, 99, 120, 145, 159, 196, 291-292, 295, 334

R

REC 13
Renewable Energy 1-2, 4, 13-14, 25, 29, 34, 37, 39-43, 52-53, 56-58, 60-61, 63, 74, 78, 102, 119-120, 126, 142-145, 202, 206-207, 210, 213-214, 218, 270, 288-292, 295, 300-301, 306, 316-318, 320, 339
Renewable Energy Resources 318, 320
Rural Semi-Arid Areas 293, 321

S

Saccharification 87, 99-100, 105-106, 108, 111, 113, 116, 167-168, 206, 212, 238, 242, 244, 246, 249, 251-252, 254, 331-334, 337
Seaweed 129, 145, 157, 162-163, 165, 167-168, 206-207, 210-212, 215, 222
Societal Gap 1, 6
Spirulina 129, 132-140, 146-147, 199, 201, 215
Superfoods 129
Sustainable 2, 6, 8-9, 11-12, 15-17, 20-22, 24, 30, 37, 41, 46-47, 51, 56-64, 73, 76-77, 89, 93, 102, 104, 107-110, 112, 114-117, 119, 127, 129, 141, 144-146, 148-150, 156, 158, 161-167, 189-191, 196, 201, 203-205, 213-215, 217-218, 221, 228-232, 253-254, 258-259, 270-271, 289-292, 295-296, 317-320, 322-323, 334, 336, 338-339
Sustainable Development 2, 11, 15, 17, 20-21, 30, 37, 51, 56, 58, 61, 115, 217, 258-259
System Boundary 22

T

Techno-Economic Analysis 113, 240, 244, 254
Thermal Stimulation 170-173, 175, 177-179, 181, 184, 186-187
Third Generation Biofuels 148, 151, 161, 230
Transesterification 25, 37, 60-61, 79, 144, 150, 156, 164-165, 167, 196, 259-260, 268, 270-272, 274, 276, 288-290

V

Viscosity 80, 87, 92, 191, 212, 242, 244, 258-261, 268, 274, 278, 287-288

www.ingramcontent.com/pod-product-compliance
Lightning Source LLC
LaVergne TN
LVHW082301150926
848178LV00051B/1329